Physical Biochemistry

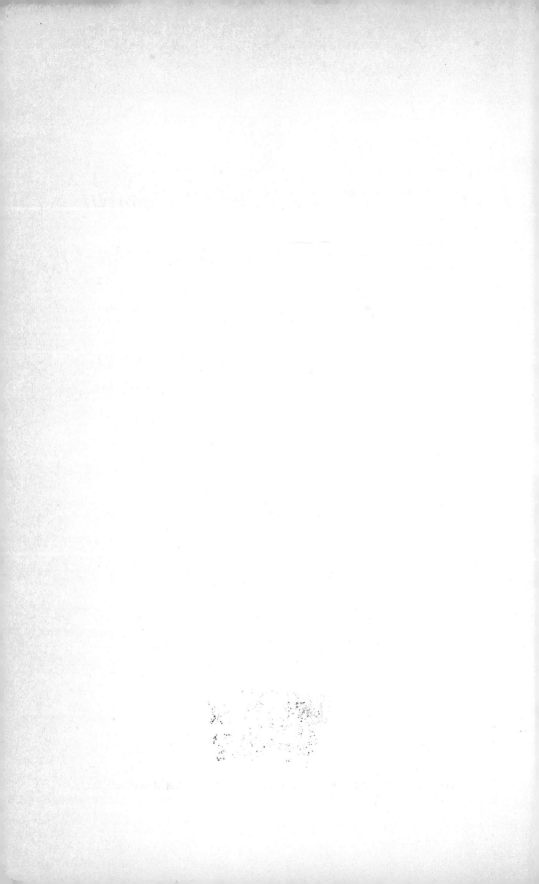

Physical Biochemistry

Applications to Biochemistry and Molecular Biology

DAVID FREIFELDER
Brandeis University

W. H. FREEMAN AND COMPANY *San Francisco*

Library of Congress Cataloging in Publication Data

Freifelder, David Michael, 1935–
Physical biochemistry.

Bibliography: p.
Includes index.
1. Biological chemistry—Technique. 2. Molecular biology—Technique. I. Title.
QP519.7.F75 574.1'9283 76-6495
ISBN 0–7167–0560–5
ISBN 0–7167–0559–1 pbk.

Printed in the United States of America

9 8 7 6 5 4 3 2

Contents

Preface

Modern biochemistry and molecular biology are concerned with the functions of biological systems. A century ago, the only means of study was by direct observation of such systems at work. Today, much more sophisticated and detailed observations can be made through electron microscopy and the specialized microscopic techniques that have been developed in the past forty years.

Scientists realized in the late nineteenth century that something was to be gained by studying the chemistry of cells. For decades thereafter, biochemists relied on the chemical methods available to them, and, indeed, great advances in understanding were achieved. Probably the most significant improvement in chemical technique was the development of the use of radioisotopic tracers; this vastly increased the sensitivity of detection and the number of kinds of biological molecules that could be identified. As it became necessary to have more sensitive methods for separating the various components of a biochemical reaction, chromatography and electrophoresis became routine procedures.

When the attention of physicists and physical chemists was directed toward biology (perhaps because of the ability of living cells to create local order even though the laws of physics state that there is a tendency toward disorder in the universe), the techniques of physics and physical chemistry—hydrodynamics, spectroscopy, scattering, and diffraction—entered the biological field.

A significant advance in biochemistry was the recognition that biological systems contain not only the small molecules with which organic chemistry is concerned, but also giant molecules, the *macromolecules,* whose molecular weights we now know can be at least 100 billion times the mass of a hydrogen atom. The importance of macromolecules to biological systems lies in the specificity that they confer both on biological reactions and in forming structural units. It is probably fair to say that in the past twenty years the greatest effort in biochemistry and molecular biology has been to characterize and understand macromolecules and their interactions with one another. This has required sophisticated methods both for separation and purification and for detailed observation of small parts of the molecules. Hence, the greater part of this book deals with techniques for the characterization of macromolecules with which to find answers to the following questions.

1. What is the precise structure at the atomic level of a macromolecule or an aggregate of macromolecules?
2. What properties of a macromolecule determine its structure and what forces participate in stabilizing it?
3. If a macromolecule binds other molecules (either small molecules or other macromolecules), what is their structure, what is the number of binding sites, and what are the physical constants (e.g., dissociation constants) for the binding?
4. Where is a particular macromolecule located within a cell or a small unit like a virus?

It should be noticed that the greatest concern is with the determination of physical parameters and structure. This derives from the (correct) belief that these characteristics determine biological function. For example, it has been said that a protein with a specialized function (e.g., an enzyme) can be thought of as an active site consisting of a few amino acids held together by α helices, β sheets, β turns, and random coils so that the structure reflects the requirements of the protein for its activity.

At present, the precise structure (at the atomic level) of a macromolecule can be determined only by x-ray diffraction analysis, and no technique has had a greater impact on the study of macromolecules than this one. (Consider, for example, the impact on genetics of the determination of the structure of DNA.) By knowing a few precise structures not only can we establish rules for determining structure, but also we have reference molecules to study using other techniques, thus explaining how to interpret the data so obtained. X-ray diffraction, unfortunately, is a very difficult and complex technique, and often years are required to determine the structure of a protein or a macromolecular complex. Traditionally, books of this sort contain general descriptions of the theory of x-ray diffraction. However, this book does not for two reasons: (1) it is very unlikely that any but a few readers will ever use the technique or even have an occasion to attempt to unravel an x-ray diffraction pattern, and (2) it is virtually impossible, in less than an entire book, to explain the principles and analytical methods so that they can be thoroughly understood. However, throughout the book, reference is made to information gained by such analysis.

Students wishing to understand modern biological thinking can find excellent texts that give biological facts and theories as we know them today. However, well-read students who attempt to read current scientific literature (i.e., journals and review articles) quickly discover that a great deal of technical information is needed. To obtain this information, they can turn to textbooks of biophysical chemistry and find excellent theoretical analyses of the available methods, together with detailed mathematical derivations describing the physics underlying the methods. Alternatively, there are manuals available that describe procedures for the use of various instruments and techniques.

In many years of teaching physical biochemistry, I have come to realize that a student who has labored through mathematical derivations rarely has achieved sufficient understanding of the techniques to read, understand, and judge the work presented in current scientific literature. Furthermore, many students lack the mathematical sophistication required for obtaining any information at all from this standard approach. For these reasons, I teach physical biochemistry in a way that completely avoids mathematical derivations. In presenting an equation, I make clear what assumptions have been made in its derivation and what conditions must be satisfied before the equation is usable. Techniques are described in detail, but mostly with words, and many examples are given as a teaching device. I feel that the approach has been successful, and this book is designed in accord with it. Derivations are left to those instructors who feel that they are necessary to round out understanding. The single aim of this book is to enable the student to read and understand the current literature.

The book is directed at advanced undergraduates or beginning graduate students who have a general knowledge of chemistry and physics. For a few of the techniques presented (especially in the sections on spectroscopic methods), a more extensive background in physics is required, and the student may have to draw on the instructor's knowledge of such techniques for a thorough grasp of them. In the course of writing this book, I had the help of many active researchers as reviewers and was pleased to discover that the book is informative to such people, filling them in on techniques developed since graduation.

I would like to thank the following people whose aid was invaluable in achieving correctness and clarity: Elliot Androphy, Dan Alterman, Carol Orr, and Jon Tumen, four undergraduates at Brandeis University, and Robert Suva, a Brandeis graduate student, who combed the manuscript for ambiguities and inaccuracies; Andrew Braun, who read the entire manuscript in search of flaws in presentation; Richard Mandel, without whom I could not have written the chapters on spectroscopy; Alfred Redfield, Helen Van Vunakis, Lawrence Levine, Robert Baldwin, Bruno Zimm, Ross Feldberg, Sherwin Lehrer, Inga Mahler, and Serge Timasheff, who read the specialized chapters; Phil Hanawalt, Paul Schimmel, and Peter Von Hippel, who reviewed the manuscript submitted for publication. Further thanks go to all of the people who gave me data and allowed the use of their illustrations and photographs. I also want to thank Mildred Kravitz and Barbara Nagy, who typed thousands of pages to bring the manuscript into its final form. My final debt is to my wife, Dorothy, who made use of her extraordinary skills as proofreader, scientist, editor, and logician, requiring perfection of me and of this book as she read the page proofs.

February 1976 David Freifelder

Physical Biochemistry

CHAPTER 1

Characterization of Macromolecules

For roughly half a century the aim of biochemistry has been to assemble a complete catalog of the chemical *reactions* occurring in living cells. The motivation for this great effort was the belief that a significant number of the biological properties of cells could be understood in terms of the reactions in which covalent bonds are formed or broken. Indeed, from the great collection of biological reactions that has been obtained, we now understand in some detail how energy is generated by chemical degradation, how biological molecules are interconverted, and how giant molecules—the *macromolecules*—are assembled from amino acids, nucleotides, sugars, and lipids.

In the past thirty years, it has become apparent that the physical *interactions* between molecules—that is, those that do not form or break covalent bonds—are at least as important as the chemical *reactions*. For example, the *regulation* of chemical reactions (i.e., the degree to which they are allowed to occur) is accomplished both by physical changes in the structure of macromolecules and by variation in the availability of active sites on macromolecules resulting from the noncovalent binding of both small and large molecules. Furthermore, the large macromolecular aggregates found either in cells or in organisms (i.e., membranes, cell walls, chromosomes, tendons, hair, etc.) derive many of their special properties from noncovalent physical interactions. Therefore, chemical reactions are only half the story; clearly, to understand a complex biological system, knowledge of the physical properties of the constituent molecules is essential. The attainment of this knowledge is the goal of *physical biochemistry*. The application of the information so obtained to biological systems is the foundation of the modern discipline called *molecular biology*. A large part of this book describes methods for characterizing macromolecules. Because the language describing macromolecules is generally unfamiliar to the student of biochemistry, the terminology and concepts used in considering the properties and shapes of macromolecules and the transitions between various forms will be explained first.

FIGURE 1-1
The amino acids and their chemical structures.

Polar amino acids (tend to be on protein surface)

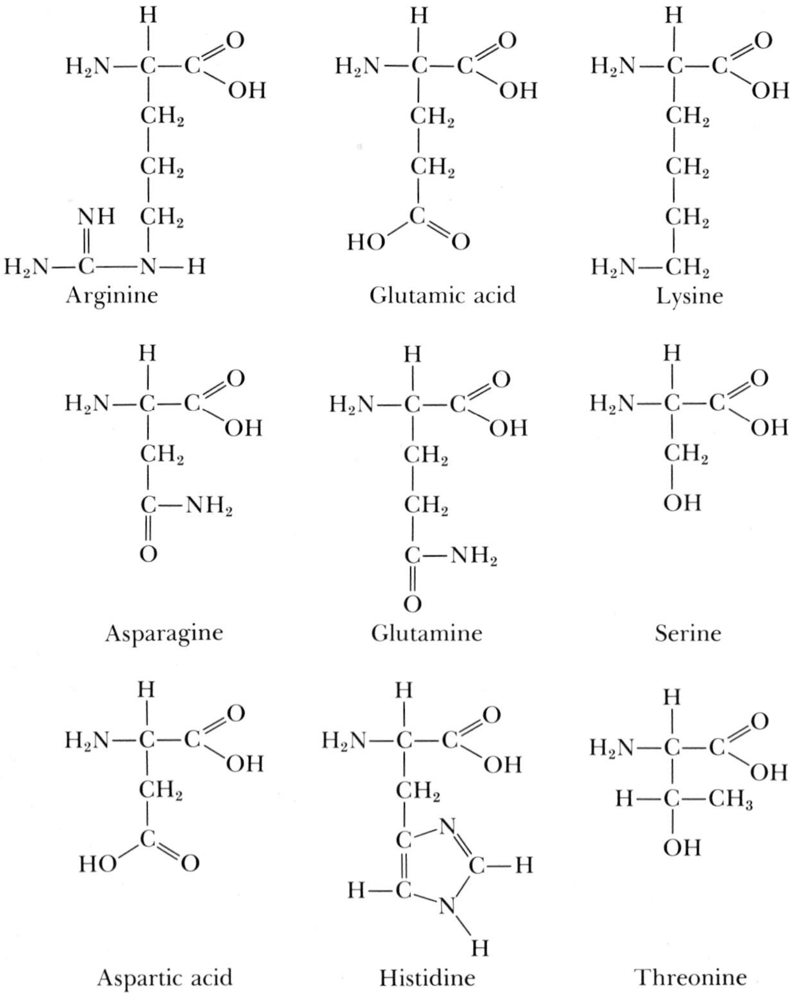

Arginine

Glutamic acid

Lysine

Asparagine

Glutamine

Serine

Aspartic acid

Histidine

Threonine

Nonpolar amino acids (tend to be internal)

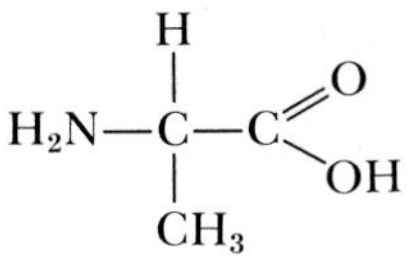

Alanine

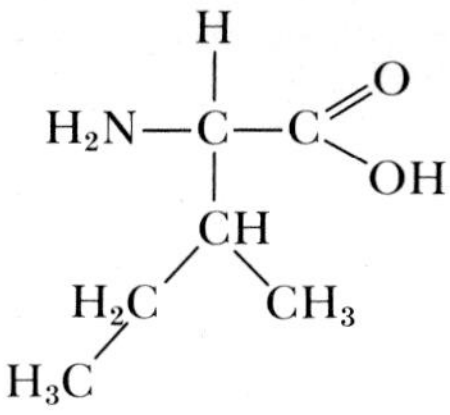

Isoleucine

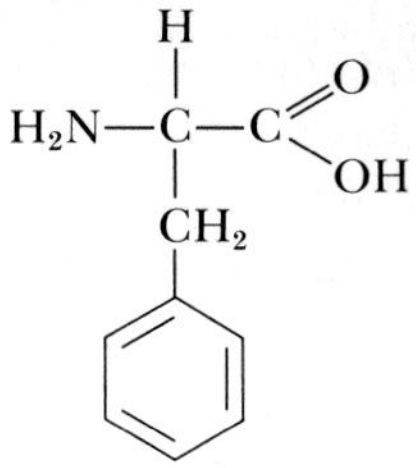

Phenylalanine

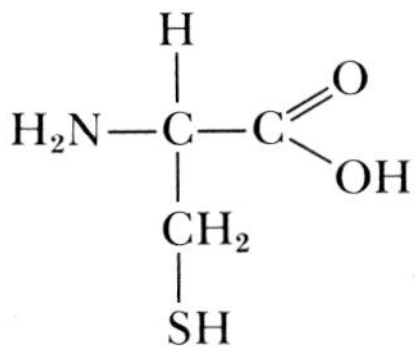

Cysteine

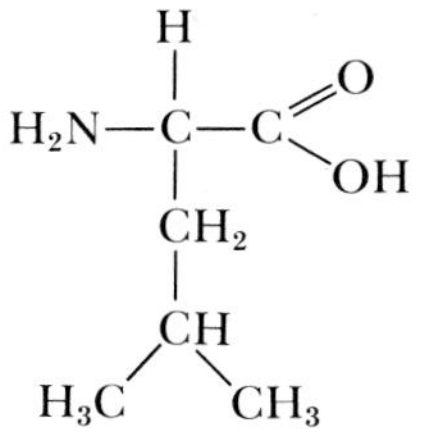

Leucine

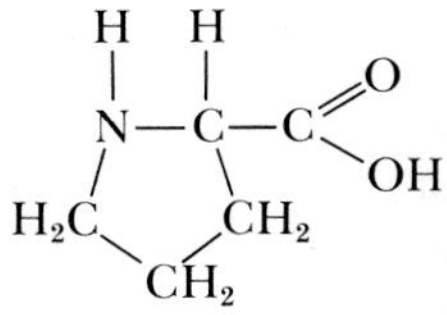

Proline

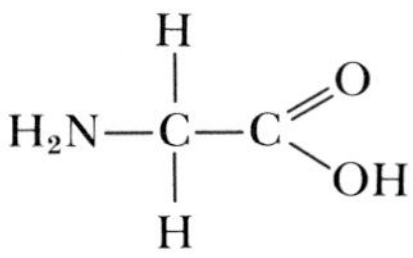

Glycine

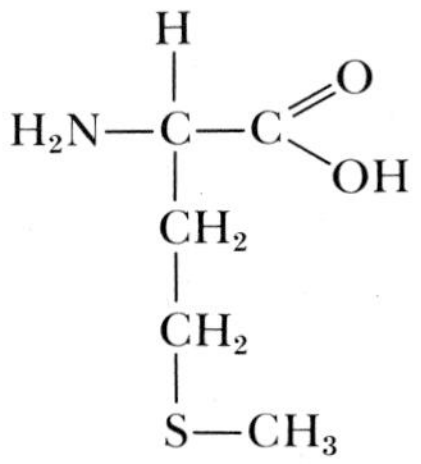

Methionine

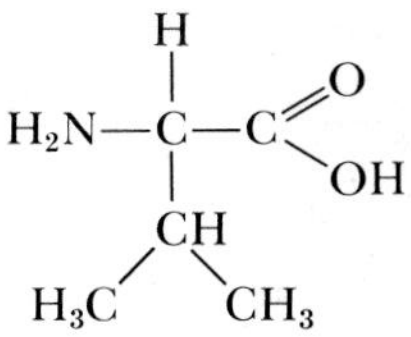

Valine

Amino acids equally frequently internal and external

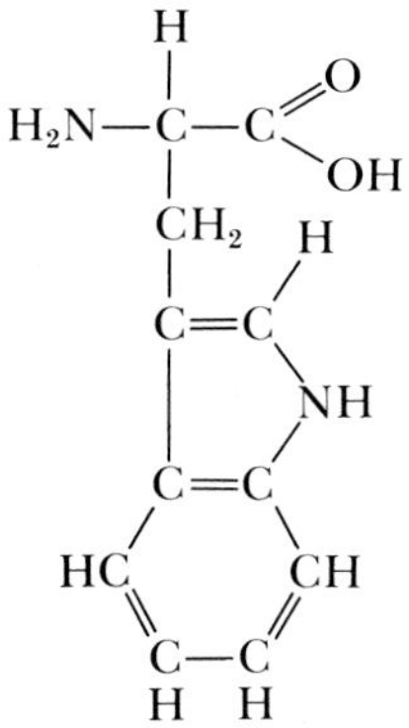

Tryptophan
(nonpolar)

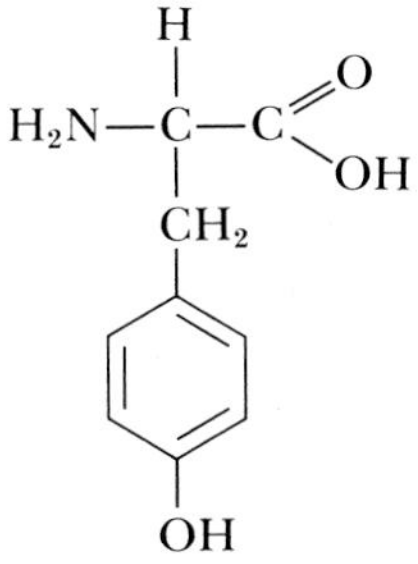

Tyrosine
(polar)

POLYPEPTIDES AND POLYNUCLEOTIDE CHAINS

The components of proteins and polypeptides are the *amino acids.* The chemical structures of the common amino acids are shown in Figure 1-1 (on pages 2 and 3), in which the amino acids are grouped to indicate their usual locations in proteins. *Polar* amino acids carry charged groups that interact significantly with water (i.e., they are *solvated*). They are also called the *hydrophilic* amino acids. Because of this strong interaction with water, polar amino acids tend to be on the surfaces of proteins, thereby maximizing contact with water. Many polar amino acids carrying opposite charges (e.g., on the negative hydroxyl and positive amino groups) tend to interact with one another to form hydrogen (H) bonds (Figure 1-2) and are therefore often in close proximity. The *nonpolar* amino acids are not charged nor easily solvated by water, and therefore tend to be internal, thus minimizing contact with water. They are also called *hydrophobic* amino acids. The sulfhydryl (SH) group of the amino acid cysteine can combine with the SH of another cysteine to form a *disulfide bridge* (—S—S—).

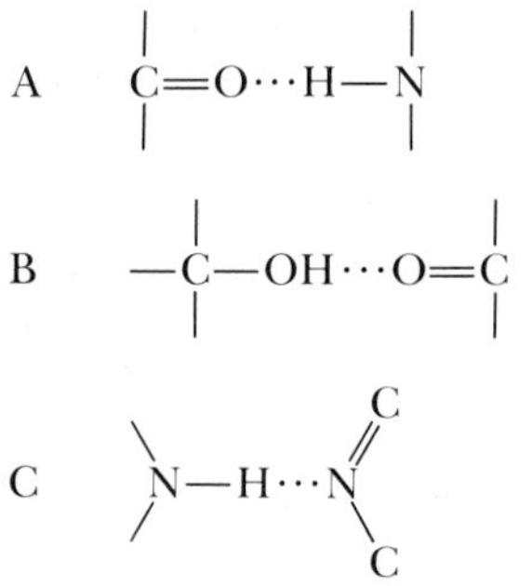

```
A      |           |
       C=O···H—N
       |           |

B      |               |
     —C—OH···O=C
       |               |

                  C
       \         //
C       N—H···N
       /         \
                  C
```

FIGURE 1-2
Structures of three types of hydrogen bonds (indicated by three dots): (A) the type found in proteins and nucleic acids; (B) a weak bond found in proteins; (C) the type found in DNA.

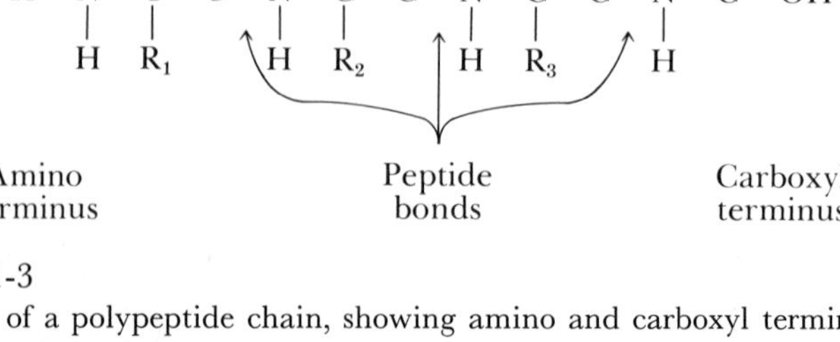

```
          H  O      H  O      H  O      O
          |  ||     |  ||     |  ||     ||
   H—N—C—C—N—C—C—N—C—C—N—C—OH
     |  |       |  |       |  |       |
     H  R1      H  R2      H  R3      H

 Amino            Peptide           Carboxyl
 terminus          bonds            terminus
```

FIGURE 1-3
Structure of a polypeptide chain, showing amino and carboxyl termini, peptide bonds, and the locations of the side chains (R_1, R_2, and R_3).

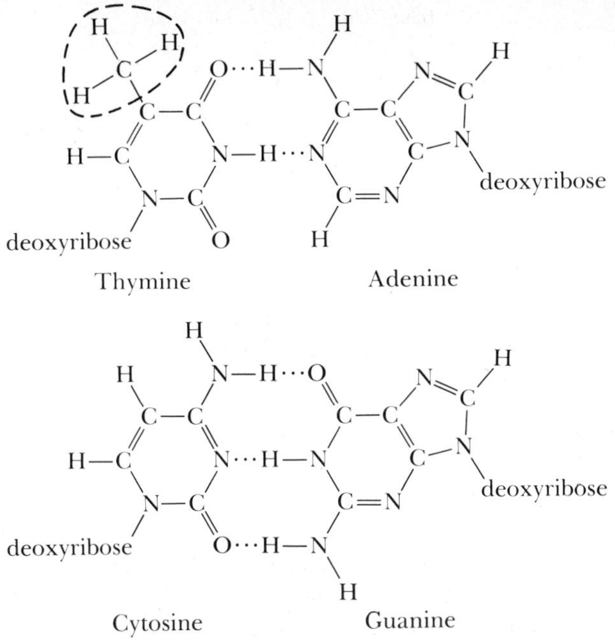

Thymine Adenine

Cytosine Guanine

FIGURE 1-4
The two common base pairs of DNA. If the encircled methyl group were replaced by a hydrogen, the result would be uracil.

Amino acids polymerize by forming a covalent bond, called a *peptide bond,* between the carboxyl group of one and the amino group of another. The resulting structure of a polypeptide or protein is shown in Figure 1-3, in which R_1 (like R_2 and R_3) represents the *side chain,* or distinguishing group, of an amino acid.

The components of nucleic acids are the bases depicted in Figure 1-4. These bases consist of relatively hydrophobic rings to which are attached charged groups that interact by means of hydrogen bonding to form the base pairs indicated. A base is covalently coupled with a sugar (deoxyribose, for DNA, or ribose, for RNA) to which a phosphate group is attached; the structures thus formed are polymerized by means of phosphodiester bonds to form a nucleic acid, as shown in Figure 1-5. When hydrogen bonds form between bases, the polar part of each base becomes less accessible to water; because the ring part is fairly hydrophobic—and considerably so compared with the sugar and the highly charged phosphate group—in a nucleic acid in which there is hydrogen bonding, the bases will be situated in a way that tends to minimize contact with water. This line of reasoning gives a rationale for the fact that, in a totally hydrogen-bonded nucleic acid such as DNA, the base pairs are internal and surrounded by the hydrophilic sugar-phosphate chains.

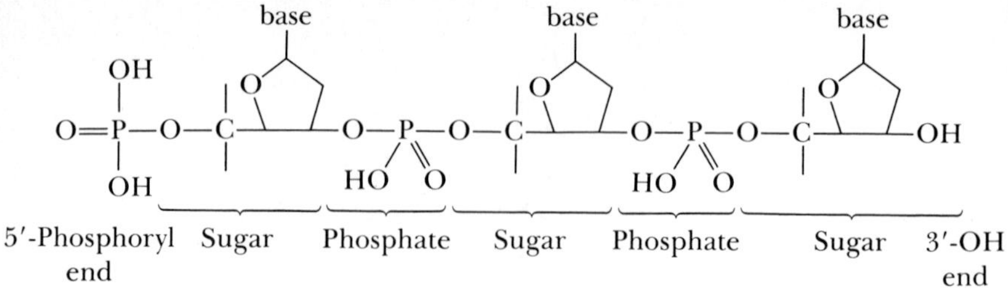

base base base

OH O O O

O=P—O—C— —O—P—O—C— —O—P—O—C— —OH

OH HO O HO O

5′-Phosphoryl end Sugar Phosphate Sugar Phosphate Sugar 3′-OH end

FIGURE 1-5
Structure of a single polynucleotide chain. The sugar (ribose or deoxyribose) and phosphate moieties alternate, a phosphate always connecting the 3′- and 5′-carbon atoms.

POLYMER STRUCTURES

The *primary structure* of a polymer consisting of different monomer types refers to the monomer sequence—for example, the amino acid sequence of a protein and the base sequence of a polynucleotide. Such sequences can be determined by chemical analysis, using many of the separation procedures described in Chapters 8 and 9. The actual chemical methods, however, will not be discussed in this book.

Because of the interactions between various amino acid side chains and between nucleic acid bases and because of the relative degree of interaction of different molecules with water, biological polymers are rarely fully extended linear chains; instead, they fold to form complex three-dimensional structures. The orientation of each monomer unit with respect to another is called the *secondary structure.* By convention, secondary structure usually (but not always) refers to the *configuration* or *conformation* of the backbone—that is, the polypeptide and the sugar-phosphate chains. The shapes commonly encountered are helices, coils, sheets, and rods, and combinations thereof.

The relative orientation of the side chains (amino acids or nucleic acid bases) is usually called the *tertiary structure.* Many biological polymers interact with one another to form complex structures such as multisubunit proteins, viruses, membranes, filaments, and so forth. This is sometimes called the *quaternary* structure. However, the use of the terms tertiary and quaternary structure is sometimes ambiguous.

The peptide bond is *planar* (Figure 1-6), which puts several constraints on the possible types of secondary structures of proteins. On the other hand, all bonds involving the α-carbon are flexible and allow a wide variety of possible structures.

The phosphodiester bonds of nucleic acids are also flexible (Figure 1-7). However, because the bases consist of planar, strongly hydrophobic ring

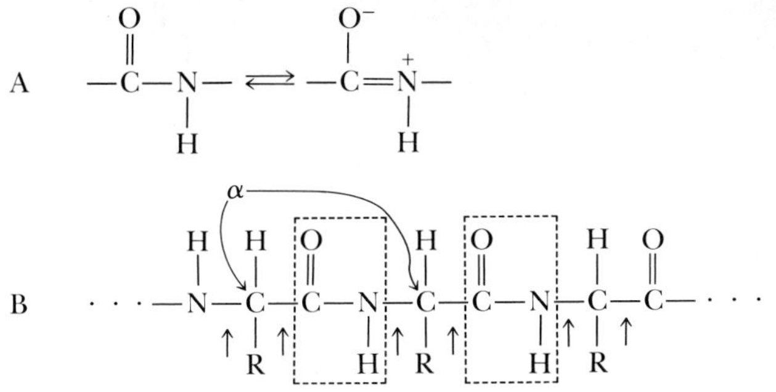

FIGURE 1-6
A. Tautomeric structure of peptide bond, showing the rigidity conferred by partial double-bond characters. This is why the peptide bond is planar. B. Part of a polypeptide chain: arrows point to bonds about which there is free rotation; the rigid peptide units are inside the boxes.

systems surrounded by only a few charged groups, they tend to stack one above the other (see Chapter 16), thereby minimizing contact with water. This tends to increase the rigidity of the structure, even in a single-stranded polynucleotide.

A linear polymer that has free rotation about all bonds in the chain and has no interaction of side groups is called a *random coil* (Figure 1-8). It does not have a unique three-dimensional structure or size because it is continually being distorted by Brownian motion. Its size can be described by an average value—the average *radius of gyration:*

$$R_G = \sqrt{\frac{\Sigma R_i^2}{N}}$$

in which N is the number of segments (or monomers) and R_i is the average distance of a segment, i, from the center of mass. It turns out that R_G is actually proportional to $\sqrt{N}$. It is also a measure of the average size of a hypothetical sphere occupied by the coil. A protein in which all hydrogen

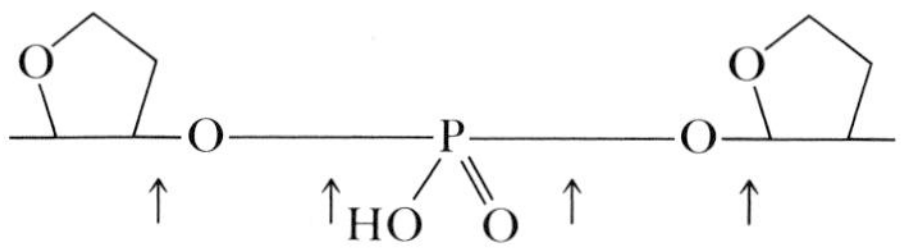

FIGURE 1-7
A phosphodiester bond; the points of possible rotation are indicated by the arrows.

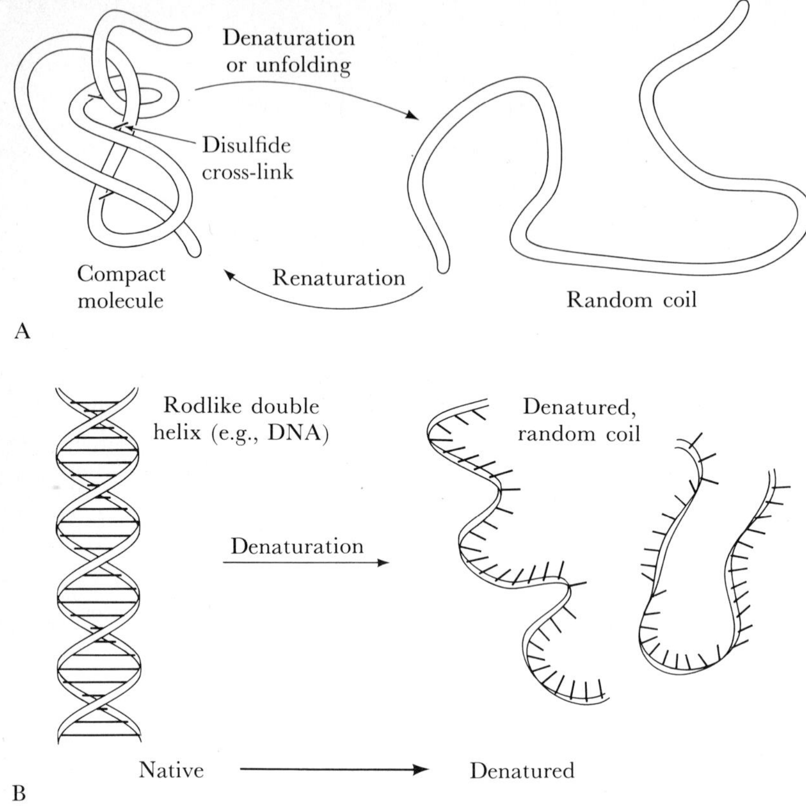

FIGURE 1-8
Several forms of macromolecules and how they are interconverted.

bonds are broken but a *few* disulfide bonds remain—which reduces the radius of gyration—is sometimes called a *near-random coil.*

Most biopolymers have substantial *side-chain interactions,* which tend to reduce the size of the molecules to less than that of a random coil. These molecules are *globular* and are said to be *compact* (Figure 1-8).

Many polymers have a *helical structure* (Figure 1-9). For proteins this means that the planar bonds are rotated about a particular point such that the rotation from plane to plane is constant. Such a structure can fit into a cylinder. A simple analogy is the structure that would be obtained if a stack of playing cards were pinned together with a straight pin at one corner but each card were rotated at a fixed angle with respect to the one below. In many proteins, rotation occurs about the α-carbon atom of the amino acid (Figure 1-6) and the structure is stabilized by bonds between carboxyl and amino groups. This particular helix is called an

α helix (Figure 1-10). In polynucleotides, the planar nucleotide bases are stacked one above another but slightly rotated. In DNA, two polynucleotide strands, each extended by this stacking, hydrogen bond to one another to form a *double-stranded helix* (Figure 1-9). Some molecules (e.g., the protein collagen) can form a *triple-stranded helix.* Helical molecules (single- and multiple-stranded) are examples of *extended* or *rodlike* molecules.

A common type of secondary structure in proteins is the β *structure,* in which two sections of a polypeptide chain (or in some cases two different chains) are aligned side-by-side and held together by hydrogen bonds. To maximize the number of hydrogen bonds, the polypeptide chains are pleated as shown in Figure 1-9. In this structure, the plane of the pleat contains the peptide group and the side chains are located alternately above and below the plane of the sheet. The figure shows only one of the strands of the β structure; normally, a second strand would be adjacent to the one shown. The alignment of the two strands may be such that the adjacent chains are running either in the same direction (*parallel*) or in the opposite direction (*antiparallel*), as shown in Figure 1-9. Such regions of a protein are called parallel and antiparallel β *pleated sheets,* respectively.

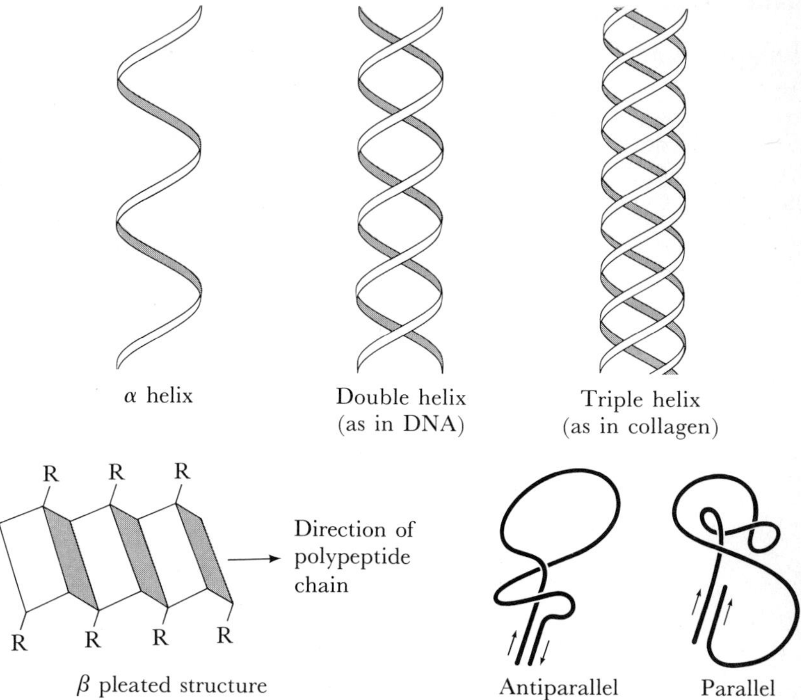

FIGURE 1-9
Several conformations of macromolecules.

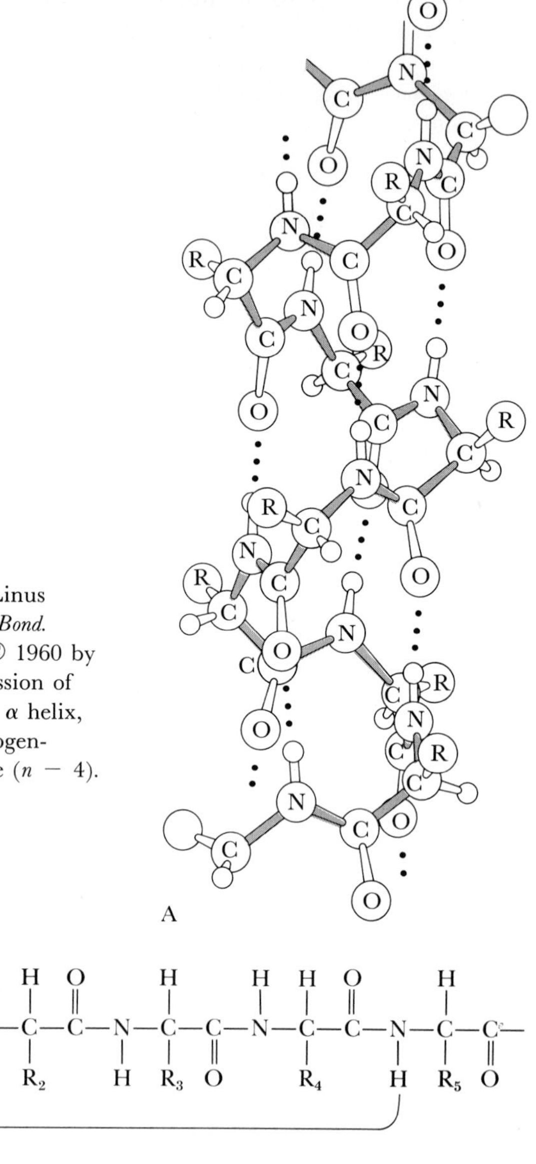

FIGURE 1-10
A. Structure of the α helix [from Linus Pauling, *The Nature of the Chemical Bond.* Copyright 1939 and 1940, 3d ed. © 1960 by Cornell University. Used by permission of Cornell University Press]. B. In an α helix, the NH group of residue n is hydrogen-bonded to the CO group of residue $(n - 4)$.

THE CONCEPT OF NATIVE AND DENATURED STRUCTURES

The term *native* structure is commonly used but is difficult to define. It can mean any of the following: (1) the structure of a macromolecule as it exists in nature; (2) the structure of a macromolecule as isolated, if it retains enzymatic activity; (3) the form of a macromolecule that has no biological activity but possesses secondary structure.

Denatured is an equally vague term usually meaning a form of a macromolecule that has less secondary structure than that which is called native. For proteins, it usually means a random (or near-random) coil. For double-stranded (native) DNA, the term specifically means single-stranded DNA that may or may not have intrastrand hydrogen bonds and may be randomly aggregated with one or more different single strands. If some of the native structure of a molecule is lost, the molecule is considered *partially denatured.*

A transition from an ordered to a disordered structure is often called a *helix-coil transition,* although the term denaturation is used with equal frequency. Helix-coil transitions are usually detected by monitoring a change in some physical property of the molecule—for example, intrinsic viscosity, optical density, sedimentation coefficient, and so forth. The usual agents used to induce a helix-coil transition are temperature, pH, salt concentration, and chemical denaturants such as urea and guanidium chloride for proteins, and formamide, formaldehyde, and ethylene glycol for nucleic acids (Figure 1-11). If temperature is used, the temperature at which the transition is 50% complete is called the *melting temperature* (T_m), or less frequently the transition temperature. Helix-coil transitions are often studied because the parameters of the transition yield valuable

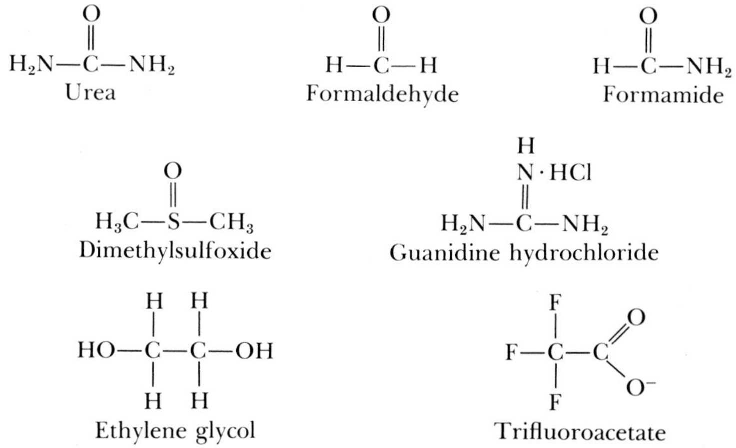

$H_2N-C(=O)-NH_2$
Urea

$H-C(=O)-H$
Formaldehyde

$H-C(=O)-NH_2$
Formamide

$H_3C-S(=O)-CH_3$
Dimethylsulfoxide

$H_2N-C(=NH \cdot HCl)-NH_2$
Guanidine hydrochloride

$HO-CH_2-CH_2-OH$
Ethylene glycol

$F_3C-C(=O)O^-$
Trifluoroacetate

FIGURE 1-11
Chemical formulas for several common denaturants. Note that many contain groups that can form hydrogen bonds.

information about the forces stabilizing the structure of a macromolecule. Examples of helix-coil transitions can be found throughout this book (e.g., Figures 14-15, 14-19, 16-11, and 17-12).

Renaturation (Figure 1-8) refers to the reformation of the native structure from a denatured form. Actually, the structure of a renatured molecule depends on the criterion used to assay restoration of the native configuration. For DNA, it specifically means reformation of a double-stranded helix from separated polynucleotide strands. For proteins, the term is less definite.

If a structure consists of subunits, separation of the subunits is occasionally called denaturation but more often *dissociation* (Figure 1-12). The latter term is used, for example, to describe a protein consisting of several polypeptide chains that has been reduced to a mixture of the separated chains. In this case, the individual subunits are not necessarily denatured by the dissociation process. On the other hand, separation of two strands of DNA is always called denaturation.

The reconstruction of a subunit-containing structure from subunits derived from *different* larger units is called *hybridization.* The mechanism may be renaturation or reassociation. For example, if DNA whose strands are labeled with the isotope nitrogen-14 is denatured and mixed with single-stranded (denatured) DNA labeled with nitrogen-15 and then subjected to renaturation, double-stranded DNA is formed, some of which consists of one ^{14}N and one ^{15}N strand (Figure 1-12). Such a ^{14}N-^{15}N-DNA is an example of *hybrid DNA.* Similarly, a double-stranded polynucleotide containing one DNA and one RNA strand is another kind of hybrid—a DNA-RNA hybrid. Proteins can also form hybrids. For example, if a protein that contains four subunits is obtained from a cow, dissociated, and mixed with the dissociated form of a similar rabbit protein, the subunits may reassociate in such a way that hybrid structures are formed, as shown in Figure 1-12.

LINEAR AND CIRCULAR POLYNUCLEOTIDE MOLECULES

Some single- or double-stranded DNA molecules (e.g., the single-stranded DNA from *E. coli* phage ϕX174 or the double-stranded DNA from *Pseudomonas* phage PM-2) have no terminal nucleotides because each nucleotide is joined covalently by means of a phosphodiester bond to the adjacent one. Such a molecule is *circular.* If there are terminal nucleotides, the molecule is *linear* (Figure 1-13). If both polynucleotide strands of double-stranded DNA are circular, the DNA is a *covalent* or *closed circle;* if there are one or more single-strand interruptions, the circle is *open* or *nicked.* Some linear DNAs (e.g., from *E. coli* phage λ) contain short, complementary single strands at each end of the molecule (i.e., strands whose base

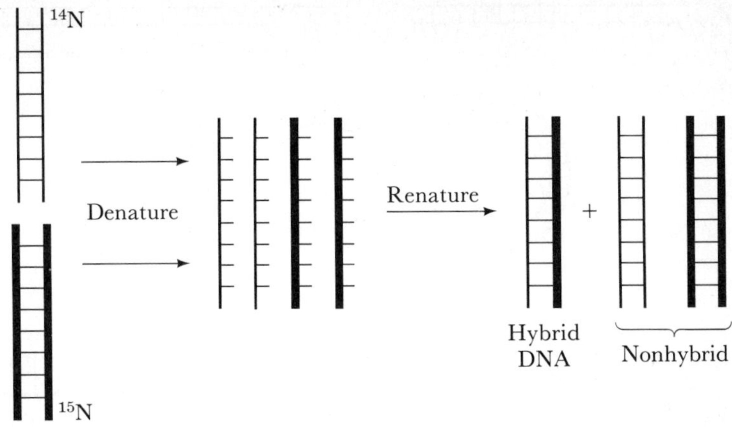

A. Formation of hybrid DNA

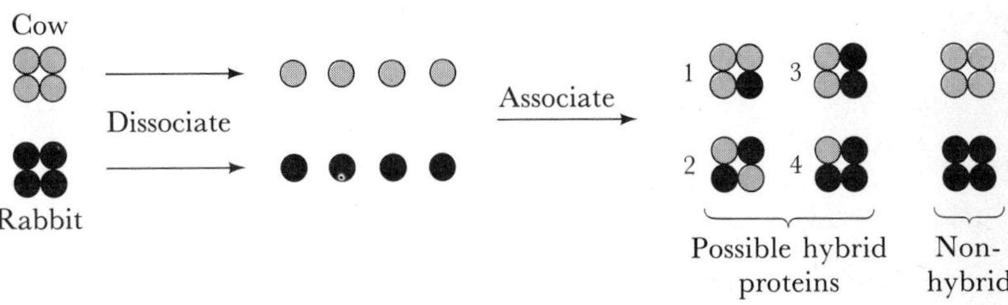

B. Formation of hybrid proteins

FIGURE 1-12
Hybridization of nucleic acids and proteins: (A) Two DNA molecules labeled with either ^{14}N or ^{15}N are denatured to produce single strands, which are then renatured. In some cases, strands having the same label renature to form the original DNA. However, hybrid DNA consisting of one strand labeled with ^{14}N and the other labeled with ^{15}N can also form. If equal amounts of ^{14}N and ^{15}N single strands are mixed, then by random association the renatured mixture will be 25% ^{14}N-DNA, 25% ^{15}N-DNA, and 50% ^{14}N-^{15}N (hybrid) DNA. (B) Formation of hybrid proteins by dissociation of subunits and reassociation. Two proteins of the same type, one from a cow and the other from a rabbit, can reassociate to form four possible hybrid types, which differ by the number and arrangement of subunits; types 2 and 3 have the same number of each subunit, but they are arranged differently.

sequences are such that the strands can be joined by the standard adenine-thymine and guanine-cytosine base pairs to form double-stranded DNA). These short strands are called *cohesive ends.* When these cohesive ends join, the resulting structure is a *Hershey circle.* If a covalent circle is twisted

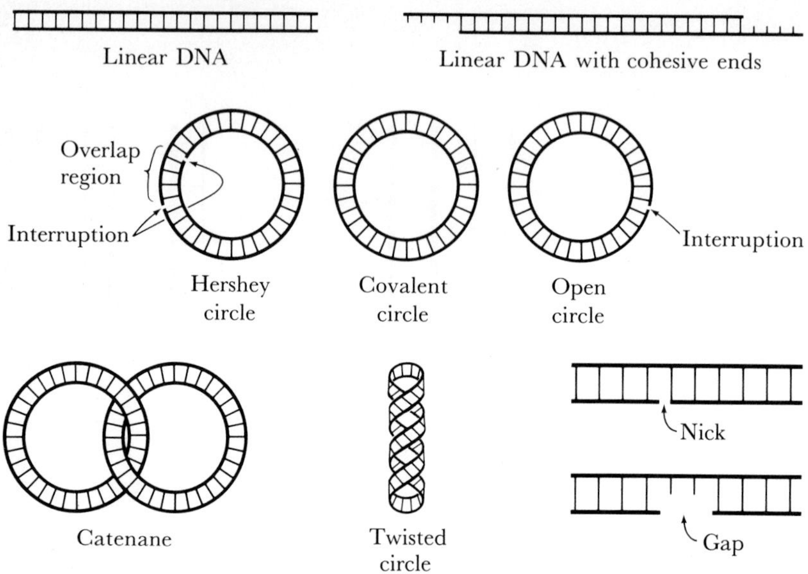

FIGURE 1-13
Various conformations of DNA.

(imagine taking a linear double-stranded DNA and twisting it before joining the ends but making sure that each strand is continuous), it is called (interchangeably) a *twisted circle,* a *superhelix,* or a *supercoil.* The number of twists per unit of molecular weight is the *superhelix density.* Note that, if the hydrogen bonds of a superhelix are broken, the individual circular strands cannot be physically separated; however, if one of the circles contains a physical interruption, the double-stranded circle can separate into two single-stranded molecules, one linear and the other circular.

In a double-stranded polynucleotide, a broken phosphodiester bond is called a *nick.* If one or more nucleotides are missing, the result is a *gap.*

CROSS-LINKS

Certain chemical and physical agents produce covalent bonds between two nucleotides. These are called *cross-links* and can be either intrastrand or interstrand: intrastrand cross-links prevent the strand from forming a random coil; interstrand cross-links prevent strand separation. A covalent bond between a polynucleotide and another molecule (e.g., a protein) is also called a cross-link. Proteins contain natural cross-links, the disulfide bonds (Figure 1-14).

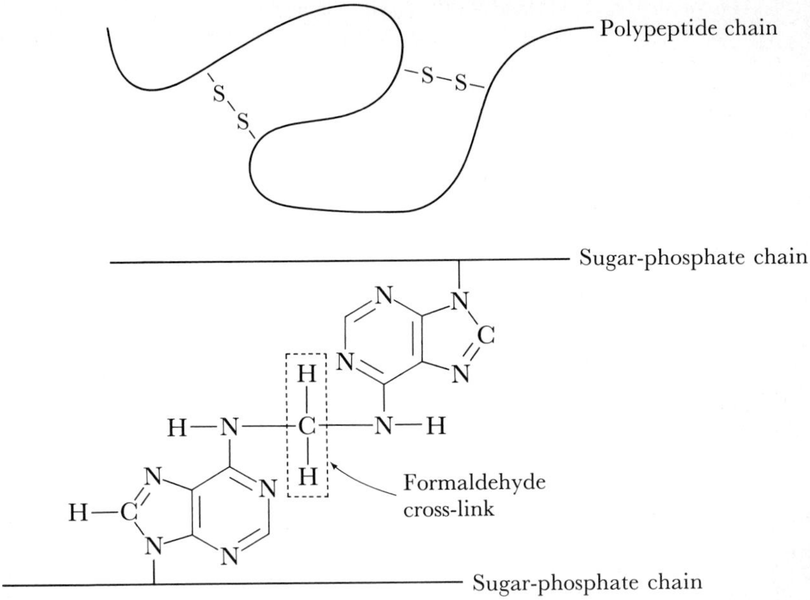

FIGURE 1-14
Two types of cross-links: (top) a disulfide cross-link (connecting two cysteines) within a polypeptide chain; and (bottom) a methylene bridge produced by formaldehyde reacting with the amino groups of two adenines in a double-stranded polynucleotide.

Selected References

Anfinsen, C. B. 1973. "Principles That Govern the Folding of Protein Chains." *Science* 181:223–230.

Bailey, J. L. 1967. *Techniques of Protein Chemistry.* American Elsevier.

Bloomfield, V. A., D. M. Crothers, and I. Tinoco. 1974. *Physical Chemistry of Nucleic Acids.* Harper & Row.

Cantoni, G. L., and D. R. Davies, eds. 1971. *Procedures in Nucleic Acid Research,* vol. 2. Harper & Row.

Cold Spring Harbor Laboratory. 1972. *Structure and Function of Proteins at the Three-Dimensional Level,* Cold Spring Harbor Symposia on Quantitative Biology, vol. 36.

Dickerson, R. E., and I. Geis. 1969. *The Structure and Action of Proteins.* Harper & Row. An excellent book for the beginner.

Flory, P. 1953. *Principles of Polymer Chemistry.* Cornell University Press. A classical theoretical book.

Haschemeyer, R. H., and A. E. V. Haschemeyer. 1973. *Proteins: A Guide to Study by Physical and Chemical Methods.* Wiley.

Mandelkern, L. 1972. *An Introduction to Macromolecules.* Springer-Verlag.

Tanford, C. 1961. *Physical Chemistry of Macromolecules.* Wiley. A great theoretical text.

Tanford, C. 1970. "Protein Denaturation." *Advan. Protein Chem.* 24:1–95.

Wold, F. 1971. *Macromolecules: Structure and Function.* Prentice-Hall.

Problems

1-1. A polynucleotide, poly AT, is a single-stranded, alternating copolymer—that is, the sequence of bases along the chain is ATATATA . . . . What kind of structure would you expect this to assume in solution?

1-2. A protein contains six cysteines. If all were engaged in disulfide bonds and if all possible pairs of cysteines could be joined, how many different protein structures would be possible?

1-3. Hershey circles are formed by heating a linear DNA that has single-stranded cohesive ends to a temperature that favors end joining. At very low DNA concentration, only Hershey circles are formed. What structures might be expected at very high DNA concentrations?

1-4. Which polynucleotide would you expect to be more compact at neutral pH, polylysine or polyvaline? Which would show the greatest dependence of shape on pH? Why?

1-5. Adenine, guanine, cytosine, and thymine have low solubility in water, whereas nucleotides are much more soluble. The solublility of the bases increases when solvents less polar than water are added to the water—for example, methanol or ethylene glycol. What would you expect to happen to DNA if it were placed in 100% ethylene glycol?

1-6. Suppose that you hybridize two proteins (as in Figure 1-12B), each consisting of three identical subunits. How many different hybrid types are possible if the subunits are arranged linearly? If arranged in a triangle?

1-7. Suppose that the proteins in Problem 1-6 are hexamers, with the subunits producing a hexagonal array. How many hybrid types are possible?

1-8. If a protein contains four identical subunits, what kinds of forces might be holding them together?

1-9. If a single strand of DNA whose sequence is AGCTAACGCGA were mixed with another strand whose sequence is TCGATTAGTCATGCGCT under conditions that favor hybridization, what type of molecule would result? Suppose that the second strand had the sequence TCGATTAGTACTGCGCT instead. Draw the resulting structure.

1-10. Calculate the fraction of ^{14}N-^{15}N (hybrid) DNA produced by denaturation and renaturation of a mixture of one part ^{14}N-DNA and three parts ^{15}N-DNA.

1-11. Write down the structure of the hexapeptide H_2N-glycine-tryptophan-proline-isoleucine-valine-methionine-COOH. How many peptide bonds are there?

PART I

Direct Observation

CHAPTER 2

Light Microscopy

Too often the biochemist or molecular biologist forgets that the chemical and physical systems and the molecules and macromolecules under study are located in cells. Not only is it worthwhile and satisfying to see the cells themselves, but also microscopic observation can yield quantitative information, especially with the important techniques of polarization, interference, and fluorescence microscopy. However, the microscope with its array of lenses, apertures, and screws can seem baffling. This chapter attempts to make it less so.

SIMPLE THEORY OF MICROSCOPY

The theory of image formation by a lens can be presented in terms of either geometric or physical optics. Geometric optics easily explains focus and aberrations; however, physical optics is necessary to understand why images are not perfectly sharp and how contrast is obtained.

The theory of geometric optics is presented in numerous texts (refer to the Selected References near the end of this chapter) and is given here only briefly. There are two rules of geometric optics from which all else follows: (1) light travels in a straight path and (2) the path bends (refracts) at an interface between two transparent media (Figure 2-1).

An ideal simple lens with two convex spherical surfaces, not necessarily having the same radii of curvature, has two focal points (Figure 2-2). If two lines representing light rays coming from a given point on an object

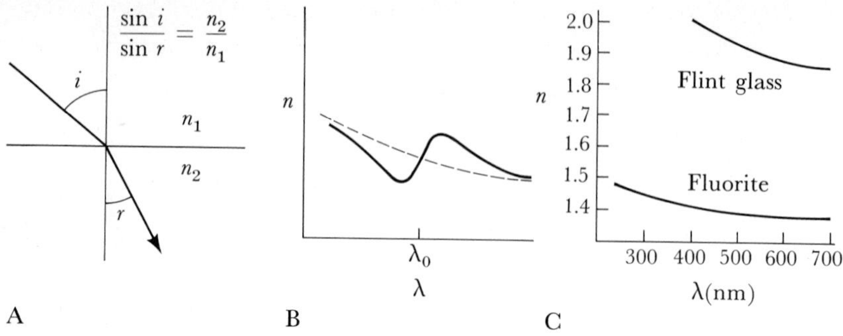

FIGURE 2-1
A. Refraction at a surface: n_1 and n_2 are the refractive indices of the medium on either side of the surface; i and r are the angles of incidence and refraction, respectively. The equation (Snell's Law) describes the relationship.
B. Dependence of n on λ: the dashed line is the kind of curve found far from an absorption maximum; the solid curve is the relation near an absorption maximum, λ_0. C. n versus λ for two materials used in the construction of microscope lenses.

(O) are drawn such that one ray is parallel to the lens axis and the other passes through the focus (F), they will emerge from the other side of the lens in a direction such that the parallel one will pass through the second focus (F′) and the other will be parallel to the optic axis, as shown in Figure 2-2. The intersection of these rays defines the image at O′. This simple construction defines an object plane and an image plane. The distances indicated in the figure obey the relation $aa' = ff'$ and the magnification of the image is $-(f/a)$, in which the minus sign indicates that the image is inverted.

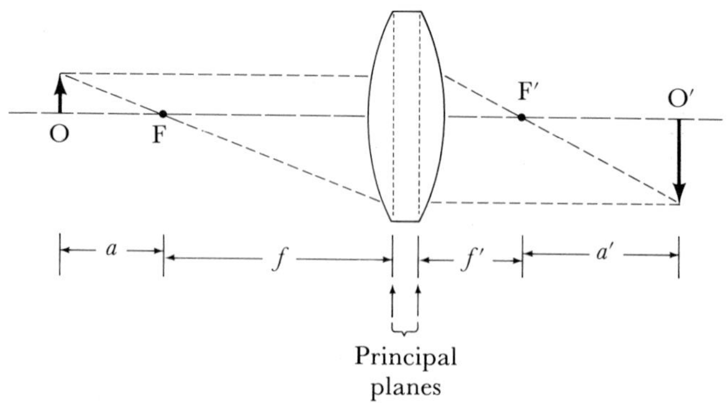

FIGURE 2-2
Image formation by a simple lens.

The simple construction just described is for an ideal lens. However, real lenses fail to bring all rays from a given point on an object to a unique focus; this is called *aberration.* The most easily understood aberration is *chromatic aberration* (Figure 2-3A), which results from the fact that, because the index of refraction of any substance depends on wavelength (Figure 2-1), the position of the focal points is wavelength dependent. Hence, with white light images will be fuzzy, being the superposition of a large number of images, not all of which can simultaneously be in focus. Chromatic aberration is simple to correct by using a lens system consisting of several types of glass for which the relations between the index of refraction (n) and wavelength (λ) balance to make n independent of λ.

Other aberrations exist even for monochromatic light. The major aberrations are called point-imaging because they result from the fact that all rays from a single point do not pass through the same image point. The principal point-imaging aberration is *spherical aberration* (Figure. 2-3B), in which rays from a single point on the optical axis of the lens are refracted by different parts of the lens and therefore do not come to focus at the same point on the axis. A second important point-imaging aberra-

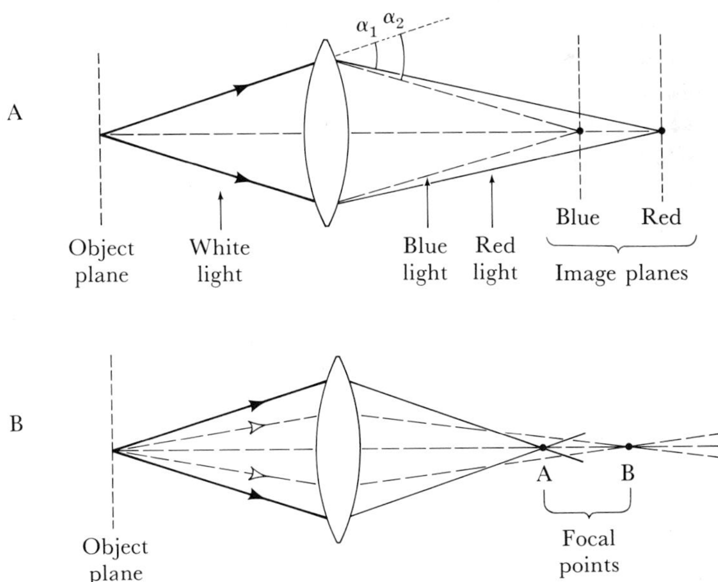

FIGURE 2-3
Two common aberrations of simple lenses: (A) chromatic aberration in which focal point depends on wavelength; (B) spherical aberration in which focal point depends on which part of the lens surface is producing the refraction.

tion is called *coma* because it gives points a comet shape so that symmetry in the object is always lost in the image. By appropriate lens construction, these aberrations are easily corrected. A lens so corrected is an *aplanatic* lens. Another aberration is *astigmatism.* Except for observation at the highest resolution (as in electron microscopy where it is extremely important), this aberration is frequently left uncorrected but, if it is corrected, the lens is called an *anastigmat.* All commercially available microscopes are corrected for aberrations before being sold.

The theory of physical optics explains image formation and resolution, which will be important in a later discussion of phase-contrast microscopy. Within the framework of physical optics, light is not thought of as rays traveling in straight lines but as electromagnetic radiation that is diffracted (bent) at edges and apertures and can interfere constructively or destructively. Because a lens is in a sense an aperture, all light passing through the lens is diffracted. What this means is that an illuminated point in the object plane appears in the image plane as a circle of light surrounded by a series of bright concentric rings resulting from constructive interference. In microscopy this pattern is called an *Airy disc* (Figure 2-4A). It can be shown that the radius of the first dark ring surrounding the central disc is $0.61\ \lambda/n \sin U$, in which λ is the wavelength of the light, n is the index of refraction on the object side of the lens, and U is the angle made by the lens axis and a line drawn from an axial object point to the edge of the aperture (Figure 2-4B). Two object points that are close together will therefore appear as two tiny discs in the image plane and the resolution of these points (i.e., the ability to state unequivocally that there are two points) is determined not only by the separation of the image points defined by geometric optics, but by the size of the tiny discs. The precise point at which resolution is lost cannot be determined (Figure 2-4C); however, the convention that the resolution limit is $0.61\ \lambda/n \sin U$ was adopted long ago. Resolution is increased by decreasing the wavelength, increasing n, and increasing U; increasing U can be accomplished either by decreasing the distance from object to lens or by increasing the lens diameter. (We will return to these points later.) The quantity $n \sin U$ is called the *numerical aperture,* or NA, and is always printed on microscope objectives. That resolution increases as angle U increases can be explained in the following way. As illuminating light falls on an object, it is diffracted by the object itself. The different orders of diffraction will leave the points of the object at various angles, which increase with the order of diffraction. If each order of diffraction is thought of as part of the information supplied by the object for forming an image, then to maximize the information all the diffracted light must be gathered. This clearly means having the maximum collection angle possible between an object and the lens aperture—that is, a large value of U.

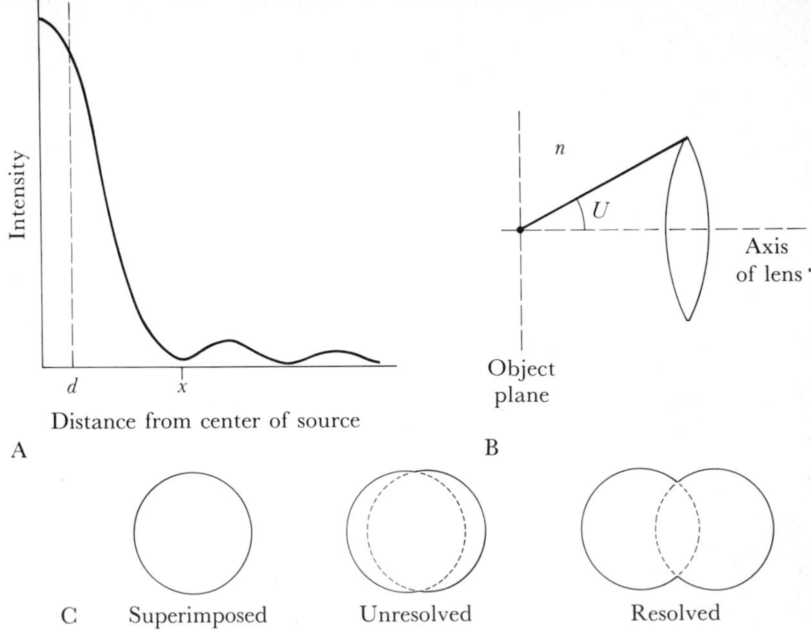

FIGURE 2-4
A. Plot of the intensity of light surrounding a light source of diameter *d*. The image is a bright central spot surrounded by bright rings and its configuration is called an Airy disc. As *d* becomes smaller, *x* increases. The size of the central region is principally responsible for the poor resolution between closely spaced point sources. As λ decreases, *x* decreases so that resolution is improved by decreasing the wavelength. B. Illustration of the angle *U*. C. Resolution of two point objects.

PARTS OF THE MICROSCOPE

A basic microscope consists of only three components: a source of light of uniform brightness, an object holder, and a magnifying lens corrected for various aberrations. However, to have very high magnification, the focal length must be very small and the eye would have to be virtually on the opposite surface of the lens. To avoid the practical problems of making a lens of very short focal length with adequate corrections while providing a reasonable distance from lens to eye, a secondary lens system called an *eyepiece* is introduced. The basic lens system (the objective lens) plus the eyepiece constitute a *compound microscope.* The basic optics of a compound microscope are shown in Figure 2-5. The object is placed just past the focal point on the object side of the lens. Thus a magnified (10–100×) and inverted image is formed far from the focal point on the image side. The eyepiece then magnifies this image somewhat (5–20×) and focuses it onto the retina of the eye, which can be at a convenient working distance.

Objective Lenses

Microscope objectives are fairly well standardized with respect to magnification and NA. Table 2-1 gives some of the parameters of typical objectives and Figure 2-6 shows a diagram of two typical objectives, a 40× "high dry" and a 100× oil-immersion lens. Note that in general NA increases with magnification so that magnification is not "empty." (Empty magnification refers to the fact that magnification can always be increased by adding more lenses but, if resolution does not increase, nothing is gained.) The NA is normally increased by decreasing the focal length because the lens diameter decreases with increasing magnification. With a lens of high NA (e.g., 1.30), this is accomplished by essentially placing the object within the lens. This is the significance of the oil in so-called oil-immersion lenses. By using an oil whose refractive index is the same as that of the glass cover slip above the object and that of the front lens element, all surfaces between object and lens are eliminated (Figure 2-6), except, of course, where the object touches the cover slip. The front lens element of an oil-immersion objective is usually greater than a hemisphere so that with the oil the object can be thought of as being within a spherical lens. This results in gathering all light diffracted in the forward direction.

Many oil-immersion lenses have a residual chromatic aberration so that the magnification of the blue image is about twice that of the red. This is objectionable for certain critical work and can be eliminated by using

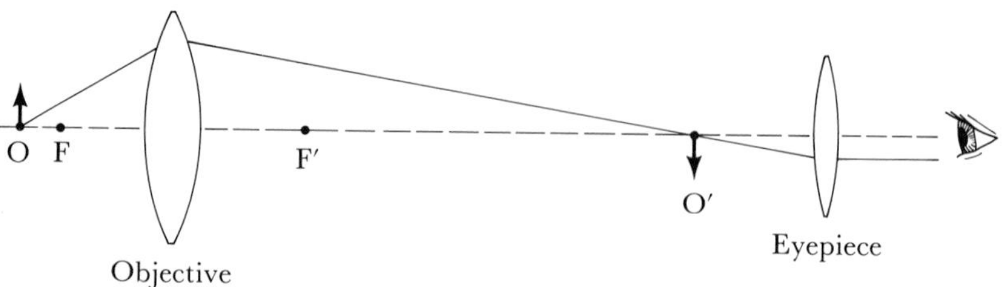

FIGURE 2-5
Optics of a compound microscope: F and F′ are the focal points of the objective.

TABLE 2-1
Properties of objective lenses.

Magnification	NA	Focal length (mm)	Working distance (mm)*	Diameter of field (mm)
10	0.25	16	5.50	2.00
20	0.54	8	1.40	1.00
40	0.65	4	0.60	0.50
40	0.95	4	0.25	0.20
95†	1.32	2	0.10	0.05

*Distance from front surface of lens to sample.
†Oil-immersion lens.

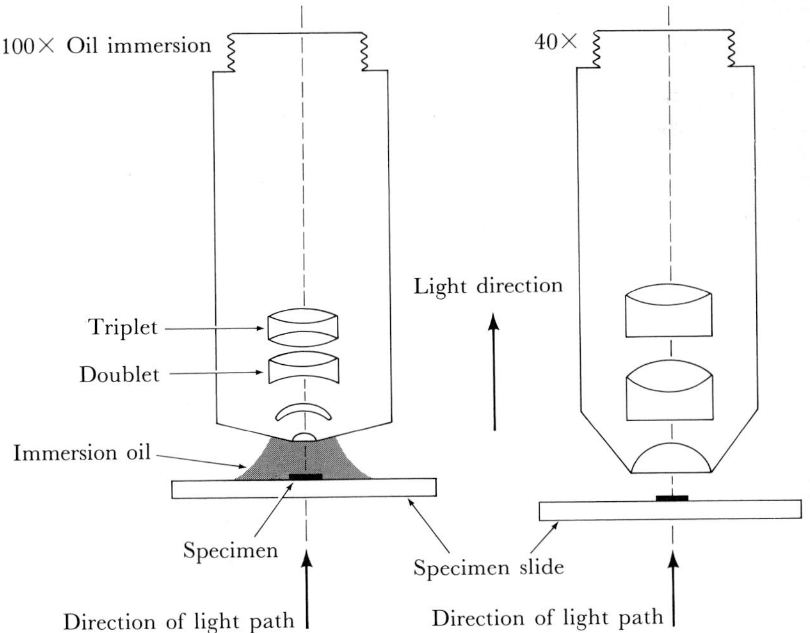

FIGURE 2-6
Typical objective lenses containing several lens elements for correcting various aberrations.

a color-correcting *compensating eyepiece* (see next section) with a partially corrected objective called an *apochromat.** The best solution for all chromatic problems is to use monochromatic illumination.

A minor aberration called *curvature of field* (long straight lines appear curved) is usually left uncorrected because normally only a small part of the field is being observed. However, curvature of field can be objectionable in photomicrography because a large area is usually examined. *Flat-field* objectives are available to correct this aberration but they are expensive and their purchase is warranted only if a microscope is used frequently for photomicrography.

Eyepieces

The principal function of the eyepiece is to deliver the image of the objective to the eye. There are various kinds of eyepieces, including Ramsden (Figure 2-7), Huygens, Kellner, and compensating. The first three are interchangeable and differ only in the manner of inserting grids, pointers, and other reference points. The compensating eyepiece is designed as a further correction of chromatic aberration. Ramsden and Huygens eyepieces each consist of two lenses at opposite ends of a tube. In the compensating and Kellner eyepieces, one of the lens elements consists of a pair of lenses cemented together (a doublet). Eyepieces come in magnifications ranging from 5 to 20. However, it should always be remembered that increasing magnification without increasing resolution is of little value and the range of magnification of eyepieces is mostly for convenience in viewing. Hence a 40× objective with a 5× eyepiece gives greater resolution than a 10× objective with a 20× eyepiece, because the 40× objective has a higher NA (the resolving power is determined solely by the objective).

Condenser Lenses

For ordinary (bright-field) microscopy, the illumination of the object must satisfy two criteria: First, it must provide a beam of light whose divergence when leaving the object plane is at least as great as the angle *U*,

*Apochromats use CaF_2 (fluorite) instead of glass for some of the lens elements because it has greater *n* and can be made with a slightly higher NA (1.35). However, such lenses cost more, they have some fluorescence (so they cannot be used in fluorescence microscopes), and the fluorite elements scatter light so that special types of illumination may be required.

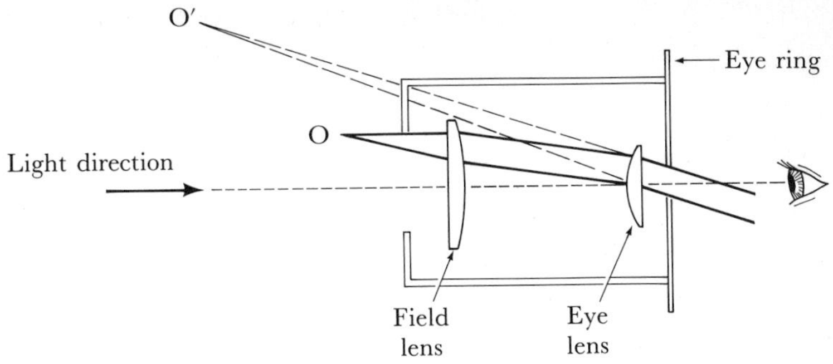

FIGURE 2-7
A Ramsden eyepiece. The image at O produced by the objective is magnified to produce the final image at O′.

to make use of the resolution capability of the objective (in jargon used by microscopists, the incident light must "fill the numerical aperture" of the objective). Second, the illumination should be uniform across the specimen and, for convenience, it should be possible to control the intensity and to eliminate stray light. To accomplish these ends, a source of light of small dimensions is employed, which is focused down to a very small area by using a set of lenses mounted in a single unit called a *condenser;* this is simply an objective operated with the light path reversed. The aberrations of a condenser are not usually corrected as stringently as are those of an objective lens because the resolving power of an objective is not significantly affected by small aberrations of the condenser. There are three basic kinds of condensers: The Abbé and aplanatic (which are for all practical purposes interchangeable) and the achromatic (which improves the color correction).

Probably the most misunderstood point in the operation of a microscope is condenser adjustment. The condenser is used to provide either *critical* or *Köhler illumination* (Figure 2-8A). In critical illumination the light source is focused by the condenser lens so that the object is illuminated by a cone of light. Operationally, this means that the position of the condenser with respect to the object is adjusted so that the spot of illuminating light is as small as possible if seen by the naked eye and as bright as possible if viewed through the eyepiece and objective *when focused on the object.* Critical illumination is very commonly used; it requires the use of a diffusing screen in front of the light source to avoid imaging the structure of the light filament on the object. This illumination method is somewhat wasteful of light but is simple to obtain. The condenser lens must be chosen so that its NA is at least as great as that of the objective in order to fill the objective aperture.

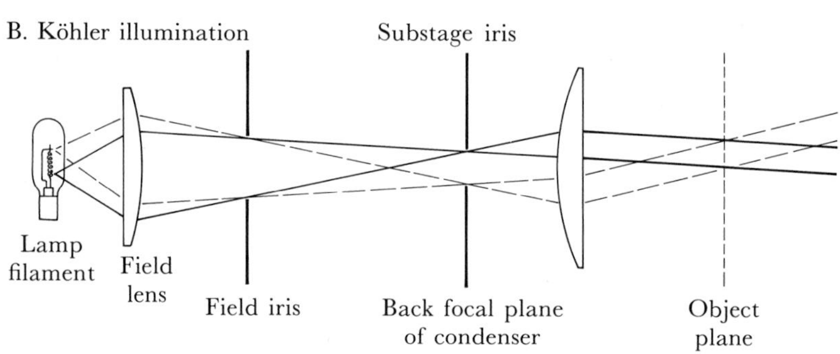

FIGURE 2-8
Types of illumination used in microscopy.

Köhler illumination is more intense and precisely controlled. An accessory lens, called a field lens, (Figure 2-8B) is placed between the light source and the condenser so that it forms an image of the source in the focal plane of the condenser (actually at the focal point). All light rays leave the condenser as parallel beams called pencils of light. They pass through the object at various angles, which increase with the distance of the source points from the optic axis. Hence the object is illuminated by a set of pencils that constitute a cone of light.* The condenser can be adjusted so that the field lens is imaged in the object plane and the lamp is imaged at the back focal plane of the objective. This is done by decreasing the field iris (Figure 2-8B) and focusing it on the object with the condenser. This provides illumination as even as that of the field lens. A condenser or substage iris is usually added to control the NA of the illuminating beam.

A complete plan of a typical microscope is shown in Figure 2-9.

---

*It is not actually a cone because the filament of the light source is not round However, because of the size and placement of the field iris, only a small circular part of the illuminated filament is used as a source.

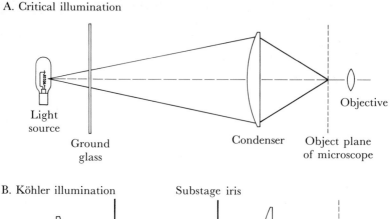

ADJUSTING A MICROSCOPE

Three adjustments must be made in setting up a microscope for use: (1) the light source and all components must be centered on the optic axis of the instrument; (2) the objective must be focused; and (3) illumination must be adjusted. In most bright-field (i.e., standard) microscopes the condenser, objectives, and eyepieces are coaxial so that only the light source must be centered. This is done by focusing on a microscope slide, removing the eyepiece, looking down the microscope tube, and moving the light source (which has adjusting screws) until the light is in the center of the objective lens. If the centering of the condenser is also adjustable, the condenser is first removed, the light source is centered as described above,

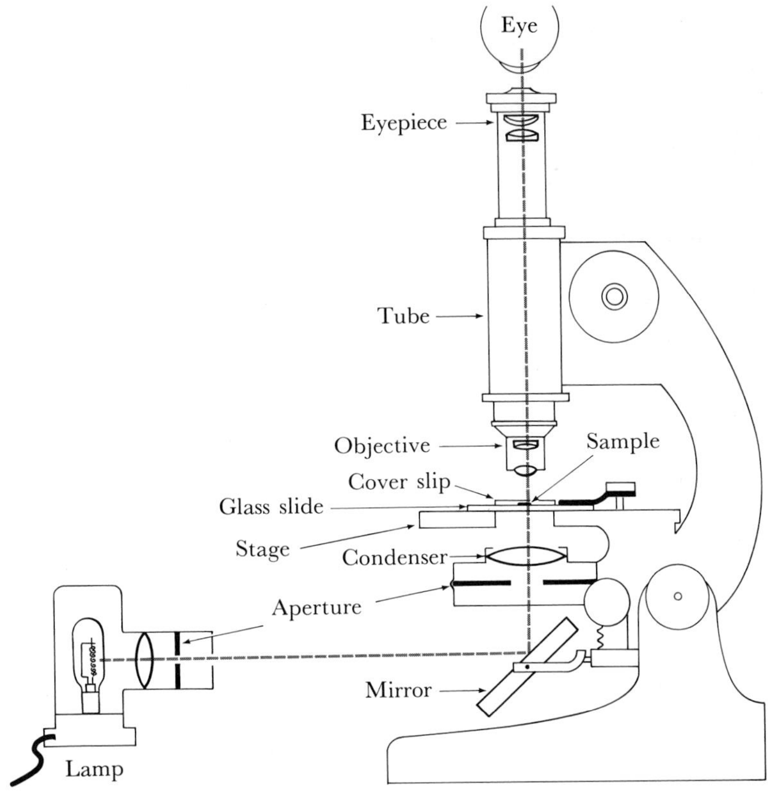

FIGURE 2-9
A complete microscope. Both tube and condenser can be separately moved along the optic axis for focusing.

the condenser is returned to the system, and the adjusting screws on the condenser are moved to center the source. The condenser is then focused on the object for critical illumination or, as discussed above, for Köhler illumination. The substage iris (Figure 2-8B) should always be adjusted so that the NA of the lens is filled, which can be determined by looking down the tube without an eyepiece and noting that the lens is fully illuminated. To eliminate stray light and glare, the field iris can be reduced so that only that part of the field containing the object is illuminated. Sometimes the field is too bright for comfortable viewing. The intensity should never be reduced by changing apertures; instead, either a neutral density filter should be inserted in front of the source or the voltage applied to the light source should be reduced.

CONTRAST

To be seen, the image of an object must differ in intensity from that of the surrounding medium. The difference in intensity between object and medium is called *contrast.* Unfortunately, most biological specimens (i.e., cells and their components) are transparent; so contrast is near zero. In the past, the solution to this problem was to stain the specimen—that is, to apply colored compounds that react with particular components of the cells. Hundreds of such substances are known, most of which have a preference (though rarely absolute) for proteins, lipids, or nucleic acids. Whether dyes exist that are really specific for particular chemical groups is a matter of some controversy. The technology of staining is described in many texts on microscopy and histochemistry (see the Selected References near the end of the chapter). Five special methods have been developed to create contrast and to obtain quantitative information from microscopy: dark-field, phase-contrast, interference, polarization, and fluorescence microscopy. We will examine them separately and give examples of the uses of each.

DARK-FIELD MICROSCOPY

Let us consider critical illumination of an object, remembering that the light coming from the condenser forms a solid cone (Figure 2-8B). If there is no object in the object plane, the field will appear to be uniformly illuminated (bright field). If an opaque disc containing a transparent annulus is inserted just below or within the condenser (Figure 2-10), the object plane will be illuminated by a *hollow* cone of light that passes

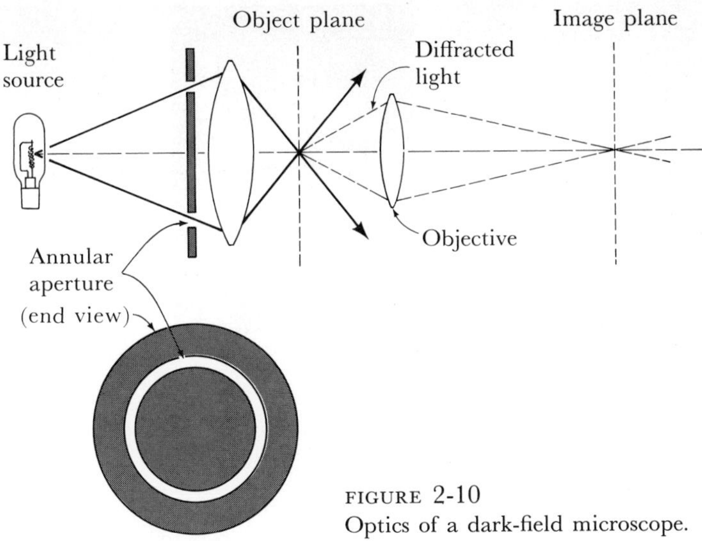

FIGURE 2-10
Optics of a dark-field microscope.

through the object plane and emerges as a second hollow cone. If the size of the annulus is sufficiently large, the hollow cone will surround the object and no light will enter the objective; in this case the field appears dark. However, if an object is present in the object plane, it will diffract light and some of the orders of diffraction will be at an angle such that they enter the objective. Even though the intensity of the diffracted light is low, a bright image will be produced against a dark background. Such an image does not have much detail (because information is lost by discarding the zero order of light), but the contrast is exceptional and governed mostly by the intensity of the light source and the degree to which internal reflections in the system have been eliminated. Hence dark-field microscopy is useful for counting small particles that are difficult to see with a bright field. Otherwise it has little use at present because of its poor resolution. If dark-field accessories are not available for a particular microscope, a dark field can be produced by gross misalignment of the light source and condenser so that little or no light enters the objective. Figure 2-11 shows a comparison of unstained bacteria seen by bright-field, dark-field, and phase microscopy.

PHASE-CONTRAST MICROSCOPY

Consider a transparent disc in a transparent medium illuminated by parallel light falling perpendicularly on the disc. For all practical pur-

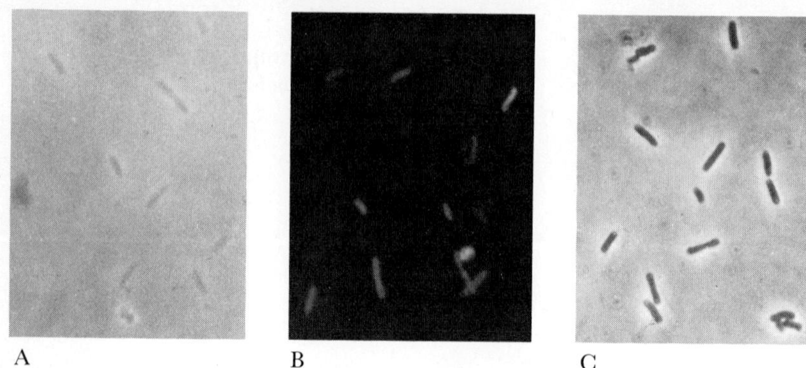

FIGURE 2-11
Bacteria (*E. coli*) seen by (A) bright-field, (B) dark-field, and (C) phase microscopy. Note that they are nearly invisible in part A.

poses, the disc is invisible because the intensity of the light passing through disc and medium is the same. However, the light transmitted by disc and medium differs in phase if the refractive indices of disc and medium are different, because a light wave is retarded when it passes through transparent matter (see Chapter 16).* Neither our eyes nor photographic film can distinguish light of different phases—hence the disc is invisible.

With small objects, the formation of an image cannot be adequately described by geometric optics; it is actually the result of two diffraction processes. The incident light is first diffracted by the object and then by the aperture of the objective. This means that every point of the object and every point of the objective contributes to the formation of every point of the image by interference of all of these diffracted waves.

Let us now consider the light that falls on the objective after passing through a point of the transparent disc. This light consists principally of zero-order diffracted light—that is, a direct beam, which might be called *undeviated light.* The diffracted light forms maxima at various angles from the zero-order wave and may be called *deviated* light; it differs in phase by one-quarter wavelength from the undeviated. The light passing through the medium is also broken down into deviated and undeviated. However, when all of these waves are recombined by the objective and projected

*It is important to understand the meaning of phase. If two sine waves moving along a single axis in the same direction have maxima and minima at the same points, they are said to be in phase. If the maxima of one and the minima of the other occur at the same point, they are one-half wavelength, or 180°, out of phase and the two waves, if superimposed, would cancel one another out.

onto the image plane, the result is an image consisting only of phase differences.

In phase microscopy the phase difference is converted into an intensity difference by the following simple trick (Figure 2-12). An annular aperture (such as in dark-field microscopy) is placed in the focal plane below the condenser so that the object is illuminated by a *hollow* cone of light. If no object is present, the hollow cone passes through the object plane and falls on the objective, forming a tiny ring of light. In the back focal plane of the objective (i.e., on the eyepiece side of the objective), an image of the condenser annulus forms. At this point a *phase plate* is introduced, consisting of a disc either with an annular groove one-quarter wavelength in depth or with an additional layer of one-quarter wavelength. For the latter arrangement, a transparent film is usually deposited directly on one of the lens surfaces in the objective. The size of this phase plate coincides exactly with that of the image of the condenser annulus. Therefore all light passing through the condenser annulus will be advanced or retarded by one-quarter wavelength, depending on whether the phase plate is a groove or a layer. If we now think of an object that diffracts the incident light, it is clear that the zero-order light (undeviated) will pass through the phase plate, whereas the diffracted light (which goes off at an angle, or is deviated) will pass through the remainder of the objective aperture. Furthermore, to make the intensity of deviated and undeviated light more nearly the same (zero-order light is always more intense than the higher orders), a thin film of aluminum or silver (neither of which alters phase but merely removes light) may be deposited on the phase plate.

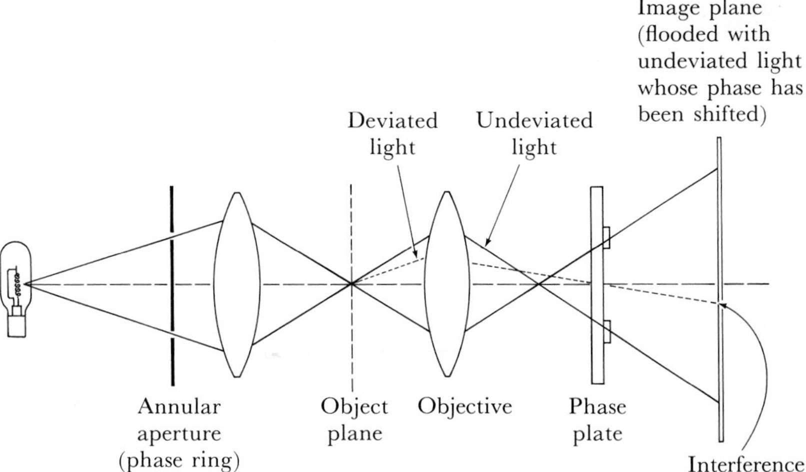

FIGURE 2-12
Optics of a phase-contrast microscope.

When the deviated and undeviated light are recombined by the objective, they differ in phase by one-half wavelength—the condition for destructive interference. Hence, the background will appear dark, the actual intensity depending on the relative intensities of the deviated and undeviated light (determined by the thickness of the aluminum in the phase plate). When an object is present, light has two components—that which passes through the object and that which passes through the medium—and the light deviated by the object will interfere destructively only with undeviated light that is at the same position in the image plane. Remember that the object itself introduces a phase difference because its index of refraction differs from that of the medium. This phase difference, when superimposed on the difference created by the phase plate, results in a loss of destructive interference. Hence the object will appear brighter than the background. (If the undeviated light were advanced by one-quarter wavelength, the object would be dark against a bright background.) Clearly, the darkness of each object depends on its particular refractive index and thickness.

In this simple way, "invisible" objects can be made visible. The only disadvantage of phase microscopy compared with interference microscopy (described in the next section) is that objects are surrounded by unavoidable halos caused by diffraction by the phase plate. Figure 2-13 (as well as Figure 2-11) shows a comparison of the same cell viewed by bright-field and phase-contrast microscopy.

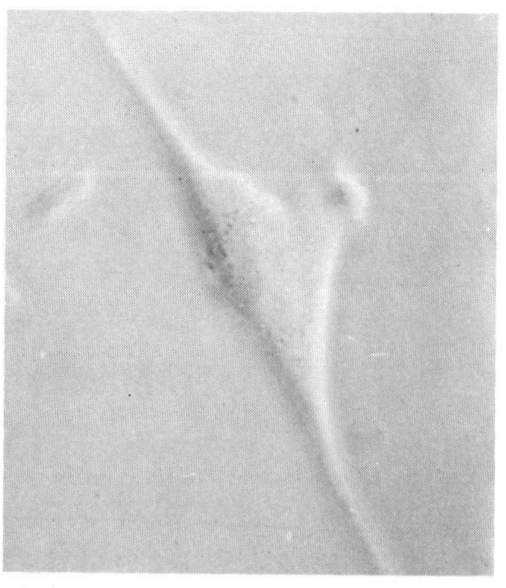

A

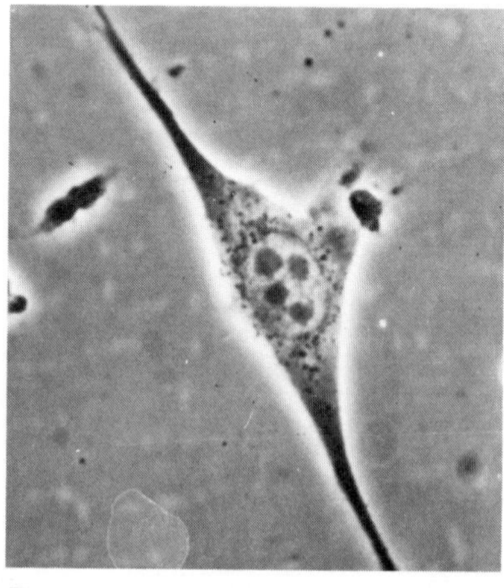

B

FIGURE 2-13
A mouse fibroblast seen by (A) bright-field and (B) phase-contrast microscopy. Note the vastly improved contrast in part B.

Alignment of a phase microscope is critical not only because the criteria for standard microscopy must be satisfied, but also because the condenser annulus must be precisely centered and focused on the phase plate. The method of alignment is described in the appendix at the end of this chapter.

Uses of the Phase Microscope

The only requirement that a sample must meet for phase microscopy is that it is nonabsorbing because the microscope is based on phase differences only. For absorbing objects (e.g., stained cells), intensity differences are added to the phase effects, often resulting in reduced contrast and poor image quality.

The phase microscope is of great value in visualizing the organelles of living cells (Figure 2-13). Nuclei contrast strongly with the cytoplasm and cytoplasmic components such as mitochondria, vacuoles, and fat droplets are easily seen. Nucleoli and chromosomes are especially visible. A membrane can be seen clearly as a dark line with a bright halo on either side although, because of the halo, unit membranes are not easily distinguished from double membranes. It is difficult to see detail in small objects such as bacteria because of the halo effect on the cell wall. However, by placing bacteria in a medium whose refractive index equals that of the cell wall or cytoplasm (e.g., concentrated solutions of bovine serum albumin), internal detail, such as nuclear bodies, become relatively easy to see.

INTERFERENCE MICROSCOPY

Like the phase microscope, the interference microscope is capable of converting phase differences into intensity differences, with the advantages that there is no halo effect and that certain quantitative measurements are possible. The most effective form of the instrument (the Dyson instrument) employs a complex system of silvered surfaces and reflectors that divides the light emerging from an object so that a fraction of it passes through the surrounding medium and a phase-shifting plate. When this fraction recombines with the other fraction of the light, that is, that which has not undergone the phase shift, interference occurs. In this way phase shifts introduced by regions in the object having different indices of refraction and/or thickness are converted into intensity differences. If white light is used, these regions have different colors, whereas with monochromatic light only intensity differences are seen.

This brief treatment is not meant to explain fully how this relatively complex instrument works. For a detailed explanation the references given near the end of the chapter should be consulted. The main point is that the student of biology should be aware of the existence of this rather uncommon instrument because it has the following capabilities:

1. Without the halo produced by the phase microscope, contrast of objects is enhanced. Small detail not easily discernible by the phase microscope can thus be seen.
2. Quantitative measurements not possible with the phase microscope can be made. The difference in optical path between a particle and the surrounding medium can be measured. Because optical path is the product of index of refraction and thickness, one can be measured if the other is known. Furthermore, the concentration of a known material can be determined if the index of refraction and the *specific refractive increment* (change in index of refraction per unit amount of solute) is known. If n_p and n_s are the indices of refraction of a protein and the solvent, respectively,

$$n_p - n_s = \alpha c$$

in which α is the specific refractive increment and c is the concentration in grams per 100 milliliters. Because n_p and α do not vary enormously from one protein to another, a rough estimate of the protein concentration within a cell can be easily made.

POLARIZATION MICROSCOPY

Note: Before reading this section, Chapter 16 should be consulted for a discussion of plane-polarized light.

Structures consisting of elongated particles in a parallel array or stacked discs embedded in a medium whose refractive index differs from that of the structure particles exhibit *form birefringence* (Figure 2-14). This means that the structures will pass plane-polarized light only if the plane of polarization is parallel to the particles. This will be true even if the particles themselves are not intrinsically birefringent—that is, if the particles transmit polarized light with equal probability for all angles of incidence. (Form birefingence is the same phenomenon encountered when fibrous molecules are subjected to flow through a tube or to the shear gradient between coaxial cylinders, as discussed in Chapter 18).

Form birefringence is easily observed in cellular material using a polarization microscope. It is of great importance not only because it can be

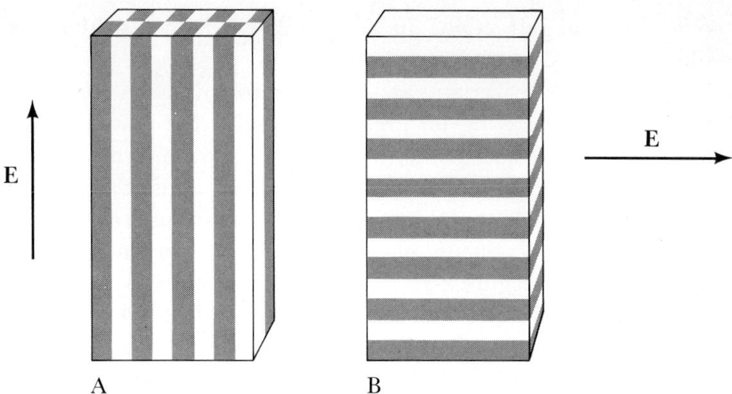

FIGURE 2-14
Structures showing form birefringence: (A) positive and (B) negative. The plane of the **E** vector passed by each is shown.

used analytically to determine orientation, but also because in some cases it is the only property that can be used to make a structure visible (e.g., cases in which the particles cannot be stained or in which their concentration or specific refractive increment is too low to generate a phase difference large enough to make it visible by phase contrast or interference microscopy).

The essential components of a polarization microscope are a polarizer located between the light source and the condenser; a rotating stage or sample holder; an analyzer, situated between the objective and eyepiece, that can be set so that its axis is perpendicular to the polarizer axis (in which case the polarizer and analyzer are said to be *crossed*); and a compensator (Figure 2-15). Because polarized light is partly depolarized by

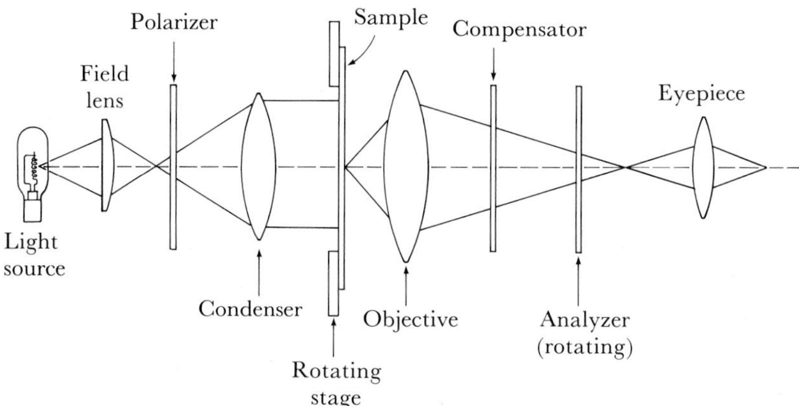

FIGURE 2-15
Optics of a polarization microscope.

reflection and refraction, the condenser is adjusted so that the object is illuminated by nearly parallel light.

When the polarizer and analyzer are crossed and either no object or an *isotropic* object (having no preferred axis for the refraction of polarized light) is in the object plane, the field appears uniformly dark. If a birefringent object is present and lying so that its axis is at an angle other than 90° with respect to the plane of polarization, it will resolve the polarized light into two components, one parallel and one perpendicular to the plane of the analyzer (Figure 2-16). Hence some light will pass through the analyzer so that the object will appear bright against a dark background. As the object is rotated, it will be invisible when lined up with either the polarizer or the analyzer and will have maximum brightness at an angle of 45°. (This is the reason for the rotating stage.)

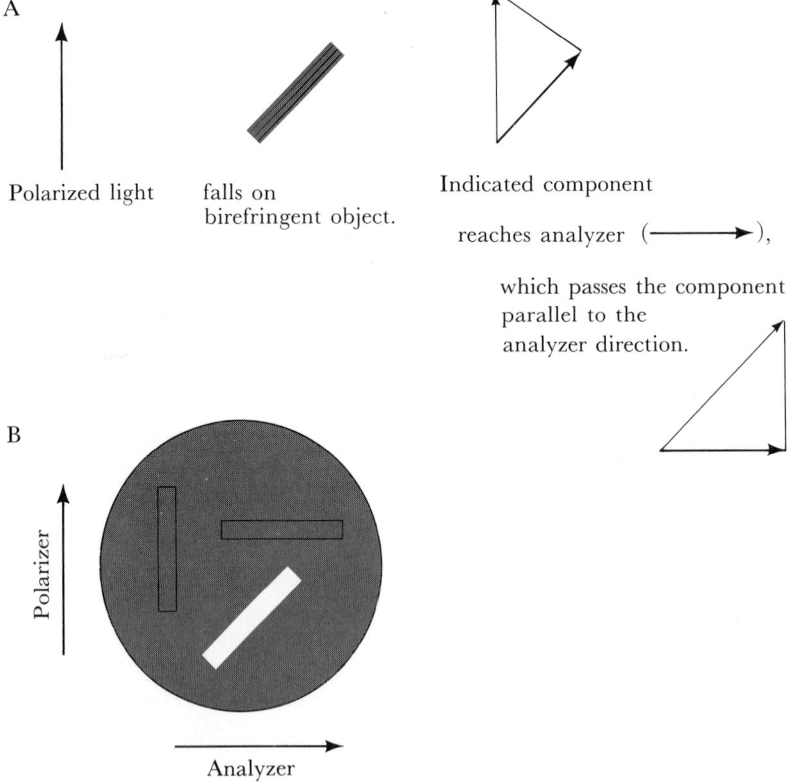

FIGURE 2-16
A. Resolution of polarized light by a birefringent object. B. The appearance of a birefringent object aligned with the polarizer, aligned with the analyzer, and at 45° to both.

Let us consider a rodlike structure with form birefringence oriented so that its long axis is at 45° with respect to the crossed polarizer and analyzer. Brightness will be maximal if it consists of either parallel elongated molecules or stacked discs, because both the axis and the plane of the discs, being perpendicular, would be at a 45° angle. The question is how to distinguish the parallel ("positively" birefringent) from the perpendicular ("negatively" birefringent) structure. This is done with a *compensator*. To understand how the compensator works, it is necessary to remember that an object is birefringent if the index of refraction of light polarized parallel to the axis differs from that for light polarized in the perpendicular direction. For positive birefringence, the parallel direction has the lower index of refraction and therefore allows the light to travel faster than in the perpendicular direction. If the velocities of the parallel and perpendicular components were equalized, the net birefringence would be zero and the object would become invisible. Consider the effect of interposing between the object and the analyzer a thin layer of a birefringent crystal (e.g., mica, gypsum) whose slow and fast direction are known and indicated. If the slow direction of the crystal is parallel to the fast direction of the birefringent sample, the velocities of the parallel and perpendicular components approach one another and the brightness of the object decreases. Therefore, if such a unit (which is the compensator) is rotated, the sign of the birefringence can be identified by means of the position of the compensator—that is, that which produces minimum or maximum brightness. The value of this capability will be described in some of the examples that follow.

Use of the Polarizing Microscope in Biology and Biochemistry

Often, the polarizing microscope can give detailed information about molecular architecture in a relatively short time. In some studies of living cells, this is the only applicable method because the more precise and sophisticated techniques described in later chapters require either a dried (i.e., dead) sample or a large volume.

Example 2-A. Orientation of molecules.

Birefringence seen in a cell indicates that there is a structure containing oriented molecules. The sign of the birefringence tells how the components are oriented. For example, studies of muscle cells and sperm tails indicated that they contain molecules arranged parallel to the fiber direction long before such structures were seen in the electron microscope.

Similarly, the birefringence of chloroplasts and the rod cells of the retina indicate that they have a lamellar (stacked-disc) structure. Oriented fibers of DNA also show the disc structure due to the purine-pyrimidine base pairs.

Birefringence measurements can also give some indication of a mixture of two oriented components. For instance, the myelin sheath of nerve cells is positively birefringent and consists of proteins and lipids. When the lipids are extracted by solvents, the positive birefringence increases so much that one may conclude that the lipid molecules must be nearly perpendicular to the highly oriented proteins.

Example 2-B. Visualization of structures not otherwise visible.
There are cell structures that are not visible by either bright-field or phase-contrast microscopy but are easily seen in a polarization microscope. An example is the mitotic spindle (Figure 2-17) and the orientation of molecules in the phragmoplast (boundary region) between plant cells undergoing cell division. By time-lapse photography and polarization microscopy, it is even possible to follow the development of these structures in living cells.

Example 2-C. Identification of helical arrays.
If positively birefringent fibers are observed looking down the fiber axis, they look dark for all rotations. However, if the fiber is helical, the fiber appears alternately dark and light, the dark corresponding to the regions where the rung of the helix points directly at the observer. In this way helical structures can be identified. For example, the inclusion bodies (orderly aggregates) of tobacco mosaic virus in tobacco leaf cells appear helical in cross section.

Use of the Polarizing Microscope to Measure Dichroism

All the objects discussed in the examples are transparent. Many objects possess a preferred direction of absorption called dichroism (see Chapter 14), which again may be intrinsic or by virtue of form. Intrinsic dichroism is most common in chemicals containing conjugated rings in which absorption is maximal when the plane of polarization is in the plane of the ring.

A polarizing microscope can be used to detect dichroism by removing the analyzer and compensator because the object itself will behave as an analyzer. Hence, if a dichroic sample is illuminated with plane-polarized light of a wavelength that can be absorbed and is rotated, there will be an angle at which it will appear darkest. If the dichroism of the molecules known to be in a given structure is known, the orientation of the molecules in that structure can be determined. This technique has not had widespread use, but the following examples should indicate the great power of the method.

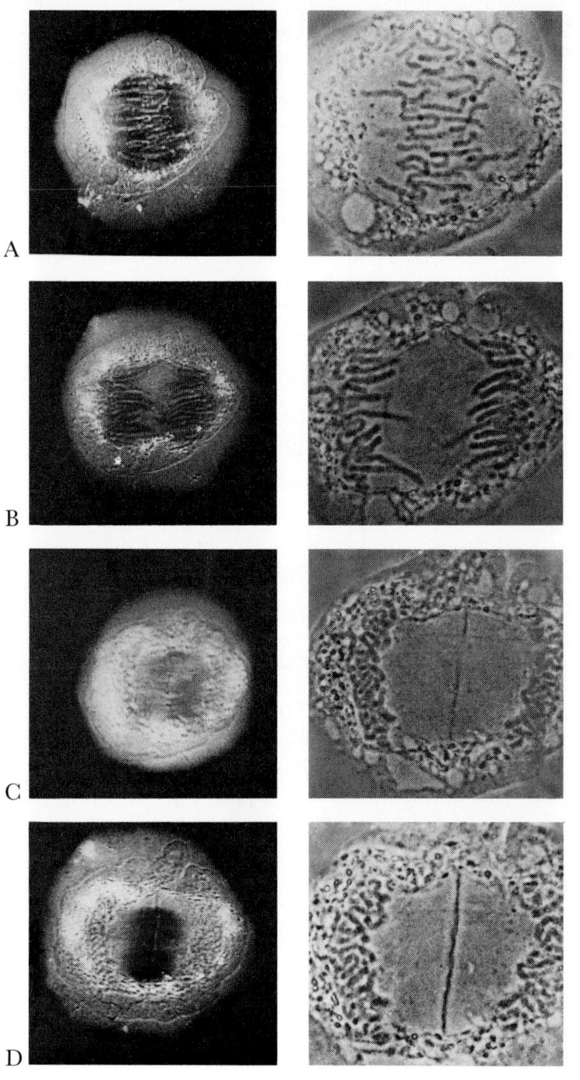

FIGURE 2-17
Cell division of *Haemanthus* endosperm seen by (left) polarization and (right) phase-contrast microscopy. The same cell has been followed through various stages of mitosis: (A) metaphase—the mitotic spindle is visible in the polarization micrograph by virtue of its birefringence and the chromosomes are visible by phase contrast; (B) anaphase—the mitotic spindle has begun to disintegrate; (C) telophase—the spindle is nearly gone and the septum separating the cells is forming; (D) the phragmoplast (an oriented array on the cell surface) is clear by polarization microscopy, whereas only the well-formed septum is seen by phase contrast. The magnitude on the right is 1.6 times that of the left. The cells on the left do not fill the illuminated area, which has been reduced by an aperture to eliminate stray light, which lowers contrast. [Photomicrographs courtesy of Robert Haynes and Raymond Zirkle.]

Example 2-D. Orientation of absorbing groups in crystals.
The orientation of the heme group in hemoglobin crystals can be determined by viewing the crystals with blue light. Because absorption along the *b*-axis of the crystal is much greater than in the perpendicular direction, the flat surface of the heme group must lie along the *b*-axis.

Example 2-E. Orientation of macromolecules in large structures.
The absorption of polarized ultraviolet (UV) light by stretched chromosomes indicates the net orientation of DNA in the chromosome because, for DNA, UV absorption is maximum perpendicular to the helical axis (i.e., parallel to the plane of the base pairs).

FLUORESCENCE MICROSCOPY

Fluorescence is an extraordinarily sensitive method for detecting minute quantities of material because, for any fluorescent material, the total intensity of the fluorescence is proportional to the intensity of the incident light (see Chapter 15). The principal problem, however, in detecting fluorescence is that of separating the fluorescence from the incident light. With a fluorescence spectrophotometer, this is accomplished by viewing the sample at right angles to the incident beam. With a fluorescence microscope, this is not easily done and optical filters are used instead (Figure 2-18). A filter that allows the exciting wavelengths but not any wavelengths in the fluorescence spectrum to be transmitted is placed between the light source and the condenser. Another filter that transmits the fluorescence wavelengths but not the exciting light is placed somewhere between the objective and the eye. Hence, in the absence of a fluorescent object, no light reaches the eye and the field is black. If an object that fluoresces is present, it will contrast strongly with the background. To analyze weak fluorescence, it is necessary to use microscope slides, coverslips, and lens elements made of nonfluorescing glass because any of these can lighten the background significantly. Apochromatic lenses must never be used because the CaF_2 elements in them are strongly fluorescent.

Because few cell components are very fluorescent (other than molecules like tryptophan, tyrosine, and riboflavin, which are found throughout the cell) and even fewer can be excited by short wavelength visible or near-UV light (because short wavelength UV requires special lenses and illumination systems), extrinsic fluors are always used—that is, a fluor that binds to particular cell components is added. Common fluors are acridine orange (for nucleic acids), fluorescein, and quinacrine.

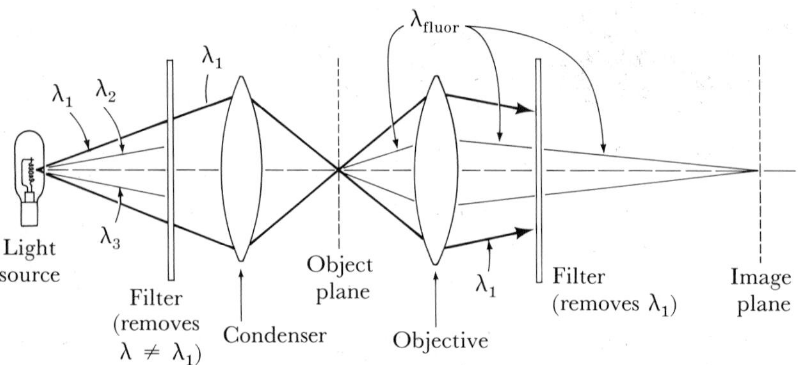

FIGURE 2-18
Optics of a fluorescence microscope.

Fluorescence microscopy is used for three purposes: to visualize components difficult to see, to localize substances by means of specific binding, and to determine orientation by means of fluorescence polarization. How this is done can be seen in the following examples.

Example 2-F. Visualization of nucleic acids in animal and plant cells. Acridine orange binds to both DNA and RNA but gives a green fluorescence with DNA and, if the dye concentration is high, an orange fluorescence with RNA (Figure 2-19). Hence, at low dye concentration, eukaryotic cells show a bright green nucleus and pale green cytoplasm; at high concentrations, the cytoplasm becomes orange. During mitosis, chromosomes glow a bright green and their morphological characteristics are easily seen *in the living cell.* In a cell infected with a DNA virus, viral inclusions can be detected and counted as bright green spots in the cytoplasm.

Example 2-G. Visualization of small organelles. The yeast nucleus is very difficult to see because of its small size. If acridine orange is added at low concentration, a bright green polar object is seen outside of the vacuole (Figure 2-19). At high concentrations, the vacuole remains colorless, the nucleus becomes a brighter green, and the cytoplasm glows orange. This was actually the first dem-

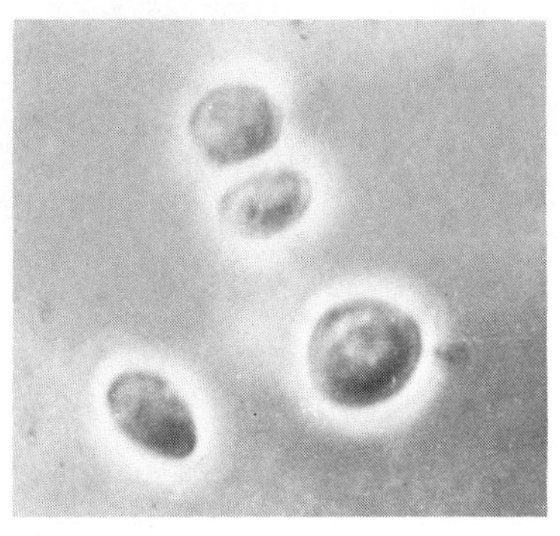

A

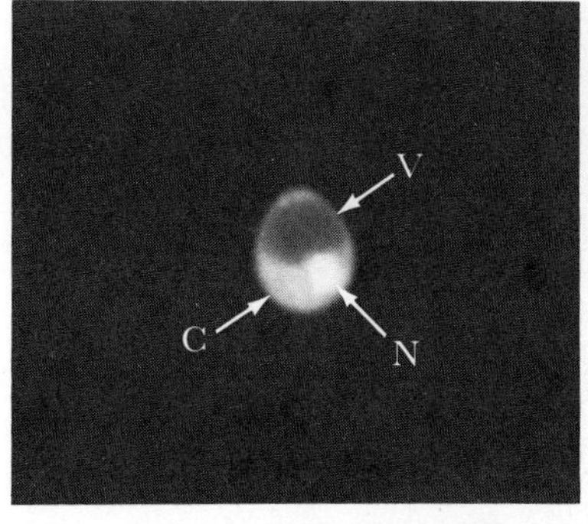

B

FIGURE 2-19
Photomicrographs of resting cells of baker's yeast: (A) as seen by phase-contrast microscopy; (B) as seen by fluorescence microscopy. Note the lack of detail and the bright halo around the cells in part A. In part B, the fluorochrome–acridine orange–has been added, causing the nucleus (N) to become bright green and clearly distinguishable from the cytoplasm (C) containing RNA, which becomes orange, and the vacuole (V), which contains no nucleic acid and is therefore nonfluorescent.

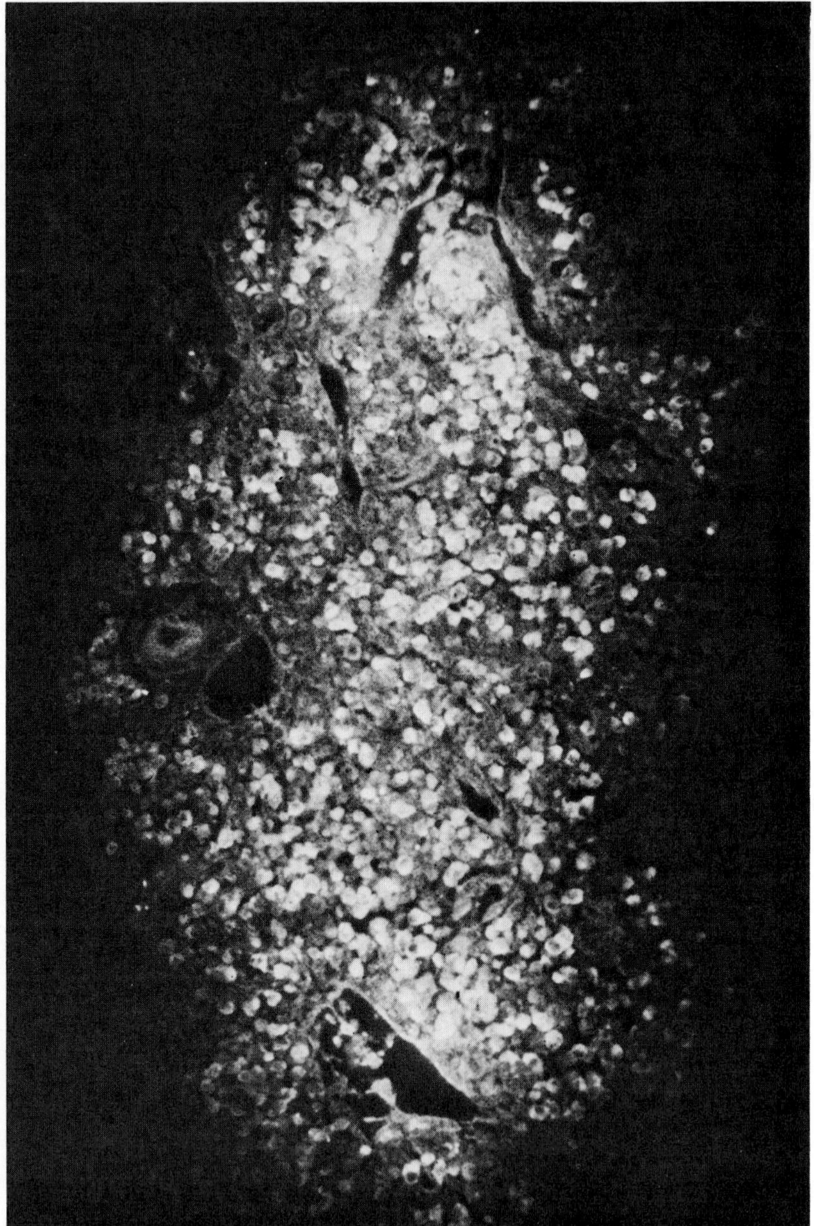

FIGURE 2-20
An example of the use of the fluorescence microscope. This cross section of a rabbit popliteal lymph node shows cells containing antibody against bovine serum albumin (BSA) on the fourth day after a booster injection. Many cells show the yellow green fluorescence of fluorescein (white) against the bluish

onstration that yeast contained DNA, because yeast has such a large RNA/DNA ratio that, in the early 1950s, chemical tests failed to show its presence.

Example 2-H. Fluorescent antibody technique.
Fluorescein can be covalently linked to antibodies prepared against various cell fractions and against proteins (Figure 2-20). Addition of the fluorescent antibody to cell sections (thin slices) or to cells made permeable to protein by treatment with acid or acetone allows localization of these substances. For example, viral antigens, cell-membrane components, histones, and many other substances have been localized in individual cells or in tissues by this method. The proteins of actin and myosin in muscle fibers were distinguished by fluorescein-labeled antiactin and antimyosin.

Example 2-I. Polarization-fluorescence microscopy.
It is known that acridine orange intercalates between the DNA base pairs. The plane of polarization of the fluorescence of acridine orange is also known and, hence, the angle of polarization of the fluorescence with respect to the DNA helix axis. Acridine orange binds tightly to chromosomes, presumably by DNA binding. Therefore, from the polarization of the fluorescence of the chromosomes, the orientation of the DNA molecule with respect to the chromosome can be determined.

Example 2-J. Identification of chromosomes by quinacrine fluorescence.
Quinacrine binds to DNA, either in solution or in chromosomes. Quinacrine mustard binds even more efficiently by reacting with free amino groups of nucleotides in DNA. When it is added to cells, chromosomes are not only fluorescent but show characteristic patterns of bright and dark bands. The cause of the pattern of bands is not yet clear; probably there is enhanced binding of quinacrine to regions high in adenine-thymine base pairs and to regions to which certain histones are prefer-

---

background of the other cells (grey). The antibody is in the cytoplasm. The tissue was fixed in 95% ethanol at 4°C overnight, then dehydrated in cold 100% ethanol, cleared in cold xylene, and finally embedded in paraffin below 60°C. Sections were then made, deparaffinized, and hydrated. Then the section was reacted with BSA (0.5 mg/ml in saline) and washed; thus, BSA was bound only in regions containing antibody, and unbound BSA was washed off. Anti-BSA labeled with fluorescein (two molecules per molecule of protein) was layered over the section, to react with whatever BSA had been bound. This revealed the antibody that was originally in the cells. In this way, it was shown that the cells in the lymph node had synthesized anti-BSA. [Courtesy of Albert Coons.]

entially bound. The value of this method is that chromosomes that are very difficult to distinguish morphologically can be easily distinguished by their patterns of bands. Furthermore, chromosomal abnormalities not otherwise recognized can be observed; therefore, this method is attaining great clinical importance. Figure 2-21 shows chromosomes stained with quinacrine mustard.

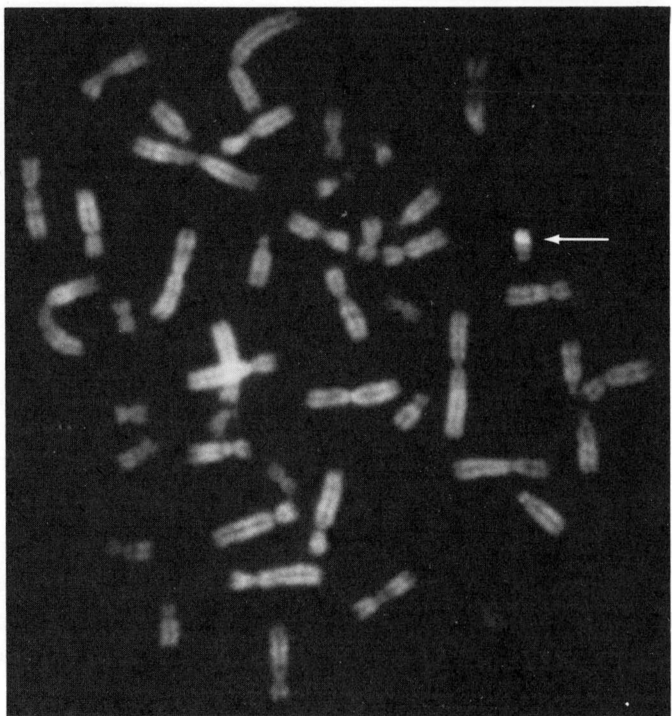

FIGURE 2-21
Chromosomes visualized by fluorescence microscopy. These are metaphase chromosomes from peripheral (blood circulation) leukocytes of a normal human male that have been stained with quinacrine mustard. Note that the chromosomes do not fluoresce with equal intensity but are banded. Chromosomes that are morphologically indistinguishable by phase-contrast microscopy or by bright-field microscopy using standard histological stains can be easily distinguished by their banding patterns. Note in particular the brightly fluorescent Y chromosome (indicated by the arrow). The background is dark because only the chromosomes are fluorescent. Details of the utility of the method can be found in T. Casperson, G. Lomakta, and L. Zech, *Hereditas* 67(1971):89. [Photomicrograph courtesy of Edward Modest.]

Appendix

Alignment of a Phase-contrast Microscope

To obtain phase contrast, it is important that the image of the condenser annulus (sometimes called the phase ring) is precisely focused on the phase plate. This is accomplished in the following way: The condenser and eyepiece are removed and the light source is adjusted so that it is roughly centered in the objective when viewed down the tube. An object is then put on the stage, the eyepiece is reinserted, and the objective with desired magnification is swung into position. The objective is then focused on the object. The condenser is replaced and focused so that an image of the lamp iris is in focus in the object plane (i.e., Köhler illumination). A special lens, called a Bertrand lens, is slid into place just below the eyepiece. With this lens the back focal plane of the objective—that is, the position of the phase plate—can be viewed. The appropriate condenser annulus is swung into position (there is a separate annulus for each objective because the size of the annulus and the phase plate must be matched). The condenser annulus is then moved by means of adjusting knobs until it is concentric with the phase plate. A slight refocusing of the condenser is usually necessary at this point so that the image of the condenser annulus is precisely superimposed on the phase plate. The Bertrand lens is then removed and the system is ready for use.

Selected References

Gurr, E. 1965. *The Rational Use of Dyes in Biology and General Staining Methods.* Williams & Wilkins.

Martin, L. C. and W. T. Welford. 1971. "The Light Microscope," in *Physical Techniques in Biological Research,* vol. 1A edited by G. Oster, pp. 2–70. Academic Press.

Oster, G. 1955. "Birefringence and Dichroism," in *Physical Techniques in Biological Research,* vol. 1, edited by G. Oster and A. W. Pollister, pp. 439–459. Academic Press.

Osterberg, H. 1955. "Phase and Interference Microscopy," in *Physical Techniques in Biological Research,* vol. 1, edited by G. Oster and A. W. Pollister, pp. 378–437. Academic Press.

Slayter, E. M. 1970. *Optical Methods in Biology.* Wiley. This excellent book explains just about everything.

Zernike, F. 1942. "Phase Contrast, a New Method for the Microscopic Observation of Transparent Objects," Parts 1 and 2. *Physica* 9:686–693; 974–985. The theory of phase contrast was developed in these two papers.

Problems

2-1. If you had a bacterial culture and you wanted to count the number of bacteria per milliliter, a reasonable way would be to count the number of particles in a measured volume using a phase-contrast microscope. But, if you wanted to count viruses, this method would not work because they are too small to be seen. Explain how such a count might be obtained using a fluorescence microscope and acridine orange. What property should the virus have to be optimally counted? Would it be easier to visualize a large DNA or RNA virus?

2-2. Would there be any difference in appearance between a stacked-disc and a stacked-ring structure in a polarizing microscope?

2-3. Explain why the image quality will be poor if an absorbing object is used in phase microscopy.

2-4. Explain how you could use a polarizing microscope to show that an object is dichroic. Would there be any significant difference between eliminating the polarizer and eliminating the analyzer?

2-5. A particular type of cell structure can be stained a light pink. However, it is barely visible with an ordinary bright-field microscope. What simple modification to the microscope could you make to increase the contrast between the structure and the surrounding medium?

2-6. If an object becomes positively birefringent when stretched, what can you guess about its molecular structure?

2-7. It is usually possible to vary the intensity of the illuminating source for all microscopes; to vary for ease of viewing and to prevent eye strain, the intensity is kept relatively low. For what type of microscopy would the maximum illuminating intensity be desirable?

2-8. Give several reasons for using cover slips in preparing samples for microscopy. What would happen to the image if the sample were in liquid but no cover slip were used?

2-9. If a cell were suspended in a medium whose index of refraction is the same as that of the cell wall, how would the cell appear?

2-10. If acridine orange (AO) is added to animal cells in mitosis (i.e., when the chromosomes have condensed), the chromosomes appear bright green when viewed with a fluorescence microscope using light that excites AO fluorescence.

We will use this technique to interpret the following experiment. An instrument exists (the UV microbeam) that allows a small part of a cell to be irradiated with an intense beam of ultraviolet light. If a part of a chromosome is irradiated with a spot whose diameter is larger than the cross section of a chromosome, five observations can be made: (1) the chromosome remains intact; (2) if observed with a phase microscope, the irradiated part of the chromosome appears pale compared with the remainder of the chromosome (which is normally black); (3) if observed with an interference microscope, using white light, the irradiated part is a different color from that of the remainder of the chromosome; (4) if AO is added, the irradiated part is not fluorescent; and (5) if the cell is stained with a dye that produces a red color wherever deoxyribose is present, the irradiated part is unstained, whereas the remainder is red.

What is the probable effect of the UV irradiation? What can you say about the probable composition of chromosomes? What can you say about the physical structure of chromosomes?

2-11. A round, flat object is observed with a polarization microscope with polarizer and analyzer perpendicular to one another. The object appears to contain a bright cross (whose arms intersect to form 90° angles), which is oriented at 45° with respect to the polarizer axis. The pattern does not change when the object is rotated. What is the probable structure of the object? What would be observed if the analyzer were removed from the microscope and the object rotated?

CHAPTER 3

Electron Microscopy

The limit of resolution of the light microscope is roughly 2000 Å (see Chapter 2), which is insufficient to visualize cell organelles, viruses, and macromolecules of current interest. This is possible, however, with the electron microscope for which the limit is less than the diameter of a uranium atom (approximately 5 Å) under special conditions.

There are few instruments that present such a bewildering array of knobs and meters as does the electron microscope; furthermore, few techniques require greater skill and attentiveness to detail. Nonetheless, for most applications the instrument is simple to use and sample preparation is not complicated. There is no doubt that every laboratory should have access to an electron microscope and that every biochemist should be proficient in its use. This chapter will not supply specific procedures or technical details but will indicate how electron microscopy (EM) is done and what its potential is.

SIMPLE THEORY OF OPERATION

An ordinary light microscope consists of a light source, a condenser for focusing the light on or near the object, an object holder (i.e., the slide and the stage), an objective lens for focusing the image, and an eyepiece for projecting the image formed by the objective onto the eye or photographic film (see Chapter 2). This is also true of an electron microscope, except

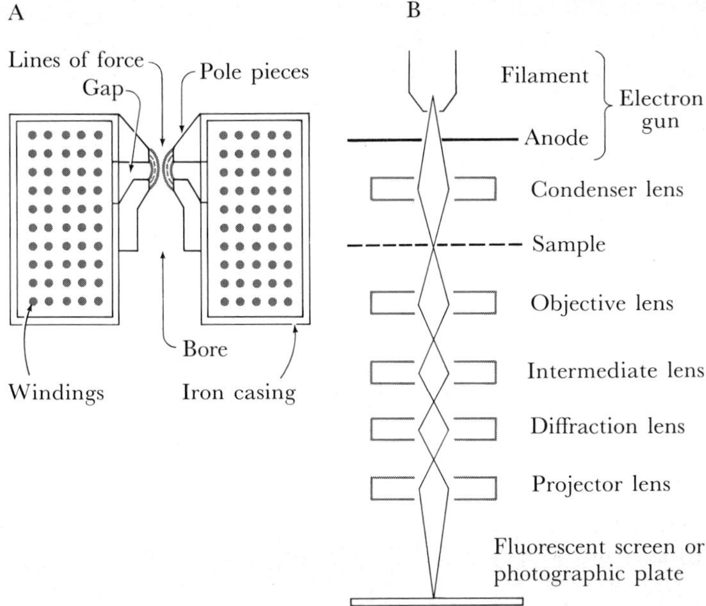

FIGURE 3-1
A. An electromagnetic lens. Although there is tremendous variation in details of design, every lens has an iron casing containing copper windings through which an electric current passes and pole pieces to concentrate the magnetic lines of force. The unit shown here is not drawn to scale; a typical lens is five or more inches in diameter with a bore having a diameter measurable in microns. B. The path of electrons through the lens system of an electron microscope. The electrons are moving downward from the filament. The components and the beam are not drawn to scale. Some microscopes have two condensers; some lack the diffraction lens.

that the light is replaced by an electron beam, the sample holder is a wire screen called a *grid,* and the lenses are electromagnets rather than glass.

A schematic of an electromagnetic lens, which basically consists of an axially* symmetric electromagnet through which the electron beam passes, is shown in Figure 3-1A. (The focusing action of a magnetic field is complex and will not be explained here. For further information see the Selected References near the end of the chapter.) A magnetic field is generated by windings of copper wire. For high magnification, lenses of

*It is frequently stated that electromagnetic lenses are *cylindrically* symmetric, meaning that they have the shape of a cylinder. Axially symmetric is a more precise term because the symmetry is with respect to the axis.

short focal length are needed and it can be shown that focal length decreases as the magnetic field increases. The magnetic field could be made larger by increasing the current in the windings, but this would generate a considerable amount of heat. Instead the field is concentrated by enclosing the wires in a soft iron casing and by inserting two conical pieces of soft iron called *pole pieces,* each of which contains a small orifice through which the beam passes. The magnetic field lines are then as indicated in Figure 3-1B.

As mentioned earlier, this brief description is not intended to explain how an electromagnetic lens works, but mainly to describe some of the parts and to introduce the terminology commonly encountered in electron microscopy. Suffice it to say that a divergent beam of electrons can be brought to focus within certain limits (i.e., there is substantial spherical aberration) at a point on the axis of the lens.

The optical system of an electron microscope is diagrammed in Figure 3-1B. The illumination source consists of a white hot tungsten filament, which emits electrons. The potential of the anode, to which electrons are drawn, is normally from 40 to 100 kilovolts greater than that of the filament. The filament and the anode together constitute the *electron gun.* The anode contains a small orifice through which some of the fastest electrons pass. This hole plus a small aperture just below it collimate the electrons to form a beam. The beam is slightly divergent because the electrons are deflected toward the edge of the orifice of the anode owing to its positive potential. The divergent beam is then made to converge onto the specimen by an electromagnetic condenser lens. The beam is rarely focused sharply on the sample because an intense beam could destroy it.

The image is formed by what is often called the subtractive action of the sample. That is, some of the electrons are scattered from the atoms of the object. The pattern of this loss of electrons generates the image pattern (in much the same way that the light intensity is reduced by an absorbing object in the light microscope). The *objective lens,* which is adjusted so that the sample is precisely at its focal point, then refocuses the beam to produce an image. This image is then magnified in several stages by three electromagnetic lenses called the *diffraction, intermediate,* and *projector lenses.* The final projector lens forms the image on either a fluorescent screen or a photographic plate.

The electron microscope differs in three respects from the light microscope: First, because electrons do not travel very far in air, the entire microscope column must be in a high vacuum; hence, the object must always be dry and therefore dead. Second, because the magnification of an electromagnetic lens is proportional to the magnetic field, which in turn is proportional to the current in the windings, the magnification can be varied continuously by varying the current through the windings of the lens. In light optics, magnification is fixed by the set shape of the glass lens; hence, many objectives are needed to cover a range of magnification. Third, none of the primary aberrations (i.e., the spherical and chromatic aberrations discussed in Chapter 2) can be corrected in standard electromagnetic lenses because the magnetic lenses are always convergent. (The Crewe microscope described in a later section partly corrects spherical aberration.) To reduce spherical aberration and thereby improve image quality, the lenses are operated at very small numerical apertures. (Numerical aperture (NA) is discussed in Chapter 2). This has the effect of severely limiting the resolution allowable by the Compton wavelength of the electron because the limit of resolution is 0.61 λ/NA. For an electron, the Compton wavelength, λ, is hc/E, in which h is Planck's constant, c is the velocity of light, and E is the energy of the electron. For a 60-Kv electron this is 0.03 Å. However, because the numerical aperture of a magnetic lens is normally about 0.0005, the practical limit is more like 4 Å. (As will be discussed later, the nature of the sample actually allows this limit to be reached only rarely.) Nonetheless, the resolution is still approximately 500 times as great as that obtained with a light microscope.

METHODS FOR PREPARING SAMPLES AND PRODUCING CONTRAST

Preparation of Specimen Supports

The great capability of all matter for scattering electrons requires that a sample be very thin—otherwise no beam will get through to form an image. In practice, the maximum thickness is approximately 0.1 micron (1000 Å) for 100 Å resolution and from approximately 50 to 100 Å for 10 Å resolution. This poses no real problem in observing viruses, fibrils, or macromolecules, but for most cells, which range from 1 to 50 microns in thickness, it is necessary to make thin sections (see Embedding, Sectioning, and Staining). This requirement clearly means that the sample support (i.e., the equivalent of a microscope slide) must also be very thin, uniform in thickness, and without obvious structure at high magnification.

The specimen support, or *grid* (Figure 3-2) for all samples consists of a disc cut from a rigid copper (in some cases, platinum) mesh with openings approximately 75 μ per side, overlaid with a thin "electron transparent" film called the support film.* Electron transparency indicates that its elec-

*The word "grid" is used by microscopists to mean both a disc cut from the mesh but not coated with a support film and the coated mesh.

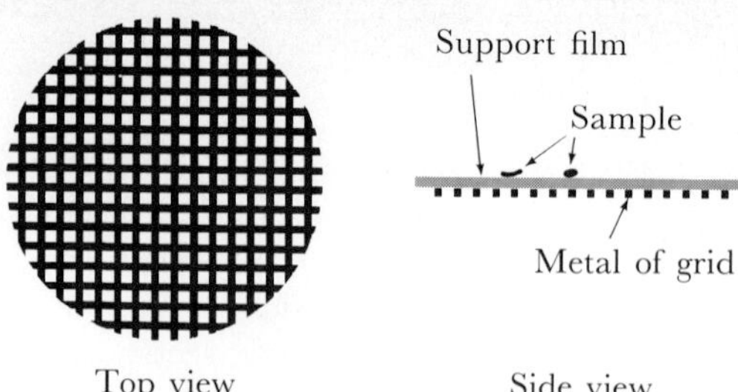

FIGURE 3-2
A specimen support or grid, which usually consists of a fine copper mesh (the grid itself) overlaid with a thin film of plastic or carbon (the support film). The support film holds the sample.

tron scattering power is both low and uniform. Commonly used films consist of layers from 100 to 200 Å of either carbon or various plastics (Parlodion, Formvar). Unfortunately, no support film is truly structureless and for high-resolution work with macromolecules whose dimensions are comparable to film thickness, variations in the intensity of the background produce a "granularity," the grains ranging from 5 to 10 Å. This seriously limits the attainable resolution. Films are prepared in one of three ways (Figure 3-3): Parlodion films are prepared by placing a drop of a solution of Parlodion in amyl acetate on a water surface. (A liquid surface is used because it is very smooth.) The droplet spreads, the solvent evaporates, and a thin film of the plastic forms. Formvar films are prepared by dipping a smooth glass microscope slide into a solution of the plastic and then removing it. When dry, the thin film on the glass will slide onto a water surface if the slide is slowly lowered into the water. (This dipping method is preferred by some microscopists for Parlodion.) Carbon films are prepared by evaporating onto freshly cleaved mica (which is a molecularly smooth surface, being a single plane of a crystal), and floating the film onto a water surface as is done in preparing Formvar films. In all cases, the film is mounted on grids in either of two ways: it can be lowered onto the grids—which had been placed on the bottom of the container beforehand—by draining the water off, or the grids can be placed on the film from above (as shown in Figure 3-3) and the entire support picked up by touching the surface with a sheet of plastic or absorbent paper.

Sample Preparation and Contrast Enhancement

The intrinsic contrast of biological material is poor because the scattering of the carbon atoms in the support film is of roughly the same magnitude as that of all the principal atoms (C, N, O, P, S) of the material. The usual

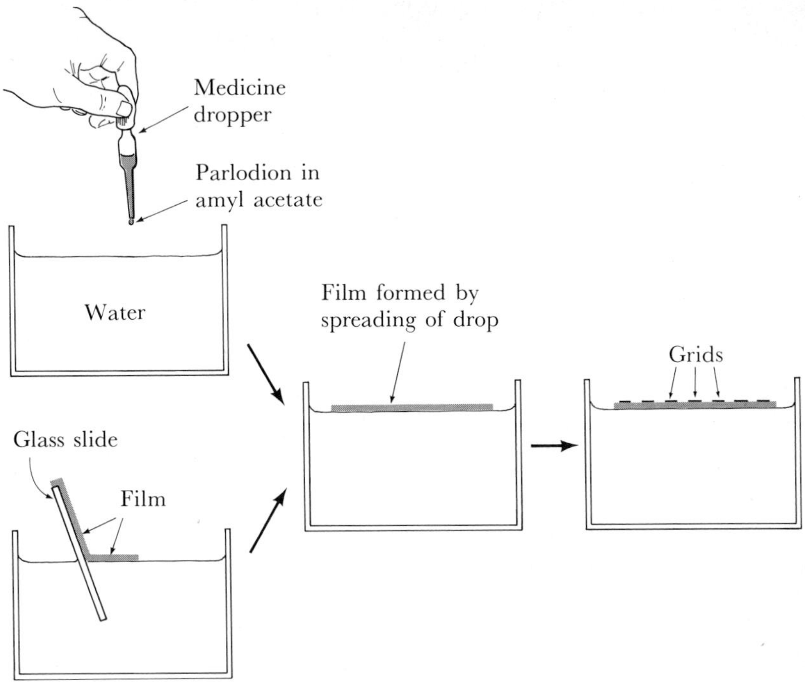

FIGURE 3-3
Preparation of a plastic support film. In one method, a droplet of Parlodion dissolved in amyl acetate is placed on a clean surface of water. The droplet spreads and the solvent evaporates, leaving a thin film of the plastic. In the other method, a glass microscope slide is coated with a thin film of either Parlodion or Formvar by dipping the slide into a solution of the plastic considerably less concentrated than the solution used with the droplet method. In some cases, the film is a thin layer of carbon that has been evaporated onto the glass or onto a mica sheet. When the slide is lowered into the water, the film comes off the glass and floats on the water. Several grids are then placed on the surface of the film and a piece of absorbent paper is placed on the grids. When the paper is lifted up, the grids adhere to it. An alternative method is to use a vessel with a bottom drain. The grids are placed on a screen platform under the water surface before forming the film. The film is then formed and the vessel is drained. As the water level drops, the film comes in contact with the grids.

method for correcting this situation is to deposit heavy metals of very high scattering power on the structure in such a way that the pattern of metal somehow indicates the features of the sample. Useful metals are osmium, platinum, lead, and uranium, although chromium, palladium, tungsten, and gold are sometimes used. Several standard methods for sample preparation and contrast enhancement follow.

EMBEDDING, SECTIONING, AND STAINING

If the material under observation is too thick for the passage of electrons, a thin slice or section must be made. To prepare a thin section, the sample must be made rigid so that it can be cleanly cut. This process, called *embedding,* consists of the gradual replacement of the aqueous material of the sample with an organic monomer (e.g., methyl methacrylate) that can be hardened by polymerization. After it has become solid, the plastic containing the supposedly undisrupted sample is sliced with an ultramicrotome (a kind of knife) into layers from 500 to 1000 Å thick. The sections are then stained (although staining is sometimes done before

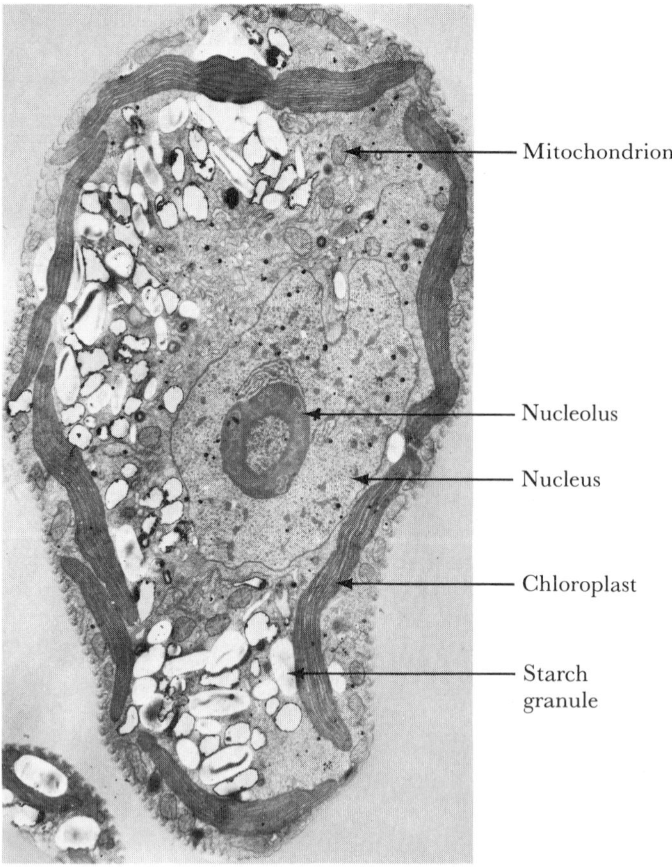

FIGURE 3-4
Electron micrograph of an ultrathin slice of *Euglena gracilis* stained with osmium tetroxide. [Courtesy of Jerome Schiff and Nancy O'Donohue.]

embedding) by exposure to solutions of salts of molybdenum, tungsten, lead, or uranium, or to the vapor of osmium tetroxide. (The word *staining* refers to the deposition of a metal by a chemical reaction or the formation of a complex with certain components of the sample, to increase the electron density.) These stains react with proteins and other macromolecules and aggregates and thereby put electron-dense material in the sample. Stained preparations are beautiful to look at (Figure 3-4) and appear to contain considerable detail, yet it must be realized that what is being observed is the distribution of metal atoms and therefore of the chemical groups that can react with a particular stain. An example of a type of artifact that can arise is the deposition of osmium on opposite sides of a thick membrane, producing two black lines separated by an unstained space, which can be mistaken for a double membrane. The embedding and sectioning procedures themselves can induce distortion because of uneven permeation of the sample by the organic monomer and because of the cutting itself.

Only a single layer through a sample is observed when looking at a thin section and this may not always be adequate. To get a picture of the entire sample, a large number of sections are normally examined. An elegant though tedious method is *serial sectioning* in which successive sections are collected in sequence and examined.

REPLICA FORMATION

The method used for observing the surface of an electron-opaque or easily-destroyed specimen is called replica formation. The specimen is coated with a thin layer of platinum and then a supporting layer of carbon (for strength), both deposited by vacuum evaporation or shadow-casting (see Figure 3-8 and the accompanying discussion). This bilayer is then floated off onto water and picked up on a grid (Figure 3-5). The replica is thus a facsimile of the surface of the object—that is, the contours are the same as those of the sample. This method has been used to study the surfaces of viruses, membranes, and certain protein crystals that are immediately destroyed by the electron beam. An example of a replica is shown in Figure 3-6.

FREEZE-ETCHING AND THE CRITICAL-POINT TECHNIQUE

In replica formation, the water in the sample must be removed before preparing the replica because the production of a film by shadow-casting must be in vacuum. This presents a problem in that structures usually collapse during air drying as a result of surface tension effects accompanying the phase changes that occur during evaporation of the solvent. Freeze-etching and the critical-point method avoid the production of artifacts

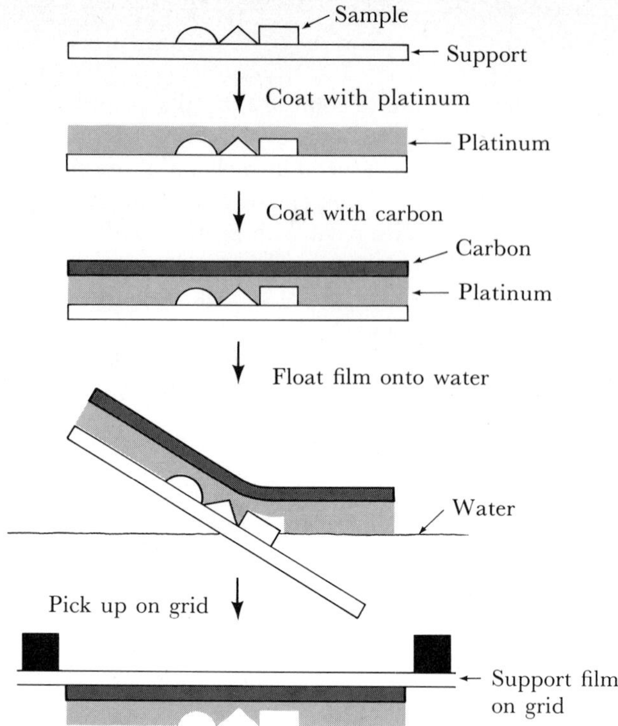

FIGURE 3-5
Formation of a carbon-platinum replica. In some cases, the support film is applied to the carbon layer before floating on water; a bare grid is then used to pick up the film (see legend for Figure 3-6).

due to drying. In freeze-etching, the sample is rapidly frozen, sectioned or fractured, and placed in a vacuum with conditions of pressure and temperature such that the water sublimes from the surface of the sample. A replica of this surface is then prepared by evaporating platinum or carbon while it is still in the vacuum. An example of a replica prepared by the freeze-etching method is shown in Figure 3-7.

The critical-point method makes use of the fact that no liquid phase can exist above a "critical temperature" characteristic of each substance. The procedure follows: First, a wet sample is soaked in ethanol. The ethanol is then exchanged with liquid CO_2 under pressure at 15°C. The temperature of the specimen is then raised above 31°C (the critical temperature) and the liquid CO_2 becomes a gas. Presumably, all three-dimensional relations are preserved. A replica can then be prepared or the sample can be observed directly in the microscope, if it had been stained beforehand. This method is especially useful in preserving macromolecular structures if molecules are deposited on a film from a solution.

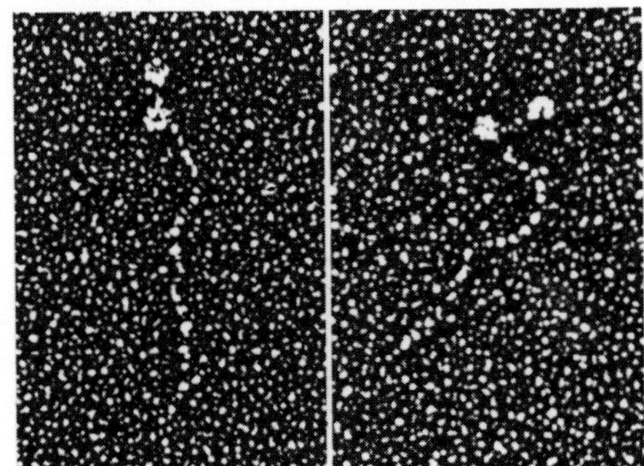

FIGURE 3-6
Replicas of myosin molecules (each panel is approximately 1200 Å wide). A solution of rabbit-muscle myosin was sprayed onto a surface of freshly cleaved mica. After the solution had dried, a thin layer of platinum was deposited by shadow casting at a low angle. A carbon layer was evaporated onto the platinum layer, and finally a Parlodion film was put on top. The mica was then slowly dipped into water at a low angle; the film came off the mica and floated on the surface. The film was then picked up on a grid. [From S. Lowey, H. S. Slayter, A. G. Weeds, and H. Baker, *J. Mol. Biol.* 42(1969):1.]

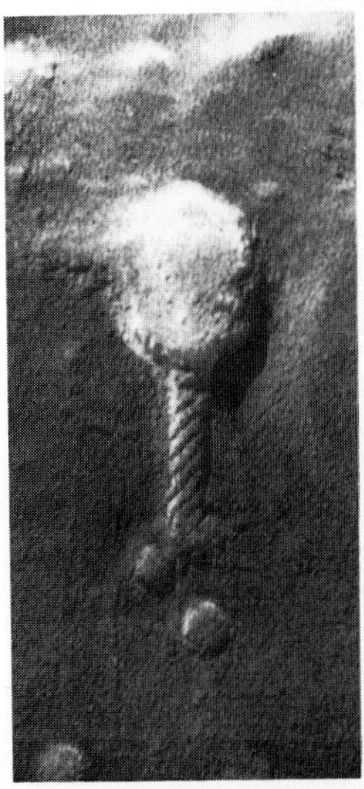

FIGURE 3-7
An *E. coli* T2 phage prepared by freeze-etching, a slight modification of the critical-point method in which a replica is made. The surface details of the phage tail are particularly clear. This should be compared with the similar T4 phages shown in Figures 3-10 and 3-12. [Courtesy of Manfred Bayer.]

SHADOW-CASTING

A great deal of electron microscopy is concerned with the structures of particles—such as viruses, phages, and ribosomes—and of macromolecules. The sizes of such objects as well as limited information about their structures can be obtained by shadow-casting. The particles (in solution or suspension) are applied to a grid overlaid by a support film by spraying. The liquid quickly evaporates, the sample is placed in vacuum, and a heavy metal is applied by evaporation. This requires boiling a metal using white hot tungsten and is typically done either by wrapping a metal wire around a tungsten wire (Figure 3-8) or by placing small lumps of metal in a tungsten container. The metal atoms are projected in all directions and, if the vacuum is good, in straight lines. If evaporation is from an acute angle (Figure 3-9), metal will pile up on only one side of the sample and will cover the grid except in the shadow of the particle. If the vertical (H) and horizontal (L) distances from the evaporation source to the specimen are

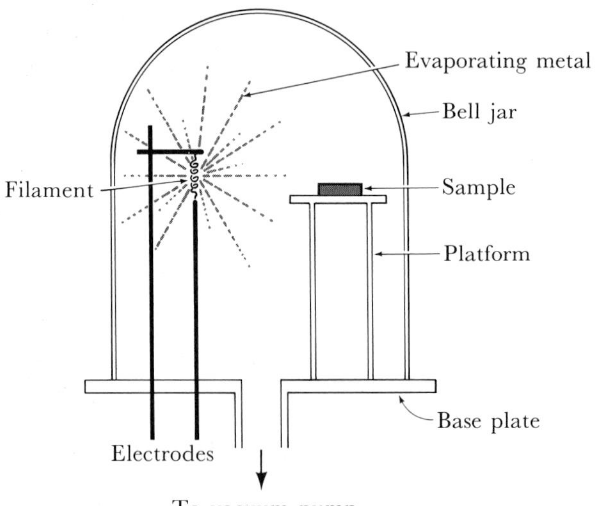

FIGURE 3-8
Apparatus for vacuum evaporation or shadow-casting. The bell jar is evacuated by vacuum and diffusion pumps. The tungsten filament, around which the metal wire to be evaporated is wrapped, is heated to the boiling point of the metal. In 10 or 15 seconds, the metal is boiled away and forms a film on the sample. The thickness of the film is proportional to the amount of metal put on the filament and inversely proportional to the square of the distance from filament to sample.

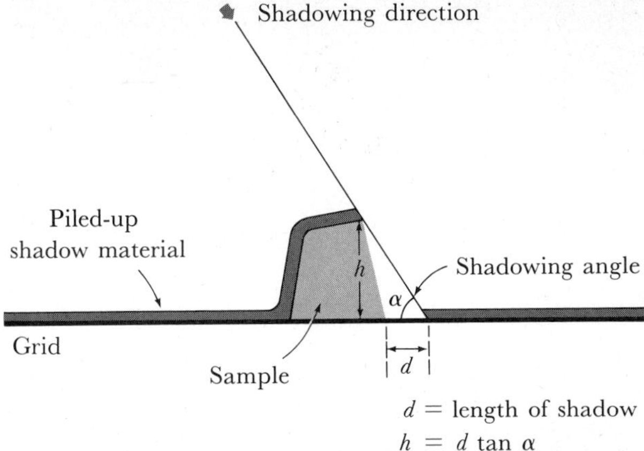

FIGURE 3-9
Determination of the height of an object from the length of its shadow. Because evaporation is not done from directly above the sample, there is a region on the support film on which there is no metal.

known, the height (h) of the particle above the surface of the grid can be calculated from the length (d) of the shadow cast by the specimen because $h/d = H/L$. Hence, the dimensions of the particle can be determined. Figure 3-10 shows a shadowed preparation of a phage.

Shadow-casting has not been too successful for smaller macromolecules because of the small size of the shadow and the granularity of the support film. Many of the problems can be avoided by using the *negative-contrast procedure* described next.

A special use of shadow-casting, the Kleinschmidt technique for observing nucleic acid molecules, will be discussed later.

NEGATIVE-CONTRAST TECHNIQUE

The negative-contrast (incorrectly and commonly called negative staining) procedure of Brenner and Horne consists of embedding small particles or macromolecules in a continuous stain or electron-opaque film (Figure 3-11). The stain penetrates the interstices of the particle but not the particle itself. The image is a result of the relative intensity of the beam at every point, which is proportional to the thickness of the opaque material at that point. Hence, contrast is achieved by virtue of the particle reducing the effective thickness of the opaque film—that is, the particles are seen in

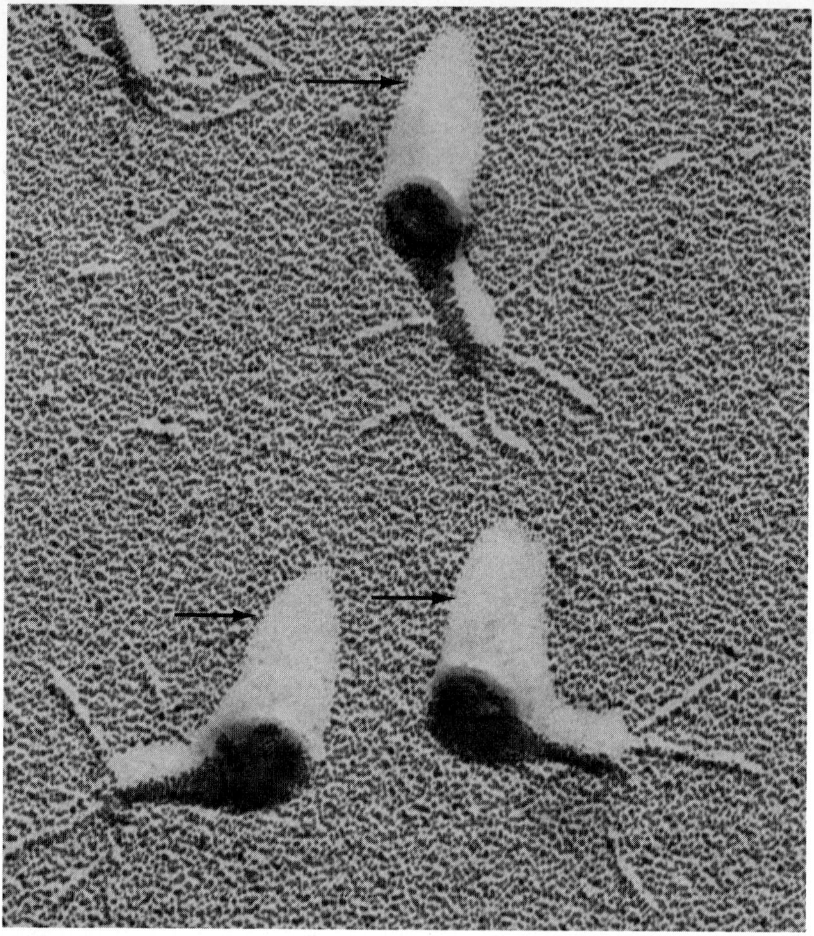

FIGURE 3-10
Electron micrograph of *E. coli* bacteriophage T4 prepared by shadowing. The shadows are indicated by arrows. [Courtesy of Jonathan King.]

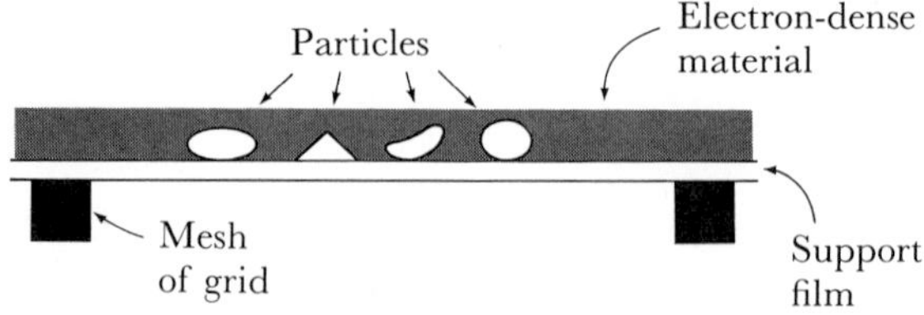

FIGURE 3-11
The negative-contrast method. Four particles are embedded in an electron-dense material. As the beam passes through the sample, the attenuation will depend on the total thickness of the stain; therefore, more electrons will pass through the regions containing each particle.

outline. Figure 3-12 shows a negative-contrast picture of a phage. In this procedure the sample is either mixed with the stain and sprayed on the grid or sprayed on the grid first and then sprayed with the stain. The interpretation of negatively stained samples is sometimes difficult because various patterns can be observed depending on (1) the thickness of the stain; (2) whether it has penetrated the interstices of the particle, (3) whether it lies above and below the particle, and (4) whether any of it has adsorbed specifically to the sample (positive staining). For this reason it is usually necessary to look at a large number of preparations and particles. None-

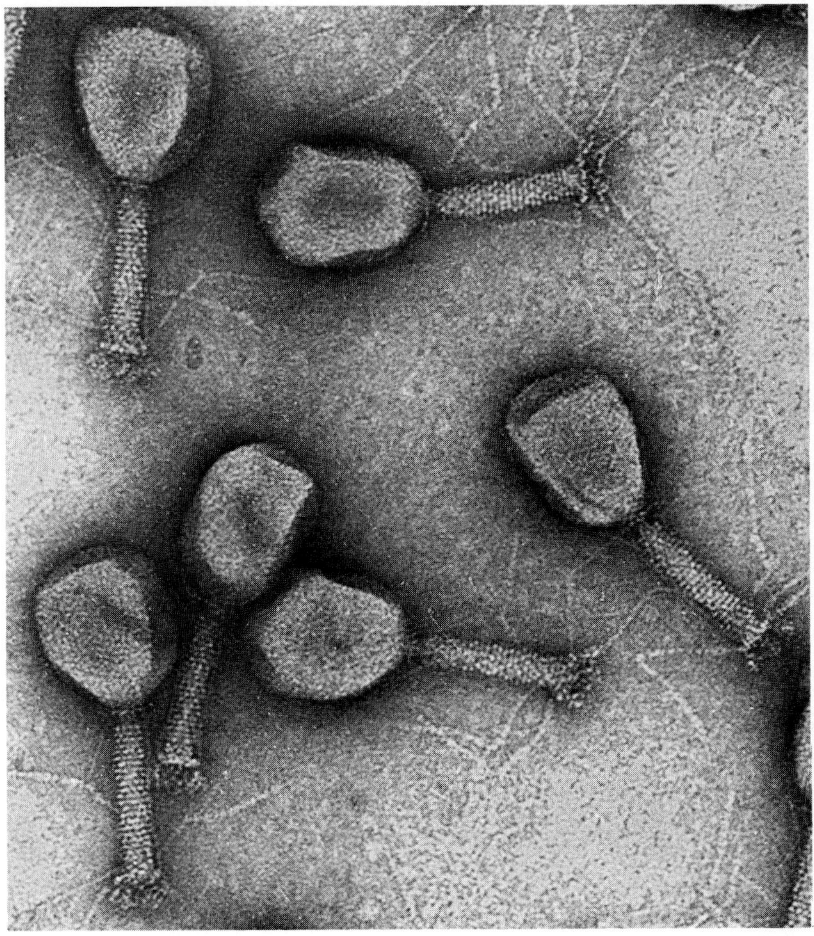

FIGURE 3-12
Electron micrograph of *E. coli* phage T4 prepared by the negative-contrast procedure. The stain is phosphotungstic acid. Compare with Figure 3-10. [Courtesy of Jonathan King.]

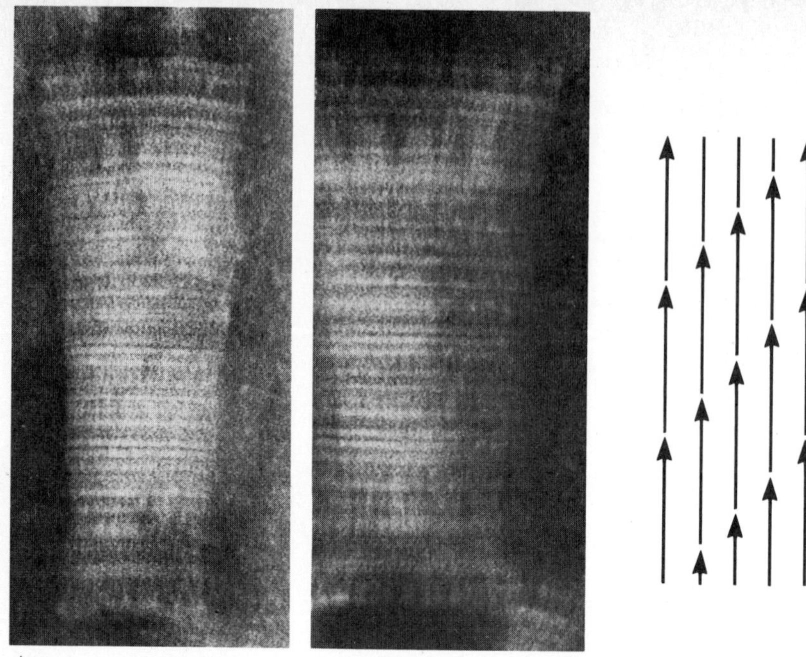

FIGURE 3-13
A. Electron micrographs of two segment-long-spacing aggregates of tropocollagen molecules positively stained by phosphotungstic acid. The length of the aggregate is approximately 2.8 μ. The bands are the parts of tropocollagen that bind the stain. [Courtesy of Peter F. Davison.]
B. A schematic representation of the collagen fiber. Each arrow represents a tropocollagen molecule; the molecules are attached end-to-end and aggregated sideways but in a quarter-staggered array. This model was derived from an analysis of the pattern of bands of this and other forms of collagen. [From A. J. Hodge, J. Highberger, G. Deffner, and F. O. Schmitt, *Proc. Nat. Acad. Sci.* 46(1960):197.]

theless, the negative-contrast method has been used successfully for a wide variety of phages, viruses, and proteins.

The most useful stain for negative contrast is a phosphotungstic acid salt (although there is some indication that this might ultimately be replaced by cadmium iodide). It should be realized of course that resolution in the negative-contrast method depends on the size of the opaque atoms (i.e., approximately 5 Å).

POSITIVE STAINING

Positive staining has not had widespread use for most macromolecules because it is not usually possible to attach a sufficiently large number of heavy atoms to obtain good contrast, although it has been possible with large molecules and structures such as ribosomes, DNA, RNA polymerase, and collagen. The collagen results are especially beautiful and deserve description because they indicate the great analytical power of this technique. When tropocollagen molecules are allowed to aggregate side-by-side to produce collagen and the resulting structure is stained with phosphotungstic acid at pH 4.2 (binding to positive groups), a pattern of bands is observed (Figure 3-13). If uranyl acetate, which binds negative groups, is used, the banding pattern is exactly the same. Therefore the sideways aggregation probably includes the conjunction of positive and negatively charged groups. Analysis of the banding pattern of several different types of collagen aggregates shows that the tropocollagen molecule can form head-to-head or head-to-tail fibers and that in the standard structure the sideways aggregation involves a displacement of one-quarter of the molecular length from one tropocollagen to the next.

KLEINSCHMIDT SPREADING WITH POSITIVE STAINING AND ROTARY SHADOWING

Probably the most spectacular method of sample preparation of the past decade has been the Kleinschmidt procedure for visualizing DNA. In a single step, artifacts due to drying are eliminated and extraordinary contrast is obtained. This technique is now used in almost all biochemical laboratories and can be learned in an afternoon. A drop of a DNA solution in 0.5- to 1.0-M NH_4 acetate containing 0.1 mg/ml of cytochrome *c* is allowed to flow down a glass slide onto the surface of 0.15- to 0.25-M NH_4 acetate (Figure 3-14). As the drop touches the surface, a film of denatured cytochrome *c* spreads across the surface. This film contains somewhat extended DNA molecules to which a thick (100–200 Å) layer of denatured cytochrome *c* binds. If a grid is touched to the denatured protein film, a drop containing a part of the film is transferred to it. When the grid with the adhering drop is immersed in alcohol, the aqueous phase is removed and the film adheres tightly to the support film on the grid. As used at present, the technique requires a preliminary positive staining with uranyl acetate; the protein adsorbed to the DNA becomes stained as well as the background film but, owing to the excess protein, good contrast is achieved. Contrast is enhanced (or created, if staining is not used) by shadow-casting a metal (usually platinum) at a very small angle while the sample is rotating. Because the DNA coated with protein projects above the protein film, metal piles up against the DNA-protein complex like snow drifting against a fence—but on both sides and on all molecules, regardless of orientation, because of the rotation. The contrast is extraordinary as shown in Figure 3-15. This method can be used to determine the length of DNA and whether it is circular or supercoiled. Under certain conditions (usually by incorporation of the denaturant formamide into

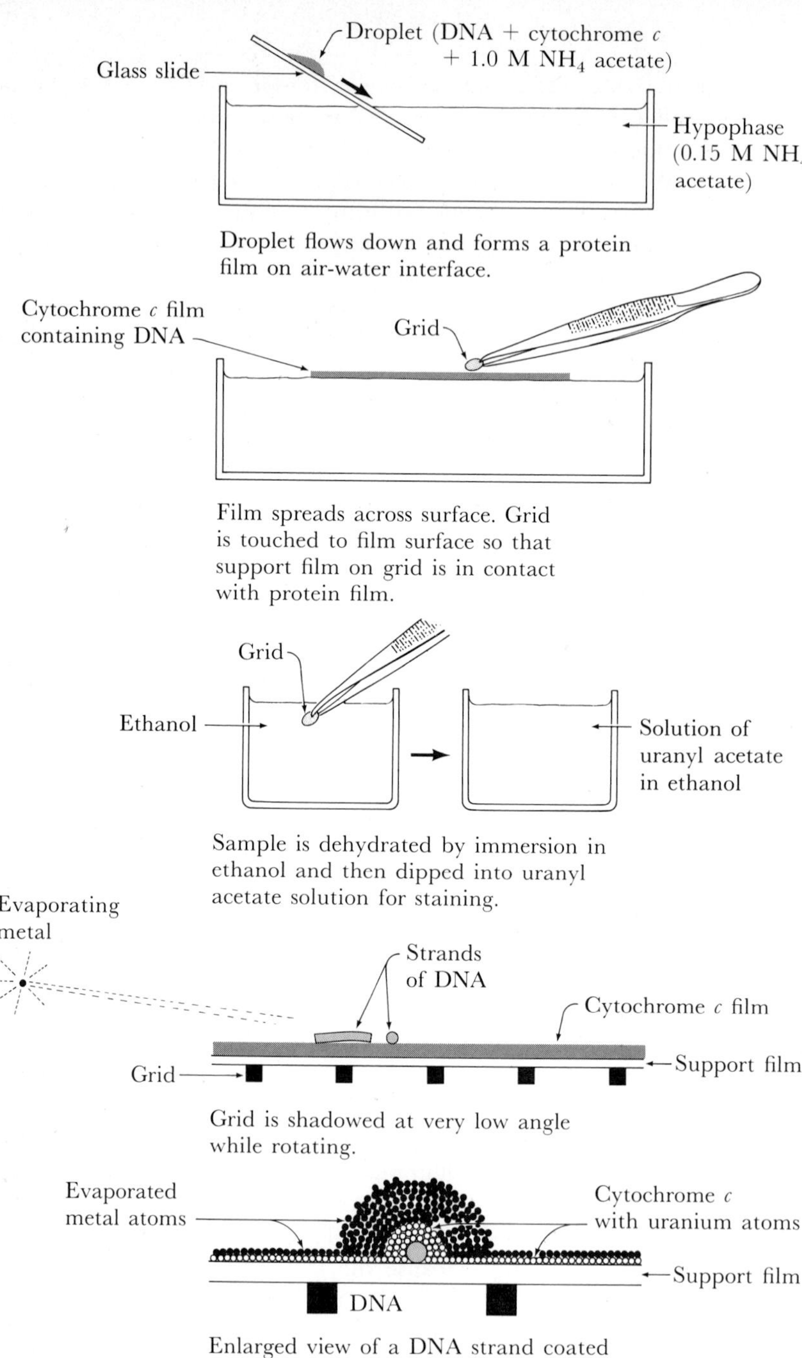
Droplet (DNA + cytochrome c + 1.0 M NH4 acetate)
Glass slide
Hypophase (0.15 M NH4 acetate)
Droplet flows down and forms a protein film on air-water interface.
Cytochrome c film containing DNA
Grid
Film spreads across surface. Grid is touched to film surface so that support film on grid is in contact with protein film.
Grid
Ethanol
Solution of uranyl acetate in ethanol
Sample is dehydrated by immersion in ethanol and then dipped into uranyl acetate solution for staining.
Evaporating metal
Strands of DNA
Cytochrome c film
Grid
Support film
Grid is shadowed at very low angle while rotating.
Evaporated metal atoms
Cytochrome c with uranium atoms
Support film
DNA
Enlarged view of a DNA strand coated with cytochrome c, stained, and shadowed.

FIGURE 3-14
Preparation of DNA for electron microscopy, using the Kleinschmidt method.

all solutions), single-stranded polynucleotides become extended and are easily visualized. Single-stranded DNA and RNA are distinguishable from native DNA by their relative thinness and kinkiness (Figure 3-15).

In a variation of this method, the diffusion method, a cytochrome *c* film is formed on a DNA solution and DNA molecules diffuse upward and adhere to the film. This is a slow process but allows much smaller DNA concentrations to be used.

One important application of the Kleinschmidt technique is seen in the *heteroduplex method.* Here single strands from two different DNA molecules are allowed to hybridize (see Chapter 18). Homologous regions (i.e., regions having complementary base pairs) show up as double-stranded DNA but nonhomologous regions remain as single-strand loops (Figure 3-16).

A recently developed variant of the Kleinschmidt technique replaces the cytochrome *c* with benzalkonium chloride, a surfactant that also forms a film to which DNA adheres. The DNA is shadowed with platinum but, because the DNA is not coated with protein, it appears much narrower than when cytochrome *c* is used. Positive staining is still used but the uranium atoms bind directly to the DNA. This procedure gives better resolution of macromolecules (e.g., RNA polymerase) bound to the DNA and will probably gain widespread use (Figure 3-17).

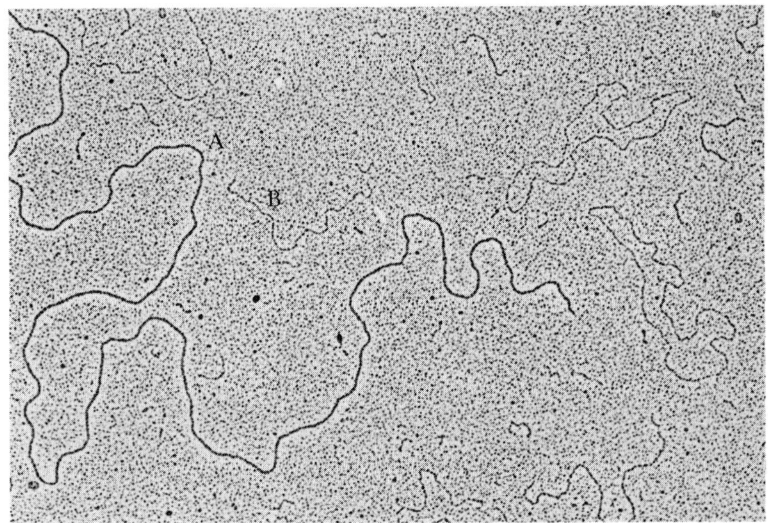

FIGURE 3-15
Electron micrograph of (A) double-stranded DNA and (B) single-stranded DNA prepared by the Kleinschmidt method. Note that single-stranded DNA is kinkier than double-stranded DNA.

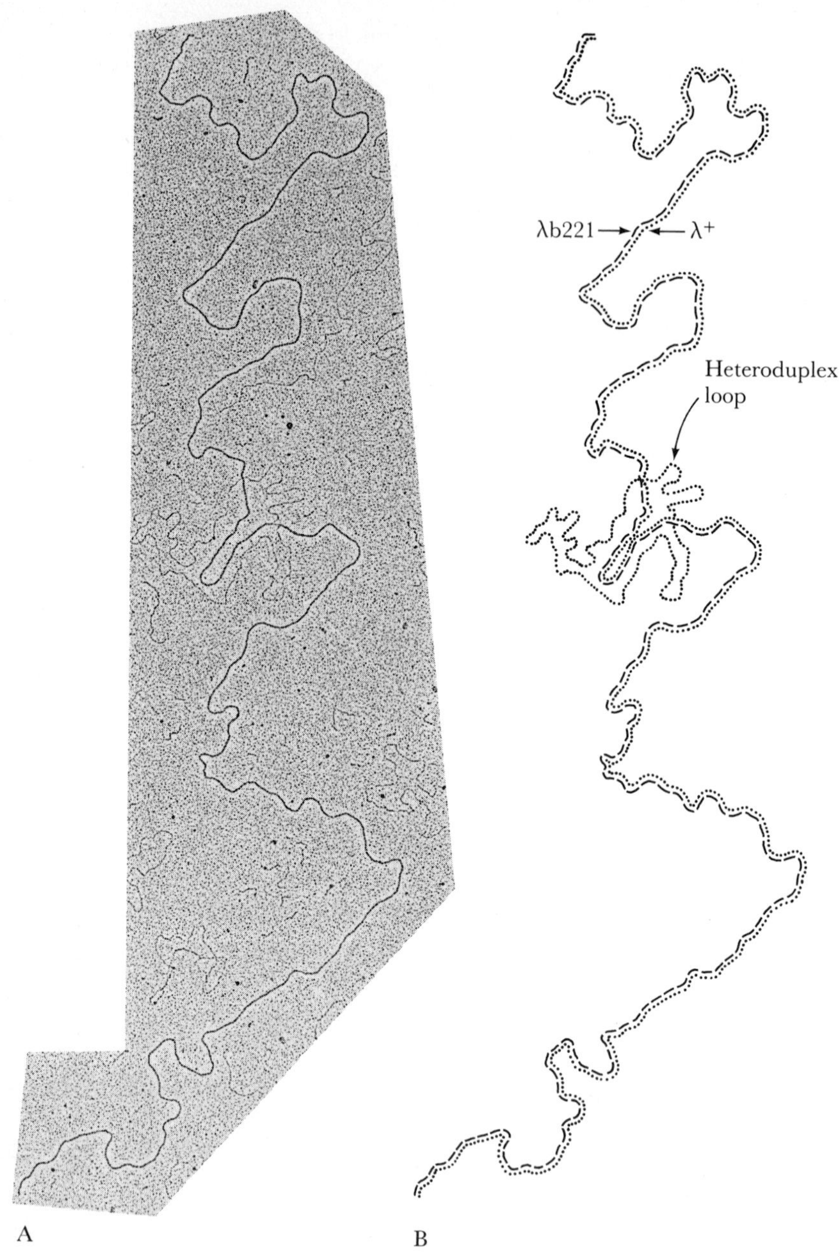

FIGURE 3-16
A. Electron micrograph of a heteroduplex of λ^+ and λb221 (a deletion phage); the DNAs were denatured and renatured. B. The dashed line in the drawing is a single strand of λb221 DNA. The dotted line is single-stranded λ^+ DNA. The heteroduplex loop is the section of λ^+ corresponding to the deletion of λb221. [Courtesy of Manuel Valenzuela.]

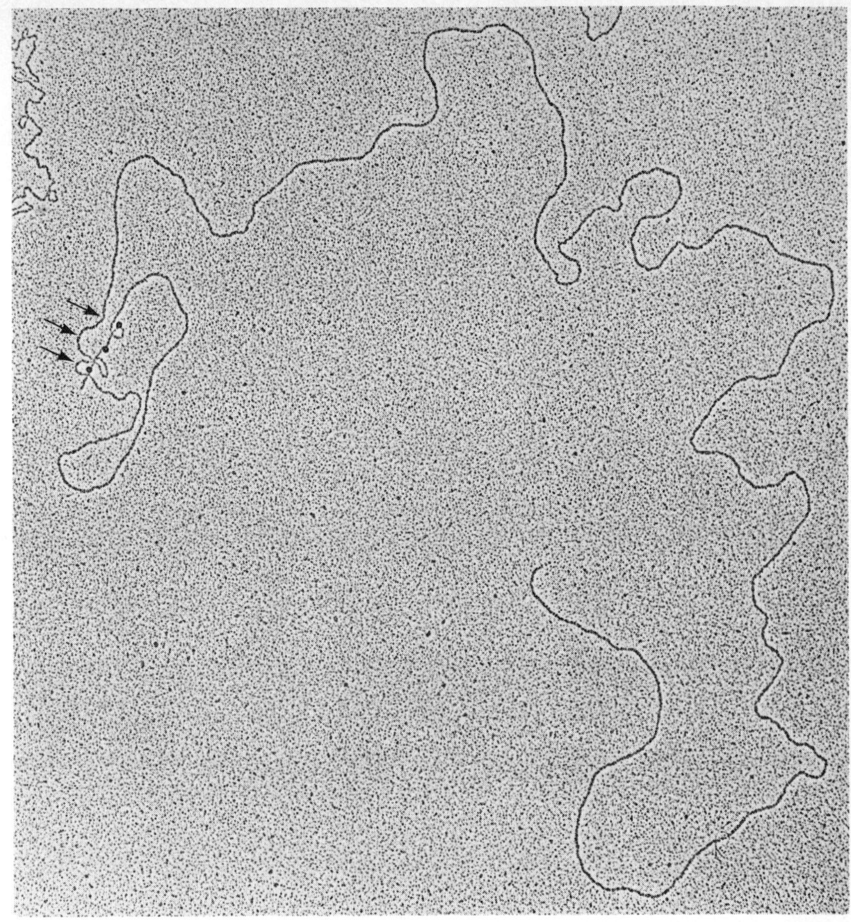

FIGURE 3-17
DNA visualized by the benzalkonium chloride method described in the *Proceedings of the National Academy of Sciences* [72(1975):83]. Molecules prepared for electron microscopy are narrower than if prepared using the Kleinschmidt method. Hence, bound protein molecules are easily seen. This micrograph shows three spherical molecules (arrows) of *E. coli* RNA polymerase bound to phage T7 DNA. The width of the photograph is approximately 3 μ. [Courtesy of T. Koller.]

SPECIAL MECHANISMS OF IMAGE FORMATION

Dark-field Electron Microscopy

Dark-field electron microscopy allows for substantially increased contrast, as is the case with the light microscope. A dark field can be obtained either by using hollow-cone illumination, as in light microscopy (Chapter

2), or by making the beam fall on the sample at an angle such that it does not enter the imaging system. The image is formed by the scattered and diffracted electrons alone. Dark-field electron microscopy has not been used very much, but for observation of macromolecules it deserves greater attention. A simple example indicates its usefulness.

Example 3-1. The binding of proteins to DNA.

DNA has a diameter of 20 Å, yet when coated with cytochrome *c*, as in the Kleinschmidt procedure, its diameter is nearly 200 Å. To observe the site of binding of a protein to DNA, the protein would have to be much larger than 100 Å, or it would be buried in the cytochrome coating. However by using ultrathin electrically charged support films, DNA or DNA plus a bound protein can be drawn from an aqueous solution without the need for the supporting protein film. Both DNA and the bound protein molecule of interest can be positively stained with uranyl acetate. However, even though the support film is not stained, contrast is poor because, in the absence of cytochrome *c*, the total amount of bound uranium is not very great. Using ordinary illumination, the DNA is nearly invisible. However, it is easily seen by dark-field electron microscopy because the background is unilluminated. A micrograph of a DNA molecule visualized by dark-field electron microscopy is shown in Figure 3-18.

It should be noted that dark-field electron microscopy does not improve the resolving power of the microscope but eliminates some of the factors in sample preparation that prevent making use of the existing resolving power. It is not yet clear what the relative merits of dark-field versus the benzalkonium chloride variant are.

The Crewe Microscope

In the ordinary transmission electron microscope, the incident beam covers the entire sample. As the electrons interact with the sample, they do so in several different ways: (1) no interaction at all (i.e., traveling through the interatomic spaces), the most abundant class of electrons; (2) elastically scattered (i.e., without loss of energy) by the orbital electrons of the atoms of the sample, the second most abundant; and (3) inelastically scattered (i.e., with loss of energy) from atomic nuclei, the least abundant. The ratio of the last two classes is a characteristic of each element, because the size of the nuclear target and hence the cross section for inelastic events increases greatly for the larger elements.

A special new microscope has been designed to take advantage of these facts. To do this, the beam is collimated into a very small (approximately

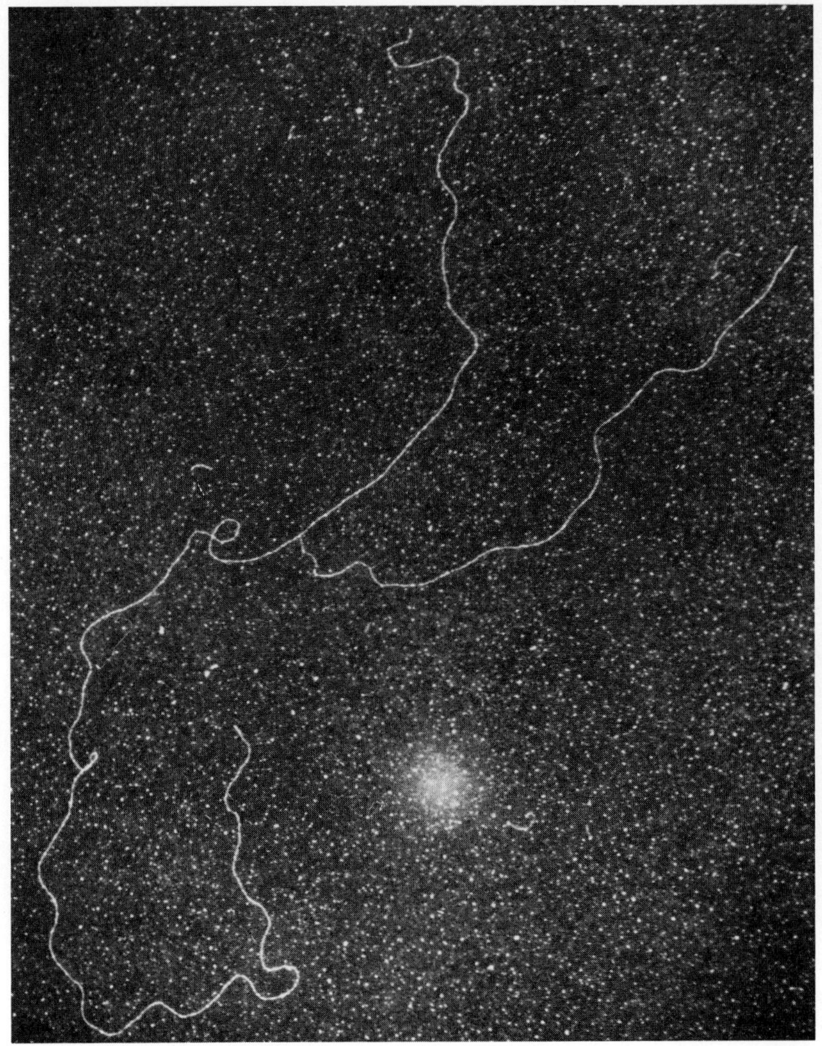

FIGURE 3-18
Dark-field electron micrograph of replicating *E. coli* T7 DNA. [From D. Dressler, in *Control Processes in Virus Multiplication,* edited by D. C. Burke and W. C. Russel, Cambridge University Press, 1975.]

5 Å) spot. The spot is swept across the sample as in a television set. As the beam moves, the ratio of the latter two classes of electrons is measured at each point by an electron-energy spectrometer. This ratio is converted into an image on a television screen by suitable electronic circuits; the information attained can also be processed by computer analysis. This new and important step in electron microscopy gives a new element of analysis to

electron microscopy because individual atoms can be identified. In some cases, the limit of resolution is improved and pictures at 2 Å resolution have been obtained. Figure 3-19 shows several examples of molecules visualized by this procedure.

The Backscatter Scanning Microscope

The scanning electron microscope (SEM) is a device that has produced the many beautiful photographs of cell surfaces seen in the past few years. This microscope is limited to about 200 Å in resolution and operates on

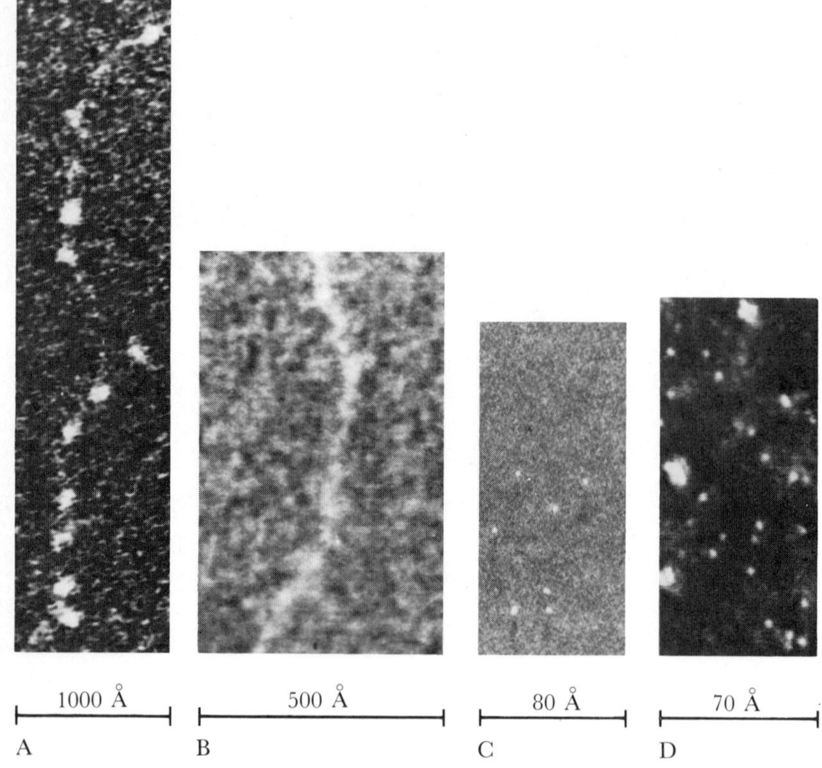

FIGURE 3-19
A. Electron micrograph of calf thymus chromatin depleted of very lysine rich histone (Hl). A fiber of DNA, 20 or 30 Å wide, is seen coated with protein particles, whose average diameter is 135 Å, spaced approximately 260 Å apart. The bar is 1000 Å. [From J. P. Langmore and J. Wooley, *Proc. Nat. Acad. Sci.* 72(1975):2691–2695.] B. Part of an unstained T7 DNA molecule showing a fiber from 20 to 30 Å thick. C. Mercury atoms. D. Uranium atoms. [All micrographs were obtained with the Crewe scanning transmission electron microscope and were kindly provided by John Langmore and Albert Crewe.]

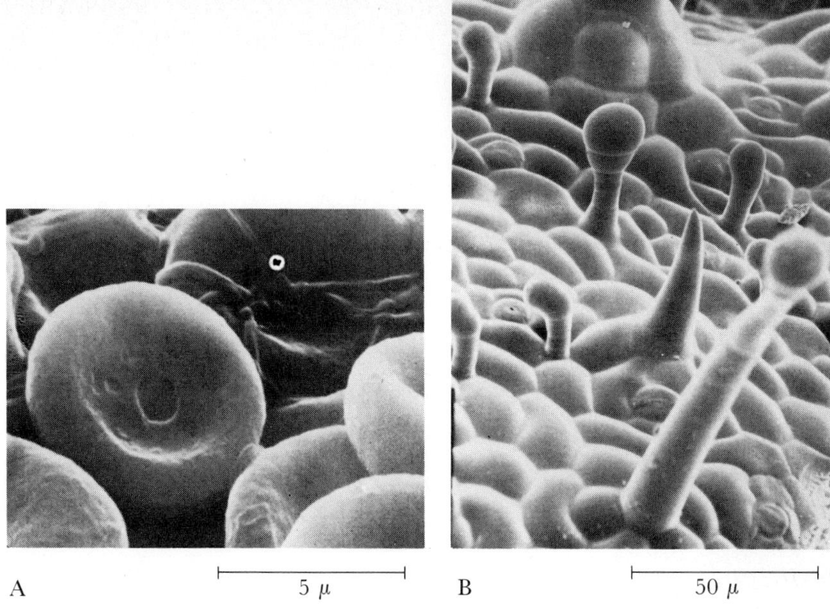

FIGURE 3-20
Scanning electron micrograph of (A) a human red blood cell and (B) the surface of a geranium leaf. [Courtesy of Thomas Hayes.]

a very different principle from that of the transmission electron microscope. Like the Crewe microscope, the beam is collimated into a small (100 Å) spot, and the spot is swept across the sample surface, which has been coated with a thick (200 Å) layer of gold or other heavy metal. As the beam impinges on the metal and penetrates into it a short distance, electrons are emitted from the gold either as secondary emissions or as directly backscattered electrons from the beam. Because of an unexpected angular relationship between the number of electrons emitted and the angle of the surface to the incident beam, which is close to, but not identical with, the way that light reflects from the surface of an object, the image formed by the collected electrons gives dramatic images of the surface being examined. An example of a scanning micrograph is shown in Figure 3-20.

Selected References

Crewe, A. V. 1971. "A High-Resolution Scanning Electron Microscope." *Sci. Amer.* 224(4):26–35.

Davis, R. W., M. Simon, and N. G. Davidson. 1971. "Electron Microscope Heteroduplex Methods for Mapping Base Sequence Homology in Nucleic Acid," in *Methods in Enzymology,* vol. 21, edited by L. Grossman and K. Moldave, pp. 413–428. Academic Press.

Greenstone, A. 1968. *The Electron Microscope in Biology.* St. Martin's Press.

Haggis, G. H. 1967. *The Electron Microscope in Molecular Biology.* Wiley.

Hall, C. E. 1966. *Introduction to Electron Microscopy.* McGraw-Hill. A classic.

Kleinschmidt, A. K. 1968. "Monolayer Techniques in Electron Microscopy of Nucleic Acid Molecules," in *Methods in Enzymology,* vol. 12B, edited by L. Grossman and K. Moldave, pp. 361–376. Academic Press.

Oliver, R. M. 1973. "Negative Stain Electron Microscopy of Protein Macromolecules," in *Methods in Enzymology,* vol. 27, edited by C. H. W. Hirs and S. N. Timasheff, pp. 616–672. Academic Press.

Slayter, E. M. 1970. *Optical Methods in Biology.* Wiley.

Problems

3-1. A sample of RNA-containing viruses is observed by the negative-contrast method. Two types of particles are observed—those that appear uniformly light against a dark background and those that have large dark centers. What is the structure of each?

3-2. A spherical virus is mixed with polystyrene spheres, 750 Å in diameter. After shadowing with gold, the length of shadow of the polystyrene spheres is 1250 Å and that of the virus 820 Å. What is the diameter of the virus? Some of the viruses have shadows ranging from 150 to 200 Å. What are these?

3-3. DNA no. 1 has a length of 10 μ; no. 2 is 9.5 μ. For most of their length, their base sequences are identical. A region from 4.0 to 4.5 μ (measured from what is arbitrarily called the left end) of no. 1 is deleted in no. 2. The region from 7.1 to 7.8 μ of no. 1 is replaced in no. 2 by sequences not found in no. 1. The two DNAs are mixed, denatured, and renatured according to the heteroduplex procedure. How will the heteroduplexes appear?

3-4. A sample of linear double-stranded DNA (all molecules identical) is digested briefly with an exonuclease attacking only the 5′P ends of DNA strands until approximately 3% of the DNA is removed. It is known that these molecules are terminally redundant—that is, the gene order is ABCD . . . XYZABC in which the length of segment ABC is 1% that of the total DNA. When these treated molecules are exposed to renaturing conditions, a new type of structure appears. Describe this structure, including the lengths of the various regions in terms of the percentage of original length.

3-5. A protein structure contains sixty spherical subunits arranged so that six are in a hexagonal array and ten hexagons are stacked one above the other. Draw the types of structures that would be observed by (A) the negative-contrast method, (B) shadowing, and (C) the replica technique.

3-6. Support films are always made of plastic or pure carbon. They are very fragile and frequently break. Stronger films could be made of metals such as chromium. Would such metal films be useful? Explain.

3-7. In measuring the length of DNA molecules, a large enough number of molecules must be measured to get a statistically significant value. Clearly, as larger molecules are studied, there are fewer molecules per grid hole. One solution to this problem is to increase the DNA concentration. However, why is this not reasonable if the molecules are longer than, say, 20 μ?

3-8. A virus-infected cell appears to be approximately 50 μ in diameter. A thin section, 100 Å thick, contains twenty-two viruses approximately 400 Å in diameter. Roughly how many viruses are there in this cell? What assumptions have you made to perform this calculation?

3-9. A particular cell is roughly cylindrical, 10 μ long and 1 μ in diameter. It contains a set of stacked discs (stacked on the cylinder axis), 1 μ in diameter and 400 Å long. A collection of cells is embedded in plastic, sliced into sections 200 Å thick, and stained with a heavy metal stain. What types of structures will be seen?

PART II

General Laboratory Methods

CHAPTER 4

Measurement of pH

Because essentially all biochemical reactions depend strongly on pH (which is defined as $-\log[H^+]$), it is important to be able to measure pH accurately. This is accomplished with a commercial pH meter by simply immersing two electrodes into a solution and reading the pH value on a dial. However, it is important to know how the instrument measures pH, because several factors can cause the observed value to differ from the actual pH.

A pH meter measures the voltage between two electrodes placed in the solution. The essence of the system is an electrode whose potential is pH-dependent. The most commonly used pH-dependent unit is the *glass electrode.* The action of this electrode is based upon the fact that certain types of borosilicate glass are permeable to H^+ ions but not to other cations or anions. Therefore, if a thin layer of such glass is interposed between two solutions of different H^+ ion concentrations, H^+ ions will move across the glass from the solution of high to that of low H^+ concentration. Because passage of a H^+ ion through the glass adds a positive ion to the solution of low H^+ concentration and leaves behind a negative ion, an electric potential develops across the glass. The magnitude of this potential is given by the equation

$$E = 2.303 \frac{RT}{\mathrm{F}} \log \frac{[\mathrm{H}^+]_1}{[\mathrm{H}^+]_2} \tag{1}$$

in which E is the potential, R the gas constant, T the absolute temperature, F the Faraday constant, and $[H^+]_1$ and $[H^+]_2$ the H^+ concentrations on the inside and outside of the glass, respectively. Clearly, if the H^+ concentration of one of the solutions is fixed, the potential will be proportional to the pH of the other solution.

A diagram of a glass electrode is shown in Figure 4-1. The glass electrode contains 0.1 *N* HCl in contact with the H^+-permeable glass. Connection to the voltmeter is by means of a silver wire coated with silver chloride, which is immersed in the HCl.

The circuit is completed by immersing into the solution a reference electrode (see Figure 4-1), which most commonly contains a Hg-$HgCl_2$ paste in saturated KCl; this is called a *calomel* electrode. If a high-temperature operation is required, Ag-AgCl is used instead of Hg-$HgCl_2$. In both cases, the KCl serves to make contact between the Hg-$HgCl_2$ or Ag-AgCl unit and the solution being measured. This unit is encased in a tube made of glass that is impermeable to H^+ ions (so that its potential is pH-*in*dependent). Electrical contact between the KCl within the electrode and the solution is by means of a fine fiber or capillary in the glass casing. (The KCl slowly flows into the sample. In cases in which the Cl^- ion is undesirable, a Hg-$HgSO_4$ reference electrode can be used.)

The voltage measured by such a system is primarily the difference between that of the glass and the reference electrodes. However, there

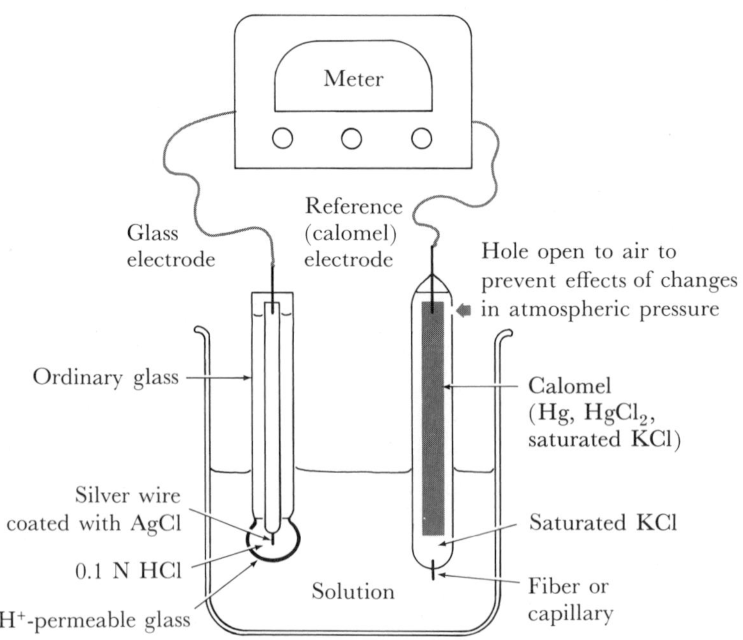

FIGURE 4-1
Glass and reference electrodes of a pH meter.

are three other potentials present in the circuit: (1) the so-called liquid-junction potential of the reference electrode resulting from the fact that K^+ and Cl^- do not diffuse at the same rate so that a charge is generated at the interface between the KCl solution in the reference electrode and the sample; (2) a poorly understood potential called the asymmetry potential, which develops across glass even when the pH on both sides is the same; and (3) the potential of the Ag-AgCl in the glass-electrode unit, which is itself an electrode because of its contact with the Cl^- of the HCl. These three potentials and that of the reference electrode itself are relatively independent of pH and of ionic strength (in the range normally encountered) and therefore can be considered to be constant. Hence, the voltage, V, measured with the total system may be expressed as the difference between the fixed potentials and that of the glass electrode:

$$V = E_{\text{fixed}} - \frac{2.303\ RT}{\text{F}} \log \frac{[\text{H}^+]_1}{[\text{H}^+]_2} \qquad (2)$$

Because the glass electrode is normally filled with 0.1 N HCl,

$$[\text{H}^+]_1 = 10^{-1}$$

and because pH $= -\log \text{H}^+$,

$$V = E_{\text{fixed}} + \frac{2.303\ RT}{\text{F}} - \frac{2.303\ RT}{\text{F}} \cdot \text{pH}$$

or

$$V = \text{constant} - \frac{2.303\ RT}{\text{F}} \cdot \text{pH} \qquad (3)$$

Therefore, the voltage generated is linearly related to the pH of the solution.

To avoid determining the constant in equation (3), and *because the concentration of the HCl in the glass electrode changes by repeated use,* a pH meter is normally standardized against a solution of known pH. (The KCl concentration of the reference electrode does not change because the solution is saturated and contains undissolved crystals.) That is, the electrodes are placed in a standard buffer (typical standards are pH 4, 7, and 10) and the meter is adjusted to read the pH of the standard. Such standardization is sufficient. However, because of minor perturbations producing slight nonlinearity, for precise determination of pH, it is advisable to use a standard whose pH is within one or two pH units of the unknown.

Note that equation (3) states that the relation between the measured voltage and the actual pH is temperature-dependent.* Hence, to determine pH it is necessary to adjust the pH meter (by means of a knob usually labeled "Temp" or "Temperature Compensation Control") to the temperature of the solution being measured. This adjustment introduces or removes a resistance in the electrical circuit so that the voltage change per pH unit increment is always the same. Clearly, because of this temperature dependence as well as that due to the effect of temperature on ionization, it is important that the measurement itself does not induce temperature changes. This point is mentioned here because, in fact, the temperature might be expected to rise during a measurement as a result of current flow through the solution. However, because the glass electrode has very high resistance, only a very small current is drawn by the voltmeter. Furthermore, the solution being measured usually has very low resistance so that the change in temperature (proportional to the resistance times the square of the current) over a short interval is very small.

COMPLICATIONS OF pH MEASUREMENT

Dependence of pH on Concentration of Ions

The pH of a buffer depends not on the concentration of the buffer ions but on a thermodynamic quantity called the activity coefficient. This parameter is strongly affected by the total concentration of ions in the solution (actually the ionic strength) so that the pH of a buffer will vary both with its own concentration and with the concentration of other salts in the solution. This is especially important in the use of commercially available pH standards because they are usually concentrated solutions to be diluted 25-fold—a dilution factor that must be rigorously adhered to. Similarly, if a buffered solution is being prepared as a concentrate (i.e., a stock solution), the pH of this stock solution should not be adjusted to the value desired for the diluted solution. Instead, the concentrated solution must be prepared so that, when the pH of a dilution (i.e., to the concentration desired for a particular experiment) is measured, it will have the required value.

---

*It is important to realize that this temperature dependence is not the same as that which makes the pH of a buffer vary with T—that effect is the result of the temperature-dependence of the dissociation constant of acid, whereas this dependence is the effect of T on the potential.

Electrode Contamination or Alteration

Any substance that can be adsorbed by the H^+-permeable glass of the glass electrode can affect the pH reading by affecting the permeability to H^+ ions. This frequently happens with protein solutions because a thin protein film can form on the glass. Fortunately, such a film can be removed by treatment with detergents or acid.

The commonly used buffer Tris (tris-[hydroxymethyl]-aminomethane) has been found to react with the components of several commercially available electrodes. Errors of as much as one pH unit have been observed. The manufacturers of electrodes usually indicate in their catalogs which electrodes are suitable for use with this buffer.

The Sodium Error

General-purpose glass electrodes are almost always somewhat permeable to sodium ions. Therefore, a potential related to the Na^+ concentration can be produced in the same way as with H^+ ions. If Na^+ is present in a solution whose pH is to be determined, the measured pH decreases as the Na^+ concentration increases, because the electrode detects the sum of the H^+ and Na^+ concentration. Hence the H^+ can *appear* to be greater, which means that the observed pH (i.e., $-\log\ [H^+$ *and* $Na^+])$ is lower than the actual pH ($-\log[H^+]$). This effect is most noticeable at high pH (when NaOH is used) and can be as high as one or two pH units in 1 M Na^+. It is important to remember this because the Na^+ ion is so ubiquitous.

Special Na^+-impermeable glass electrodes are commercially available and can be used to prevent the sodium error if the presence of the Na^+ ion is necessary. Alternatively, because the permeability of the glass to other alkaline earth cations (e.g., K^+) is very low, if present at all, the Na^+ error can be prevented by use of potassium salts. If work at high pH is necessary, it is best to use KOH rather than NaOH, if possible.

It should be noted that the frequently encountered statement that general-purpose electrodes read low at high pH is not strictly true, because it is not the OH^- but the Na^+ that causes the difficulty. The so-called high-pH electrodes are simply glass electrodes made with Na^+-impermeable glass.

A Special Problem with Tris Buffer

Tris buffer is probably the most widely used buffer in biochemistry because of its high buffering capacity, its low toxicity, its low interference with most biochemical reactions, its lack of interaction with metal ions, and its availability in very pure form. However, three drawbacks must be kept in mind: (1) it reacts with certain electrodes, as mentioned earlier; (2) its pH varies more with temperature than does that of most buffers—pH increases approximately 0.03 pH units per degree Celsius from 25°C to 5°C; (3) its concentration dependence is also greater than that of most buffers—the relative pH of a buffer at 0.01, 0.05, and 0.1 molar is x, $x + 0.05$, and $x + 0.1$, respectively.

TYPES OF ELECTRODES

A manufacturer's catalog of pH-meter electrodes can be bewildering. However, careful perusal indicates that there are actually only a few types and combinations thereof.

Glass ("H^+ permeable") electrodes are of four types: general purpose, high temperature, low Na^+ error ("high pH" electrode) and Na^+ ultrasensitive (used to measure Na^+ concentration).

Reference ("H^+ impermeable") electrodes are of three types: general purpose (usually calomel), high temperature (usually Ag-AgCl), and chloride free (uses Hg-$HgSO_4$ instead of chloride). Some are constructed with a hole in the tip covered with a ground glass sleeve to allow rapid equilibration with samples of very high viscosity or with slurries or emulsions.

Combination electrodes consist of a glass and a reference electrode in a single unit and are used almost exclusively in biochemistry. Their particular advantage is that, with a single unit, smaller volumes of solution can be measured. Their disadvantages are higher cost and the fact that they must be discarded if one of the elements fails.

There also exist electrodes for measurements other than of pH. For example, metallic electrodes can be used to measure redox potential and the concentration of specific ions. Platinum is usually used for oxidation-reduction measurements because it is resistant to chemical attack. Silver electrodes are used for direct potentiometric determination of Cl^-, Br^-, I^-, S^{2-} and SH^- (i.e., any ion whose silver salt has low solubility). These measurements are performed with the millivoltmeter attachment of most commercial pH meters. For example, in measuring Cl^-, solutions having different chloride concentrations are measured and the reading on the millivoltmeter is plotted against $[Cl^-]$. The concentration of a sample can then be determined from this standard curve. Redox potentials are similarly measured by using standard solutions of known redox potential.

Electrodes also come in a variety of shapes and sizes for various purposes.

THE USE OF pH PAPER

A variety of color indicator papers exist for pH determination, some supposedly sensitive to 0.2 pH units. They are somewhat useful for rough testing, but it is important to know that errors of several pH units can be introduced by high salt concentration, protein, and certain chromatographic materials such as DEAE-cellulose. In some cases, the dyes in the paper can react with organic substances in the solution and this can result in color changes that are unrelated to pH. Finally, in weakly buffered solutions, the dyes in the paper can act as a buffer and change the pH or give a false reading.

Selected References

Bates, R. G. 1964. *Determination of pH: Theory and Practice.* Wiley.

Dole, M. 1941. *The Glass Electrode.* Wiley.

Brochures supplied by the manufacturers of Beckman, Corning, and Radiometer pH meters are good sources of information.

CHAPTER 5

Radioactive Labeling and Counting

Many biochemical analyses require the detection of minute (10^{-14}–10^{-6} moles) quantities of material. However, chemical tests are rarely responsive to less than 10^{-7} moles. This limitation has been alleviated by the development of radiotracer technology through which extraordinarily sensitive detection of radioisotopically labeled material has allowed studies of many substances in quantities of 10^{-12} moles to become routine. In addition, the use of radioactivity has permitted the development of powerful experimental approaches to various types of problems. Such approaches employ the *double-labeling* technique for following two substances simultaneously or for distinguishing two identical substances synthesized at different times; the *pulse-chase* method for following a substance at a time after its synthesis without the interference of material concurrently synthesized; and *exchange analysis* for measuring participation in reactions.

In this chapter, these and other techniques will be described in some detail. The methods for detecting and measuring radioisotopes will also be presented herein. Autoradiography has been excluded from this chapter because its technology and applications are rather different from the subject of this chapter. It is presented in Chapter 6.

TYPES OF RADIATION USED IN BIOCHEMISTRY

Nuclear radiation is a result of the spontaneous disintegration of atomic nuclei. Of the several kinds of emitted radiation those of importance in isotopic labeling are β particles (emitted electrons) and γ rays (photons).*

Beta Particles

For a particular β-emitting nucleus, β particles are emitted having a continuous range of energy from zero to a maximum value (E_{max}) characteristic of the particular isotope. A plot of the relative probability of emission of a β particle as a function of energy is called a β spectrum. The spectra for several commonly used radioisotopes are given in Figure 5-1. The energy of a given β emitter is traditionally described by stating E_{max}, even though the fraction of particles with energy near E_{max} is very small. A better description is probably given by E_{mean}, the mean energy, because a large fraction of the particles have energy near this value. E_{mean} is roughly $\frac{1}{3}$ E_{max}. These spectra are important in discriminating between different isotopes in the same sample and will be discussed again in the section on liquid scintillation counting.

When β particles pass through matter, their energy is dissipated mostly by ionization and/or excitation of the atoms with which they collide. These interactions are detected by Geiger-Müller (ionization) and scintillation (excitation) counters, which will be described in detail in a later section.

Gamma Rays

A γ ray is a form of electromagnetic radiation and for a given radioisotope is emitted with one or more discrete energy values rather than over a continuous range, as with β particles. A γ ray is uncharged and therefore does not directly ionize atoms in its path. However, it can interact with an orbital electron of an atom and eject it from the orbit, or with a nucleus to produce an electron-positron pair (Figure 5-2). In both types of interaction, the secondary electrons produced by the absorption of a γ photon are like β particles and can ionize and excite other atoms. Hence, the detection of a γ ray is ultimately accomplished in the same way as is a

---

*Alpha particles are rarely used for two reasons: (1) they are difficult to detect because they are strongly absorbed by the samples themselves; and (2) there are few α-emitting isotopic labels that can be satisfactorily used for biological materials.

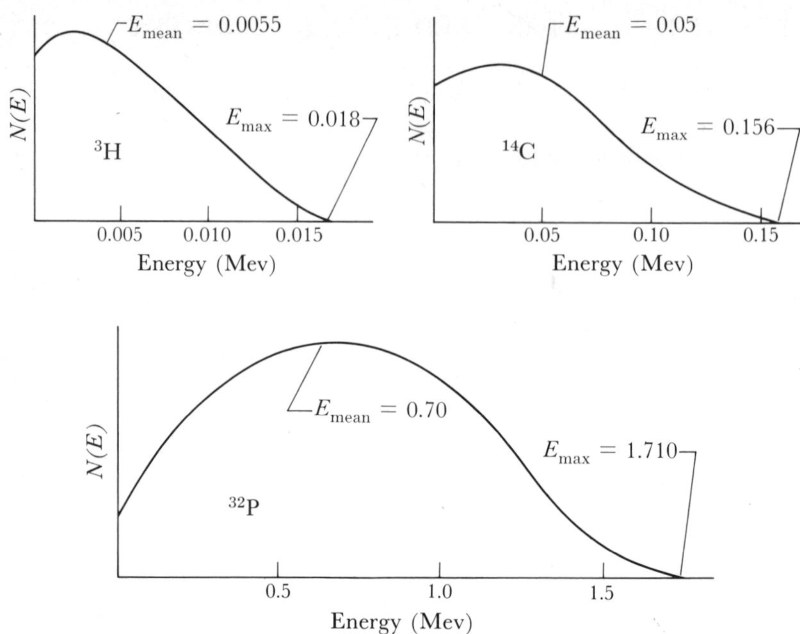

FIGURE 5-1
Beta spectra for ^{3}H, ^{14}C, and ^{32}P: E refers to the energy of the particles and $N(E)$ is the number of particles emitted with energy E. This is in fact a measure of the probability of emission of a particle at that energy.

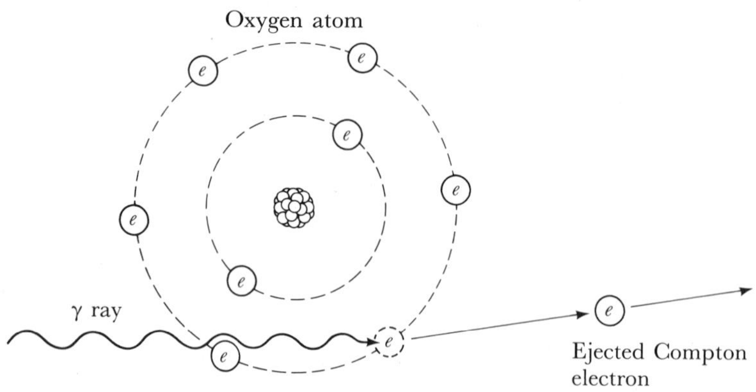

FIGURE 5-2
Ejection of an electron by the interaction of a λ photon with an oxygen atom.

β particle. In practice, as will be seen in a later section, because of the low probability of interaction (often spoken of as the high penetrability of γ rays), special detectors are needed for γ counting.

PROPERTIES OF THE RADIOACTIVE DECAY OF CHEMICAL COMPOUNDS

At any particular time, the number of atoms of a radioactive material decaying per unit time is proportional to the number of atoms present at that time. That is, if N is the number of atoms present at time t, and dN the number of atoms disintegrating in the interval dt, then

$$-\frac{dN}{dt} = \lambda N \tag{1}$$

in which λ is the decay constant. Thus, if N_0 is the number of atoms at $t = 0$,

$$N = N_0 e^{-\lambda t} \tag{2}$$

An exponential decay equation states that, in any given time interval, the radioactivity (disintegration rate) will decrease by the same fraction; hence, it is convenient to express the decay constant as the half-life, $\tau_{1/2}$, the time required for the activity (decay rate) to decrease by one-half. Hence, $\tau_{1/2} = -(\log_e 2)/\lambda = +0.693/\lambda$. This simply means that after one half-life, one-half of the initial activity remains, after a second half-life, one-quarter of the activity, and so forth. Half-lives for some of the isotopes commonly used in biological studies are given in Table 5-1. (Note the enormous range in half-lives.) The principal reason for knowing the half-life of an isotope being used in an experiment is that, if the half-life is short compared with the time it takes to do the experiment, the amount detected will depend on the time at which the measurement is made; hence, if radioactivity is being used to determine the amount of material present at a given time, a correction for the changing count rate must be made. The half-life also determines the maximum specific activity (see page 92) that is obtainable and is therefore a factor in selecting an isotope for a particular experiment.

Radioactivity is expressed in units of *curies.* One curie (C) is defined as the number of disintegrations per second per gram of radium and equals 3.70×10^{10} disintegrations per second. For most biological applications, quantities much less than one curie are normally used and the milli- (mC) or microcurie (μC) is employed. Furthermore, in practice, a minute is the standard time unit—hence, 1 μC $= 2.22 \times 10^6$ disintegra-

TABLE 5-1
Characteristics of commonly used isotopes.

Isotope	Particle emitted	E_{max} (Mev)	Half-life
^{3}H	β	0.018	12.3 years
^{14}C	β	0.155	5,568 years
^{24}Na	β	1.39	14.97 hours
	γ	1.7, 2.75	
^{32}P	β	1.71	14.2 days
^{35}S	β	0.167	87 days
^{40}K	β	1.33, 1.46	1.25×10^{9} years
^{45}Ca	β	0.254	164 days
^{131}I	β	0.335, 0.608	8.1 days
	γ	0.284, 0.364, 0.637	

tions per minute. For reasons that will be discussed later, it is fairly difficult to determine the absolute number of disintegrations because radiation counters detect only a fraction of them; hence activity is usually stated as detected counts per minute (cpm). For highly quantitative experiments in which the amount of radioactivity is used to calculate the absolute amount of material present, the value of cpm is converted into disintegrations per minute (dpm) by dividing by the efficiency of counting.

Radioisotopes and isotopically labeled compounds are rarely isotopically pure—that is, the radioisotope is usually diluted by the presence of chemically identical nonradioactive isotopes. (A few isotopes—e.g., ^{32}P—can be prepared in a pure, or *carrier-free,* state, but they are rarely used as such.) The relative abundance of a radioisotope is described by the *specific activity*—that is, the disintegration rate per unit mass. This is normally expressed as activity (C, mC, μC) per millimole or micromole (mmol, μmol). Unfortunately, specific activity can be expressed in different ways: for example, the specific activity of ^{14}C-glycine (which contains two carbon atoms) may be stated as 50 μC/μmol or 25 μC/μmol *C atom.* Hence, it is important to note the units of specific activity.

Radiochemical purity is a complication often difficult to deal with. For example, when an isotopically labeled biochemical is synthesized, the reaction mixture usually contains a variety of labeled products, which are separated to produce radiochemically pure compounds. Manufacturers of radiochemicals usually supply an assay of the radiochemicals present in the sample and thereby indicate the degree of purity. However, radiochemical purity can change in time for two reasons: First, radiation emitted by one molecule can alter an identical labeled molecule either directly by ionization followed by chemical rearrangement or indirectly by ionizing the solvent (radiolysis) and producing reactive species that can attack the originally pure compound. This is especially true of radiochemicals stored in aqueous solution. Second, if a substance contains two isotopically labeled atoms (which is uncommon except in the so-called uniformly labeled compounds), the decay of one to produce a new atom will result in a molecular rearrangement. For example, consider a substance with two ^3H atoms in each molecule. Beta decay of one ^3H nucleus converts the hydrogen into helium. Because helium does not participate in chemical-bond formation, it will fall off and a chemical rearrangement must occur, thus forming a new radioactive compound containing the second ^3H atom. Similarly, ^{14}C decays to form ^{13}N and, because the valences of carbon and nitrogen differ, a chemical rearrangement must occur. For the most part these problems can be minimized by following the manufacturer's directions for storage, by using freshly made compounds, and, when necessary, by purifying the compounds by chromatography or electrophoresis before use.

MEASUREMENT OF BETA ACTIVITY BY METHODS EMPLOYING GAS IONIZATION

A β particle that passes through a gas may dislodge an orbital electron from one of the atoms of the gas, which results in the production of an ion pair—the dislodged electron plus the remaining positively charged ion. The β particle often has sufficient energy to ionize several atoms successively. If the gas is contained in a chamber in which there are two charged electrodes, the secondary electrons and the positive ions will be attracted to the anode and cathode, respectively, and this can be recorded as a tiny pulse of charge or current. A medium-energy, single β particle will produce from 10^2 to 10^3 ion pairs per centimeter, but not all of these ions can be collected by the electrodes because the ion pairs rapidly recombine in the wake of the particle. The efficiency of collection of the ions depends on the voltage between the electrodes. For example, with very low voltage most ions recombine before reaching the electrodes but, as the voltage is increased, a greater fraction of the ions is captured before they have recombined. Ultimately, a voltage is reached at which all ions are collected and *saturation* is achieved. This property of ion chambers is shown in Figure 5-3. At a much higher voltage (well above that required for saturation), the dislodged orbital electrons are accelerated toward the anode at such high velocities that they also cause ionization of the gas atoms, producing what is called an *avalanche of ions* or *gas amplification.* At

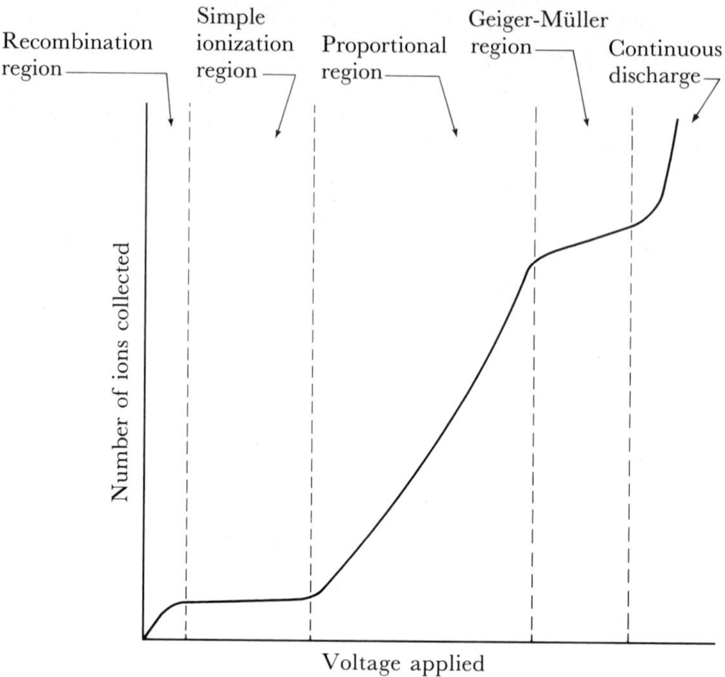

FIGURE 5-3
Output (number of ion pairs) of a Geiger-Müller tube as a function of the voltage between anode and cathode.

even higher voltage, the number of secondary ions collected becomes proportional to the number formed in the original ionization (the proportional region). This is followed by a second region of saturation (the Geiger-Müller region) in which all possible ion pairs, both primary and secondary, are collected; ultimately, the voltage is so high that the gas is ionized by the applied voltage even when no β particles are present (the continual discharge region). The complete dependence of the system on voltage is shown in Figure 5-3.

Ionization detectors can be operated in either the proportional or the Geiger-Müller region. Proportional counting has the advantage that particles of different energies can be distinguished by pulse-height analyzers (see page 101), because the size of the current pulse received by the electrode is proportional to the energy of the original charged particle entering the chamber. In this way, two different radioisotopes could be present in the same sample and the relative amounts of each could be determined from their β spectra by the procedure described on page 103 for scintillation counting. The disadvantage of the proportional counter is that extraordinarily stable, high-voltage power supplies are necessary because small fluctuations in voltage can produce large fluctuations in current. Proportional counting has been replaced by liquid scintillation counting (see next section) and will not be discussed further.

In the Geiger-Müller (G-M) region, maximal gas amplification is achieved and voltage fluctuations have little or no effect. This results in a reliable, sensitive, relatively inexpensive, and stable counting device—the Geiger-Müller counter.

A diagram of the most commonly used type of Geiger-Müller counter is shown in Figure 5-4—the so-called *end-window counter.* The detector consists of a cylinder, the inner wall of which is metallized and serves as the cathode, an axial wire anode, and an end window (typically made of mica or a plastic called mylar) through which β particles enter the gas-filled chamber. An electronic system measures the current produced by the capture of electrons by the anode. The tube is usually filled with an inert gas such as helium, neon, or argon to which has been added a small amount of a *quenching agent*—usually butane, propane, ethanol, chlorine, or bromine. The quenching agent prevents continuous ionization, which would cause the detector to fail to respond to any other than the first in-

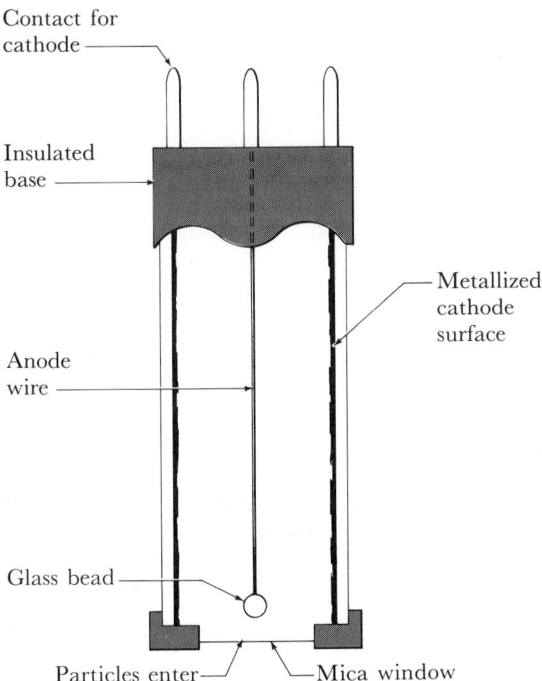

FIGURE 5-4
An end-window Geiger-Müller tube.

coming particle. When a positive ion nears the cathode, it sometimes pulls off an electron and becomes a neutral atom again. This recombination results in the production of both x rays and short wavelength ultraviolet radiation, either of which can ionize the gas molecules and form secondary electrons, producing the subsequent avalanche of electrons. It is this that causes a self-generating, continuous ionization. However, if a quenching gas (e.g., butane) is present that has a lower ionization potential than the major gas (e.g., helium), the positive ions of the helium will collide with the butane and acquire an electron from the butane to form a neutral helium atom and a positively charged butane ion. These butane ions move to the cathode and pick up an electron to become neutral again; however, because of the physical and chemical properties of the butane, the excess energy of recombination is not converted into electromagnetic radiation but instead breaks the chemical bonds of the butane resulting in the destruction of the molecule. After a certain number of discharges no quenching gas will be left. Actually, for most commercial G-M tubes, from 10^8 to 10^{10} pulses are possible before the tube becomes useless. This limitation can be eliminated by adding an entrance and an exit port and passing the gas continuously through the tube. This is called a *flow tube.* The sensitivity of the tube can also be improved by eliminating the window; many of the β particles either fail to pass through the window or lose sufficient energy during the passage that they are no longer able to ionize the gas. To do this, the tube is mounted so that it is in close contact with the sample holder. Because the gas would be lost whenever the sample was inserted or removed, the gas is continually flushed through the tube. Hence, this modification is known as a *windowless flow counter.*

MEASUREMENT OF RADIOACTIVITY BY LIQUID SCINTILLATION COUNTING

The ideal instrument for measuring radioactivity would detect all decays but no such instrument exists. Geiger-Müller windowless counters are very efficient detectors of high-energy β particles such as those from ^{32}P but are very inefficient for low-energy particles such as those from ^{3}H. The two major factors that limit efficiency are that not all emitted particles reach the detector and, of those that do, not all are counted. The main reason for failing to reach the detector is that the geometry is usually such that some particles are emitted in a direction that misses the detector. For example, even if a sample were flush with the front surface of a G-M detector, one-half of the particles would still be emitted in a direction away from the tube. The principal reason for failing to detect a particle that does enter the G-M tube is that the particle may have insufficient energy to cause ionization of the gas. By looking at the β spectra in Figure 5-1, it can be seen that the particle energy can be as low as zero (there is no E_{min}), so that some fraction will always be less than the energy required to ionize the gas. It can also be seen that the shape of the spectra is such that, with decreasing E_{mean}, there is an ever-increasing fraction of particles having energy in this very low range. This is in fact the main problem in detecting low-energy β particles with all counters.

The geometric problem could be solved if the sample were contained *within* the detector. In this way, losses due to failure to reach the detector would be limited to those particles whose range is so short that they fail to leave the sample itself ("self absorption"). Such a solution to the geometric problem is provided by the technique of *liquid scintillation counting.*

In liquid scintillation counting, the sample is dissolved or suspended in a solvent containing one or more substances that are fluorescent (Table 5-2). In brief, the emitted particle causes a pulse of light, which is detected by an optical device (a photomultiplier tube) that converts it into an electrical pulse that can be counted.

TABLE 5-2
Solvents and fluors commonly used in liquid scintillation counting.

Solvents

Toluene

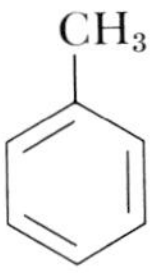

1, 4-Dioxane

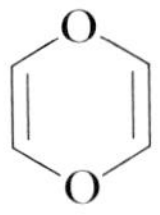

p-Xylene

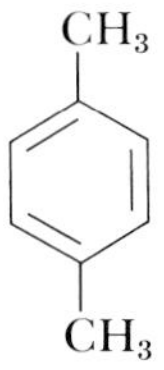

Fluors

PPO

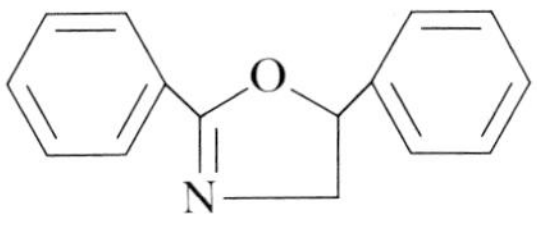

POPOP

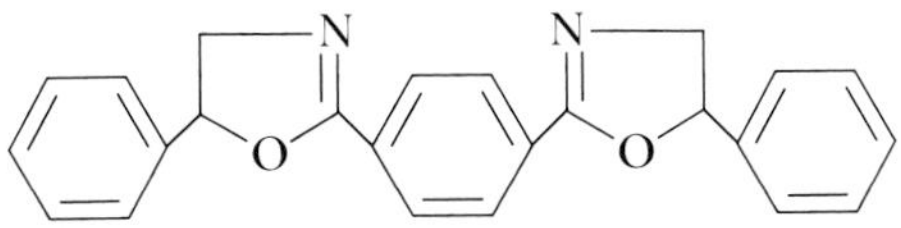

The Scintillation Process

Let us first understand how the decay produces detectable fluorescence. Consider a β particle that leaves the sample and enters the solvent in which the sample has been placed. In most solvents, the energy of the particle either would be dissipated as heat or would cause a chemical alteration (e.g., ionization or dissociation). However, in certain solvents, the energy is absorbed by the solvent molecules that are raised to an excited state. The excited molecule then returns to the ground state and gives up its energy by emission of a photon of light (see Chapter 15 for a discussion of the fluorescence process). The wavelengths of these emitted photons are very short and are not detectable by most photodetectors and so a fluorescent substance (a *fluor*) at a fairly low concentration must be added; this substance efficiently absorbs the photons emitted by the excited solvent molecules and then reemits photons at a longer wavelength. The photons emitted by the fluor must be detected by a photomultiplier tube. However, if the wavelengths of the photons emitted by the fluor are not in the region of highest sensitivity of the photomultiplier, an electrical pulse will not be generated. Therefore, a second fluor is almost always added. The secondary fluor absorbs the photons emitted by the primary fluor and reemits them as fluorescence at a longer wavelength, which the photomultiplier can detect with high efficiency. It should be noted that neither primary nor secondary fluors absorb much of the original energy of the β particles (because they are present at exceedingly low molarity) and are simply wavelength shifters.

This sequence of events converts the energy of the emitted particle into a flash of light—that is, a collection of photons emitted in an extremely short interval ($\sim 10^{-9}$ seconds). Note that the light need not come from a single point because of the multiple transfer of energy. The efficiency of detection of the original decay depends on (1) the properties of the solvent (i.e., the fraction of the absorbed energy that is converted into excitation and the fraction of the excitation energy that is transferred to the primary fluor rather than being dissipated as heat); (2) the number of photons produced by the primary and secondary fluors; (3) the geometry of the photomultiplier tubes (i.e., the efficiency by which the photons are gathered); (4) the signal-to-noise ratio of the photomultiplier (*noise* refers to pulses produced in the absence of light); and (5) the circuitry that converts the charge of the photomultiplier into a voltage.

Background Noise

Noise is a significant problem in the design of liquid scintillation counters—a problem that can be readily understood by observing the magnitude of the electrical pulse resulting from a single β decay. To give a rough idea of what happens, a 50 Kev β particle (e.g., from ^{14}C) will yield a few hundred photons in the commonly used solvent-fluor systems. These photons result in roughly 50 photoelectrons (10%–20%) being produced by the photocathode of the photomultiplier. This is amplified approximately 10^6-fold by the multiplier to yield 5×10^7 electrons, or a charge ranging from 0.5×10^{-11} to 1.0×10^{-11} coulomb, which can be converted into an electrical pulse of 0.1 or 0.2 volts. For ^3H, E_{max} is 5.5 Kev and the pulse is about 0.018 volts.

The low voltage of these pulses causes the problem because the thermal noise of photomultipliers produces pulses on the order of 0.005 volts at room temperature. Hence, a substantial fraction of emitted β particles (a fraction that increases with decreasing E_{max}) produces pulses of lower voltage than does thermionic noise.

This problem has been attacked in two ways: The first was to place the photomultiplier and the samples in a freezer at a temperature ranging from 0° to 5°C, which reduced the noise level by a factor of about 4; now with the advent of improved low-noise photomultipliers, room temperature ("ambient") operation is possible. Thermionic noise results in a single pulse, whereas β particles usually result in the production of several photons. Hence, the second way of attacking the problem requires the use of two photomultipliers for viewing the sample; the outputs of the photomultipliers are fed into a *coincidence circuit,* which registers a count only when two pulses are *simultaneously* (i.e., within seconds) received by the photomultipliers. In this way, the single noise pulses are discarded. The timing of the coincidence circuit is such that it will only very rarely be the case that one noise pulse will follow another so closely in time that it will be recorded; only when two photons have been produced as a result of a *single* β particle will the coincidence circuit register a pulse. Coincidence circuitry reduces the background count rate from about 10^5 to 15 cpm (i.e., the number of pulses that pass the coincidence circuit).

The necessity of using coincidence circuitry eliminates the possibility of 100% efficiency of detection of low-energy β particles because a minimum of two photons (one for each photomultiplier) is required and a certain number of the decays do in fact yield only one. However, this disadvantage is more than offset by the tremendous efficiency resulting from the huge reduction in background.

Several other sources of background counts exist although they are of a lesser order of magnitude. They are, however, the ones encountered by the user of the counter rather than the designer. One source is the radioisotope ^{40}K found in glass. If the sample is a solution, it must be in a container, which is usually a glass vial fabricated from "low potassium" glass. This is necessary because of the large amount of the radioisotope ^{40}K found in ordinary glass. However, even low-potassium glass contains significant amounts of ^{40}K and contributes a background of about 15 cpm—the same amount contributed by thermionic noise. For some operations requiring very low backgrounds, polyethylene vials can be used to eliminate the ^{40}K background, but they are not resistant to all solvents used in scintillation counting and are therefore not in general use. Another source of background is Čerenkov radiation, the wavelength of which is in the region of efficient response of the photomultipliers. Čerenkov radiation is produced by the responses of the scintillation solvent (without added fluor) and the vial itself to cosmic rays. It produces pulses in the low-to-mid-energy range and contributes another 10 cpm. Yet another source is environmental activity. When the sample vial is filled with solvent and fluors, environmental radioactivity adds about 40 cpm to the background. Taken together these effects, including thermionic noise, produce a background ranging from about 80 to 90 cpm. However, because emitted β particles always have a maximum energy, it is always possible to require (by means of pulse-height analysis and by the use of discriminators—see page 101) that pulses with energies above that value be rejected. This results in a reduction of background to about 35 cpm with no loss of counting efficiency.

Quenching

The ability to detect radioactivity depends on the signal-to-noise ratio. So far, the discussion of background has been concerned with maximizing this ratio by reducing noise. Now the problems associated with *maximizing the efficiency of production of photons* (i.e., maximizing the signal) will be considered.

Quenching refers to a reduction in the efficiency of transferring energy from the β particles to the photomultiplier (by any means). Quenching results in a decreased (perhaps to zero) number of photons per β particle and therefore the production of a pulse of reduced voltage. The three most common mechanisms are *chemical, color,* and *dilution* quenching.

CHEMICAL QUENCHING

When a sample is added to a scintillation solution, the sample itself may contain substances that either absorb some of the energy of the β particles without emitting any photons or absorb the photons emitted by the excited solvent molecules without fluorescing. The most common chemical quenchers are water, acids, salts, and dissolved oxygen.

COLOR QUENCHING

Colored samples can absorb some of the photons emitted by the secondary fluor so that they never leave the counting vial. It is important to realize that a lack of visible color in the sample is no criterion for the absence of color quenching because the human eye cannot determine whether a substance absorbs in the near ultraviolet. It is, though, generally true that substances that appear yellow, red, or brown probably introduce severe color quenching.

DILUTION QUENCHING

The dilution of the solvent and fluor by the sample reduces the probability of a scintillation event. If the sample is a liquid, there is no way to avoid this; however, it can be corrected for in the analysis of data and will be described in a later section.

Proportional Counting and Pulse-height Analysis

The greatest value of scintillation counting is the ability to determine the ratio of two radioisotopes present in a mixture. This is possible because the voltage pulse produced as a result of a decay is proportional to the energy of the emitted particles. The resolution of the two isotopes is accomplished with a *pulse-height analyzer* equipped with *discriminators.* A pulse-height analyzer is an electronic instrument that can sort fluctuations (pulses) in current or voltage. As each pulse is detected, its magnitude (pulse-height) is also registered. The instrument is used in circuitry that is designed to count pulses in different voltage *intervals.* For example, it can be adjusted to detect pulses either greater than zero or some other value, less than a certain value, or between any pair of values. The controls that determine the voltage levels defining the voltage range are called *discriminators.* Two discriminators are normally employed to define an interval that excludes low-energy photomultiplier noise and high-energy environmental noise. With more than two discriminators, it is possible to count two intervals simultaneously. Counting in two ranges simultaneously requires two counting circuits, called *scalers.* Each scaler is said to be counting in a *channel* defined by the particular voltage levels, frequently designated L1, L2, L3, and so forth. Decay rate can be plotted against energy or pulse height by counting pulses between a variety of pairs of voltage levels. Figure 5-5 shows a plot for ^{3}H and ^{14}C. Using this ability to determine the decay rate in a given pulse-height interval, the ratio of two isotopes can be determined. How this is done can best be seen by example.

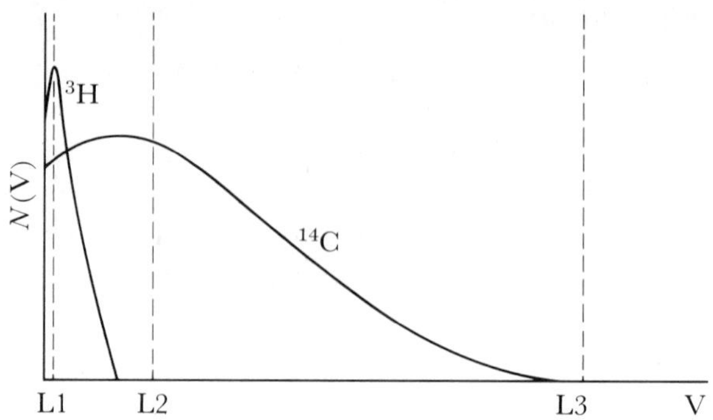

FIGURE 5-5
Simultaneous counting of ^{3}H and ^{14}C. The ^{3}H and ^{14}C spectra are plotted on a single scale: L1 is the lower limit for most counters, below which thermal noise becomes severe; L2 and L3 are the voltage levels used in Example 5-A. The y-axis is an arbitrary scale giving the count rate in a small interval centered on a particular voltage, V.

Example 5-A. Ratio of ^{3}H to ^{14}C activity in a single sample. Consider a radioactive sample containing both ^{3}H and ^{14}C labels and counted in the two voltage ranges defined by L1, L2, and L3 of Figure 5-5. Data obtained for the sample are given in Table 5-3. By counting ^{3}H- and ^{14}C-standard samples using the same instrument settings, the amount of ^{3}H and ^{14}C present in each channel can be calculated as shown in Table 5-3.

What is done follows. The distribution of cpm of the ^{3}H- and ^{14}C-standard samples in channels A and B tells how the ^{3}H and ^{14}C cpm are distributed in the experimental sample. Line 1 shows that only 0.0015 of all ^{3}H cpm are in channel B, which is approximately zero for the sample. Hence, all cpm in channel B are due to ^{14}C, as indicated in line 2. The data for the ^{14}C-standard sample shows that the cpm in channel A is 0.416 times the cpm of ^{14}C in channel B. Because line 2 shows that channel B contains only ^{14}C cpm, the calculation of line 3 can be carried out. Hence, of the 500 cpm in channel A, 333 are ^{14}C and the remainder are ^{3}H, as shown in line 4. Because there is virtually no ^{3}H in channel B, the amount in channel A is all of the ^{3}H, as stated in line 5. The total ^{14}C is then the sum of the ^{14}C in channels A and B, as shown in line 6. The ^{3}H-to-^{14}C ratio can then be calculated as in line 7. Note that the simplicity of the calculation comes about by selecting L2 so that there are, for all practical purposes, no ^{3}H counts in channel B. It should also be noted that, if there were a great deal more ^{14}C, the ^{3}H determination would be inaccurate. For

TABLE 5-3
Calculation of ^3H / ^{14}C ratio in a doubly labeled sample counted in two channels of a scintillation counter.

Sample	L1–L2 (channel A)	L2–L3 (channel B)	B/A	A/B
Experimental	500 cpm	800 cpm		
^3H-standard	16,840	25	0.0015	—
^{14}C-standard	4,250	10,200	—	0.416

Calculations for sample

1. ^3H cpm in channel B = (0.0015)(500) = 0.75
2. ^{14}C cpm in channel B = 800 − 0.75 ≅ 800
3. ^{14}C cpm in channel A = (0.416)800 = 333
4. ^3H cpm in channel A = 500 − 333 = 167
5. Total ^3H cpm = 167
6. Total ^{14}C cpm = 333 + 800 = 1,133
7. ^3H / ^{14}C = 167/1,133 = 0.147

example, suppose that there were twenty times as much ^{14}C as ^3H. Then the ^{14}C in channel A would be 20 × 333 = 6,660 and the cpm registered in channel A would be 6,660 + 167 = 6,827. The ^3H value would then be obtained by subtracting 6,660 from 6,827.

To understand why subtracting large numbers is a problem, consider the statistics of counting. According to the theory of statistics, the standard error of a measurement is $N^{-1/2}$ in which N is the total counts (not cpm). This means that there is a 68% probability that the "true" value is in the range $N \pm N^{1/2}$. If the samples in the example with 20× ^{14}C were counted for one minute, the ^3H would be obtained from the difference between 6,827 ± 82 and 6,660 ± 81—which could range from 5 to 430 cpm. Clearly, the error in the difference would be huge. If the samples were counted for 10 minutes, the difference would be (68,270 ± 261) − (66,600 ± 259), which could range from 115 to 219 cpm, an improvement but still a large error. With the $^3H/^{14}C$ ratio used, the range is from 164 to 180 for 10-minute counts.

To indicate the value of double-label analysis, consider an extension of this example. Suppose that the ^3H is in leucine at a specific activity of 5 C/mmol and the ^{14}C is in tyrosine at a specific activity of 4 C/mmol. Let us assume that the counting efficiency for ^3H and ^{14}C is 30% and 90%, respectively. Therefore the actual ^3H and ^{14}C cpm is (measured ^3H cpm/^3H efficiency) and (measured ^{14}C cpm/^{14}C

efficiency) or 167/0.30 = 556 cpm and 1,133/0.90 = 1,259 cpm, respectively. The amounts of leucine and tyrosine are therefore (amount ^{3}H/sp. activity of leucine) and (amount ^{14}C/sp. activity of tyrosine). The molar ratio of leucine to tyrosine is therefore (^{3}H/^{14}C) · (specific activity of tyrosine/specific activity of leucine) = (556/1,259) · (4/5) = 0.353.

Example 5-A assumes that the distribution of cpm between channels A and B for ^{3}H is the same in the experimental sample as in the ^{3}H-standard sample and that this is also true for ^{14}C. This means either that there is no quenching or that the quenching in the sample is the same as in the standards. If this is not the case, the relative degrees of quenching would have to be determined. As will be seen in the section on sample preparation, this can be a formidable problem in doubly labelled samples. Hence, it cannot be overemphasized that the samples used as standards *must* be prepared in an identical manner as the samples containing both labels. A common error made by novices is to use the standards provided by the manufacturer of the counter for determining the ratio of cpm in the channels. These standards are typically radioactive toluene in a toluene-based solvent in an argon atmosphere—an ideal unquenched situation almost never encountered in practice.

SAMPLE PREPARATION

The method of sample preparation can have a great effect on counting efficiency.

In G-M counting, the sample is usually deposited on an aluminum disc called a *planchet* and dried, if it is in liquid. Frequently, insoluble samples are collected on the surface of a thin membrane filter (Chapter 7) and cemented onto the planchet. The principal problem in sample preparation for G-M counting is self-absorption—that is, if the sample is too thick, many of the β particles fail to reach the counter because they are absorbed by the sample itself, or they lose sufficient energy that they cannot ionize the gas. For β particles as energetic as those from ^{32}P, self-absorption is rarely a problem; for ^{14}C, it can be a problem if the mass of the sample exceeds 1 mg; for ^{3}H, it is so severe that G-M counting is rarely possible. The problem of self-absorption is circumvented by using samples of various thickness (i.e., various volumes), determining the count rate per unit volume for each, and determining whether the cpm is proportional to volume. If this is not the case, a graph is made of cpm versus volume and is used to correct the data. Figure 5-6 illustrates how this is done. It is

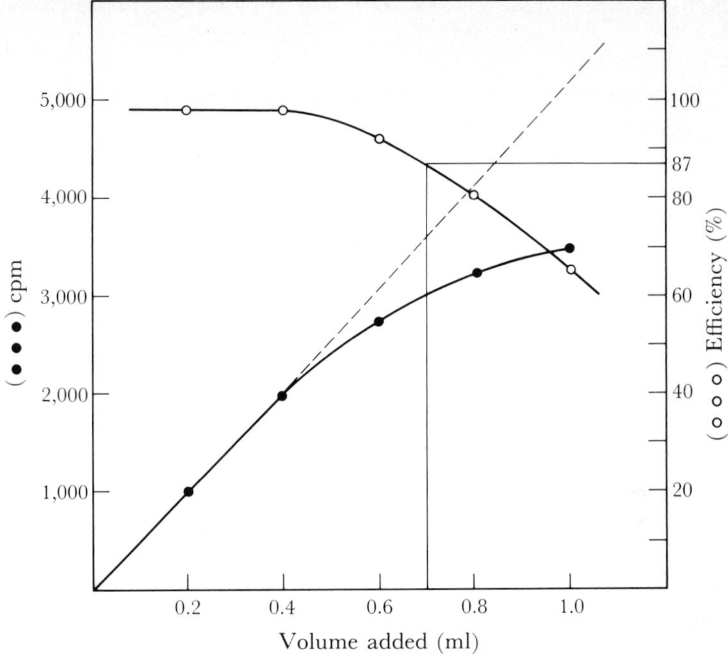

FIGURE 5-6
Evidence of self-absorption: the measured radioactivity is not proportional to the volume added; the dashed line indicates the cpm that would be expected if there were no self-absorption; the ratio of the solid curve to the dashed line gives the efficiency (open circles). Hence, if a similar 0.7-ml sample yielded 4,500 cpm, the corrected value would be 4,500/0.87 = 5,184 cpm.

often necessary to add a separate compound of high activity to an identical sample in order to have a count rate that is high enough to obtain a statistically significant correction curve.

For scintillation counting, the ideal sample is one that is dissolved in the counting solution because then every radioactive atom is in intimate contact with the solvent. If the sample is insoluble, it will exist in the form of small particles or aggregates in the solvent and the energy of some of the beta electrons will be reduced by passage through the aggregate—that is, there will be self-absorption. Again, this is not a significant problem for high-energy particles but must be considered for ^3H. In general, sample solubility is a problem in biological experimentation because the fluors used are usually soluble only in nonpolar solvents in which most biological molecules are insoluble.

There are three basic methods for preparing biological samples for scintillation counting: (1) the scintillation solvent is altered in a way that permits the addition of aqueous solutions so that soluble material remains in solution; (2) insoluble substances are chemically or physically converted into a soluble form; (3) insoluble substances are collected on nitrocellulose, glass fiber, or paper filters and then dried. The first two methods introduce quenching and this must be dealt with by methods described next; the third method is rapid and convenient but also presents several problems.

Alteration of Counting Fluid to Permit Water Uptake

The usual counting fluids are toluene-based and cannot take up water without severe quenching. Dioxane-based solutions can accept water (as much as 29%), but dioxane is less efficient in energy transfer than toluene. Dioxane-based fluids can be improved by the addition of naphthalene to increase efficiency plus a variety of alcohols or ethers to maintain naphthalene solubility when water is added. A particularly common dioxane-based fluid is called Bray's solution. In each system the additives produce quenching, but this can usually be ignored if relative amounts of radioactivity are being measured because it will be the same for all samples. However, water added to the sample and the salts dissolved in the water produce quenching, the degree of which increases as more water is added. The usual recourse is to add water or a solution to each sample so that the total amount of added material is the same for all samples in a set. An alternative is to determine the degree of quenching in each sample, which will be described shortly.

The mean range of ^3H β particles in water is approximately 0.5 micron. Hence, if a precipitate or insoluble material could be kept dispersed so that the particle size were much less than 0.5 μ, self-absorption losses would be small. This can be done in several ways. If a surfactant such as Triton-X100 or an emulsifier such as Cab-O-Sil (silica gel) is added to toluene, water can be added up to 10% by volume and the water droplets will be dispersed in the emulsion as particles <0.1 μ in diameter. Other emulsifying mixtures are PCS and Aquasol (New England Nuclear Corporation). An alternative is to filter the sample onto finely divided diatomaceous earth (Celite, Johns-Manville Corporation) and count in any emulsifying mixture. It is not clear, though, why the latter method works.

Solubilization of Sample

Acidic substances such as proteins and polypeptides, acidic polysaccharides, and nucleic acids can be solubilized by reaction with organic bases—primary examples are Hyamine 10-X-hydroxide and primene 81-R (high-molecular-weight quaternary and primary amines, respectively). A technique that is sometimes used with ^{14}C-labeled material is to incinerate the sample and collect the radioactive CO_2 by bubbling through Hyamine.

Insoluble bases, such as metallic ions, are solubilized by 2-ethylhexanoic acid or dialkyl hydrogen phosphates. Most of these methods are technically complex and introduce significant quenching.

Collection on Membrane Filters

Insoluble samples are commonly filtered or dried onto paper filters—a convenient way to remove water. But this popular method has two disadvantages: (1) the efficiency of counting depends on the orientation of the paper in the counting vial (because the paper is opaque in the solvents and intercepts some of the photons) and (2) very small molecules (e.g., amino acids and purines) penetrate the paper and become inaccessible to the solvent.

A great improvement is to collect fine precipitates on nitrocellulose membrane filters (e.g., Millipore, Schleicher and Schuell, and Gelman) or, even better, on fiberglass papers as described in Chapter 7. This is probably the most common method of sample preparation in modern biochemistry because of speed and convenience, although self-absorption can be a problem, especially if the concentration of material is >1 mg/cm^2. This will be discussed in more detail in the next section. Nitrocellulose filters present a special problem that is not commonly recognized. After a sample has been collected, the filter is usually dried in a 100°C oven to remove trace amounts of water to prevent chemical quenching. Sometimes there is a slight yellowing (or charring) caused by the heating. The yellow material leaches out into the scintillation solution and absorbs the blue-violet photons emitted by the fluor. This yellowing often goes unnoticed and is a frequent cause of the lack of reproducibility in duplicate samples. This problem is best avoided either by washing the sample in an organic solvent to aid in rapid drying or by reducing the drying temperature to 80°C. The replacement of nitrocellulose filters with fiberglass filters eliminates the problem of yellowing of the support. Fiberglass is superior because, also for unknown reasons, it gives a higher efficiency in toluene-based solutions and, as indicated earlier, lacks the problem of variable color quenching resulting from the yellowing of nitrocellulose filters in the course of drying at a high temperature.

DETERMINATION OF THE EFFICIENCY OF COUNTING BY PULSE-HEIGHT ANALYSIS

An evaluation of the efficiency of counting is clearly necessary if the work being done requires that the *absolute* number of disintegrations be known, although in most experimental situations this is not the case. In the more common situation in which radioactivity in several samples is compared, it is necessary to know only that the efficiency of counting is the same in all samples. This is not always so, because the samples may have different chemical compositions, which might lead to different degrees of chemical quenching; different colors; or variable amounts of dilution quenching. Furthermore, it is often necessary to compare a soluble sample with an insoluble one, or two insoluble samples with different physical properties.

Chemical quenching is the type encountered most frequently. The *channels ratio,* the *external standard ratio,* and the *internal standard ratio* methods are the most common procedures used to evaluate the degree of quenching. The first two are based on the fact that quenching reduces counting efficiency by shifting the spectrum to a lower energy range (Figure 5-7A) so that counts are apparently lost because a larger fraction have an energy that the scintillation counter fails to detect—that is, they are in the one-photon range that the coincidence circuit rejects or they are below the L1 level needed to reduce photomultiplier noise. However, it should be noted that, if the quenched and unquenched spectra are examined between two pairs of levels, L1 → L2 and L2 → L3, quenching causes a *relative increase* in the number of counts in channel A compared with B (i.e., the channels ratio changes), although the total count rate *decreases.* Hence, in the channels ratio method, a set of standards is prepared with various amounts of quenching (e.g., by the addition of such quenching agents as water or chloroform) and a curve is prepared relating the observed cpm to the ratio of cpm in channels A and B (Figure 5-7B). Hence, for a sample with suspected quenching, the sample is counted in the same pair of channels used for the standard, the ratio is calculated, and the percentage of quenching is determined from the curve shown in Figure 5-7B. For example, if a particular sample had 500 cpm and the channel

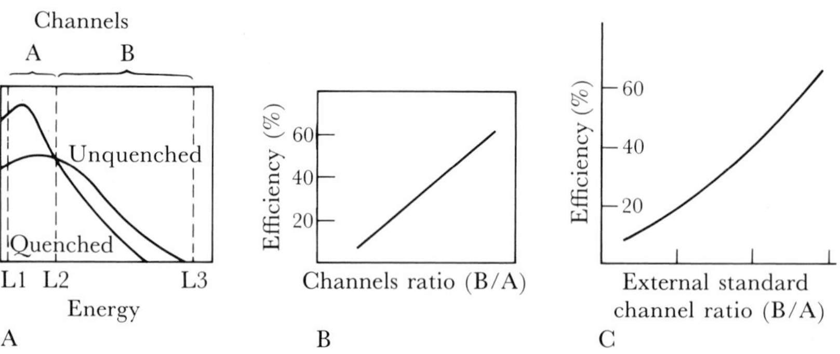

FIGURE 5-7
A. Shift in spectrum caused by quenching. The area remains constant.
B. Correction curve for channels ratio method of quench correction.
C. Correction curve for external standards method of quench correction.

ratio showed 20% efficiency, the cpm in an unquenched sample would be 500/0.20 = 2,500.

The *external standard ratio* procedure requires the use of a radiation source (typically ^{137}Cs) external to the counting vial but contained within the instrument and bombarding the sample with γ rays. When the γ rays pass through the solvent, some electrons are produced, which in turn are detected in the same manner as β particles. In a quenched sample, the energy spectrum of these electrons is also shifted and a similar plot of counting efficiency versus the channel ratio can be made (see Figure 5-7C). The advantage of the external standard ratio method over the channels ratio method is that the γ source provides a very high count rate ($\sim 10^5$ cpm) and therefore great statistical reliability (remember the $N^{1/2}$ rule of Example 5-A). This method is especially valuable if the activity of the sample in question is too low to use the channels ratio method.

Modern instruments with associated computers are available that utilize the external standard method and automatically correct for quenching before printing out the data.

If a sample is quenched and contains two labels, special techniques are required for quench correction (when it is possible). Descriptions of such techniques are usually given in instrument manuals or in the specialized texts listed near the end of this chapter.

The third method of quench correction—the *internal standard ratio* method—is carried out in the following way: After counting a sample, a known amount of a nonquenching, labeled compound is added and the sample is recounted. From the ratio of the observed activity to the amount known to have been added, the efficiency of counting can be calculated. Although this method seems perfect, it has two problems: (1) the volume of added material is usually so small that there is a significant error (3%–5%) in volume measurement and (2) the sample is destroyed and cannot be recounted if desired. This method is not often used because it is considered tedious by most workers.

Both the internal and external standard methods can be used to correct for color and dilution quenching. However, colored samples are more easily dealt with by oxidizing them with H_2O_2, which usually eliminates the color. Dilution quenching is best dealt with by successive dilutions of the sample with the fluor solution. The cpm/unit volume of counting solution can then be plotted against sample concentration and extrapolated to zero concentration. This method is time-consuming but satisfactory.

It is important not to confuse self-absorption with quenching. The quench corrections discussed above correct only for changes in the β spectrum associated with the responsiveness of the scintillation fluid and give no information about the extent of self-absorption. With insoluble samples or particles collected on a filter, losses are caused by the physical properties of the sample itself.* The problem of self-absorption is difficult to solve and often the principal cause of low counting efficiency. This is especially true in studies of the synthesis of macromolecules such as proteins and nucleic acids because the most common means of sample preparation is to precipitate the macromolecule with acid (hydrochloric, perchloric, or trichloracetic) and collect the precipitate on a filter. If self-absorption is taking place, an increase in efficiency can sometimes be achieved by reducing the sample size if the sample is very large and on a filter because the precipitate may have formed a thick layer. However, dilution does not usually eliminate self-absorption because the precipitate of most substances has a minimum size that is not reduced by dilution. Unfortunately, in low-energy β particles such as those from ^{3}H, self-absorption due to these microparticles is already severe. In general, self-absorption cannot be eliminated in any way other than by the use of solubilizing agents.

The best way to *count relative amounts* of radioactivity in different insoluble samples is to make all samples as nearly identical as possible. One method of accomplishing this is to add an excess and fixed amount of a precipitable material (e.g., DNA or bovine serum albumin) to each sample so that *variation* in the amount of precipitate from one sample to the next is reduced. If a soluble sample is to be compared with an insoluble one (e.g., to relate the specific activity of a protein to that of an amino acid used in the synthesis), the only way to obtain high precision is to solubilize the samples and count them in identical mixtures.

A word of caution is in order concerning solubilization and samples on filters. If an insoluble sample on a filter is placed in a solvent in which the filter material is soluble (e.g., dioxane for nitrocellulose filters), the sample may appear to dissolve but in fact it is usually the case that only the filter has dissolved and the sample is still in particulate form. Hence, this procedure does not eliminate the self-absorption problem.

GAMMA-RAY DETECTION

Gamma emitters have recently become of great value in biochemistry, especially in radioimmunoassay (see Chapter 10) in which antibody or other proteins are labeled with the γ emitter ^{126}I. This radiation is easily

*At first glance, it may seem that the problem would be greater on a filter because of the attenuation of the β particles passing through the filter material itself. However, this is surprisingly unimportant because the efficiency of counting an insoluble sample is little affected by its being on a filter. Presumably, the particles are seated so high on the filter that they are nearly surrounded by the solvent.

detected and counted by means of commercially available γ counters, using an external sample-scintillation detector, which work in the following way. The sample is placed in a glass, or plastic, tube and inserted into a large thallium-activated NaI crystal [NaI(Tl)]; this is called "well" counting (Figure 5-8). Because γ rays have very high energy, they leave the sample and the container virtually without loss of energy. With an efficiency of a few percent, they produce electrons while traversing the crystal. These electrons excite adjacent parts of the fluor crystal and thereby produce fluorescence, which is detected by a photomultiplier tube. As in liquid scintillation counting, the appropriate circuitry converts the light into pulses, which are counted. Coincidence circuitry is not necessary in γ counting because the γ energy is so high compared with the background noise of the photomultipliers that the voltage defining the lowest level of the pulse-height analyzer (i.e., L1 of liquid scintillation counting) can be set above that of the background pulses. In fact, the system usually uses many photomultipliers *acting independently* to increase the efficiency of the counter.

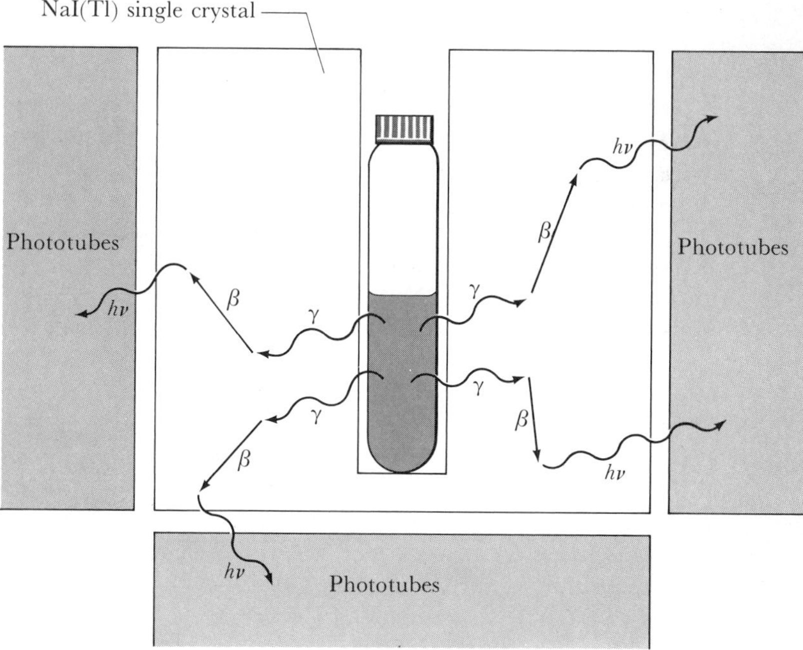

FIGURE 5-8
Well counting of γ rays. The γ rays interact with the crystal and produce β particles that excite the crystal, which then emits photons ($h\nu$) detectable by the phototubes.

As in liquid scintillation counting, some double-label experiments are performed using different γ-emitting sources. Again, because the pulses produced are related to the energy of the γ ray, pulse-height analysis can be carried out to distinguish different isotopes—for example, ^{126}I and ^{131}I. Detailed discussions of γ spectrometry can be found in the references given near the end of the chapter.

EXAMPLES OF THE USE OF RADIOACTIVE MATERIALS

This section includes many examples of the use of radioactivity in biochemistry and molecular biology. Radioisotopes can be used (1) to test material in quantities that are too small for direct chemical testing, (2) to distinguish molecules that are identical but are in different chemical locations, (3) to analyze mixtures that are too complex for traditional chemical analysis, and (4) to demonstrate participation in a reaction in which the products are chemically indistinguishable from the reactants. The first situation is illustrated in Examples 5-C and 5-H; the second in 5-B and 5-E; the third in 5-F and 5-G; and the fourth in 5-I.

Example 5-B. Reactions of *Escherichia coli* DNA polymerase I.
If purified DNA polymerase I is added to a reaction mixture containing a buffer, Mg^{2+}, DNA, and the four deoxyribonucleotide-5′-triphosphates (of adenine, thymine, guanine, and cytosine) labeled with ^{32}P in the α position (nearest to the deoxyribose) and incubated at 37°C, the ^{32}P becomes insoluble in 10% trichloroacetic acid (TCA) and can be collected on a membrane filter—an indication of polymerization to DNA because DNA, but not the nucleotide triphosphates, is insoluble in TCA. Note that newly synthesized DNA is distinguishable from the template DNA by virtue of the radioactivity of the new DNA. If only three deoxynucleotide-5′-triphosphates are present, or if Mg^{2+} or DNA are not present, or if the nucleotides are ribo- instead of deoxyribo-, the ^{32}P remains soluble. Hence, polymerization requires all four *deoxy*ribonucleotides, Mg^{2+}, and DNA. If any of the triphosphates is replaced by mono- or diphosphate, ^{32}P remains soluble (i.e., there is no polymerization); hence, the triphosphate group is necessary in the substrate. If the deoxyribose nucleotide-5′-triphosphates carry the ^{32}P in the β or γ position, and ^{14}C in the deoxyribose moiety, the ^{14}C, but not ^{32}P, becomes acid insoluble. (The ^{14}C and ^{32}P can be distinguished by liquid scintillation counting.) Hence, the β and γ phosphate groups are removed in the reaction; only the α phosphate remains in the polymer. Thus, by the use of radioactivity, the different phosphate atoms can be distinguished.

Example 5-C. Molecular weight of a DNA molecule by end-group labeling.

The enzyme *E. coli* polynucleotide kinase transfers a γ phosphate from adenosine-5′-triphosphate (ATP) to a 5′-hydroxyl terminus of DNA. DNA normally contains two 5′-phosphoryl termini (one for each polynucleotide strand), which can be converted into 5′-hydroxyl groups by the enzyme alkaline phosphatase. In the laboratory at Brandeis University, 4.7 μg of purified, homogeneous (meaning that all molecules are the same size) phage λ DNA with 5′-hydroxyl termini was reacted with γ-^{32}P-ATP at a specific activity of 3 mC/μmol, using polynucleotide kinase. When acid-precipitated and counted, the sample showed 1,870 cpm, using a counter that detected ^{32}P at 85% efficiency. Hence, 1,870/0.85 or 2200 dpm of ^{32}P were precipitated. In units of microcuries, this is $2{,}200/(2.2 \times 10^6) = 10^{-3}$ μC. From the specific activity, 10^{-3} μC is equivalent to 3.3×10^{-7} μmol of ATP, or 1.98×10^{11} molecules. Because 4.7 μg of λ DNA was added, the weight of a single λ DNA molecule is $4.7/9.9 \times 10^{10} = 4.7 \times 10^{-11}$ μg or 31.0×10^6 atomic weight units. Here, the radioactivity served to measure phosphorylation of 5′-hydroxyl groups with very high sensitivity. New terminal phosphoryls could also be distinguished from old 5′-phosphoryl groups.

Example 5-D. Identification of "buried" tyrosines in a hypothetical protein.

Tyrosine is the only amino acid that can be iodinated. A hypothetical protein is known from amino acid analysis to contain six tyrosine residues. When 1 μg of purified protein reacts with ^{131}I, 2,550 cpm can be precipitated with trichloroacetic acid and collected on a fiberglass filter. If the reaction takes place in the presence of a denaturing agent—so that the protein is totally unfolded—3,820 cpm are acid-insoluble. Hence, $(2{,}550/3{,}820) \times 6 \simeq 4$ tyrosines are available for iodination when the protein is in the native configuration. Hence, $6 - 4 = 2$ are unavailable and therefore "buried" in the three-dimensional structure. In this case, radioactivity allows the detection of iodine in the protein when the amount of iodine is too small for chemical detection.

Example 5-E. Permeability of a bacterium to adenylic acid (AMP).

Not all radioactive chemicals added to a bacterial culture can penetrate the bacteria; some are excluded. If a radiochemical enters the bacteria and the bacteria are collected by filtration, radioactivity will appear on the filter. If it is excluded, the radioactivity will pass through the filter and any residue can be washed away. For example, if either ^{32}P-AMP or ^{3}H-AMP (with ^{3}H in the purine ring) is added to a culture of the bacterium *E. coli* and after a period of growth the bacteria are collected on a membrane filter and washed thoroughly with a buffer, neither ^{3}H nor ^{32}P is found on the filter. Hence, AMP is not taken up by the bacterium. In this case, radioactivity allows the added AMP to be distinguished from the AMP already present in the cell. If there were an uptake of an amount equal to 0.1% of that already within the bacterium, the increase by that amount would be undetectable by chemical procedures; using radioactive material, the 0.1% could be detected against a zero background because the internal material is unlabeled.

Because both ^{32}P-AMP and ^{3}H-AMP are being tested in this example, the lack of uptake of either label also shows that the cells do not cleave external AMP to form adenosine and phosphate in that it is known that these two substances can enter the cell. This is an important point because there are examples, such as thymidine, in which cleavage in this case to thymine) occurs before uptake.

Example 5-F. Identification of a substance by precipitation with antibody.

If the bacterium *E. coli,* growing in a medium containing ^{14}C-glucose as the sole carbon source, is infected with phage T4 and the infected cells are allowed to grow, phage-mediated proteins are synthesized. If the infected cells are broken open and antibody specific to purified T4 tail fibers is added, ^{14}C is precipitated. If the cells are broken at various times, the rate of synthesis of tail fibers can be determined from the amount of precipitated ^{14}C as a function of time. A similar analysis can be made with cells infected with phage T5 and antibody specific to T5 tail fibers. However, if the cells are simultaneously infected with T4 and T5 and are later disrupted, ^{14}C can be precipitated by anti-T4 tail fibers but not by anti-T5 tail fibers. Hence, T5 tail fibers are not synthesized in such a dual infection. Such an experiment would be virtually impossible without radioactivity because of the small quantities and the difficulty in identifying tail fibers in a mixture as complex as a cell lysate plus antiserum.

Example 5-G. Purification of a protein for which there is no biological or chemical assay.

Proteins are usually detected by virtue of having a measurable biological activity (e.g., enzymatic or inhibitory to some reaction) or a distinguishable physical property (such as the spectral property of hemoglobin). But detection by these means is not always possible, in which case the double-label method described herein is used.

If a phage-infected bacterium capable of making a particular protein is grown in a medium containing ^{3}H-leucine, all proteins will be ^{3}H-labeled. If a mutant bacterium that makes only a fragment of the protein is grown in the presence of ^{14}C-leucine, all proteins will be labeled, but there will be no ^{14}C-labeled protein corresponding to the intact protein of interest. If two cultures thus labeled are mixed and the total protein is isolated and fractionated by chromatography, most fractions will have a fixed ratio of ^{3}H to ^{14}C. The protein of interest will be in a fraction that has a high ratio of ^{3}H to ^{14}C (Figure 5-9). Hence, purification schemes can be based on maximizing this ratio until a fraction is obtained that contains ^{3}H but no ^{14}C. If all proteins

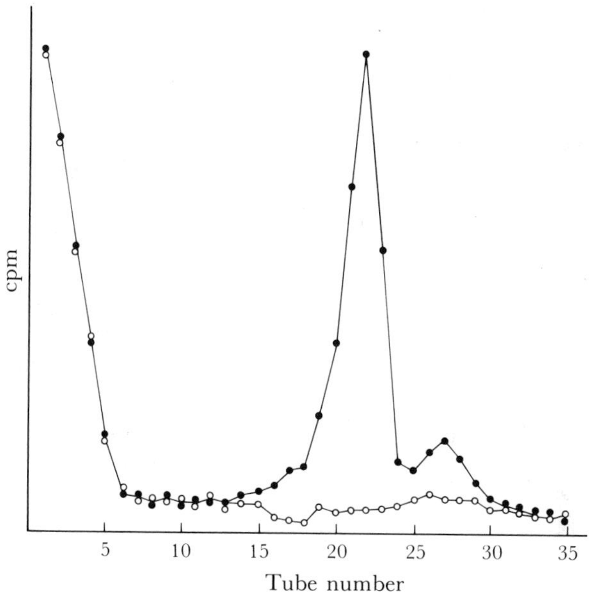

FIGURE 5-9
The use of double labeling in protein purification. A noninducible lysogenic strain of the bacterium *E. coli* was UV-irradiated to suppress protein synthesis. The cells were then infected either with phage λ wild-type in the presence of ^{3}H-leucine or with λcIsus10 (which makes a short fragment of the λ repressor) in the presence of ^{14}C-leucine. Proteins were isolated from each culture, mixed, and chromatographed on DEAE-cellulose (see Chapter 8). The elution pattern of the chromatographic column is shown as the amount of ^{3}H and ^{14}C in each fraction eluting from the column. A peak of ^{3}H activity appears for which there is no corresponding ^{14}C peak. This represents a protein produced in the λ wild-type infection but not in the λcIsus10 infection—that is, the λ repressor. This was the first means of assaying the λ repressor. Solid circles, ^{3}H; open circles, ^{14}C. [Courtesy of Mark Ptashne.]

are labeled, the protein will be chemically pure (which is, of course, not the case in a phage-infected cell because most bacterial proteins are nonradioactive).

Example 5-H. Sedimentation analysis of bacteriophage DNA structures in a superinfected lysogenic bacterium.

If a stable lysogen of the *E. coli* bacteriophage λ is infected with another λ phage, no phage are produced. The fate of the incoming DNA can be investigated by isolating all intracellular DNA and analyzing it by zonal centrifugation through a preformed density gradient (see Chapter 11). The analysis could be based on the fact that the λ phage DNA has a different molecular weight (3×10^7) from that of *E. coli* DNA (2.6×10^9) were it not for the fact that *E. coli* DNA usually breaks into pieces of many sizes during isolation. The question then is how to distinguish the λ and the *E. coli* DNA. This is easily done if the λ DNA is radioactive because, if radioactivity is used as an assay, all radioactivity will be in λ DNA. Therefore, ^{3}H-λ can be used to infect an *E. coli* lysogen. The total DNA is isolated and analyzed by centrifugation. After sedimentation, the centrifuge tube is fractionated and the radioactivity in each fraction determined. The data obtained in an experiment of this type is shown in Figure 5-10A and indicates that the major fraction of the DNA is converted into a nonreplicating twisted circle. Figure 5-10B shows the result that would have been obtained had radioactive phage not been used. Because the ratio of bacteria to phage DNA is about 150, it would have been difficult to recognize any of the peaks seen in Figure 5-10A.

Example 5-I. Studies of exchange reactions.

Because exchange reactions result in the production of a substance chemically identical with that of the starting material, traditional chemical analysis fails to give any evidence for a reaction. However, exchange–or, more general, participation–is easily studied with radioisotopes, as the following two examples will show.

Pyridoxal phosphate is a cofactor for many enzymatic reactions. If ^{3}H-pyridoxal phosphate is used, it is found that if the ^{3}H is in certain positions it will appear in water as ^3H_2O. The labeled enzyme-bound pyridoxal phosphate can be removed from the water by gel-exclusion chromatography (see Chapter 8) and the ^{3}H in the water can be counted by the direct addition of the water to a scintillation mixture that accepts water. If ^{3}H is found in the water, direct evidence is provided that pyridoxal phosphate participated in the reaction by means of a proton transfer.

DNA is a double-stranded polynucleotide held together by hydrogen bonds. If DNA is added to ^3H_2O and then removed at various times

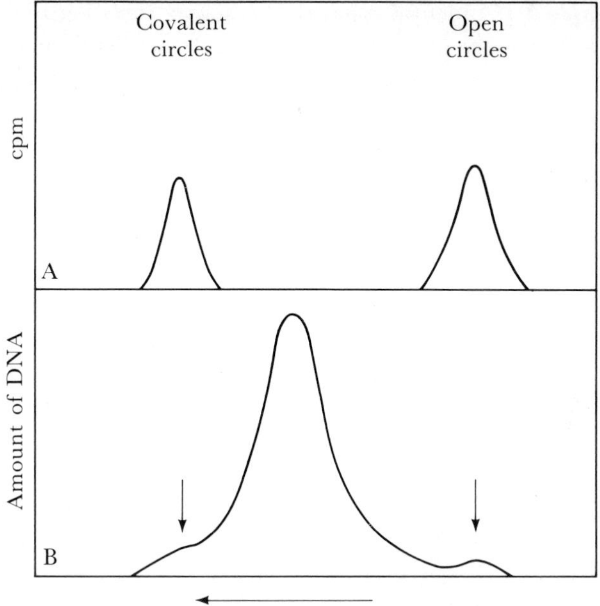

FIGURE 5-10
Comparison of experiments done with and without the use of radioactivity. A strain of the bacterium *E. coli*, lysogenic for phage λ, is infected with λ phage. It is known that the injected DNA is converted into a mixture of open and covalent circles that are easily distinguished by sedimentation in an alkaline sucrose gradient (Chapter 11). The graphs illustrate two such sedimentation runs. In graph A, the phage are labeled with ^{3}H-thymidine and the sedimentation pattern shows radioactive DNA only. The two forms are easily distinguished. In graph B, the phage is nonradioactive and the DNA concentration throughout the gradient is plotted. Because the bacterial DNA is in great excess, the phage DNA appears as barely visible blips (see arrows).

by gel-exclusion chromatography, ^{3}H appears in the DNA. From the kinetics of exchange and the effects of various denaturing agents, it can be shown that hydrogen bonds are continually breaking and reforming. This phenomenon is called "breathing" and has also been used to determine the rate of protein folding and unfolding (see Chapter 18).

Example 5-J. Identification of the active sites of enzymes.
Radioisotopes play an important role in identifying the active sites of enzymes because enzymatic reactions require such low enzyme concentrations that huge volumes would be necessary (if, in fact, sufficient enzyme were available) to obtain enough material for traditional chemical analysis. The following is a simple example of the determination of the number of binding sites.

That 1-amino-2-bromoethane covalently binds to the active site of bovine plasma amine oxidase can be easily demonstrated if the second carbon is labeled with ^{14}C. Measuring the ratio of moles of ^{14}C bound to mole of enzyme tells the number of binding sites. By hydrolyzing the protein to amino acids and by identifying the type of chemical linkage, the site of binding (i.e., the amino acid) can be identified. If the amino acid sequence of the enzyme being studied is known, hydrolysis to peptides identifies the location of the amino acid in the protein.

Selected References

Birks, J. B. 1964. *Theory and Practice of Scintillation Counting.* Pergamon.

Kobayashi, Y., and D. V. Maudsley. 1969. "Practical Aspects of Liquid Scintillation Counting," in *Methods of Biochemical Analysis,* vol. 17, edited by D. Glick, pp. 55–133. Interscience.

Millipore Corporation. *Multiple Sample Filtration and Scintillation Counting.*

Some of the best information available can be obtained from the manuals for various scintillation counters and the brochures supplied by manufacturers of radiochemicals.

Problems

5-1. The specific activity of methyl-^3H-thymidine is 6 C/mmol. What fraction of the thymidine molecules is radioactive? If DNA having a molecular weight of 25×10^6 is labeled with this thymidine and if 50% of its base pairs are AT, how many ^3H nuclei are there per DNA molecule? If a counter has 52% efficiency, what weight of DNA will give 1,000 cpm?

5-2. If ^3H, ^{14}C, and ^{32}P could each be used with equal simplicity in a double-label experiment, which pair of isotopes would you choose? Why?

5-3. In an experiment to measure DNA synthesis, would ^3H-thymidine or ^{32}P be a better choice? Why?

5-4. A scintillation counter is adjusted such that the ratio of counts in channels A and B is 1,000 for ^{3}H and 0.2 for ^{14}C. Using the same levels, a sample is counted and has 1,450 cpm in channel A and 1,620 in channel B. What is the ratio of ^{3}H to ^{14}C in this sample?

5-5. Referring to Problem 5-4, if the sample contains ^{3}H-thymidine at 6 C/mmol and ^{14}C-uracil at 0.5 C/mmol and the counting efficiencies are 25% for ^{3}H and 82% for ^{14}C, what is the molar ratio of thymidine and uridine in the sample? Suppose that the thymidine is in DNA that is 43% GC and the uridine is in RNA that is 28% uridine. What weight of DNA and RNA are in the sample?

5-6. Because ^{32}P causes strand breakage in DNA, to study only unbroken DNA it is advisable to incorporate only one ^{32}P per DNA. In that way, after a ^{32}P has decayed, the DNA is no longer radioactive and hence not detectable. If the minimum concentration of phosphate in a growth medium that supports the growth of a DNA-containing bacteriophage is 10^{-3} M, how much pure ^{32}P must be added per milliliter of solution to obtain one ^{32}P per phage if the DNA has a molecular weight of 20×10^6?

5-7. A sample counted for one minute shows a count rate of 752 cpm. For how many minutes should it be counted to have a 1% probable error?

5-8. In a scintillation counter, the sample is observed by two photomultiplier tubes on either side of the sample vial. The sample vials can hold as much as 20 ml of solvent. Do you think the count rate will be affected by the volume of the solvent if the sample is collected on a filter? What about the background?

5-9. A set of quenched standard samples is counted in two channels. The following cpm are observed even though all samples have the same amount of added radioactivity (i.e., 1,000 cpm).

Sample	Channel A	Channel B
1	502	500
2	478	425
3	445	354
4	410	290
5	379	227
6	332	170

An unknown sample is counted. It has 1,822 cpm in channel A and 1,211 in channel B. How many cpm are actually in this sample?

5-10. A particular kind of cell must be grown in a complex growth medium containing a huge number of nutrients. The medium has never been analyzed. You wish to label the protein with ^{14}C-leucine. Add 100 μl of ^{14}C-leucine at a specific activity of 50 μC/μg and a concentration of 25 μg/ml to 10 ml of growth medium. After allowing the cells to grow in this medium for 1 hour, collect 0.1 ml, acid precipitate, and count in a counter with 70% efficiency for ^{14}C. Suppose that the counter shows 1,251 cpm. The 0.1-ml sample is known to contain roughly 25 μg of protein and the protein is 4% leucine. How much ^{14}C-leucine must be added per milliliter of medium to increase the amount of collected radioactivity to approximately 5,000 cpm?

5-11. Suppose that you have in your laboratory a Geiger counter with an efficiency of 22% for ^{14}C and a background of 6 cpm and a scintillation counter with an efficiency of 72% for ^{14}C and a background of 38 cpm. In an experiment with ^{14}C, you expect your sample to have very low activity—that is, from 75 to 100 cpm. Which counter should you use? Why?

CHAPTER 6

Autoradiography

Autoradiography is a method by which radioactive material can be localized—for example, within a particular tissue, cell, cell part, or even molecule. In this technique, a sample containing a radioactive substance is put in direct contact with a thick layer of a photographic emulsion specially designed for autoradiography. Radioactive atoms decay in the sample and the emitted radiation activates individual silver halide grains in the emulsion and renders them susceptible to conversion into metallic silver by a photographic developer.* On chemical development, the resulting pattern of grains shows the distribution of radioactive material within the specimen (Figure 6-1). Observation is by microscopy. The image gives two specific bits of information: the location of the radioactive material with respect to the object or its parts, and its intensity, which is related to the amount of radioactive material present. This chapter describes how autoradiography is performed, how the appropriate isotope and preparative technique is selected, the problems arising in its use, and how the kinds of information obtained can be used to supply answers to particular questions.

---

*All photographic film consists of a suspension of silver halide crystals in gelatin. The crystals have the unusual property that, on exposure to light or radiation, they are activated in the sense that various reducing agents (developers) become capable of chemically converting the silver halide into metallic silver. In the absence of such activation, they are resistant to chemical reduction.

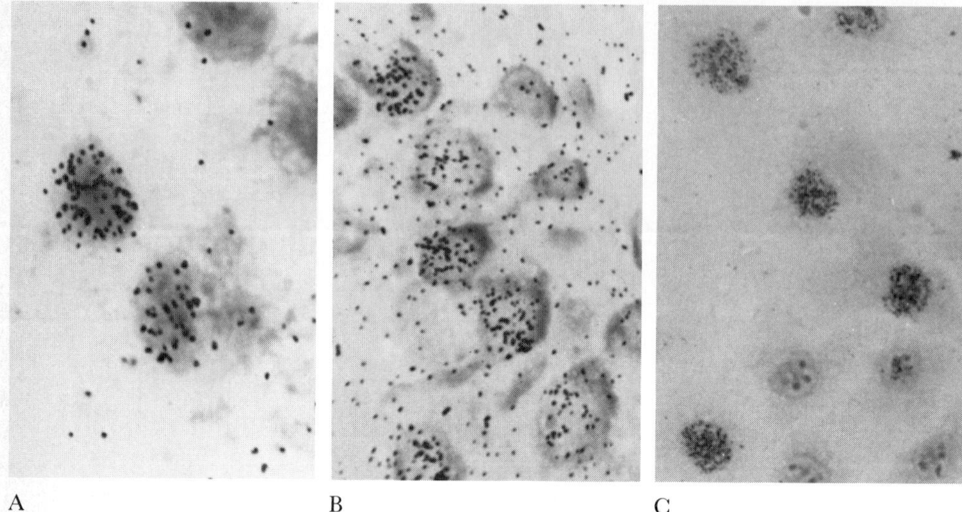

FIGURE 6-1
A. Autoradiogram showing concentration of radioactivity in nuclei isolated from a guinea pig uterus that had been exposed for 15 minutes to ^{3}H-1,2,6,7 progesterone. The nuclei were placed on a slide and overlaid with nuclear emulsion. After eleven days the autoradiogram was developed to make the silver grains visible and then stained with methyl green-pyronin to make the nuclei visible. This shows that progesterone or a compound derived from progesterone is rapidly localized in the nuclei of uterine cells. Magnification 1100×. [From W. Stumpf, in *Methods in Cell Biology,* vol. 3, edited by D. M. Prescott, Academic Press, 1976.] B. Autoradiogram of a section of a rat hippocampus showing nuclear concentration of ^{3}H-corticosterone in neurons one hour after intravenous injection. After the thin section was mounted on a slide, nuclear emulsion was applied and the autoradiogram was exposed for ninety-five days, developed, and then stained with methyl green-pyronin to make the nuclei visible. The high concentration of grains over the nucleus shows that a great deal of the corticosterone is in the nucleus. There are grains between the nuclei also: to determine whether these are due to ^{3}H in the cytoplasm or to background, it is necessary to determine the number of grains per unit area of emulsion in a region far from the tissue section. [From W. E. Stumpf and M. Sar, in *Methods in Enzymology,* vol. 36, edited by B. W. O'Malley and J. G. Hardman, Academic Press, 1975.] C. Autoradiographic detection of DNA synthesis in animal cells. African green monkey kidney cells were grown for six hours in a growth medium containing ^{3}H-thymidine. The cells were washed, dehydrated, and covered with autoradiographic film. After exposure and development of the film, the nuclei (gray regions) were stained. Such an experiment allows the identification of the cells that replicated their DNA during the six-hour period (i.e., those whose nuclei are covered with grains). Hence, the fraction of replicating cells can be measured. [Courtesy of James A. Robb.]

NUCLEAR EMULSION USED IN BIOLOGICAL STUDIES

Nuclear emulsion differs from standard photographic film principally in the high ratio of silver halide to gelatin—roughly equal volumes in nuclear emulsions—and in the small size of the grains (0.02–0.3 μ). The emulsions in common use are of three types: premounted, liquid, and stripping. A premounted emulsion is a relatively thick (250 μ) layer of emulsion that has been mounted on a glass microscope slide. A liquid emulsion is supplied as a gel, which must be melted; the sample is dipped into the molten gel and withdrawn, and the emulsion hardens and forms a film whose thickness depends on the concentration of gelatin in the liquid. Stripping film is supplied as a thin (about 5 μ) film mounted on glass. It is removed from the glass with a knife and then placed on a water surface. The sample, which has been premounted on a glass microscope slide, is placed under the floating film and lifted up into the film, thus transferring the film to the microscope slide. This is allowed to dry and the thin emulsion adheres tightly to the slide. These processes are shown diagrammatically in Figure 6-2.

ISOTOPES COMMONLY USED IN BIOLOGICAL AND BIOCHEMICAL STUDIES

The radioisotopes most commonly used are of three energy types—high (e.g., ^{32}P), medium, (e.g., ^{14}C and ^{35}S) and low (e.g.,^{3}H); almost all are β emitters. On occasion, α emitters such as polonium and thorium are used. The autoradiographic properties of these isotopes are given in the next section.

TRACK LENGTH OF VARIOUS EMITTED PARTICLES

As a particle emitted by a radioactive source passes through a nuclear emulsion, it continually loses energy by collisions with nuclei and orbital electrons. Some of this energy produces defects in the silver halide crystals and thereby renders them developable (i.e., they are exposed). The pattern of grains in the emulsion is called a *track,* which has three parameters—length, grain density (either grains per unit length or grains per total track length), and shape (e.g., linear, curved, angled, etc.). These parameters are determined by the mass of the particle, the particle energy, the emulsion, and the development of the emulsion. The effect of mass and energy are described next.

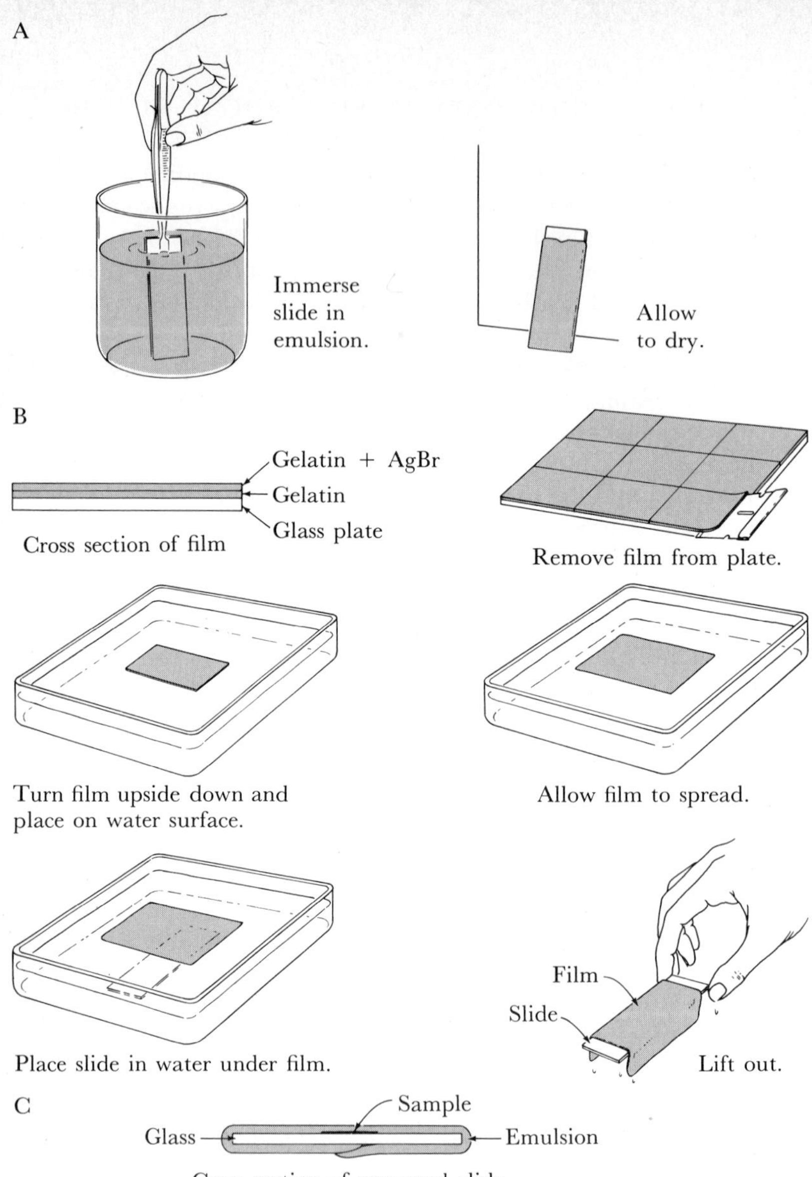
A
Immerse
slide in
emulsion.
Allow
to dry.
B
Gelatin + AgBr
Gelatin
Glass plate
Cross section of film
Remove film from plate.
Turn film upside down and
place on water surface.
Allow film to spread.
Place slide in water under film.
Film
Slide
Lift out.
C
Sample
Glass
Emulsion
Cross section of prepared slide

FIGURE 6-2
Methods of putting emulsions onto a sample: (A) Dipping method. A glass microscope slide to which the sample is affixed is dipped into molten emulsion. A thin layer adheres to the glass and solidifies at room temperature. (B) Stripping method. Commercially available glass plates on which stripping film has been made are scored with a razor blade to make rectangular pieces approximately 1 inch by 3 inches. The film is stripped from the plate by forcing a razor blade under the edge of the film. The film is inverted, floated on water, and picked up as shown. (C) Cross section of a finished slide with sample and emulsion prepared by the stripping method.

Alpha Particles

Alpha particles are heavy, have two positive charges, and usually have an energy between 4 and 8 Mev.* These massive particles are relatively unaffected by collision with electrons, tend to maintain a straight path following such a collision, and have a tremendous disrupting effect on orbital electrons as they pass through an emulsion. This results in excitation of just about every silver halide crystal that they traverse and therefore produces a very high grain density. Because an α particle interacts with a very large number of electrons per unit distance, it loses energy rapidly and has a relatively short track length (usually between 15 and 40 μ). Figure 6-3A is a photograph of alpha tracks in a thick emulsion.

Beta Particles

Beta particles are electrons and are therefore easily scattered by orbital electrons. As β particles collide with other electrons, they rapidly lose energy and are sharply deflected in each collision. The magnitude of the deflection depends on the energy of the particle; at very high energy, the momentum is so great that the particle has a greater tendency to move in a straight line and be minimally deflected. This means that, as energy is lost in each interaction with an orbital electron, the probability of greater deflection in the next interaction increases. Hence, because the energy of the particle continuously decreases, the encounters with other electrons cause the path of the particle to become more and more tortuous. Because the electron density of matter is very great, these sharp deflections tend to balance out so that over a short distance the track remains fairly straight (occasionally a β particle will pass close enough to a nucleus to be both accelerated by the positive charge and sharply deflected). The grain density (a measure of the number of interactions per unit distance along an *apparent* path) increases as the particle loses energy, which means that the grain density will always be greater at the end of a track than at the beginning (Figure 6-3B); this is the principal way of determining the direction of movement of a particle in an emulsion.

The same considerations apply to α tracks—that is, increasing grain density toward the end of a track. The principal difference between α and β tracks is that, because of the charge and great mass of the α particle, it interacts with more electrons per unit distance than does a β particle;

---

*The energy of emitted particles is measured in electron volts, the energy acquired by an electron falling through a potential difference of one volt, or 1.602×10^{-19} joule. The usual units are kev and Mev: 10^3 and 10^6 electron volts, respectively.

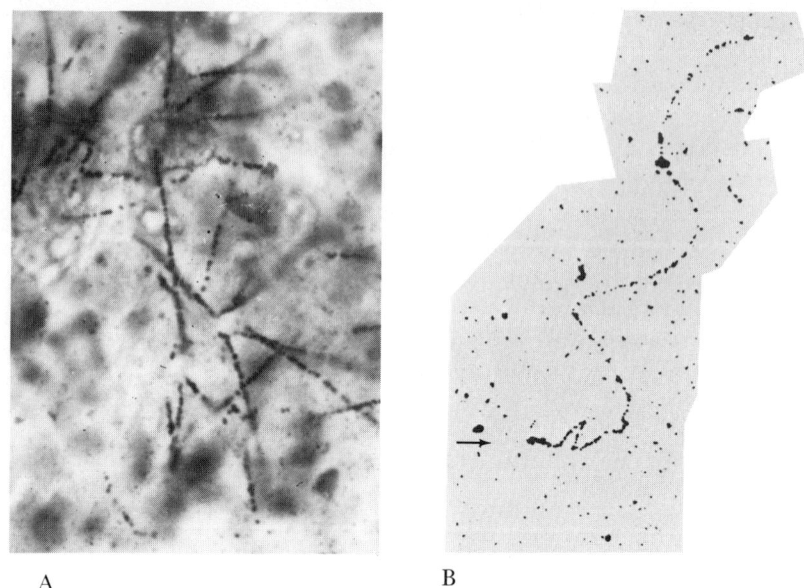

FIGURE 6-3
A. Alpha particle tracks emitted from thorotrast (thorium dioxide) particles in a thin section of rabbit spleen. The thin section was overlaid with liquid emulsion, which was then allowed to harden. After exposure, the emulsion was developed and the nuclei stained. Note the high grain density and the straightness of the tracks and compare them with part B. The tracks are from 40 to 45 μ long. [Autoradiogram supplied by Hilde Levi.] B. A track produced by a 300-kev (medium energy) β particle showing that grain density increases at the end of the track (arrow). The track is approximately 250 μ long. [From R. H. Herz, *Photographic Action of Ionizing Radiations,* Wiley-Interscience, 1969.]

it therefore loses energy at a greater rate per unit distance and for a given energy has a shorter track length than a β.

As described in Chapter 5, Figure 5-1, each isotope has a wide range of energies. Hence, the track lengths for a particular isotope will also show a great range of values.

PHYSICAL ARRANGEMENTS BETWEEN EMITTING SOURCE AND EMULSION

In this discussion, *emitter* means a nucleus that is decaying and *source* refers to a collection of potential emitters.

There are three basic source-emulsion relations in common use (Figure 6-4): (1) the source is embedded in the emulsion and the emulsion is

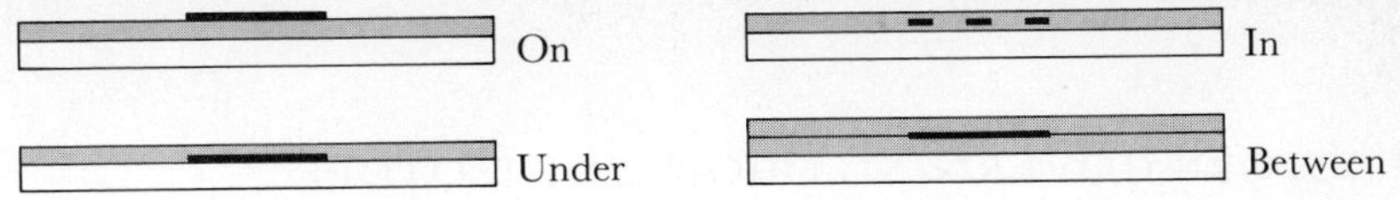

FIGURE 6-4
Four source-emulsion relations. The one labeled "under" is the arrangement for stripping film or for dipping. The "between" arrangement is used if a very thick emulsion is required.

thicker than the maximum track length; (2) the source is on a surface—usually a glass microscope slide—and is covered with an emulsion whose thickness is greater than the maximum track length; and (3) the source is on a surface and the emulsion thickness is much less than the maximum track length. With types 1 and 2, the entire track length can be seen—although with type 2 only one-half of the number of tracks are seen (the other half never entering the emulsion). With type 3, only the very beginning of the track is seen. We will see later how these three types have different applications.

Figure 6-5 shows schematic diagrams of tracks of low-, medium-, and high-energy isotopes prepared in each of the three ways. Carefully note the enormous differences.

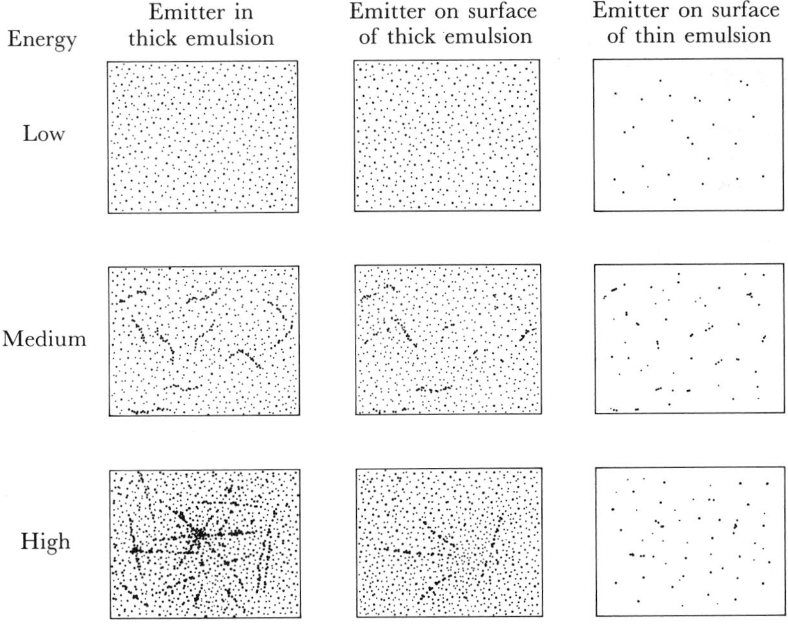

FIGURE 6-5
Drawings of low (^{13}H), medium (^{14}C), and high (^{32}P) energy tracks *in* thick emulsions and *on* thick and thin emulsions. Note that high-energy tracks become dots on a thin emulsion and low-energy tracks are totally obscured by background with thick emulsions.

FACTORS GOVERNING CHOICE OF ISOTOPE AND CHOICE OF EMITTER-EMULSION RELATION

In any experiment, high resolution, high efficiency, and low background are requisite and how they can be attained is discussed in this section.

Resolution

Resolution can mean the ability to determine the position of the emitting source, the ability to separate the individual grains to get an accurate grain count, the ability to separate two emitting sources, and more. This discussion is concerned primarily with the first two points.

The isotope itself affects resolution. As the energy of the emitter is increased, the tracks will be longer and have fewer grains near the point of emission (see Figure 6-5). This in general decreases the ability to localize this point because the grains in the low density region near the source cannot be readily distinguished from the background grains. Hence, isotopes such as ^{14}C, which give curved tracks, cannot be used for high-resolution localization of a source. If the energy is great enough that the tracks are long and straight, as is the case with ^{32}P, resolution is improved if the specific activity of the source is great enough to produce many tracks. In this case, the tracks can be easily extrapolated to a well-defined origin. This will be described later in Example 6-A.

If either the source itself is very thick or the source is far from the emulsion (instead of being in direct contact with it), resolution will be lost because grains will again be further from the source.

The size of the grains also affects resolution because a single grain can be exposed by an interaction anywhere within the crystal. Therefore, large grains can have their centers considerably off the path of the particle, thus giving lower resolution in localizing the source. Furthermore, if a source consists of a set of emitters that are very near one another and are an isotope such as ^{3}H for which a decay can expose only one grain, to count decays (e.g., to determine the number of emitters) each grain must be sufficiently small that exposed grains will be separated by unexposed ones. Another problem encountered in grain counting with low-energy isotopes is that a single grain can be exposed only once. Hence, if the grain size is too large (larger than the space between the emitters), there will no longer be correspondence between the number of decays and the number of grains.

The sensitivity of the emulsion—that is, the energy required to activate a crystal—also affects resolution. To understand this, one must remember that the grain density is always greater at the end of the track than at the beginning because more energy is lost per unit distance when the particle energy is low. Hence, if the emulsion is insensitive and requires a large energy loss to expose a grain, then those particles emitted at high energy will not expose a grain until near the end of the track. This results in a loss of grains near the source and a comparable increase far from it—which decreases resolution.

Efficiency

Ideally, every decay should produce a track. However, because a particle must reach the emulsion, this does not always happen. There are two basic problems: first, if the sample is on the surface of the emulsion only one-half of the decays will enter the emulsion; and, second, even if the sample is embedded in the emulsion, there will be self-absorption of the energy by the sample owing to its finite thickness. For high-energy isotopes such as ^{32}P, self-absorption is not a severe problem—for example, a 5-μ thick sample (about the maximum thickness for most cells or tissue sections) absorbs less than 1% of the energy. For the less energetic isotopes such as ^{14}C and ^{35}S, there is 82% and 70% transmission at a thickness of 5 and 10 μ, respectively. For low-energy isotopes such as ^{3}H, self-absorption is severe; at 0.5 μ the efficiency is only 16% and at 5 μ, 4%. In small cells such as a bacterium, ^{3}H can be detected at an efficiency of 27%.

Because most grains are produced near the end of a track, emulsion thickness also affects efficiency. Indeed, with ^{32}P, there is a direct proportionality between efficiency and emulsion thickness of as much as 50 to 100 μ. For ^{14}C, maximum efficiency is reached at an emulsion thickness between 3 and 5 μ. Because the ^{3}H β particle has a range of only 1 μ, nothing is gained by using thicker emulsions (in fact, matters will be worse because of increased background).

Background

Developed emulsions that have not been exposed to a radioactive sample contain dark grains called *background.* How, then, can a track be identified? For high-energy isotopes, this is no problem because tracks are dense with grains. For ^{14}C, the grain density of tracks ranges from high to low. The convention adopted by autoradiographers is that a minimum of four grains in a straight line is required to define a track because such a configuration has a low probability of occurring in an unexposed emulsion. However, for ^{3}H, a track is usually just one large grain, and so it is necessary to reduce background to a very low level if this isotope is used.

Background has many causes: accidental exposure to light, the presence of chemicals and metal ions (especially Cu) in the sample, mechanical pressure (slight pressure on undeveloped emulsion produces heavy blackening on development), certain conditions of development, electric sparks, and background radioactivity (e.g., ^{40}K in glass, ^{14}C in gelatin, and cosmic rays). There are two types of background—long tracks and individual grains. Long tracks are seen mostly in thick emulsions. For individual background grains, the grain density (i.e., grains per unit volume) is constant and the number of grains per unit area increases with emulsion thickness. Therefore, there is never any reason to use an emulsion thicker than is necessary to achieve maximum efficiency; furthermore, it is often advisable to sacrifice efficiency and use very thin emulsions to improve the ability to recognize tracks—this is usually the case with ^{3}H.

Background can form before the sample is applied or during exposure (i.e., the time elapsed between placing sample and emulsion in contact and development) and development. A prior background of tracks is easily dealt with if liquid emulsions (see page 133) are used because the tracks (but not the individual grains) are destroyed when the emulsion is melted. The prior background of stripping film (see next section) is reduced because the film contains a latent image fader (a chemical that reverses the effect of exposure of a grain), which is removed at the time of applying the sample because the fader dissolves in the water used to float the film. Background subsequent to sample application is normally minimized by scrupulous cleanliness and by appropriate storage and development.

TECHNIQUES FOR PLACING EMULSION AND SAMPLE IN CONTACT

Temporary Contact Method

The sample is placed in contact with the emulsion, held in place by pressure during exposure, and then removed before development. A thin protective sheet may be placed between the sample and the emulsion to prevent exposure due to chemicals in the sample. This method is used to identify spots in chromatograms and electrophoresis gels and to localize radiochemicals in large biological samples such as leaves, bone sections, and large tissue sections. The emulsion used for this technique is usually x-ray film. Semiquantitative information can be obtained using this method if certain conditions are met. However, because a grain cannot be exposed twice, it is important that the exposure be set so that blackening of the film is proportional to the amount of radioactivity; if exposure is extended to the point that areas of very low activity produce blackening of the film, the high-activity regions may have passed the point of linear

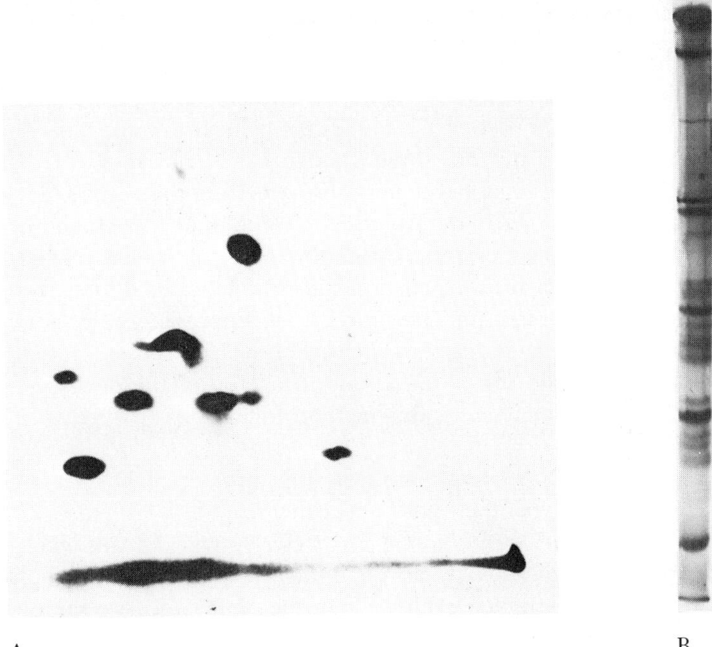

FIGURE 6-6
A. Autoradiogram of a two-dimensional chromatogram-electrophoregram. The major protein subunit of polyoma virus was iodinated with ^{125}I (which reacts principally with tyrosine) and digested with trypsin, spotted on paper, chromatographed, and then electrophoresed in the direction perpendicular to the chromatography. The paper was dried and placed in contact with x-ray film. The blackening of the film corresponds to the positions of the tyrosine-containing peptides. [Courtesy of William Murakami.]
B. Autoradiographic detection of the bands of a gel electrophoregram of proteins labeled with ^{3}H-leucine. [Courtesy of Junko Hosoda.]

response of the emulsion. Two examples of the use of this method are shown in Figure 6-6. Note that, with samples of this type, observation is not made by light microscopy because neither grains nor tracks are being counted.

Permanent Contact Methods

The three permanent contact methods in use are: mounting on a pre-formed emulsion, coating with or dipping into melted emulsion, and applying stripping film.

MOUNTING ON A PREFORMED EMULSION

Either a drop of the sample is dried onto the emulsion or a tissue section is floated on water and the emulsion (which comes premounted on a glass slide) is brought up under it so that the tissue lies on the film. This is a very simple method but has three disadvantages: development may be nonuniform because the developer must penetrate the sample, the grains and tracks must be viewed through the sample, and background with preformed emulsions is high. On the other hand, sensitivity is high. If the sample consists of radioactive particles that can be diluted so that resolution is no problem and the purpose of autoradiography is to count the particles, this method is excellent. For such purposes, sensitivity can be doubled by pouring liquid emulsion on top of the deposited sample. A classic experiment using this method follows.

Example 6-A. Determination of the number of DNA molecules in a bacteriophage—the "star" experiment.

The DNA of *Escherichia coli* bacteriophage T2 was labeled with ^{32}P to very high activity—approximately 100 ^{32}P atoms per phage. DNA was then extracted from some of the phage and two samples were applied to the emulsion: (1) an aliquot containing a certain number of phages and (2) an aliquot containing *all* of the DNA from the same number of phages. A liquid emulsion was then poured on top so that the sample was embedded. After exposure and development, heavy tracks coming from a single point were observed (see Figure 6-7). These configurations represented ^{32}P β particles emitted by a single phage or a DNA molecule and were called "stars." It was found that the average number of stars per droplet was the same for both DNA and phage, suggesting that there was only one DNA molecule per phage. If there had been two DNA molecules of equal size per phage, the number of stars in the DNA sample would have been twice that of the phage sample. However, if the phage had consisted of, for example, one piece of DNA containing 80% of the DNA of the phage and ten 2% pieces, the 2% pieces would not have given stars (they would not have been recognized because they would rarely have had more than one track or ray) and the number of stars in phage and DNA would have been equal. However, the number of rays per star was also counted and found to be the same for both DNA and phage. Hence, both DNA and phage contained the same number of ^{32}P atoms, indicating that there was only one DNA molecule per phage. It was, of course, necessary to determine the number of rays per star with high statistical precision so that an 80% piece could be distinguished from a 100% piece.

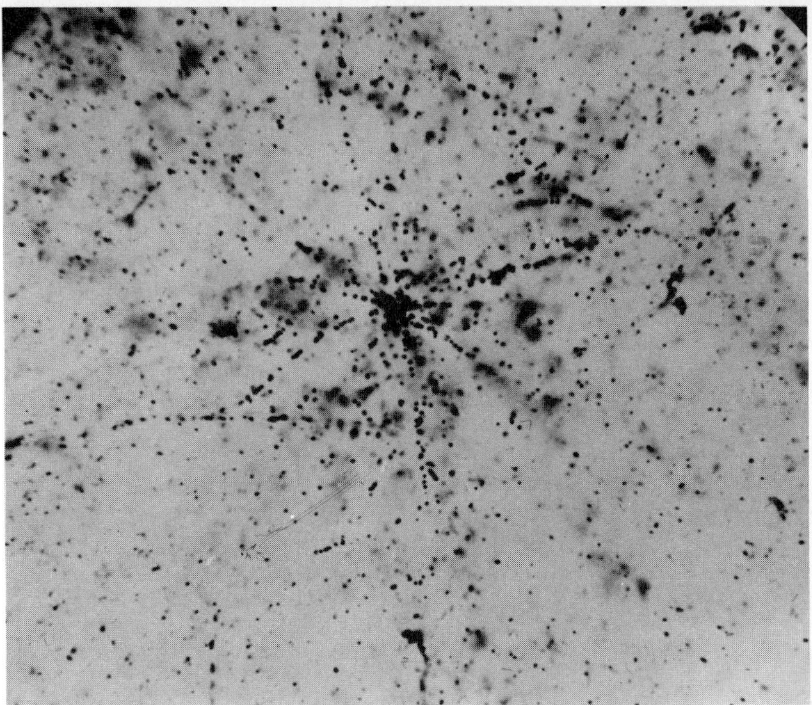

FIGURE 6-7
A star. A bacteriophage labeled with ^{32}P is embedded in emulsion. All ^{32}P β tracks originate from a single point, which allows the number of tracks (rays) per star to be counted. To count all rays it is necessary to focus the microscope up and down in order to recognize those that are not in the plane of the source. [Courtesy of Charles A. Thomas.]

DIPPING METHOD

The sample is mounted on a microscope slide and the slide is dipped into a melted emulsion (see Figure 6-2). This method gives the most intimate contact between sample and emulsion, thus maximizing efficiency for low-energy emitters. Dipping has the advantages that very thin emulsions can be obtained by appropriate dilution and choice of temperature, sample preparation is very simple and rapid, and liquid emulsion can be prepared with silver halide grains of minimum size, thus improving resolution of single grains. Its disadvantage is nonuniform thickness, but this is no problem if ^{3}H is used; as indicated earlier, ^{3}H β particles rarely penetrate past the first micron of crystals.

The ability of the dipping method to provide resolution of individual ^{3}H β particles is excellent. Hence, the dipping method has been used to determine the relative amount of DNA (as ^{3}H-thymidine) in a cell grown in various ways, as shown in the following example.

Example 6-B. Demonstration that amino acid starvation prevents reinitiation of a round of DNA synthesis in bacteria.

If a strain of the bacterium *E. coli* requiring thymine and leucine for growth is grown in a medium containing ^{3}H-thymidine (^{3}H-dT), which is a DNA precursor, and the amino acid leucine, ^{3}H-dT is incorporated into DNA as replication proceeds. If the leucine is removed from the medium, incorporation of ^{3}H-dT continues for about one generation and then stops. The residual amount of DNA synthesized during the leucine-starvation period is exactly what is calculated for a *population* of cells randomly distributed in all stages of DNA synthesis in which all cells engaged in replication complete synthesis of only that molecule whose replication was in progress at the start and do not reinitiate synthesis. However, to confirm this interpretation it is necessary to show that the distribution of incorporation *per cell* does agree with the proposed distribution of ages of the individual cells because it is possible that these distributions are quite different—for example, most cells might make no DNA and a few might make more than one copy and the agreement might be fortuitous.

To test this, a culture of bacteria was divided into two parts. One was grown in medium containing ^{3}H-dT and leucine for many generations; the other was grown in nonradioactive thymidine and leucine for many generations and then was shifted to a medium containing ^{3}H-dT but no leucine and grown until ^{3}H-dT incorporation stopped. Dilutions of each culture were applied to microscope slides and dried and both cultures were prepared for autoradiography using the dipping method. After exposure for several days and development, the entire slide with the cells and emulsion was dipped into a cellular stain that made the bacteria visible with the light microscope. Figure 6-8 shows an example of stained cells with ^{3}H grains. The number of cells with 0, 1, 2, 3, . . . grains was tabulated. The first culture showed the distribution of grains

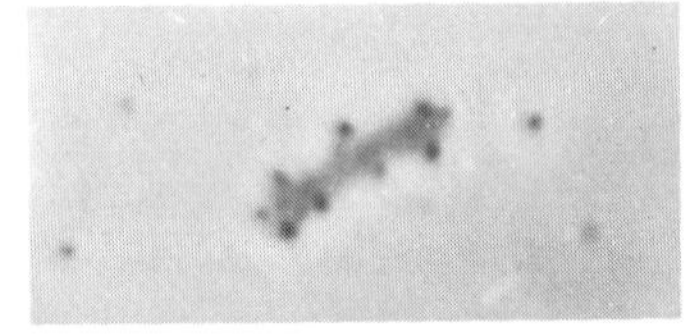

FIGURE 6-8
Autoradiogram of a bacterium labeled with ^{3}H-thymidine. Each grain represents the decay of a single ^{3}H nucleus. The bacterium and the grains are at different levels in the sample making it difficult to get both cells and grains in sharp focus. In this picture, the grains are in focus so that the bacterium appears somewhat fuzzy. [Courtesy of William Howe.]

per cell, representing a full complement of DNA. The second culture showed the amount of DNA synthesized in each cell during the leucine-starvation period. A statistical analysis of the distribution obtained for this culture agreed with the interpretation. This experiment confirmed the important hypothesis that protein synthesis is necessary for initiating DNA synthesis.

STRIPPING-FILM METHOD

Stripping film consists of a 5-μ layer of nuclear emulsion on a 10-μ gelatin layer with the gelatin in contact with a sheet of glass. If a cut is made through the gelatin with a scalpel, the film can be pulled or stripped from the glass. In the stripping method the film is placed on a water surface, the emulsion side on the water, where it swells for 2 or 3 minutes to about 140% of its initial area. A slide containing the sample is dipped in the water under the film and lifted up with the film draped on it. As it dries, it shrinks and adheres tightly to the slide (see Figure 6-2). Both emulsion and gelatin remain in contact with the sample during exposure, development, and microscopic observation. The sample is usually stained after development to make it visible.

Stripping film is very easy to work with and gives high resolution in the range of 0.5 to 5.0 μ. The emulsion thickness is more uniform than in any other method, so that valid comparisons of the radioactivity of different structures or parts of cells can be based on grain counts. However, it does have a few disadvantages: it has lower sensitivity than other emulsions; if stripping is done at low humidity, there are flashes of static charge that increase background; and the procedure is fairly slow compared with dipping. It is certainly the most widely used autoradiographic method, although some workers argue that the dipping method has all of the advantages of stripping film plus that of speed; probably the choice between stripping and dipping is a matter of preference.

One of the classic experiments of molecular biology was done with stripping film.

Example 6-C. Demonstration of the number of conserved subunits of chromosomes.

In 1957, Matthew Meselson and Franklin Stahl showed that DNA replicated semiconservatively (see Chapter 11, Example 11-K); that is, each strand served as a template for the replication of the other so that the two parental strands segregated into separate daughter, double-stranded helices. The chromosomes of root cells of the bean *Vicia faba* can be labeled with ^{3}H-thymidine by growth for a long period in a medium containing ^{3}H-dT. Autoradiograms of these cells in mitosis show that all of the grains are localized above the chromosomes (see

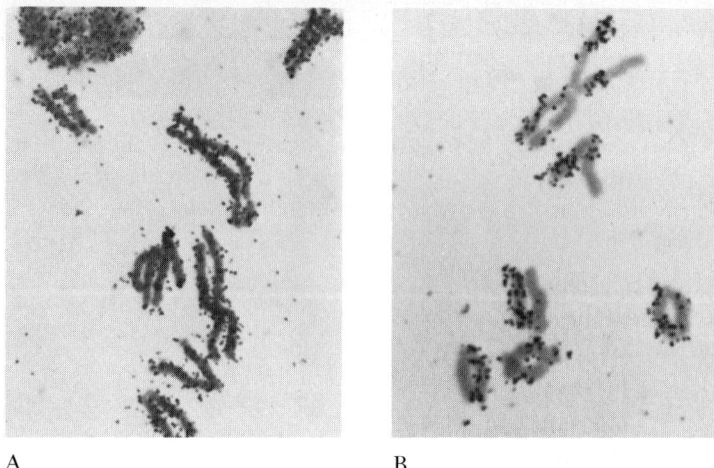

FIGURE 6-9
Autoradiograms showing chromosomes of *Vicia faba* at the (A) first and (B) second divisions after labeling with ^{3}H-thymidine during one synthetic phase. In part A, both sister chromatids are labeled. In B, only one of each pair is labeled. Note that in the uppermost X-shaped configuration there is label in both upper arms but distal and proximal to the point of intersection (the centromere). This is an example of sister chromatid exchange. [From J. Herbert Taylor, *Molecular Genetics,* part 1, Academic Press, 1963. Autoradiograms courtesy of J. Herbert Taylor.]

Figure 6-9). If, after extensive labeling, the cells are grown in a non-radioactive medium for a period, autoradiograms of samples taken at various times show that after one generation, in the cells that are in mitosis both daughter chromosomes contain radioactivity but the grain count per chromosome is half that of the original chromosome. After many cell divisions, the grain count per chromosome remains constant at this value of one-half and there are never two labeled sister chromosomes separating during anaphase. In this way, J. Herbert Taylor showed that a chromosome contains two subunits that segregate at the first division after synthesis and are henceforth conserved. This result agrees with the Meselson-Stahl experiment if the subunits are the individual polynucleotide strands of the DNA.

MOLECULAR AUTORADIOGRAPHY

The principal use of autoradiography has been to *localize* radioactivity in cells or tissues. However, it is also possible to use autoradiography to *visualize* or determine physical parameters of molecules and this is called molec-

FIGURE 6-10
Autoradiogram of a replicating *E. coli* chromosome. [Courtesy of John Cairns.]

ular autoradiography. (The first example of this use was given in Example 6-A in which the molecular weight of the DNA molecule of a phage was determined.) This powerful method, which has not had widespread use, has been used in several studies of DNA structure to obtain information not available by any other method. A well-known example is given below.

Example 6-D. Visualization of a replicating *E. coli* DNA molecule. John Cairns grew the bacterium *E. coli* for many generations in a medium containing ^{3}H-dT; he isolated unbroken DNA, extended it by adsorption to nitrocellulose filters (see Chapter 7), and prepared stripping-film autoradiograms. The beautiful pictures that he obtained are exemplified in Figure 6-10. The autoradiograms showed that the *E. coli* DNA molecule is a circle and gave the length of the molecule. (Even though DNA is easily visualized by electron microscopy [see Chapter 3], electron micrographs of such large molecules [$M = 2.6 \times 10^9$] in extended forms have never been obtained because their preparation for electron microscopy causes breakage of the DNA. If conditions producing breakage are avoided, the molecules are usually tangled.)

In a variant of this experiment it was possible to demonstrate which part of the molecule had replicated (information necessary to formulate a model of replication). This was done by growing the cells for a little *more* than one generation in a medium containing ^{3}H-dT. After one generation only a single strand of the DNA is labeled; however continued replication for an additional short period produces a short section of DNA, *both strands of which contain ^{3}H-dT.* Hence, the DNA replicated in the second generation will have *twice the grain density* (grains per unit length) of the remainder. In this way, doubly replicated DNA was identified autoradiographically and as expected it was always connected to the replication forks. A remarkable finding was that there were two regions of double grain density—one attached to each fork—a result that showed that both forks were growing parts and that DNA replicates bidirectionally (Figure 6-11).

FIGURE 6-11
A. An autoradiogram of an entire *E. coli* chromosome obtained by labeling, with tritiated thymidine, a synchronized culture of a temperature-sensitive initiation-defective strain while it was engaged in a synchronous round of replication. The intensity of labeling was increased near the end of this round of replication and kept at this high level while the culture was allowed to reinitiate the next round of replication. This strategy of labeling will heavily label those parts of the chromosome replicated at the beginning and at the end of the replication cycle. If replication is bidirectional, then these two regions of labeling should be symmetrically disposed on the chromosome 180° apart, as is the case. [From R. L. Rodriguez, M. S. Dalbey, and C.I. Davern, *J. Mol. Biol.* 74(1973):599–604.] B. Another way of looking at the replicating region of a DNA molecule. Unlabeled *E. coli* were grown for 10 minutes in a medium containing ^{3}H-thymidine with low specific activity and then for 5 minutes with very high specific activity. The DNA was isolated and prepared for autoradiography in a way that produces

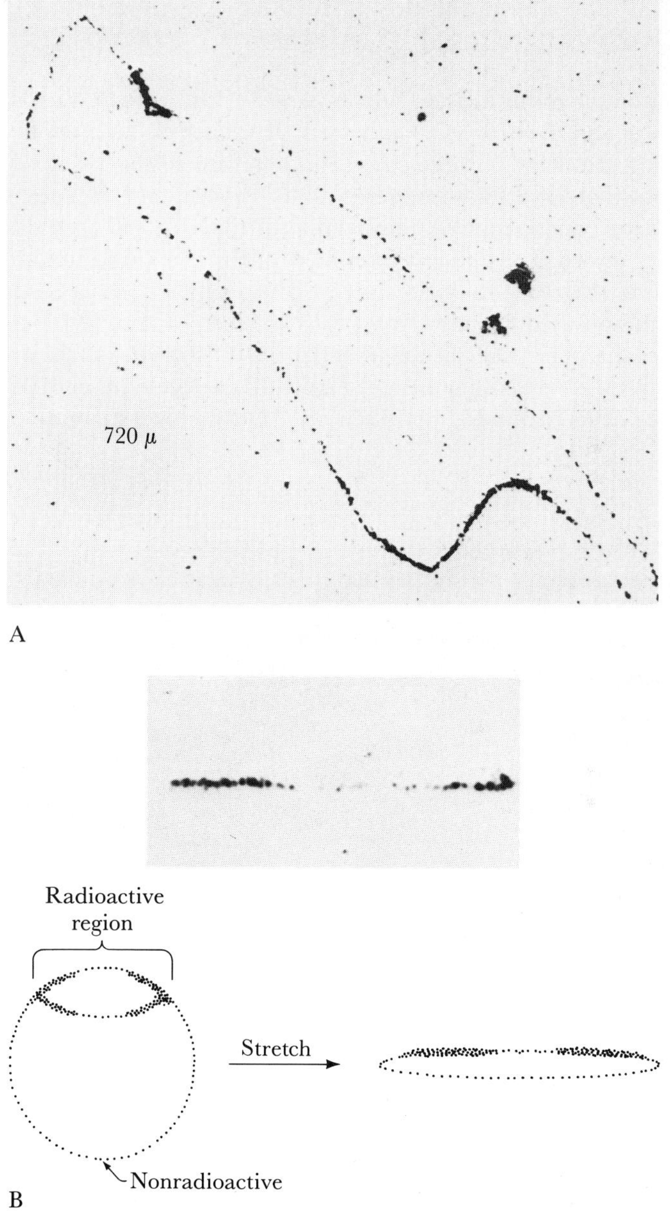

stretching. Because of this stretching, the pattern observed is the result of a superposition of two DNA strands. An interpretation of the pattern is shown in the diagram under the autoradiogram. The fact that the region of low grain density is flanked by two high-density regions also demonstrates bidirectional replication. [From R. Rodriguez and C. I. Davern, *J. Bacteriol.*, in press.]

ELECTRON-MICROSCOPIC AUTORADIOGRAPHY

Autoradiograms are normally observed with the light microscope. Lucien Caro and Robert van Tubergen developed a high-resolution method using the electron microscope. This necessitated the use of very thin emulsions that would not interfere with observations of the specimen with the electron microscope. A thin section of tissue or cell embedded in plastic and adsorbed to an electron-microscope grid is dipped in an emulsion so dilute that only a single layer of silver halide crystals is deposited. The emulsion used has very tiny ($<0.1\ \mu$) crystals. After exposure and development, the tissue section is stained with uranyl acetate and examined with the electron microscope (Figure 6-12). A resolution of $0.1\ \mu$ is obtainable. This method is not often used but deserves more attention.

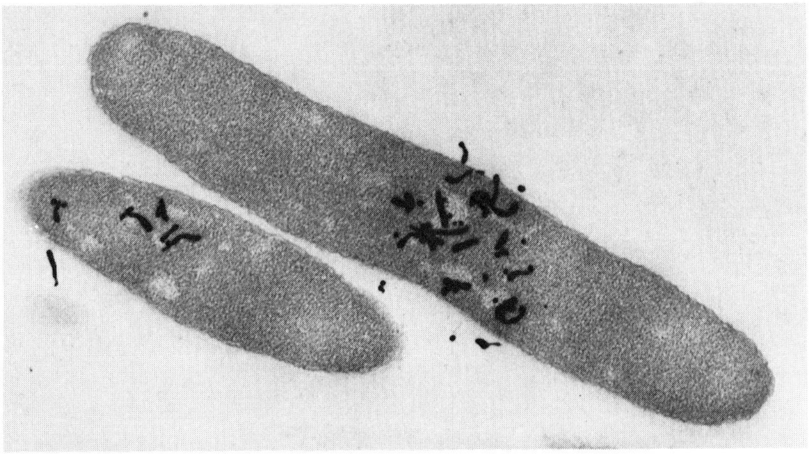

FIGURE 6-12
Electron-micrographic autoradiogram of *E. coli* bacteria labeled with ^{3}H-thymidine. Note that by electron microscopy the grains appear as threads of silver rather than dots. [Courtesy of Lucien Caro.]

Selected References

Caro, L. and Van Tubergen, R. P. 1962. "High Resolution Autoradiography." *J. Cell. Biol.* 15:179–188. The method of electron-microscopic autoradiography.

Gabran, P. B. 1972. *Autoradiography for Biologists.* Academic Press.

Gude, W. D. 1968. *Autoradiographic Techniques: Localization of Radioisotopes in Biological Material.* Prentice-Hall.

Levinthal, C., and C. A. Thomas. 1957. "Molecular Autoradiography: The β-ray Counting from Single Virus Particles and DNA Molecules in Nuclear Emulsion." *Biochim. Biophys. Acta* 22:453–465.

Problems

6-1. When bacteriophages adsorb to bacteria, the phages inject all of their DNA into the bacterium. If the phages are x-irradiated, they lose viability. It is known that the x rays produce double-strand breaks in the DNA. Design an autoradiographic procedure to determine whether an x-irradiated phage injects all of its DNA or only a fragment.

6-2. A phage is labeled with forty-two atoms of ^{32}P (on the average) per phage. The efficiency of the detection of ^{32}P by autoradiography in a thick film is 94%. How many days should the film be exposed to get an average of twelve rays per star?

6-3. Suppose that you suspect that a large cell (50 μ in diameter) incorporates a particular compound only into its cell wall and that other workers have argued that it is incorporated entirely into the cytoplasm. Could these alternatives be distinguished by autoradiography? How?

6-4. A strain of *E. coli* is claimed to have four complete chromosomes if grown in a particular very rich growth medium. How could this be proved by autoradiography? (Hint: It is possible to allow a cell to divide and form a microcolony on agar and then autoradiograph the microcolony when only a few hundred cells are present.)

6-5. You wish to identify radioactive spots on a paper chromatogram by contact with x-ray film. You could use ^{3}H or ^{14}C. Which would you use? Why?

6-6. A bacterial mutant has the property of synthesizing protein at 42°C at 1/10 the rate that it does at 37°C. Design an autoradiographic experiment to determine whether all of the bacteria in a population share this property or whether there is a normal and an abnormal fraction of the population. Indicate which biochemical, isotope, and emulsion you would use. Why would it be necessary to use different exposure times for the control and the mutant populations?

6-7. A curious mutant of *E. coli* has the property of dividing asymmetrically and producing a normal cell and one having approximately 1/10 the volume at a fairly high frequency. The small cells never divide again. How could you determine whether they have the normal amount of DNA? of RNA?

6-8. Nucleoli are dense bodies in the nuclei of many eucaryotic cells; in animal cells, they are usually about 1/10 the diameter of the nucleus. Can whether they contain DNA be determined by autoradiography? Explain.

6-9. Cells of male strains of *E. coli* form pairs with cells of female strains and transfer DNA from male to female to produce genetic recombinants (i.e., females that have lost some of their own genes and replaced them with male genes.) Genetic tests indicate that no recombinants are obtained if males are mixed with males or females with females. This could be due either to lack of homosexual pairing or to lack of transfer. Design an autoradiographic experiment to determine whether homosexual pairing occurs.

CHAPTER 7

Membrane Filtration and Dialysis

A common operation in chemistry and biochemistry is the separation of one substance from another. Before the era of centrifugation, electrophoresis, chromatography, and so on, this was accomplished by filtration, when possible. Originally, filters were of fine cloth (e.g., cheesecloth was used to separate curds from whey) and in fact cheesecloth is sometimes still used for preliminary clarification of tissue extracts. Later, porous paper replaced cloth and ultimately, to control the size of the particles retained, papers were developed having different pore sizes. To retain particles smaller than those retained using the finest papers, filters were developed consisting of a cellulose acetate, nitrocellulose, or fiber-glass matrix. An even finer membrane is dialysis tubing, which passes *small* molecules and ions but retains *macro*molecules and macromolecular aggregates. A variant of dialysis tubing is the so-called molecular filter, which separates *small* macromolecules from *large* macromolecules.

In the course of using nitrocellulose membranes, it was found that they also have useful adsorptive properties (unrelated to their ability to filter), which enable them to bind particular macromolecules (e.g., single-stranded polynucleotides and some proteins) that are actually smaller than the pores of the filter; this property has contributed significantly to modern analytical techniques, and will also be described in this chapter.

NITROCELLULOSE FILTERS AND MEMBRANE FILTRATION

A nitrocellulose filter consists of a close network of nitrocellulose fibers.* The method of manufacture allows control of the maximum size of particles passing through the filter. The pores of the filter are not circular but irregularly shaped and account for roughly 80% of the surface area (Figure 7-1A–C). The filters are sufficiently thin that retained particles do not penetrate the filter but tend to remain on the surface; hence the particles can be easily washed off the surface. The best-known manufacturers are Millipore, Schleicher and Schuell, and Gelman. The range of maximum particle size passed is from 0.01 to 8.0 μ. The most commonly used filter has a pore size of approximately 0.45 μ.

Because of both the small pore size and the surface tension, liquids do not easily pass through these filters with gravity as the driving force so that pressure or suction is usually employed. A typical suction apparatus is shown in Figure 7-2. With the vacuum developed by a water aspirator, the flow rate through a 0.45 μ filter is approximately 65 ml/min/cm^2. A second consequence of the small pore size is that the filters can be used only for small quantities of material because they easily clog; the amount of material retained before there is a significant decrease in flow rates is approximately 250 μg/cm^2 of area. To avoid the clogging problem, the filters are sometimes used with a thick paper or fiberglass "prefilter" to remove large particles that might easily produce clogging. These filters are, however, excellently suited for the collection of microprecipitates.

Two little-known facts about nitrocellulose filters are important. First, nitrocellulose is hydrophobic so that, in order for the filters to be wettable, many contain a detergent or surfactant, the identity of which is not disclosed by the manufacturers. Second, the filters are very brittle when dry

*Although commonly called nitrocellulose filters, some are mixtures of cellulose nitrate and cellulose acetate (e.g., those made by Millipore Corporation).

FIGURE 7-1
A. Photograph of Millipore membrane filters, which are available in a variety of sizes and are sometimes ruled. B. Scanning electron micrograph of a Millipore filter, type HA, whose average pore size is 0.45 μ. C. Pore-size distribution of the filter shown in part B. D. Scanning electron micrograph of a Millipore Microfiber Glass filter (picture width, 20 μ.) Note the greater variety in the size of the pores than found in the membrane filter part B. [Photographs courtesy of Millipore Corporation.]

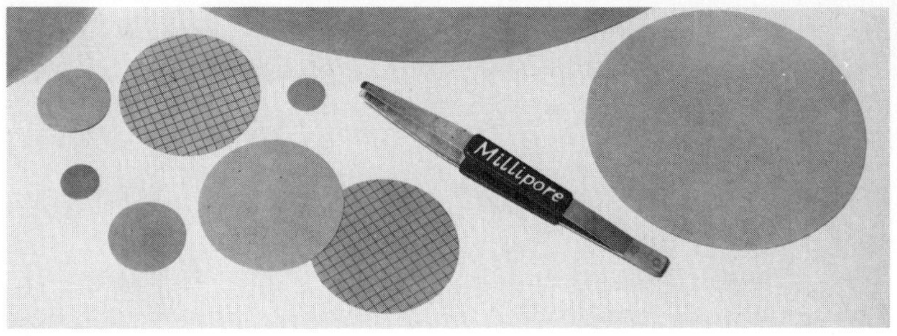

A

B

Relative pore incidence

0.35 0.40 0.45 0.50

Pore diameter (μ)

C

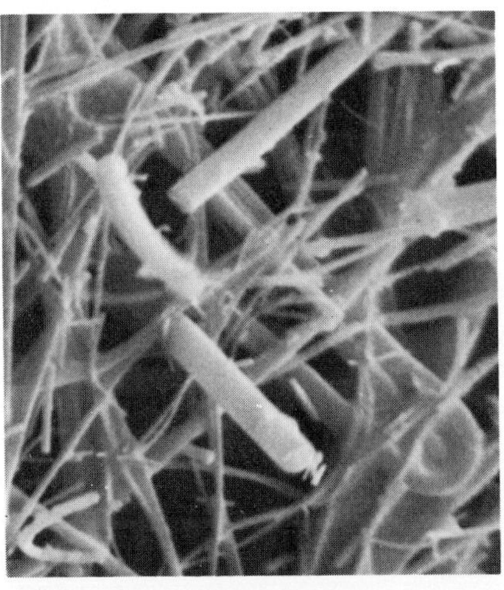

D

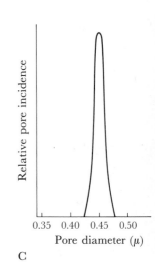

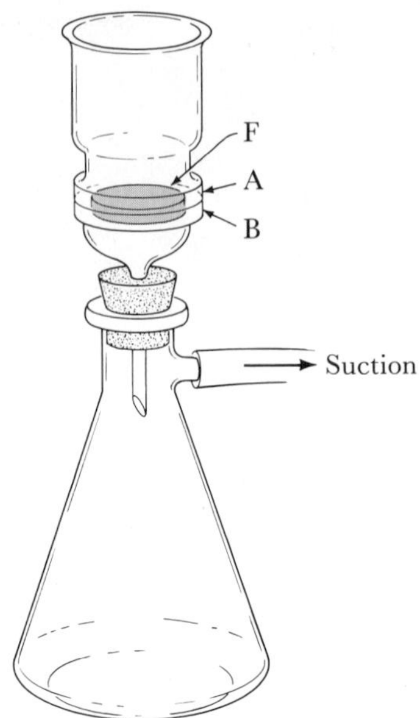

FIGURE 7-2
Typical apparatus for suction filtration: Rims A and B are held together by a spring clamp; F is a fritted glass support on which the filter is placed.

and many contain small amounts of glycerol to increase pliability. If the filters are not prewashed, these materials may contaminate the filtrate—a considerable disadvantage if it is the filtrate that is wanted.

Nitrocellulose filters come in white, black, or green circles or sheets, with or without ruled grids to aid in particle counting. The colors are used for the microscopic identification of some microorganisms. Because nitrocellulose is soluble in a variety of solvents and attacked by certain chemicals, the manufacturers supply similar membrane filters of cellulose acetate, nylon, and polyvinylchloride for use in certain solvent systems.

FIBERGLASS FILTERS

Fiberglass filters consist of a network of fine glass fibers (Figure 7-1D). This network cannot be made as fine as that of nitrocellulose filters so that, to have strength and the ability to retain particles of small size, fiberglass filters are relatively thick—approximately 0.25 mm. The advantages of fiberglass filters are very high flow rate (120 ml/min/cm^2), great capacity before clogging, resistance to almost all solvents, ability to be heated to high temperatures for rapid drying, and low cost (about one-third the cost of membrane filters per unit area). The disadvantages are that sometimes tiny pieces of glass fiber are found in the filtrate, only a small number of porosities are available, and, because fine particles penetrate the relatively thick filter, quantitative removal is not always possible.

In the following section, the uses of nitrocellulose and fiberglass filters are described and, when possible, which of the two types to use is indicated (in some cases it is a matter of preference).

FILTRATION WITH NITROCELLULOSE AND FIBERGLASS FILTERS

Clarification of Solutions

For a variety of chemical and biological applications, it is necessary to remove particulate matter from a liquid. If the particles are very fine and if there is not too much material, either a nitrocellulose or a fiberglass filter can be used unless the particle is too small to be retained by fiberglass. If there is a great deal of coarse material, prefiltration with a large-pore fiberglass filter is necessary to avoid clogging (most prefilters are made of fiberglass because of the high capacity).

Some proteins, viruses, and bacteriophages adsorb to nitrocellulose so that the filter must be pretreated in some way if larger particles are to be removed without losing any of these materials. Passing 0.1% bovine serum albumin (BSA) in H_2O through the filter followed by extensive washing with buffers of low ionic strength usually saturates the adsorption sites and eliminates the binding of other proteins. Single-stranded polynucleotides bind to both nitrocellulose and fiberglass filters in moderate ionic strengths and this is not prevented by the BSA wash. However, at low ionic strength this is not a problem.

If the solution is to be clarified for special studies such as measurement of absorbance, fluorescence, optical activity, or nuclear magnetic resonance spectroscopy, either type of filter is satisfactory. However, if light scattering or viscometry is to be done (in which dustfree solutions are absolutely necessary), fiberglass filters are probably unsatisfactory because tiny glass particles sometimes enter the filtrate.

Collection of Precipitates for Counting Radioactivity

As described in Chapter 5, for counting macromolecular radioactivity, it is almost always necessary to separate radioactive macromolecules from small radioactive molecules and to concentrate the sample to a small volume to avoid diluting the scintillator or adding water. This is usually accomplished by acid precipitation, using either perchloric or trichloracetic acid (hydrochloric acid can be used for polynucleotides but it is ineffective with protein). Following acidification, the precipitates can be collected rapidly on either a nitrocellulose or a fiberglass filter and washed, using the apparatus shown in Figure 7-2. For liquid scintillation counting, fiberglass should be used because it produces slightly higher counting efficiency and it can be dried at high temperatures without the charring that sometimes produces color quenching (see Chapter 5). For Geiger counting, the two filter types are equivalent.

Media Transfer for Growing Bacteria

In some experiments with growing bacteria, it is necessary to change the growth medium very rapidly—for example, if a radioactive compound or essential nutrient is to be removed. For this purpose, nothing is superior to membrane filtration. Only two problems must be considered: (1) nitrocellulose filters contain glycerol, which must be washed out before use (otherwise it is added to the second medium when the cells are eluted); and (2) many bacteria adsorb to nitrocellulose (approximately 2×10^6 cells/cm^2), obviating quantitative transfer at low cell density (this can usually be prevented by passing 0.1% bovine serum albumin—approximately 0.25 ml/cm^2—through the filter before collecting the cells). Following filtration, cells are removed from the filter by pipetting growth medium onto the filter or by immersing the filter in the medium and agitating gently. Although this technique has been used extensively for bacteria, using nitrocellulose filters, it has recently been applied to animal cells, using fiberglass.

NITROCELLULOSE FILTERS IN BINDING ASSAYS

Nitrocellulose filters bind proteins and single-stranded DNA under certain conditions. This binding can be used as a basis for a large number of enzymatic and physicochemical assays and as a means of purification of various materials. The binding properties of two commercially available nitrocellulose filters are shown in Table 7-1. It should be noted that the binding of proteins is independent of ionic strength, but single-stranded DNA adheres to these filters only at relatively high ionic strength. The poor binding of single-stranded DNA to the Millipore brand is not understood, although it could be because these filters are not pure nitrocellulose. RNA does not bind to these filters.

The use of these binding properties is best seen by the following examples.

Example 7-A. Purification of covalently closed, circular DNA.
Some strains of bacteria and some animal cells contain covalently closed, circular, double-stranded DNA molecules whose polynucleotide

TABLE 7-1
Binding various substances to nitrocellulose filters.

Substance	Schleicher and Schuell	Millipore	Ionic strength*
Proteins	+	+	0.01–1.0
Single-stranded DNA	+	Poorly	0.15
Single-stranded DNA	–	–	0.01
Double-stranded DNA	–	–	>0.05
Double-stranded DNA	+†	+†	0.001–0.005
Single-stranded RNA	–	–	0.01–1.0
Complex of protein and double-stranded DNA	+	+	0.01–1.0

NOTE: Some of the data has been provided by Dr. Andrew Braun.
*Ionic strength has been tested only in the range indicated.
†Binding is inefficient.

strands contain no interruptions (see Chapter 1, page 14). They frequently account for only a small fraction of the total DNA. If DNA is treated with alkali, hydrogen bonding is destroyed and the single strands separate; however, if the DNA is a covalently closed circle, the strands remain physically entangled. When the DNA is returned to neutrality and incubated at a slightly elevated temperature, the covalently closed circles rapidly reform; the single strands derived from linear DNA will ultimately rejoin, but this is a very slow process and strongly dependent on DNA concentration. A circular DNA molecule containing one interruption in one of the strands will separate to form a linear and a circular strand; reformation of a double-stranded circle has the same requirements as linear DNA. If alkali-treated and neutralized DNA is passed through a Schleicher and Schuell filter at high ionic strength (Table 7-1), only double-stranded covalent circles will pass through the filter (because all other components are single strands that bind to the filter) and will thereby be purified (Figure 7-3). For large-scale preparations, columns of powdered nitrocellulose are used.

Example 7-B. Assay of messenger RNA.
Filters containing bound, single-stranded DNA are prepared by passing a solution of single-stranded DNA through the filter and then drying it in vacuum. This DNA remains bound even if the filter is immersed in water. If radioactive RNA and the filter are placed in a small vial under renaturing conditions leading to the formation of a hydrogen-bonded complementary DNA-RNA hybrid (see Chapter 1) and then

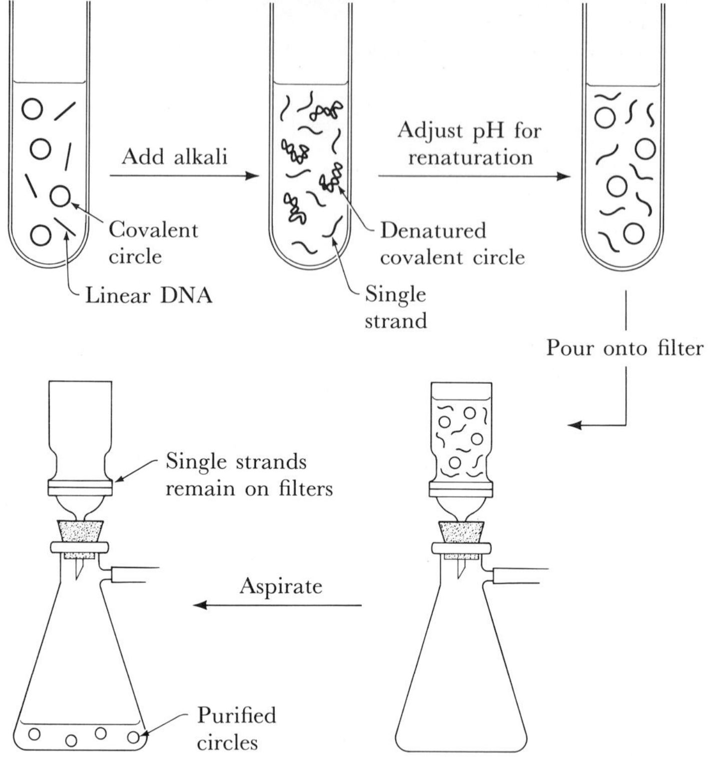

FIGURE 7-3
Plan for the purification of covalent circles with nitrocellulose filters. A DNA sample is denatured by treatment with alkali. The pH is then adjusted to a value at which covalent circles rapidly reform but separated strands do not. The mixture is filtered. The double-strand circles pass through the filter; the single strands adhere to the filter surface. The circles are thereby purified.

the filter is placed under suction and washed, most of the unbound RNA will be washed through the filter (Figure 7-4). If treated with the enzyme pancreatic ribonuclease (which hydrolyzes single-stranded RNA but fails to attack RNA in a double-stranded structure), the remaining unpaired RNA is digested. The filter is then extensively washed and only specifically bound (hybridized) RNA remains, which can be detected by its radioactivity.

If the washed filter is then heated to above the melting temperature for the DNA-RNA hybrid (Chapter 1), the specifically bound RNA is released; this is the best way to purify specific messenger RNAs.

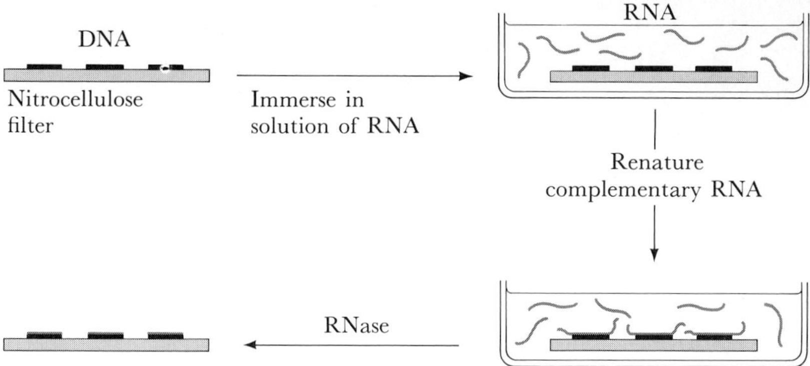

FIGURE 7-4
Method of hybridizing RNA to nitrocellulose filters containing bound, single-stranded DNA.

Example 7-C. Assay of complementary single-stranded DNA. Filters containing nonradioactive, bound, single-stranded DNA, prepared as in Example 7-B, can be used to assay complementary single-stranded DNA. However, if radioactive single-stranded DNA is added to such filters, it will, of course, also bind, whether or not there is prebound DNA (see Table 7-1). However, by washing a filter already containing nonradioactive DNA with bovine serum albumin and polyvinylpyrollidone, the remaining single-stranded DNA binding sites become saturated so that no radioactive single-stranded DNA can be bound. By immersing such a pretreated filter in a solution containing radioactive DNA and subjecting it to conditions of hybridization as in Example 7-B, the complementary radioactive DNA will anneal to the previously bound DNA. Unbound DNA can then be washed away and the bound radioactivity counted. This can be used to identify a particular species of DNA. For example, if the bacterium *E. coli* is infected with phage λ and incubated in a medium containing ^{3}H-thymidine, both *E. coli* and λ DNA will be synthesized. If the radioactive DNA is converted into single-stranded DNA (by thermal or alkaline denaturation) and annealed to filters to which are bound either *E. coli* or λ DNA, the relative amounts of radioactivity bound to each filter will indicate the fraction of each in the original radioactive mixture.

Example 7-D. Assay of proteins that bind to double-stranded DNA. If a mixture of proteins is incubated under appropriate conditions with radioactive double-stranded DNA, some DNA-protein complexes will form from special DNA-binding proteins (Figure 7-5). Because protein binds to the filter, any radioactive DNA complexed with pro-

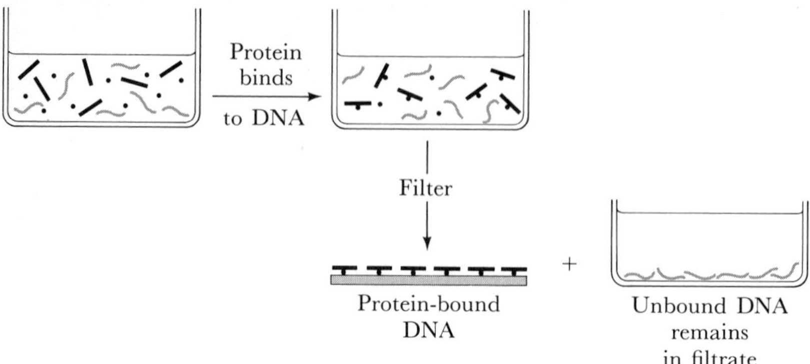

FIGURE 7-5
Filter-binding assay of protein-bound DNA. The sample contains protein (solid circles) and two types of DNA. The protein binds to the DNA shown as a heavy line but not to that indicated by the light line. When filtered through nitrocellulose, all protein binds. DNA molecules bound to the protein remain on the filter. All other DNA molecules pass through. The amount of bound radioactive DNA can be a measure of the amount of binding protein.

tein will adhere by means of the protein. These assays are best done with Millipore filters because the background due to unbound DNA is very low (see Table 7-1). This simple assay has been used in the purification of repressor proteins because they are detectable only by virtue of their ability to bind to specific sites on double-stranded DNA. It has also been used to detect DNA-enzyme intermediates in certain reactions and to measure the number of binding sites for particular proteins (e.g., RNA polymerase) on a DNA molecule.

MISCELLANEOUS USES OF MEMBRANE FILTERS

A membrane filter can also serve as an immobilizing agent. For example, in matings between chromosome-transferring male (Hfr) strains of the bacterium *E. coli* and female recipient strains, Brownian motion disrupts the mating pair. In the construction of bacterial strains with special properties, it is often necessary to transfer a gene that is normally passed from male to female only very late in the mating event. However, the transfer of "late" genes is rare owing to the disruption described above. If the females are collected on a membrane filter and the males are filtered on top of them, mating proceeds normally if the filter is placed on the surface of a nutrient agar plate. Because the cells are immobilized there is little

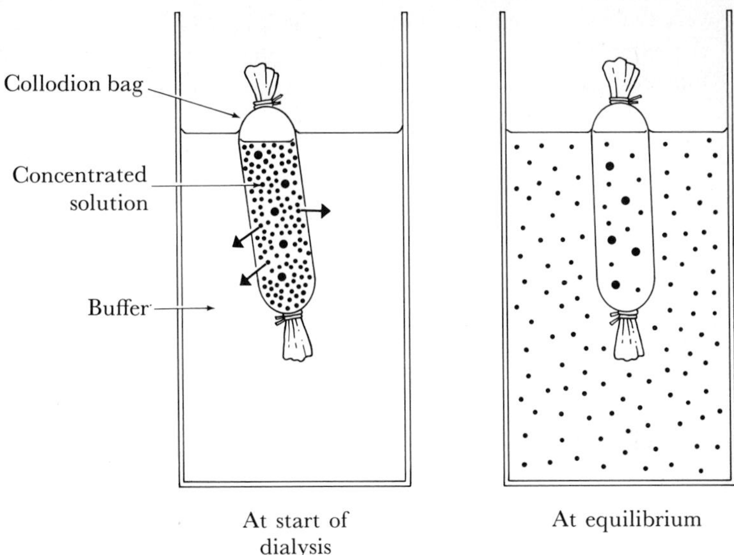

FIGURE 7-6
Dialysis. Only small molecules (dots) diffuse through the collodion membrane. At equilibrium, the concentration of small molecules is the same inside and outside the membrane. Macromolecules remain in the bag.

disruption and the frequency at which late genes are transferred is increased considerably. At the end of the mating period, the cells can be washed from the filter and plated onto appropriate agar plates.

DIALYSIS AND MOLECULAR FILTRATION

Conventional filtration separates *particulate* matter from fluids by passing suspensions through porous material. However, with so-called semipermeable membranes it is possible to separate *dissolved* molecules by virtue of their molecular dimensions.

The best-known method is *dialysis,* in which an aqueous solution containing both macromolecules and very small molecules is placed in a collodion bag, which is in turn placed in a large reservoir of a given buffer (Figure 7-6). Small solute molecules (except those that are highly charged) freely pass through the membrane until equilibrium is reached.* To con-

*Equilibrium in this case means that the concentrations inside and outside the bag are the same.

vert the solute composition within the bag into the composition wanted, the external fluid must be repeatedly changed to maintain the required final composition. Water also passes freely through the bag, thus causing concentration or dilution, depending on whether the internal solution is less or more concentrated than the external (osmotic effect).

A variation called *reverse dialysis* (Figure 7-7) can be used to concentrate the material in the bag. The filled bag is packed in a dry, water-soluble polymer (which cannot enter the membrane) such as polyethylene glycol. Water then leaves the bag to equilibrate with the dry external phase. Sucrose can also be used but, because it is a dialyzable substance, it will enter the bag as water is removed.

Semipermeable membranes that allow small but not large macromolecules to pass through are also commercially available. An example is Spectropor tubing (Spectrum Medical Industries, Inc.), which is used in much the same way as dialysis tubing. It is available in three grades, with molecular weight cutoff (i.e., maximum size passed) ranging from 6,000 to 14,000.

An important modification of dialysis tubing is the Diaflo (Amicon Corporation) or the Pellicon (Millipore Corporation) membrane. This is a very thin (0.1–1.0 μ) polymer membrane having a pore size that ranges from 2 to 100 Å mounted on a thicker (50–250 μ) supportive layer of an

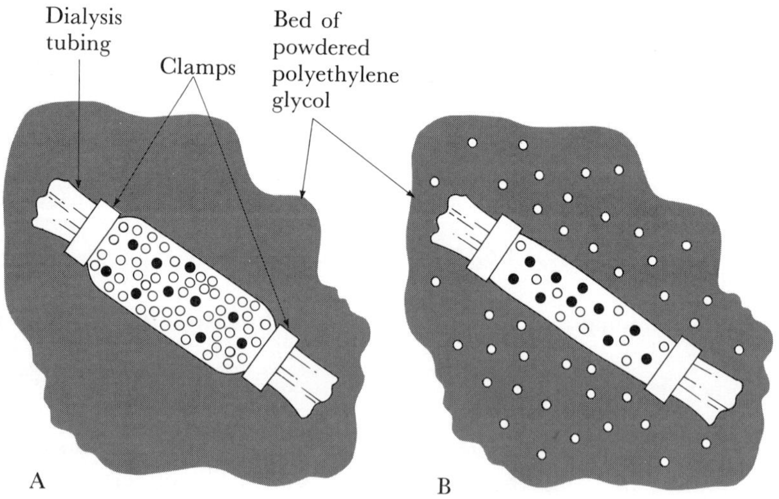

FIGURE 7-7
Reverse dialysis: (A) a solution of macromolecules (solid circles) and solvent molecules (open circles) is placed in a dialysis bag that is packed in dry polyethylene glycol (PEG) powder; (B) the solvent molecules leave the bag and enter the PEG phase. Neither the PEG nor the macromolecules can pass through the membrane and so the solution is concentrated.

open-celled "sponge;" it comes with a wide variety of molecular weight cutoffs, ranging from 500 to 100,000. Such membranes can be used for concentration (using a membrane through which only water passes), desalting in preparation for chromatography, and fractionation by molecular size. The flow rate through these membranes is so low that they are operated under pressure. Special holders are usually required for their use (Figure 7-8). Modifications of these instruments are available for handling especially large or small volume or for multiple samples.

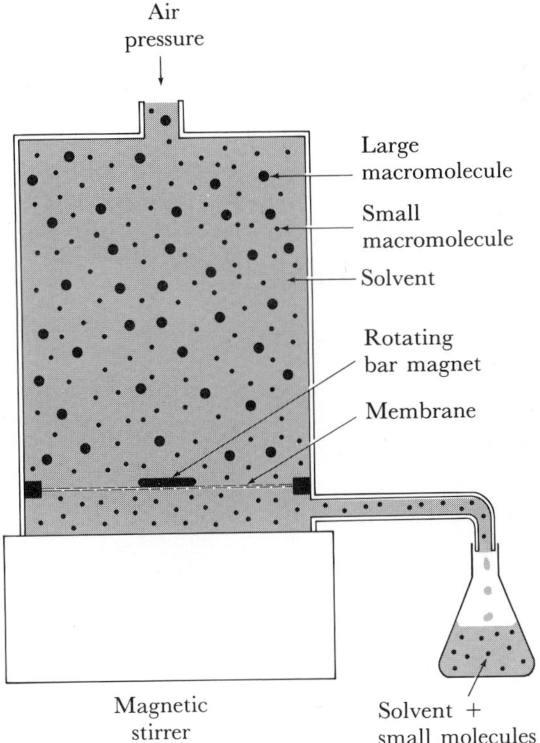

FIGURE 7-8
Apparatus for molecular filtration. A solution containing small and large macromolecules is forced against the molecular membrane. Molecules larger than the cutoff size fail to pass through the membrane. Smaller molecules and the solvent do pass through. A rotating bar magnet is used to prevent the membrane from clogging. Either small or large molecules can be purified in this way. The large molecules are also concentrated.

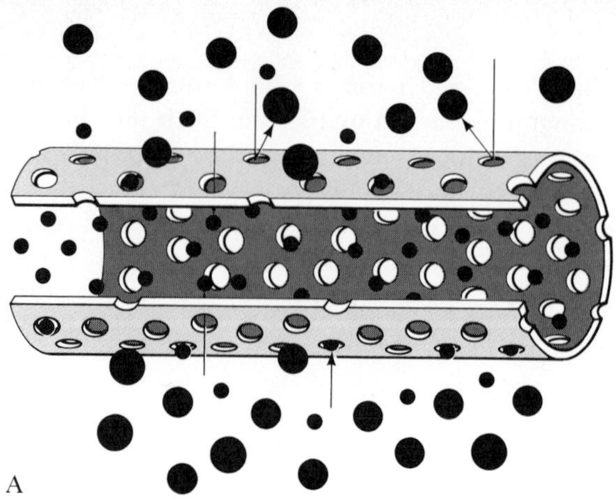

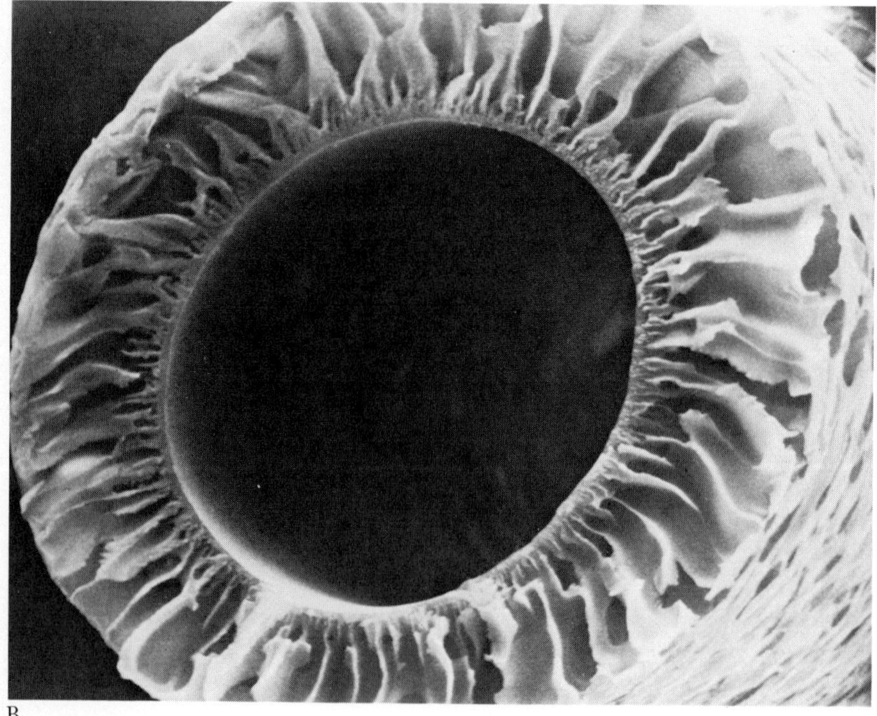

FIGURE 7-9
A. The operation of a hollow fiber. The fiber is initially immersed in a solution of large and small molecules. The small molecules pass through the pores of the fiber to the inner channel and are swept away by flowing water. The large molecules cannot pass through the pores. [Courtesy of Bio-Rad Laboratories.] B. Electron micrograph of a hollow fiber. [Courtesy of Amicon Corporation.]

GLASS FIBER DIALYSIS

Semipermeable glass fibers are valuable devices for both dialysis and concentration. They are hollow-bore fibers whose glass walls contain pores of controlled size. Molecules smaller than the pores pass freely through the wall of the fiber (Figure 7-9A). These fibers are usually used in bundles, thus providing a very large surface area. They are normally used in a unit of the type shown in Figure 7-10. To change the buffer in a sample containing a macromolecule to a second buffer, the sample is placed in the vessel and a large volume of the second buffer is allowed to flow through the fibers (Figure 7-10A). The small molecules of the two buffers rapidly exchange through the pores of the fibers; because the buffer within the fibers is in excess, the first buffer is replaced by the second. The macromolecules fail to penetrate the pores of the fibers and remain outside. If desalting is required, water is passed through the fibers. For concentrating samples, the arrangement shown in Figure 7-10B is used. The sample is again in the vessel and suction is applied to the fibers. The pressure differential then forces the solvent and solute molecules into the fibers,

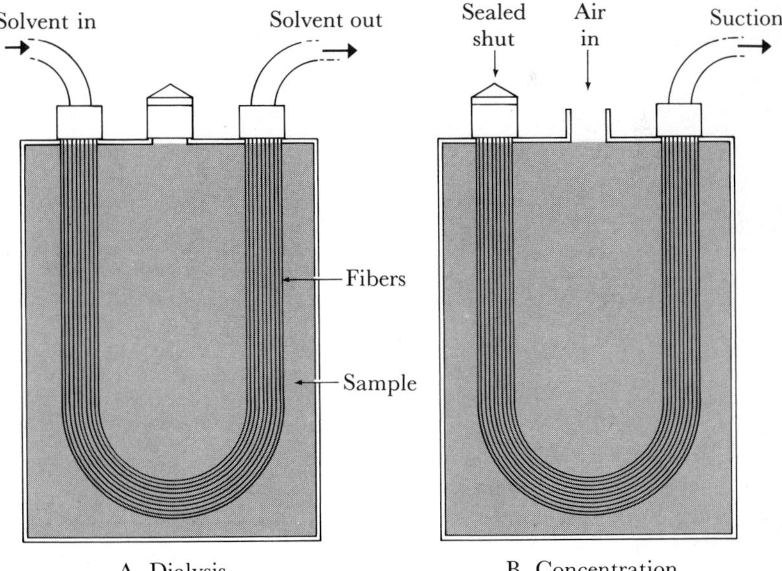

FIGURE 7-10
Two ways of using hollow fibers: (A) dialysis, in which a solvent flows through the fibers and small molecules enter the fibers, thus reducing the concentration of small molecules in the sample; and (B) a concentration vacuum is applied to the fiber bundle and the solvent and small molecules enter the fiber, thus concentrating any macromolecules in the sample.

thus concentrating the solution. Other physical arrangements that allow the sample to pass through the fibers are possible but are usually not done because the fibers easily clog.

Selected References

Bøvre, K. and W. Szybalski. 1971. "Multi-step DNA-RNA Hybridization Techniques," in *Methods in Enzymology,* vol. 21, edited by L. Grossman and K. Moldave, pp. 350–382. Academic Press.

Kennell, D. 1971. "Use of Filters to Separate Radioactivity in RNA, DNA, and Protein," in *Methods in Enzymology,* vol. 12A, edited by L. Grossman and K. Moldave, pp. 686–692. Academic Press.

Millipore Corporation. *Molecular Filtration.*

Excellent brochures are supplied by H. Reeve Angel, Inc. (properties of paper and fiberglass filters), Amicon Corporation (Diaflo membranes), Millipore Corporation (nitrocellulose filters and Pellicon membranes), and Schleicher and Schuell (nitrocellulose filters).

Problems

7-1. Suppose that you are collecting a precipitate from an acetone solution. Would a nitrocellulose or fiberglass filter be better? Why?

7-2. If you were collecting a precipitate from an aqueous solution, what criteria would you use in deciding between filtration and centrifugation?

7-3. Particles presumably smaller than the pores of a particular nitrocellulose filter are efficiently collected on the filter. They might adsorb or they might aggregate and form large clusters. How could you distinguish these alternatives?

7-4. Linear and circular DNA molecules having the same molecular weight can be separated on nitrocellulose filters if the flow rate is very high. At low flow rate there is no separation. Which would remain on the filter? Why?

7-5. Growth medium for biological material must always be sterile. Sterilization is normally done by raising the temperature of the solution to 135°C. However, this is not always possible because some of the components may be destroyed by heating. Explain how filtration could be used for sterilization. Describe what components must be sterilized and how you would do it.

7-6. In preparing protein or nucleic acid polymerization reaction mixtures for scintillation counting, a common procedure is to add trichloroacetic acid to the mixture. The polymer is insoluble in acid, whereas the monomer remains soluble. When this mixture is filtered, the polymer remains on the filter and the monomer passes through. However, a small fraction of the monomer often adsorbs tightly. If 10^6 cpm of monomer is in the mixture, 0.02% is polymerized, and 0.01% adsorb, then the background radioactivity will be too high for reliable counting. Given that proteins and nucleic acids are soluble in alkali, design a procedure for reducing the background. State the type of filter that will be used and estimate the background resulting from the improved procedure.

7-7. Suppose that you are collecting bacteria on a filter in order to transfer them to a new growth medium. Transferring is done by placing the filter in the new medium and agitating violently to wash off the bacteria. In a series of experiments, you have checked the recovery of bacteria from the filter as a function of the total number of bacteria collected and have obtained the following results:

Number of bacteria collected	Percentage washed off filter
5×10^8	98
3×10^8	100
1×10^8	89
5×10^7	50
3×10^7	35

What would the recovery be (approximately) if 10^7, 10^6, and 10^5 bacteria were collected? Explain.

7-8. That dialysis tubing adsorbs both proteins and nucleic acids is well established. Explain how you might determine whether this is occurring for a given protein or nucleic acid. Enzymes lose biological activity on dialysis. How could you distinguish adsorption from inactivation for a given enzyme?

PART III

Separation and Identification of Materials

CHAPTER 8

Chromatography

A goal of biochemistry is to separate and identify chemical compounds. Chromatography is one of the most effective techniques for accomplishing this. Although it is generally acknowledged that the method was developed in 1906 by the Russian botanist Mikhail Tswett, who separated plant *pigments* (hence the name), it was described in 1855 by Karl Runge, a German chemist who separated inorganic materials by paper chromatography; in fact Pliny the Elder reported the separation of dyes on papyrus and devised a papyrus chromatographic test for iron. However, chromatography did not become a serious technique until the work of Archer Martin and John Synge in 1944, who later received a Nobel Prize for developing the methodology of partition chromatography.

Today there are many kinds of chromatography—adsorption, partition, ion-exchange, and molecular-sieve—and many specialized techniques for using them—column, paper, thin-layer, and gas chromatography. Technical modifications can be introduced if chromatography is to be used for large-scale work (i.e., for producing large quantities of a relatively pure material) rather than as an analytical procedure. This chapter describes each of these procedures and many of the modifications.

PARTITION CHROMATOGRAPHY

Simple Theory of Partition Chromatography

Grossly dissimilar substances like iron filings and glass particles can be easily separated with a magnet; sand and sugar can be separated by dissolving the sugar in water. However, if the substances are similar in physical and chemical properties, separation procedures become more complex and subtle. In chromatography, substances are placed in a system consisting of two physically distinguishable components—a *mobile phase* and a *stationary phase*—and molecular species separate because they differ (many of them only slightly) in their distribution between these two phases. The relative movement of each molecule is a result of a balance between a driving force (i.e., the movement of the mobile phase) and retarding forces. The retarding forces that we will be considering first are partition and adsorption.

In discussing chromatography, the following standard terminology is used. The stationary phase is the *sorbent.* If the sorbent is a liquid held stationary by a solid, the solid is called the *support* or *matrix.* The mobile phase is the *solvent* or *developer* and the components in the mixture to be separated constitute the *solute.*

The theory of *partition chromatography* is that, in general, if two phases are in contact with one another and if one or both phases contain a solute, the solute will distribute itself between the two phases. This is called partitioning and is described by the *partition coefficient,* the ratio of the concentrations of the solute in the two phases.

Partition chromatography will be described in terms of the operation of a *column*—that is, a tube filled with a sorbent and a solvent. A solution containing the solute is layered on top of the sorbent and allowed to enter the sorbent. The solvent is then allowed to pass continually through the column. Although the sorbent and solvent within the column are certainly continuous from the top of the column to the bottom, the column can be thought of as consisting of a large number of individual layers ("theoretical plates"), each containing the two phases. Consider 256 *identical* molecules that distribute themselves equally between the two phases, one stationary and one mobile, and a column with eighteen plates (Figure 8-1). In the uppermost plate (the *origin*) the 256 molecules are distributed so that 128 are in each phase. When the mobile 128 molecules from plate 1

FIGURE 8-1
Principle of operation of a column in which separation is based on partition. The column has arbitrarily been divided into eighteen theoretical plates. Five hundred twelve molecules are loaded onto the column. Two hundred fifty-six of these (bold type) distribute equally between mobile phase (roman type) and stationary phase (italic); they are the 1:1 class. The 1:3 class (light type) distribute so that 25% are in the mobile phase and 75% in the stationary phase. A transfer means that all molecules in the mobile phase advance to the next plate. Following each transfer the number of molecules of each class redistribute according to the 1:1 or 1:3 rule. The graph shows the distribution of each after twenty transfers. Note that the 1:3 class moves more slowly through the column.

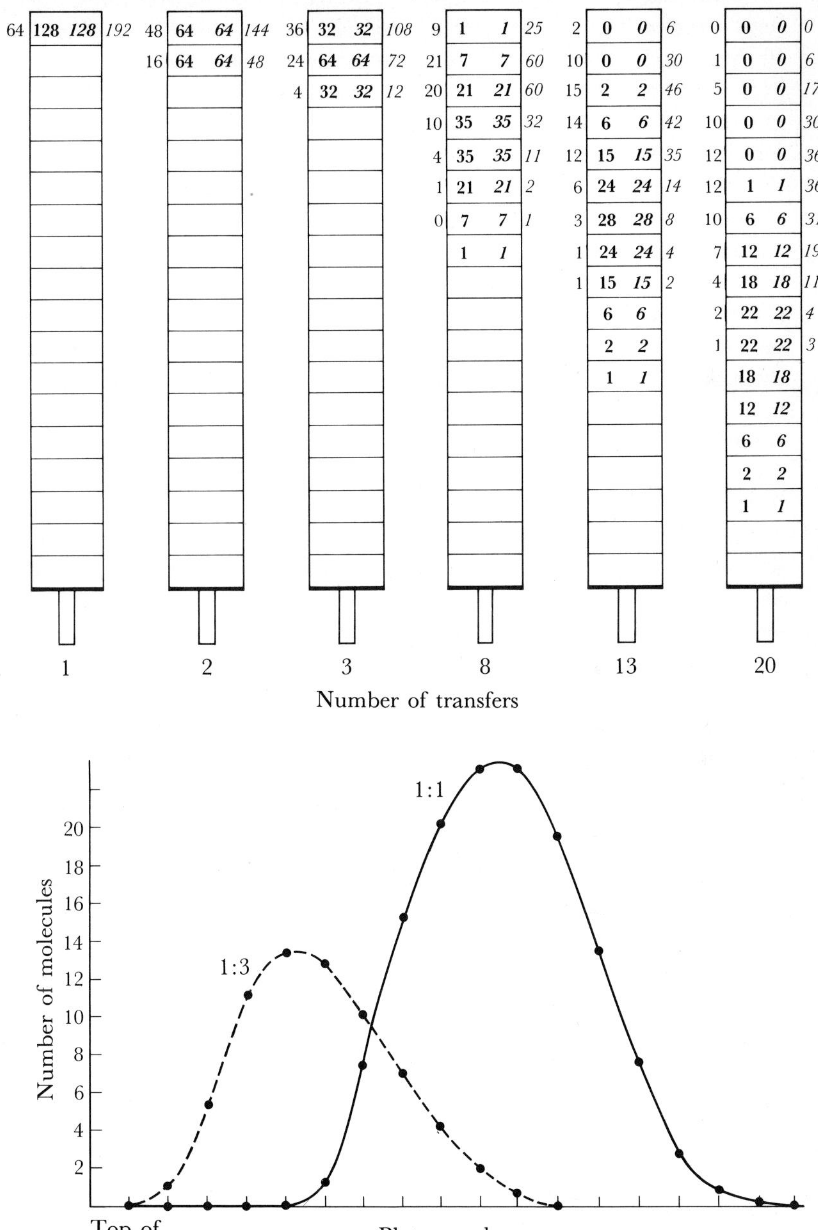

1
2
3
8
13
20
Number of transfers
1:1
1:3
Number of molecules
Top of column
Plate number

enter the second plate, they redistribute 64 and 64 in that plate; the 128 remaining in plate 1 also redistribute 64 and 64, as shown in Figure 8-1. If the mobile phases each advance again by one plate, redistribution again occurs. After twenty successive transfers, the situation shown in the graph is achieved. Suppose at the origin there also were 256 identical molecules *of a different type* (shown in italics), which distribute with three times as many in the stationary phase. Figure 8-1 shows the distribution of these molecules after twenty transfers also. The distributions of these two kinds of molecules are different and a substantial fraction of the molecules have separated. As the number of theoretical plates is increased (i.e., if the column length is increased), greater separation will result. It should be noted that, as the number of plates increases, material is spread throughout a greater part of the column. In a real situation in which the total number of molecules is huge (i.e., $>10^{16}$), the number in plate 1 would not reach zero. On the other hand, the degree of spreading would decrease in the sense that a *greater percentage* of the material is in a small fraction of plates. For example, after eight transfers, 40% of the molecules illustrated in Figure 8-1 that distribute 1:1 are contained in 4/8 of the plates, whereas, after twenty transfers, 40% are in 6/16, or 3/8, of the plates. These considerations should make it clear that in the ideal case, in which the distribution is determined only by partition, the *resolution of two substances will improve as the length of the stationary phase increases.*

Examples of Partition Chromatography

The two most common types of partition chromatography are *paper* and *thin-layer chromatography.* In both cases, the matrix contains a bound liquid: water molecules are bound to cellulose in paper chromatography, and the solvent used to form the thin layer is bound to the support in thin-layer chromatography (see pages 177–180). (These techniques are sometimes thought of as types of adsorption chromatography, because adsorptive effects do enter into the degree of separation; however, the principal mode of action is by partitioning.) Other examples of partition chromatography are *gas-liquid* and *gel chromatography,* which are described in detail on pages 180 and 185.

Partition chromatography may also be carried out in columns by using a matrix that does not adsorb the solutes. Common supporting materials are diatomaceous earth (e.g., Celite), silica gel, cellulose powder, and certain cross-linked dextrans (e.g., Sephadex LH20). The stationary phase is created by suspending the support or washing the column with the appropriate sorbent. In this way the sorbent either coats the particles of the support and is retained by adsorption or simply penetrates the interstices of the particles and is held there by capillarity. It is important to realize that the stationary phase does not fill the spaces between the particles—that space is to be occupied by the mobile phase. Typical stationary-phase materials are hydrophobic solvents such as benzene for the separation of nonpolar materials or hydrophilic solvents such as an alcohol, for polar materials. Typical mobile phases are alcohols or amides for the nonpolar material or water for polar substances. Note that the stationary-phase materials are liquids.

Partition chromatography is used primarily for molecules of small molecular weight. To reduce diffusional spreading (i.e., the broadening of peaks), very small starting zones and rapid separation is necessary. How this is done will be described in the sections on paper and thin-layer chromatography.

ADSORPTION CHROMATOGRAPHY

Simple Theory of Adsorption Chromatography

Consider a solid surface containing a wide variety of binding sites—for example, regions that are electron-rich (negatively charged), electron-poor (positively charged), nonpolar, and so forth—and a liquid containing solute in contact with the surface. If binding is reversible, the number of molecules bound to the surface will depend on the solute concentration. This dependency (of which there are three types) is shown in Figure 8-2. Curves of this sort are called *adsorption isotherms.* The most common is the convex curve—that is, binding sites with high affinity are filled first so that additional amounts of solute are bound less tightly. The binding isotherm is a characteristic of a particular molecule and sorbent. If a given concentration of a molecule is applied to the surface (which in practice is usually a collection of particles in a column or on a solid support) and solvent is

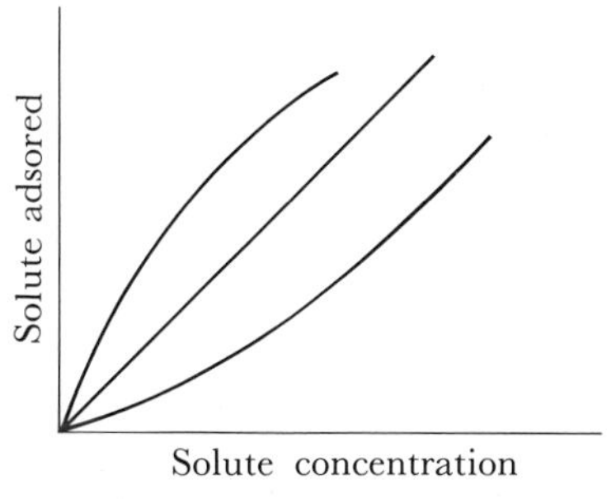

FIGURE 8-2
Three common distributions of solute between adsorbent and solvent as a function of solute concentration. These curves are called adsorption isotherms. Because of the concentration dependence of the fraction bound, the distribution as a function of the number of transfers is more complicated than that of Figure 8.1.

TABLE 8-1
Common materials used in adsorption chromatography.

Material	Substances separated
Alumina	Small organic molecules, proteins
Silica gel	Sterols, amino acids
Activated carbon	Peptides, amino acids, carbohydrates
Calcium phosphate gel	Proteins, polynucleotides
Hydroxyapatite	Nucleic acids

allowed to flow across the surface, a fixed amount will bind and the remainder will move along. The advancing material will bind with two differences: (1) the retarding force is binding, or adsorption, and (2) the fraction bound is not a constant fraction but decreases with decreasing concentration. The rate at which the substance moves is related to the strength of binding—that is, the tighter the binding, the slower the movement. Clearly then, molecules can be separated if they have different adsorption isotherms because they will be retarded to different extents.

Types of Adsorption Chromatography

Adsorption chromatography uses a mobile liquid phase and a solid stationary phase with the one exception of gas-liquid chromatography (page 180). Separation is either in columns or on thin layers. Table 8-1 gives the common adsorbents and their uses.

An important variation of adsorption chromatography is ion-exchange chromatography (page 195). This differs mainly in that the composition of the mobile phase is such that, as the material is being applied to the adsorbent, the solute becomes immobilized. Migration does not begin until a new mobile phase is added. This is not different in principle but is a special case in which adsorption is very strong.

It is important to realize that partition and adsorption chromatography are rarely exclusive in that adsorptive effects may be present in paper, thin-layer, and sometimes gel chromatography.

OPERATION OF COLUMNS

Probably the most common way to hold the stationary phase or support is in a column. In column chromatography, a tube is filled with the material constituting the stationary phase, plus a solvent. Then a small volume (i.e., a thin layer) of sample is placed on the stationary phase and allowed to enter the column (this is called *loading* the column). The chromatogram is then developed by flowing a solvent (the mobile phase) through the column (Figure 8-3). The latter process is called *eluting* the column. As different substances move through the column, they separate and appear in the effluent when particular volumes of liquid have passed through the column. The total volume of material, both solid and liquid, in the column is called the *bed volume.* The volume of the mobile phase is the *void, retention,* or *hold-up* volume. The amount of liquid that must be added to produce a peak of a particular solute in the effluent is the *elution volume.*

The manner in which the *bed* is formed in a column is called *packing.* It is important that the bed be homogeneous and free of bubbles, cracks, or spaces between the walls. The uneven flow resulting from inhomogeneity is called *channeling.* The usual effect of channeling is to distort the elution pattern so that single substances appear in multiple peaks. This is a result of the rapid movement of the mobile phase down through the column, reestablishing partitioning or adsorption at a new point.

The liquid leaving the column (the eluent) is usually collected as discrete fractions, using an automatic collector such as that shown in Figure

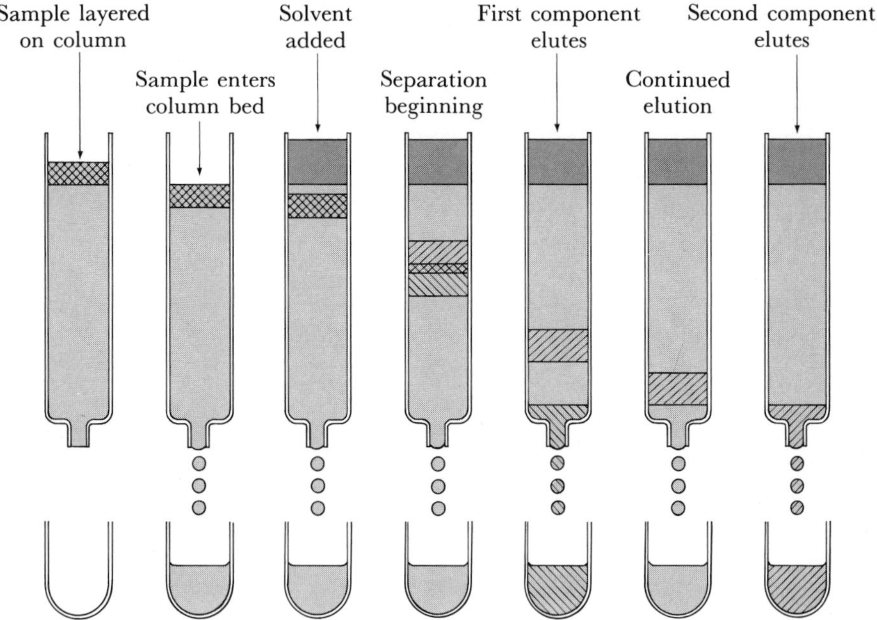

FIGURE 8-3
Operation of a column showing the loading of the column and various stages of elution.

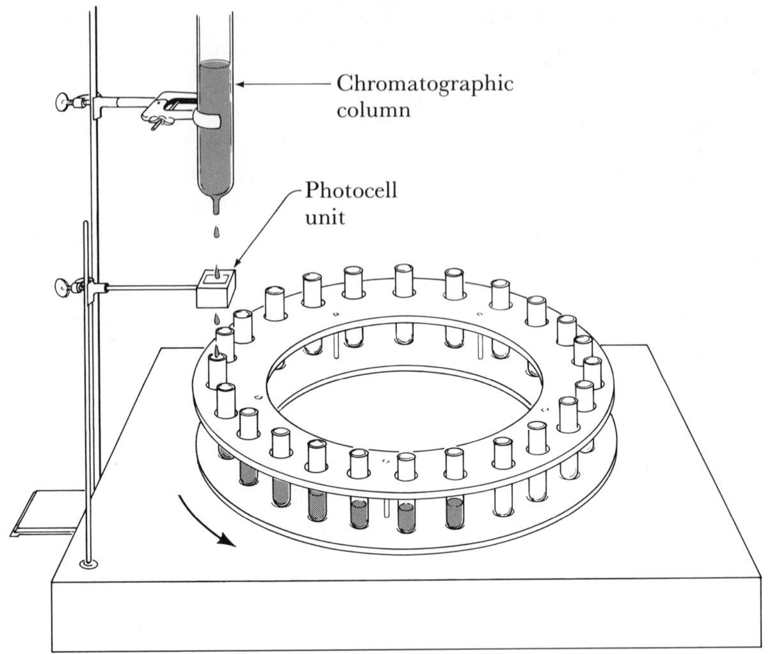

FIGURE 8-4
A simple fraction collector. A measured number of drops or a measured volume falls into one of the tubes; the tubes then advance until the next tube is centered under the liquid outflow. The drops are detected by a photocell unit containing a small light and a photocell; each drop interrupts the light beam and is counted.

8-4. The separated components are then found and identified by testing aliquots of each fraction—for example, spectral measurements, chemical tests, radioactivity, and so forth. In cases in which analysis is by the absorption of light, an automatic, continuously recording spectrophotometer is used. The sample passes through a tube before fractionation and the optical density at an appropriate wavelength is measured continuously and plotted on a chart recorder.

Columns are eluted in one of three ways: In the simplest method a single solvent flows *continually* through the column. This is a common method in ion-exchange chromatography (see page 200) and in gel chromatography.

Stepwise or batch elution is commonly used if the column is being operated for preparative purposes. The column is eluted with one solvent until a predetermined volume has been applied. Then a second solvent is added. This method has the advantage that conditions can be arranged so that a particular material can be eluted in a rather small volume. For example, consider a mixture of substances, one of which (X) is strongly retarded and the others weakly retarded when solvent A is used; X is weakly retarded by solvent B. Hence, if the column is extensively washed with solvent A, most of the material is removed from the column but most of X remains bound. If the column is then washed with solvent B, X is rapidly eluted in a relatively small volume.

The third method is *gradient elution,* which consists either of changing the ratio of two solvents or of increasing the concentration of one or more of the components in the solvent (e.g., the salt concentration). The latter is the most common way of eluting adsorption and ion-exchange columns. Gradients are prepared by means of apparatus of the type shown in Figure 8-5. A gradient maker consists of two vessels, a *reservoir* and a *mixing*

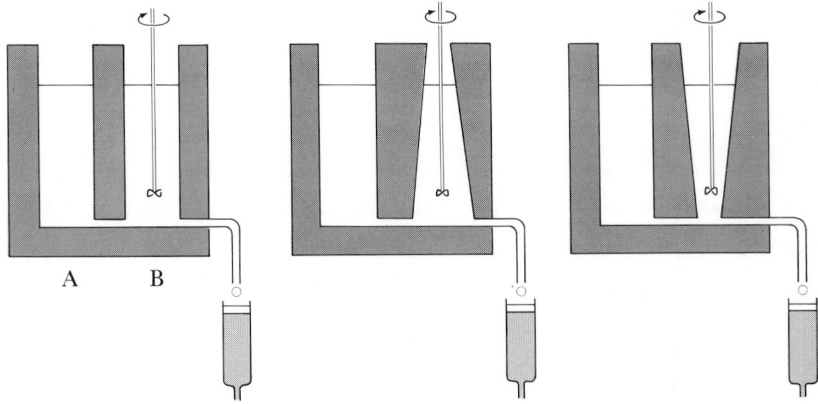

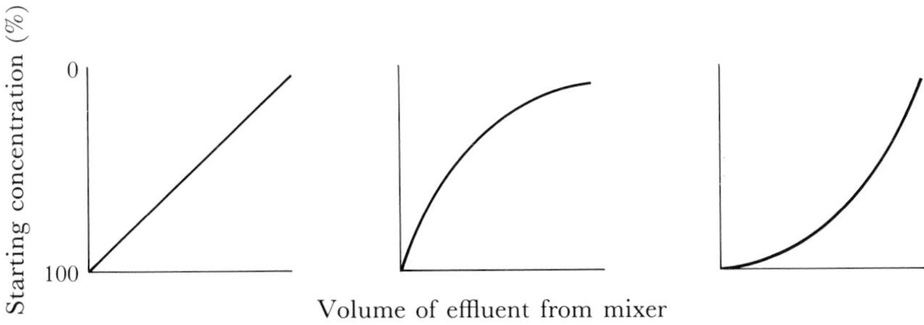

FIGURE 8-5
Three types of gradient-making apparatus and the resulting gradients. As liquid drains from the mixing chamber, liquid from the reservoir flows into the chamber to maintain constant hydrostatic pressure. If the densities of the two liquids are the same, the heights in the two chambers will be identical. If the fluid in B is denser than that in A, the level in B will be lower than that of A. A linear gradient results only if the volumes of the reservoir and mixing chamber are equal.

chamber, which are connected at their bases. The reservoir contains the more concentrated solvent. Liquid leaves the mixing chamber and enters the column. Because the hydrostatic heads (not necessarily the heights, because the densities of the liquid in the two vessels may differ) must be equal, liquid simultaneously flows from the reservoir to the mixing vessel. If the chambers have the same shape, the gradient is linear; concave and convex gradients can also be prepared, as shown in Figure 8-5.

PAPER CHROMATOGRAPHY

Partition chromatography can be performed on columns of cellulose. Paper chromatography is a variant of this procedure in which the cellulose support is in the form of a sheet of paper. Cellulose contains a large amount of bound water even when extensively dried. Partitioning occurs between the bound water and the developing solvent. Frequently, the solvent used is water itself so that it is reasonable to ask whether the mode of action in this case is adsorption. Certainly, some adsorptive effects are present but, because the physical structure of bound water is very different from that of "free" water, partitioning can still occur.

Experimental Procedure for Paper Chromatography

Accompanying the change in the form of the support (from column to paper) is a change in methodology. In paper chromatography there is no effluent and *substances are distinguished by their relative positions in the paper after the solvent has moved a given distance.*

A tiny volume (approximately 10–20 μl) of a solution of the mixture to be separated is placed at a marked spot on a strip or sheet of paper (Figure 8-6) and allowed to dry. This spot defines the *origin.* The paper is then placed in a closed chamber and one end is immersed in a suitable solvent (the mobile phase). Capillarity draws the solvent through the paper, dissolves the sample as it passes the origin, and moves the components in the direction of flow. (Note that, because the sample must be dissolved before it can migrate, a factor in determining the separation of the components is the rate of solubilization into the mobile phase.) After the *solvent front* (see Figure 8-6) has reached a point near the other end of the paper, the sheet is removed and dried. The spots, which may or may not be visible, are then detected and their positions marked. The relative distance traveled by a spot and by the solvent front is called the R_f. Values of R_f depend on the substance, the paper, and the solvent.

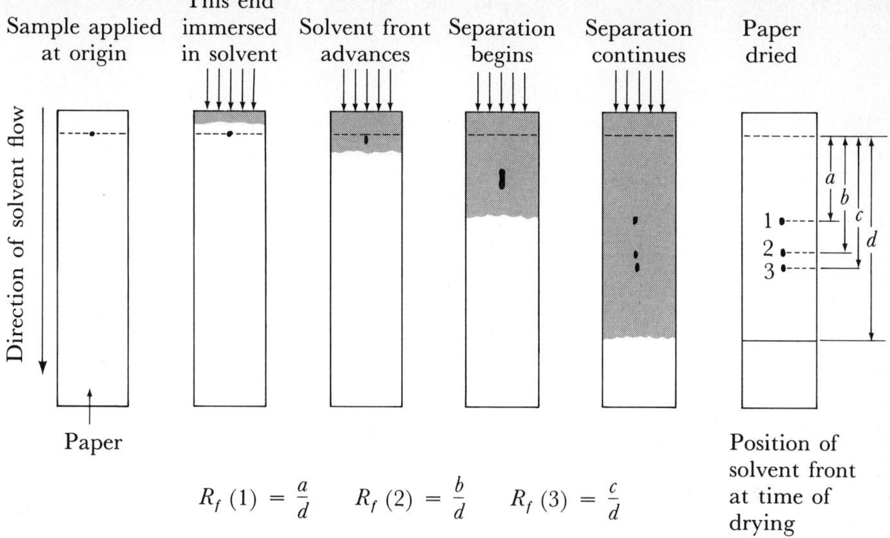

FIGURE 8-6
Spotting and developing a paper chromatogram with a sample containing three components.

Paper chromatograms can be developed by either ascending or descending solvent flow. The experimental apparatus for each is shown in Figure 8-7. There is little difference in the quality of the chromatograms and the choice is usually a matter of personal preference. Descending chromatography has two advantages: (1) it is faster because gravity aids in the flow and (2) for quantitative separations of materials with very small R_f values, which therefore require long runs, the solvent can run off the paper. Its only disadvantage is the care with which the apparatus must be assembled because dirt or poor contact where the paper passes over the support bar can result in inhomogeneous flow and consequent streaking.

A particularly useful variant is two-dimensional paper chromatography. In this method, after chromatography has been carried out in a single direction, the paper is dried and then rechromatographed at right angles to the original direction of flow, using a different solvent system (Figure 8-8). In this way, substances that fail to separate in the first solvent can often be separated in the second.

Detection and Identification of Spots

Spots in paper chromatograms can be detected by their color, by their fluorescence, by the chemical reactions that take place after the paper

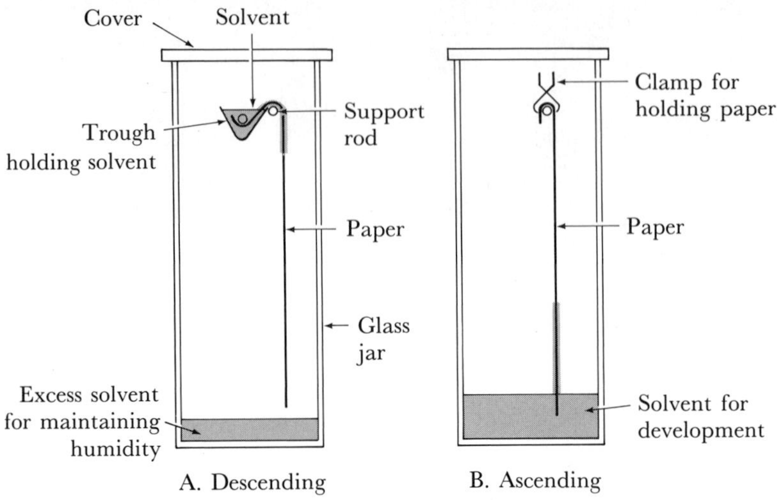

FIGURE 8-7
Experimental arrangement for descending and ascending paper chromatography.

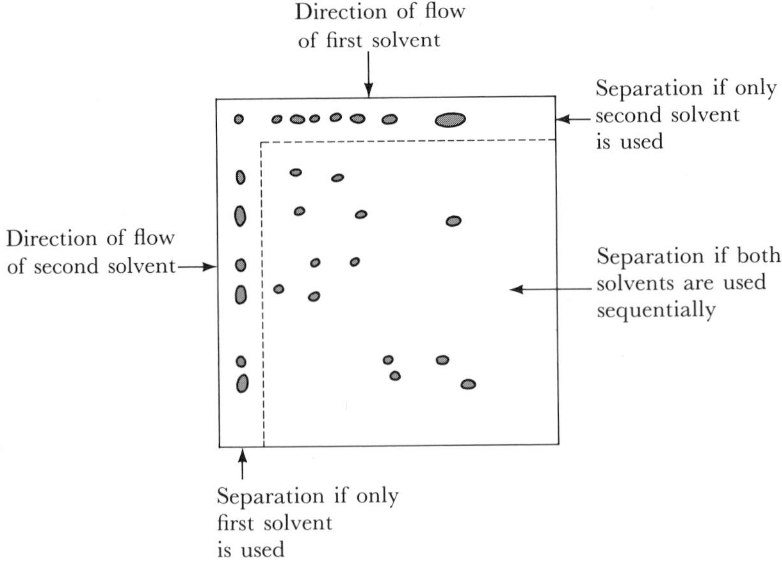

FIGURE 8-8
Two-dimensional paper chromatography.

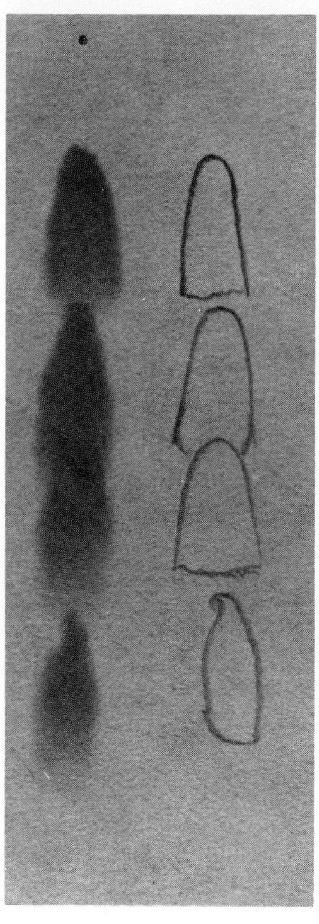

FIGURE 8-9
A paper chromatogram in which four amino acids have been separated. After the paper had dried, it was sprayed with ninhydrin, which produces a colored product with amino acids. The dot shows the origin. The column on the right shows a drawn outline of each spot.

has been sprayed with various reagents, or by radioactivity (Figure 8-9). The autoradiographic detection of spots with x-ray film is described in Chapter 6. Identification is usually based on comparison with standards of known R_f or by elution. Elution is accomplished by cutting out the spot and soaking the paper in the appropriate solvent; this can often be done in a quantitative way.

Fingerprinting: A Special Application of Paper Chromatography to the Study of Proteins

An important problem in molecular biology is to identify an isolated protein or to identify the site of an amino acid change in a mutant protein. The fingerprinting technique developed by Vernon Ingram allows this to be done relatively simply.

If a protein is digested under defined conditions with various proteases (enzymes that break peptide bonds), small peptides are produced. The number and types of peptides depend on the particular protein and protease used (e.g., for a given protease, cleavage may occur only next to a particular amino acid or class of amino acids). The peptides can be separated by two-dimensional paper chromatography, giving a distribution of spots rarely, if ever, duplicated by a different protein. Such a peptide map is called a fingerprint—an example is shown in Figure 8-10.

The fingerprint of a mutant protein with a single amino acid change differs from that of a wild-type protein (Figure 8-10). If the amino acid change does not affect the site of cleavage by the protease, a single spot in the fingerprint will disappear and one new spot will appear. If it does affect the site of cleavage, several spots will be altered. By eluting the original and mutant spots and determining the amino acid composition of each

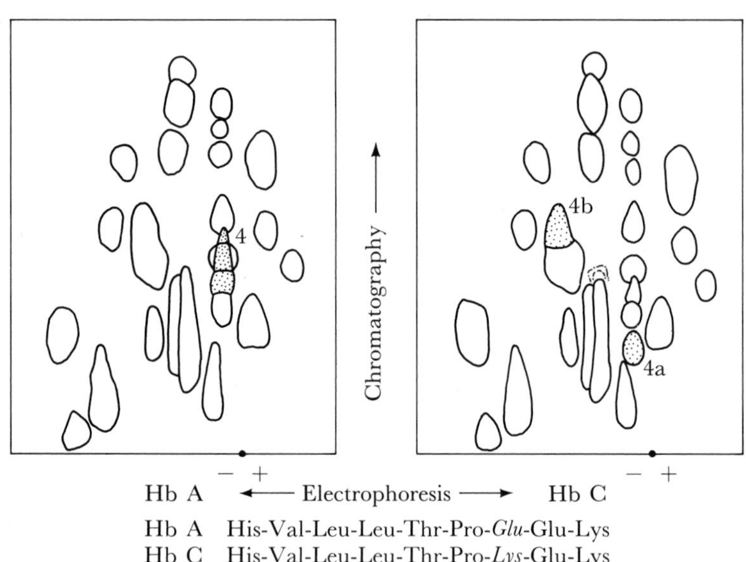

FIGURE 8-10
Tracings of fingerprints of trypsin digests of two different human hemoglobins. The peptides in the digests were first chromatographed and then electrophoresed. Note that the shaded peptide in normal hemoglobin A is missing in the variant, hemoglobin C, in which two new peptides, 4a and 4b, appear. The amino acid sequence of peptides 4, 4a, and 4b are also shown. The conversion of glutamic acid (Glu) into lysine (Lys) in the variant creates two new peptides because trypsin cannot cleave between two molecules of glutamic acid but can cleave the peptide bond between lysine and glutamic acid. [From J. A. Hunt and V. M. Ingram, *Nature* 181(1958):1062–1063.]

(this can be done by complete enzymatic hydrolysis to amino acids followed by paper chromatography), the amino acids exchanged can be determined.

The fingerprinting technique has had several important applications in molecular biology, among which are:

Proof That a Protein (A) Is a Product of Cleavage of a Larger Protein (B). If all spots in a fingerprint of protein A are found in protein B, it is likely that protein A is a part of the amino acid sequence of protein B. Hence, either protein B is made from protein A by the addition of amino acids or protein A is derived from protein B by hydrolysis. Usually, kinetic studies of the synthesis of A and B can distinguish these possibilities.

Identification of the Amino Acids Inserted by Various Suppressor Transfer RNAs. Certain mutations cause the premature termination of amino acid sequences in proteins. This termination is reversed by the presence of suppressor tRNA molecules, which insert an amino acid at the site of premature termination and thereby allow continued synthesis to the natural terminus. A protein results that has a single amino acid replacement. Fingerprints of such proteins allow the identification of the amino acid inserted by each of the known suppressor tRNAs.

Identification of Base Changes Produced by Particular Mutagens. Because the genetic code is well known, it is possible to identify base changes that produce particular mutations. For example, suppose that a mutagen is found to produce frequent changes from phenylalanine to isoleucine. The DNA triplets corresponding to phenylalanine are AAA and AAG and to isoleucine are TAA, TAG, and TAT. Because most mutations change only a single base, the particular mutagen would have to lead to frequent replacement of A with T.

THIN-LAYER CHROMATOGRAPHY

Thin-layer chromatography (TLC) originally developed from a need to separate lipids. Although paper chromatography was faster than column chromatography, papers could be prepared only from cellulose products, which were not of great value for nonpolar materials. Thin-layer chromatography has the advantages of paper but allows the use of any substance that can be finely divided and formed into a uniform layer. This includes inorganic substances—such as silica gel, aluminum oxide, diatomaceous earth, and magnesium silicate—and organic substances—such as cellulose, polyamide, and polyethylene powder.

In thin-layer chromatography, the stationary phase is a layer (0.25–0.5 mm) of sorbent spread uniformly over the surface of a glass or plastic plate. The plates are prepared in the following way (although prepared plates of many different sorbents are now commercially available and in common use).

A slurry of the sorbent is made in a solvent specified for the particular sorbent. For very small chromatograms, a microscope slide is coated with the sorbent by dipping it into the slurry. For larger chromatograms, several layers of narrow tape are put along the two edges of a plate of glass that will be vertical during the chromatography. The number of layers of tape determines the thickness of the final sorbent layer. The slurry is poured onto the glass at an untaped edge and is spread evenly by sliding a glass rod that extends from one tape to the other along the plate. The plate is then dried and the tapes are removed. Henceforth, the plates are treated as in paper chromatography. The sample is then applied with a micropipette and dried. The TLC plate is placed in a chamber containing the solvent and developed by ascending chromatography (Figure 8-11). After the solvent front has almost reached the top, the plate is removed from the chamber and dried. If desired, the dried plate can be rechromatographed at right angles with a second solvent for two-dimensional work. Spots are usually located as in paper chromatography by natural color, by fluorescence, or by spraying various reagents that react with the substances in the spots to produce color. Commonly used sprays are ninhydrin for amino acids; rhodamine B for lipids; antimony chloride for steroids and terpenoids; sulfuric acid plus heating for almost any organic substance (produces charring); potassium permanganate in sulfuric acid for hydrocarbons; anisaldehyde in sulfuric acid for carbohydrates; bromine vapor for olefins; and so forth. Material can be eluted from the chromato-

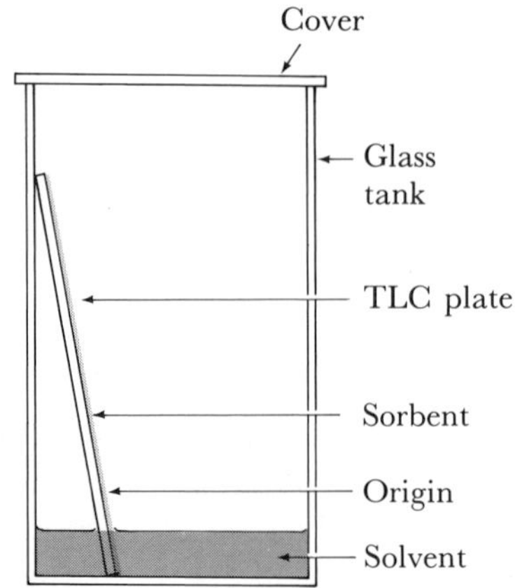

FIGURE 8-11
Typical arrangement for ascending development of a thin-layer chromatographic plate.

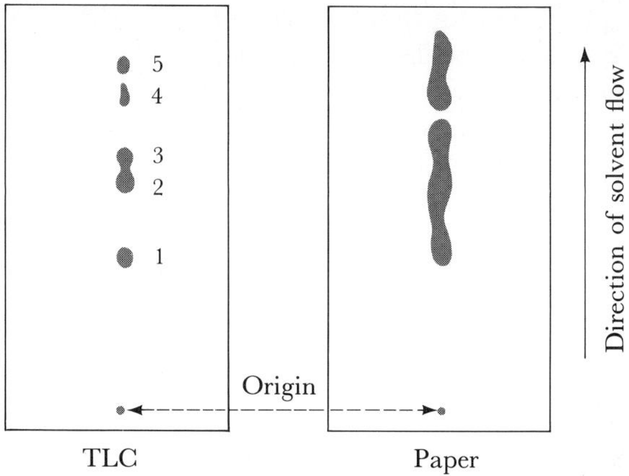

FIGURE 8-12
Comparison of separation by thin-layer and paper chromatography of adenine (1), adenosine (2), hypoxanthine (3), inosine (4), and uridine (5), both on cellulose and developed by water.

gram by scraping off the sorbent and eluting the powder with a suitable solvent. A typical thin-layer chromatogram is shown in Figure 8-12.

Thin-layer chromatography is widely used because, compared with paper or column chromatography, it offers the following advantages: greater resolving power because spots are smaller, greater speed of separation, a wider choice of materials as sorbents, easy detection of spots, and easy isolation of substances from the chromatogram.

Two factors are responsible for the great resolving power of thin-layer chromatography: First, the weight ratio of sorbent to solute is from 10^3 to 10^4:1, whereas in column chromatography the ratio is normally about 50:1. Second, because the particles can be made very fine (<0.1 mm), the surface-to-volume ratio is very high, yielding a large active area for a given amount of sorbent. Particles of such small size cannot be used in columns because the weight of the material causes compacting, clogging, and extremely slow flow rates. This is detrimental because, with very slow flow rate, diffusional spreading increases.

Paper chromatography is limited to cellulose products and those few materials that can be made into paper. Thin layer chromatography is not limited in this way. However, even for chromatography with materials that can be made into paper, thin-layer chromatography is advantageous because it is faster and has higher resolution. The problem with papers is that the fibrous structure and associated capillarity of the fibers tends to increase spot size. The materials used in thin-layer chromatography can be pulverized to eliminate the fibrous structure. With smaller spots, running time can be decreased because it takes less time for the spots to separate. Often, thin-layer chromatography can be completed in a matter of minutes. This is a particular advantage if the solvent required for optimal separation is not known, because a large number of solvent systems can be tested in one day.

The smaller size of the spots also means that the material in a spot is more concentrated; thus, a smaller amount of material is detectable. In fact, for many substances, thin-layer chromatography is from fifty to one hundred times as sensitive as paper chromatography and samples as small as one nanomole can be detected. For amino acids, the detectable amount of material is about one-tenth that required for paper chromatography, which can be of great value if only small samples of purified proteins are available for amino acid or peptide analysis. For nucleotides, about one-hundredth of the amount required for paper is needed.

GAS-LIQUID CHROMATOGRAPHY

In the types of partition chromatography described so far, the sample is carried in a liquid mobile phase past a liquid stationary phase, the liquid being immobilized by adsorption to or absorption by a supporting solid. In gas-liquid chromatography (GLC), the mobile phase is a gas; the stationary phase is again a liquid adsorbed either to the inner surface of a tube or column (*open-tubular,* or *capillary, operation*) or to a solid support (*packed-column operation*), such as diatomaceous earth, Teflon powder, or fine glass beads. The liquid is usuallly applied as a solid dissolved in a volatile solvent such as ether. For example, beads are dipped into a solution of polyethylene glycol in ether. When the ether evaporates, each bead is coated with polyethylene glycol. At the temperatures used for gas-liquid chromatography, the polyethylene glycol melts and remains on the bead as a liquid film. The sample, which may be any compound capable of being volatilized without decomposition, is introduced as a liquid with an inert gas—such as helium, argon, or nitrogen—and then heated. This gaseous mixture passes through the tubing, which is arranged as shown in Figure 8-13. For packed-column operation, the tubing is approximately 0.5 cm in diameter and from 1 to 20 meters long. For the capillary method, the length is from 30 to 100 meters. For very high resolution, capillary systems with 2 kilometers of tubing are used. The vaporized compounds continually redistribute themselves between the gaseous mobile phase and the liquid stationary phase, according to their partition coefficients, and are thereby chromatographed. At the end of the column a suitable detector is used.

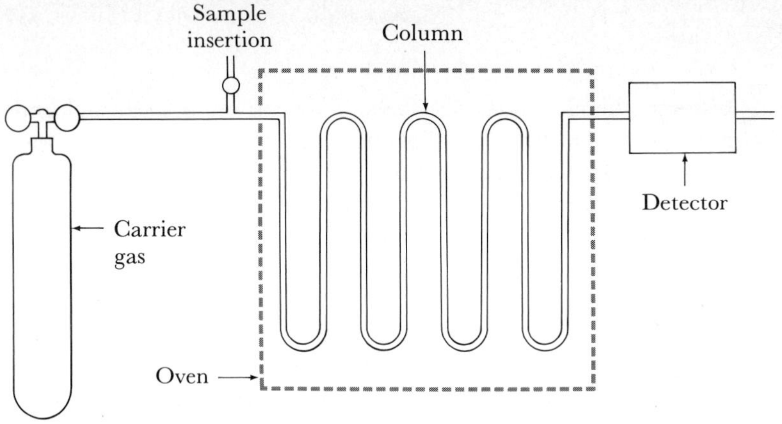

FIGURE 8-13
Apparatus for gas-liquid chromatography.

Detection of Substances in the Effluent

Three types of detectors are commonly used in analytical work: the thermal conductivity cell (sensitivity, approximately 10μg), the argon ionization detector (sensitivity, 10^{-5} μg), and the flame ionization detector (sensitivity, approximately 10^{-5} μg). The action of the thermal conductivity cell is based on the fact that the electrical resistance of a wire is temperature-dependent. If a gas flows at a constant rate past a hot wire, the wire will be cooled to a temperature determined by the flow rate and thermal conductivity of the gas. Hence, at constant flow rate, the temperature, and therefore the resistance, is a characteristic of a particular gas. If the composition of the gas changes, there will be a change in the thermal conductivity of the gas and therefore of the resistance; this is usually plotted on a chart recorder as a function of time of flow of the gas.

With an argon ionization detector, the carrier gas, argon, passes through a detection chamber (at the exit port of the column) similar to a Geiger-Müller counter and is ionized by being bombarded with β particles. As explained in Chapter 5, when the positive argon ion nears the cathode, it pulls off an electron and is neutralized. X rays are produced by this recombination and they ionize more argon atoms. This causes a self-generating, continuous ionization so that a constant current is produced by the Geiger-Müller tube. If a compound is present whose ionization potential is less than argon, it can collide with the argon ions and transfer an electron. The compound is then positively charged. When it approaches the cathode, it picks up an electron and is also neutralized. However, with most organic compounds, the excess energy of recombination does not lead to the production of electromagnetic radiation but breaks chemical bonds. Hence, whenever such a compound is present, the current between the electrodes decreases. This current is measured and plotted as a function of time. A calibration is also necessary, as with the thermal detector. The sensitivity of this type of detector is 0.1 μg.

The flame ionization detector is shown schematically in Figure 8-14. Hydrogen gas is burned with air in the presence of the column effluent. Any carbon in the sample is burned and converted into carbon dioxide. For reasons that are not clear, electrons and negative ions are produced and detected as a current, which is then converted into voltage differences. This detector operates linearly from 0.01 μg to 5 mg and essentially counts carbon atoms. A typical gas chromatogram is shown in Figure 8-15.

Identification of Components from the Detector Output

Each of the detectors indicates the amount of material emerging from the column as a function of time. However, there is no direct indication of the identity of the material producing a particular peak nor of the amount in the peak. There are various means of identifying peaks. For example,

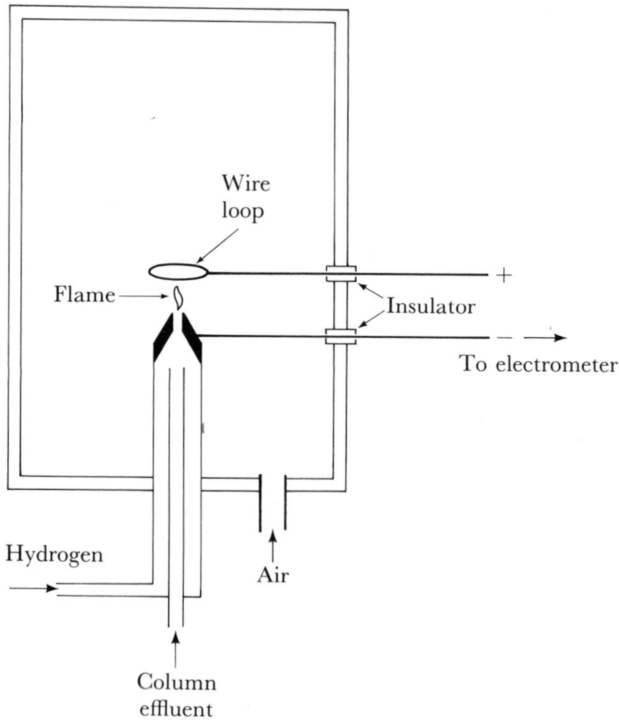

FIGURE 8-14
A flame ionization detector.

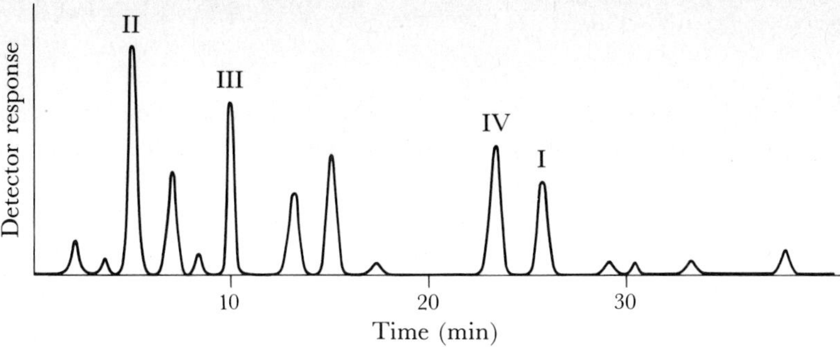

FIGURE 8-15
A gas chromatogram of methyl esters of fatty acids prepared from a rat liver that had been perfused with 25 mM of glucose and 2 mM of arachidonate (I). Pentadecanoate (II) was added before homogenizing and extracting the liver to obtain a measure of the recovery of fatty acids from the original sample; methyl heptadecanoate (III) was added to the mixture before injection into the gas chromatographic column to quantitate the amount injected. A flame ionization detector was used. The peaks represent normal components of rat liver; peak IV appears only if animals are fed diets free of essential fatty acids. [Courtesy of John Lowenstein.]

if the substances present in a mixture are known in advance (and the purpose of the chromatography is to determine the amount of each), peaks may be identified by preparing a duplicate sample containing a small amount of an added known substance to the mixture and rechromatographing this mixture. If the known substance is the same as one of the components in the mixture, the size of that peak will increase (Figure 8-16). If the peaks are part of a homologous series (e.g., ethyl esters of fatty acids),

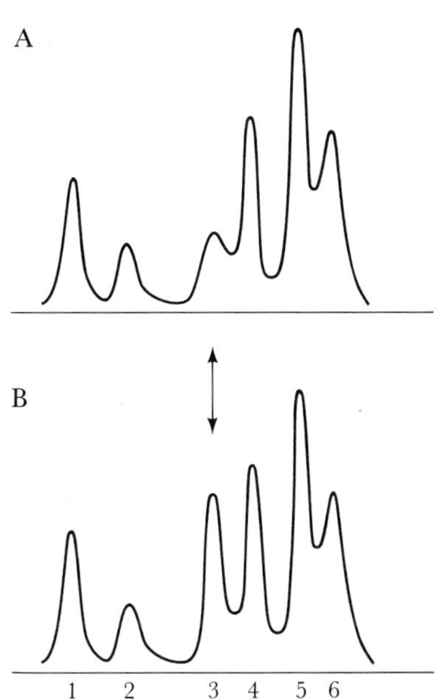

FIGURE 8-16
Identification of a peak in a gas chromatogram by the addition of a known compound. The mixture contained (1) phenol, (2) *o*-cresol, (3) *p*-cresol, (4) *m*-cresol, (5) 2,4-xylenol, and (6) 2,5-xylenol. The sample was run (A) and then *p*-cresol was added and the sample was rerun (B). Note that peak 3 has increased.

the logarithm of the retention time (i.e., the time between injection of the sample and its appearance in the detector) can be plotted as a function of the number of carbon atoms (Figure 8-17) and usually a straight line will result. The addition of one or two of the standards of the homologous series then provides a calibration for the chromatogram. Of course, this method can be used only if it is known that all substances are part of a homologous series.

The identification of unknown substances is difficult and there is no general method. Sometimes the effluent gas is bubbled through test solutions that give particular colors with specific functional groups (e.g., aldehyde, alcohol, etc.). Another method consists of chemical alteration and rechromatographing against various standards or analysis by other physical techniques. This is often done if what the substances are can be surmised. Of course, in identifying substances, it is always necessary to use a nondestructive detector (e.g., thermal conductivity cell).

The quantitative determination of the amount of material in a peak is not a straightforward procedure. For any given compound, the amount of material is proportional to the area of the peak. However, the proportionality constant varies with each substance. Hence, it is first of all essential to be able to identify the material in the peak. Quantitation then requires that a standard curve be prepared; that is, various quantities of the particular substance are chromatographed and a graph is drawn to relate peak area to amount of material.

Advantages of Gas-Liquid Chromatography

The separation in gas-liquid chromatography is excellent. Sensitivity and speed are extraordinary, with 10^{-12} gram being detectable for many substances. Because the rapidity of development of chromatograms depends on the rate of diffusion between the mobile and stationary phases and because the diffusion rate of gases is much greater than that of liquids, the gas chromatogram can be run approximately one thousand times as fast as that produced by liquid column chromatography. Hence, separations are frequently achieved in less than a minute. Furthermore, by using a nondestructive detector and condensing the samples at the collection end, it is also possible to use gas-liquid chromatography preparatively. Large preparative instruments can purify gram quantities of material.

Uses of Gas-Liquid Chromatography

Gas-liquid chromatography can be used with any substance that can be volatilized. This includes thousands of organic compounds. Nonvolatile

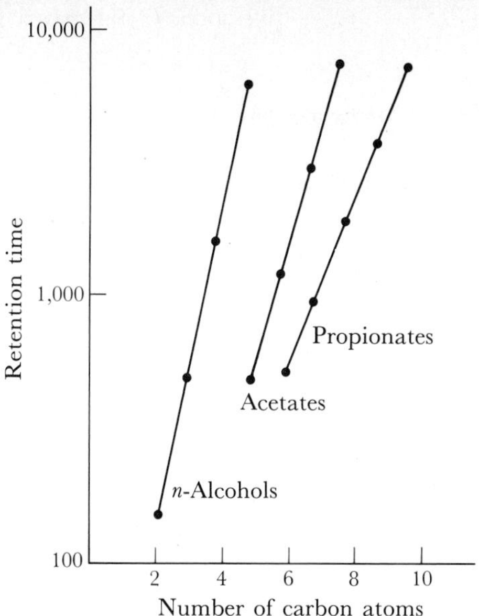

FIGURE 8-17
Curves showing that the logarithm of the retention time varies linearly with the number of carbon atoms for homologous series.

substances can also be examined if converted into volatile ones by oxidation, acylation, alkylation, and so forth.

The principal use in biological samples has been the separation of alcohols, esters, fatty acids, and amines. Hence, it is especially valuable in the study of intermediary metabolism and in working out enzyme reaction mechanisms. It has been extensively used to identify the components of flavorings and wines and for detecting pesticides in biological material.

Gas-liquid chromatography plays an important role in organic analysis also. For instance carbon and hydrogen can be determined to 0.5% and 0.1% accuracy, respectively, by burning the sample in a dry stream of oxygen free of CO_2 and H_2O and determining the amount of CO_2 and H_2O. Functional groups can also be identified—for example, alkoxy groups are detected by iodination to form an alkyl iodide, which is easily identified. The position of a double bond can be determined also by cleaving the bond by oxidation or ozonolysis and chromatographing the products.

GEL CHROMATOGRAPHY

Gel chromatography (or molecular sieve chromatography, as it is sometimes called) is a special type of partition chromatography in which separation is based on molecular size.

Simple Theory of Gel Chromatography

The basis of gel chromatography is quite simple. A column is prepared of tiny particles of an inert substance that contains small pores. If a solution containing molecules of various dimensions is passed through the column (Figure 8-18), molecules larger than the pores move only in the space between the particles and hence are not retarded by the column material. However, molecules smaller than the pores diffuse in and out of particles with a probability that increases with decreasing molecular size; in this way, they are slowed down in their movement down the column. As long as the material of which the particles are made (i.e., the gel) does not adsorb the molecules, the probability of penetration is the principal factor determining the rate of movement through the column. Hence, molecules are eluted from the column in order of decreasing size or, if the shape is relatively constant (e.g., globular or rodlike), decreasing molecular weight.

In detailed analysis of the mechanism of gel chromatography, it is clear that this steric effect, although the principal factor, does not alone explain the chromatographic behavior of all molecules. Another important factor is the charge of the molecule, although this is only manifested at very low ionic strength when highly charged small molecules seem to be excluded

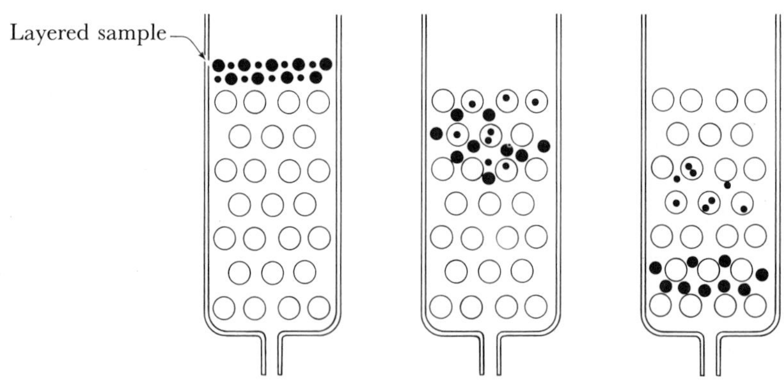

FIGURE 8-18
Separation of two molecules by passage through a column containing particles of a porous gel. The molecules larger than the pores move more rapidly than the smaller ones because the smaller ones move in and out of the pores.

from the pores even though the size is sufficient. This is probably due to electrostatic repulsion between the molecules, thus limiting the number of molecules in a pore at any given time. At very low ionic strength, there are also apparently adsorptive effects with some types of gels.

Materials in Gel Chromatography

A gel is a three-dimensional network whose structure is usually random. The gels used as molecular sieves consist of cross-linked polymers that are generally inert, do not bind or react with the material being analyzed, and are uncharged. The space within the gel is filled with liquid and this liquid occupies most of the gel volume.

The gels currently in use are of three types: dextran, agarose, and polyacrylamide. They are used for aqueous solutions.

Dextran is a polysaccharide composed of glucose residues and produced by the fermentation of sucrose by the microorganism *Leuconostoc mesenteroides.* It is prepared with various degrees of cross-linking to control pore size and is supplied in the form of dry beads of various degrees of fineness that swell when water is added. Swelling is the process by which the pores become filled with the liquid to be used as eluant. It is commercially available under the trade name Sephadex (Pharmacia Fine Chemicals, Inc.).

Agarose (which is obtained from certain seaweeds) is a linear polymer of D-galactose and 3,6-anhydro-1-galactose and forms a gel that is held together without cross-links by hydrogen bonds. It is dissolved in boiling water and forms a gel when cooled. The concentration of the material in the gel determines the size of the pores—which are much larger than those of Sephadex. This makes it useful for the analysis or separation of large globular proteins or long, linear molecules such as DNA. Agarose is useless as a solid gel because the flow rate is too low; so it is supplied as wet beads called Sepharose (Pharmacia Fine Chemicals, Inc.) and Bio-gel A (Bio-Rad Laboratories).

Polyacrylamide gels are prepared by cross-linking acrylamide with *N,N′*-methylene-bis-acrylamide. Again, the pore size is determined by the degree of cross-linking. These gels differ from dextran and agarose gels in that they contain a polar, carboxylamide group on alternate carbon atoms, but their separation properties are much the same as those of the dextrans. Polyacrylamide gels, which are marketed as Bio-gel P (Bio-Rad Laboratories), seem to be as useful as the dextrans, although they have been used less frequently. They do have an advantage over the dextrans in that they are commercially available in a wider range of pore sizes.

Porous glass beads (Bio-Glas, Bio-Rad Laboratories) have also recently been used for aqueous solutions, but to date experience with them has been limited.

The gels heretofore described swell in water and a few organic solvents: glycol, formamide, and dimethylsulfoxide. They fail to swell in pure alcohols, hydrocarbons, and most polar and nonpolar organic solvents. However, substances such as lipids, steroids, and certain vitamins are more easily handled in such solvents. Several gels have been developed for this purpose. For gel chromatography in nonpolar organic solvents, a cross-linked polystyrene gel (Styragel, Dow Chemical Co.; Bio-beads S, Bio-Rad Laboratories) has been used successfully. For polar organic solvents, there is a methylated Sephadex. A hydroxypropyl derivative of cross-linked dextran (Sephadex LH) can be used for both polar and nonpolar solvents.

In general, the pore size determines the range of molecular weights in which fractionation occurs. Table 8-2 gives representative values for some of the materials. There is usually no difficulty in selecting the appropriate material. However, the gel beads come in various sizes: coarse, medium, fine, and superfine. The rule is the coarser the bead, the more rapid the flow rate and the poorer the resolution. Hence, superfine is used if maximum resolution is required—for example, for analytical work. Fine is recommended for most preparative work in which columns are fairly large and flow rate is of some concern. The coarser grades are for very large preparations in which resolution is less important than time.

TABLE 8-2
Materials commonly used in gel chromatography.

Material and trade name	Fractionation range* (molecular weight)
Dextran	
Sephadex G-10	700
Sephadex G-25	1,000–5,000
Sephadex G-75	3,000–70,000
Sephadex G-200	5,000–800,000
Polyacrylamide	
Bio-gel P-2	200–2,000
Bio-gel P-6	1,000–6,000
Bio-gel P-150	15,000–150,000
Bio-gel P-300	60,000–400,000
Agarose	
Sepharose 2B	2×10^6–25×10^6
Sepharose 4B	3×10^5–3×10^6
Bio-gel A-0.5 M	30,000–500,000
Bio-gel A-15 M	30,000–15×10^6
Bio-gel A-150 M	5×10^6–150×10^6

*Globular proteins or polynucleotides.

Advantages of Gel Chromatography

For the separation of molecules whose molecular weights differ, gel chromatography is unsurpassed for the following reasons:

1. Because the chromatographic behavior of almost all substances in gels is independent of temperature, pH, ionic strength, and buffer composition, separations can be carried out under virtually all conditions. For very labile materials (e.g., enzymes), this means that the conditions for maximum stability can be maintained.
2. Because there is virtually no adsorption, very labile substances are not affected by the chromatography. For example, some enzymes are inactivated or altered by binding to adsorbent surfaces or ionic-exchange resins.
3. There is less zone spreading than with other chromatographic techniques (for reasons that are unclear).
4. The elution volume is related in a simple manner to molecular weight.

ESTIMATION OF MOLECULAR WEIGHT

It has been observed for a variety of gel types that a plot of a parameter, K, versus log M (molecular weight) yields a straight line except for very small and very large molecules (see Figure 8-19). To date, the only deviations are small aromatics, highly charged molecules in very low ($\ll 0.01$) ionic strength, and a few rodlike molecules such as collagen and fibrinogen. The parameter K is defined as $(V_e - V_0)/V_s$, in which V_e is the volume of solvent required to elute the molecule of interest, V_0 is the void volume or the volume required to elute a molecule that never entered the stationary phase, and V_s is the volume of the stationary phase. In gel chromatography, the volume of the gel (which is mostly the volume of its pores) is

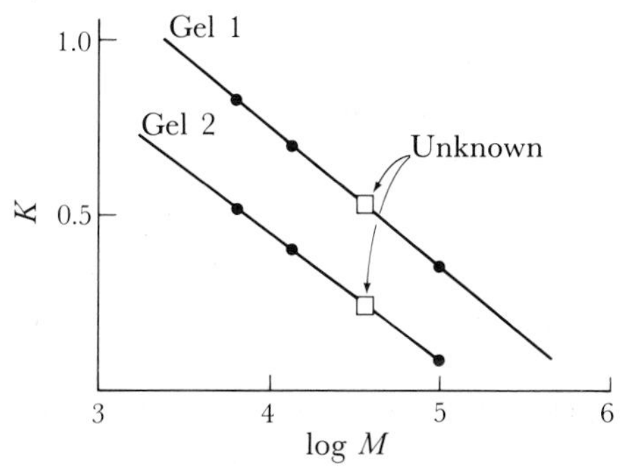

FIGURE 8-19
Typical data obtained for determining the approximate molecular weight of proteins by gel chromatography, using three known proteins (solid circles) and two different types of gels (1 and 2). The open squares represent the sample (K is defined in the text). Using two different gel types is valuable because usually, if a protein violates the K–log-M proportionality (because of its shape), the measured values of M will differ for the two gels.

V_s, and V_0 is the volume for a substance completely excluded by the gel; V_0 is usually measured by passing the excluded high-molecular-weight material, dextran blue, through the column, and V_s is measured as $V_t - V_0$ in which V_t is the total volume of the column.

For globular proteins and for many carbohydrates, this technique is very reliable when various types of Sephadex and Bio-gel P are used. Most important, the estimation of M can be made from unfractionated cell extracts as long as the protein can be assayed; almost every other method requires highly purified samples. Hence, to determine M simply requires the inclusion of several molecules of known M to define the straight line and M of the sample can be calculated by interpolation. The precision of these determinations is about 10%. A somewhat more precise variation of this technique will be described in the section on thin-layer gel chromatography.

The determination of a molecular-weight distribution is frequently important in the characterization of natural and synthetic polymers. Again, with physical methods such as centrifugation, relatively pure material is needed. However, with gel chromatography, it is possible to determine the distribution by merely measuring the amount of the substance as a function of elution volume. This is done by preparing a calibration curve and using a relatively simple mathematical correction for zone spreading.

Applications of Gel Chromatography

Gel chromatography is basically a separation procedure and has its most widespread use in purifying enzymes and other proteins and in fractionating nucleic acids. The dextran gels are especially valuable for unstable proteins as explained earlier. In preparative work, both dextrans and polyacrylamides have been used in volumes ranging from a few milliliters to several liters. For industrial purposes, 1,000-liter columns have even been employed. Various species of RNA and viruses have been successfully fractionated and purified, using agarose gels.

In large-scale purifications of macromolecules in which various types of fractionation procedures are needed, it is often necessary to remove salts, change buffers, concentrate, and remove substances such as phenol and the detergents used in the isolation and purification of nucleic acids. This is often time-consuming (e.g., by precipitation or dialysis) and for this reason can result in the loss of unstable samples. However, gel chromatography provides a very rapid way to accomplish this. For example, salts and small molecules can be rapidly removed because they are retarded by all of the gels. Buffer exchange can be accomplished merely by passing a solution of macromolecules through a column previously equilibrated with the desired buffer. Because the macromolecules move more easily through the gel than the components of the original buffer, they will be eluted in the buffer with which the gel was equilibrated. Concentrating macromolecules is easily accomplished by the addition of dry gel particles, using a type whose pore size is smaller than the molecules being examined. As the beads swell, they imbibe water but not the macromolecules, thus concentrating the macromolecules. With most procedures for concentrating macromolecules, there is a problem in that the salts are also concentrated and excessive salt concentrations can alter (sometimes dissociate or denature) some macromolecules. This is no problem if gels are used because they imbibe the salts as well as the water.

Specialized examples of gel chromatography in preparative work (e.g., purifications) follow:

1. In the chemical synthesis of various reagents, it is usually necessary to separate the product from the reactants. For example, in preparing fluorescent antibodies (see Chapters 2 and 10) by reacting antibody with fluorescein isothiocyanate, the conjugated protein must be separated from unreacted dye. This can be done with Sephadex, using gels that pass large proteins in the void volume. Of course, the unreacted protein is not separated from the conjugated protein but, for the fluorescent antibody technique, this is usually unnecessary.
2. In the assay of enzymes or the determination of cofactor requirements, the enzyme preparation sometimes contains inhibitors of small molecular size or the cofactors themselves. Also, in physical studies of some molecules (e.g. in fluorescence spectroscopy), interfering substances may be present. Such small molecules are easily removed with the dextran or polyacrylamide gels.
3. Similarly, there are frequently contaminants of large molecular size in mixtures being assayed for small molecules. The small-pore dextrans are useful in such cases. Also, proteins must often be freed of nucleic acids; this can sometimes be done by using an agarose gel, which impedes all proteins and passes nucleic acids in the void volume.
4. For most physical analyses of nucleic acids, the sample must be free of protein. This is also easily done. An interesting example from my own laboratory is the preparation of DNA from crude cell extracts for electron microscopic analysis, using the Kleinschmidt cytochrome *c* method (see Chapter 3). Small (20–50μl) samples of cell extracts containing DNA are passed through tiny Sepharose columns that have been equilibrated with the solution needed for electron microscopy. The DNA comes through the column in the void volume in the appropriate buffer, whereas all other material is retained. Following the addition of cytochrome *c*, the sample is ready to prepare for microscopy.
5. The most common use of gel chromatography is in the purification of proteins (see Appendix, step 7). To purify a protein from a cell extract,

it is usually necessary to use a sequence of separation procedures based on such parameters as solubility in certain solutions, charge, molecular weight, and so forth. The step in which size separation takes place almost invariably uses gel chromatography.

Gel chromatography is also a valuable analytical tool. The determination of molecular weight mentioned previously is an important example of this. Other examples are:

1. In studying RNA metabolism, various fractions of RNA are usually distinguished by zone centrifugation or even better by polyacrylamide-gel electrophoresis. Gel chromatography with agarose is also of great use. An example is shown in Figure 8-20.
2. Plasma protein fractions must often be determined quantitatively in the diagnosis of certain human diseases. This can be done directly with dextran gels and has been developed as a reliable test for macro- and hyperglobulinemia.
3. The tritium-exchange (see Chapter 18) method for examining protein or DNA structure requires that the macromolecule be very rapidly separated from ^3H_2O. This can be done in about 10 seconds using charged gels because the ^3H_2O is strongly retarded in all gels. If the smallest pore size is used, the macromolecule comes through rapidly in the void volume or shortly thereafter. An example is shown in Figure 8-21.

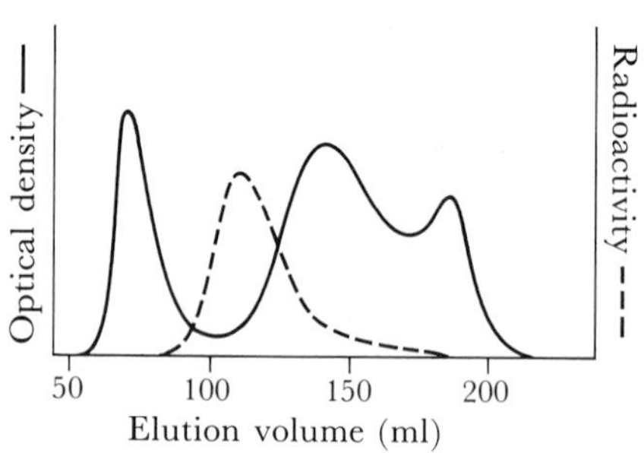

FIGURE 8-20
Separation of various species of nucleic acids by chromatography on Sepharose. KB cells were infected with ^{32}P-labeled poliovirus. The cells were lysed shortly after infection, and nucleic acids were isolated and chromatographed. The total cellular nucleic acid was detected by measuring the optical density at 260 mμ. The polio RNA was revealed by its radioactivity. The peaks in order from left to right are KB-cell DNA, polio RNA, KB ribosomal RNA, and KB transfer RNA. [Courtesy of Pharmacia Fine Chemicals, Inc.]

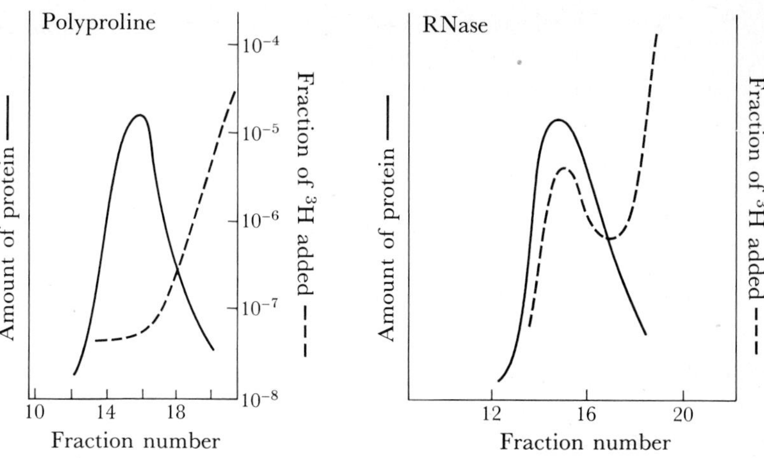

FIGURE 8-21
One of the earliest ^{3}H-exchange experiments. Polyproline and RNase were incubated in ^3H_2O for several hours to allow maximum exchange. The samples were then diluted in H_2O and at various times applied to a Sephadex G-25 column to separate the protein and ^3H_2O. In the graph at the left, there is no ^{3}H associated with polyproline, from which it can be concluded that either it never became tritiated or exchange is very rapid. In the graph at the right, it is clear that the RNase has ^{3}H bound to it. [Redrawn with permission from W. Englander, *Biochemistry* 2(1965):798–807. Copyright by the American Chemical Society.]

4. Gel chromatography can be used to study binding between proteins and small molecules either by separating the product and the reactants or by passing protein through a column equilibrated with the small molecule. A simple calculation allows the determination of binding constants. The great value of gel chromatography in studies of chemical equilibrium is that a gel column can be operated over a wide range of concentrations, pH, ionic strength, and temperature because the pore size of the gel is unaffected by these factors.

Thin-layer Gel Chromatography

As indicated earlier, thin-layer chromatography has the advantages of quick separation, high sensitivity, simple equipment, and ready elution and is rapidly replacing paper chromatography as an analytical method. Thin-layer gel chromatography (TLG) offers these same advantages plus the opportunities afforded by the gels.

Thin-layer gel chromatography is similar to thin-layer chromatography in that a thin layer of material is spread on a glass plate, the sample is spotted, and a mobile phase traverses the layer. However, there are important differences:

1. In thin-layer chromatography, the support is dried before applying the sample. However, because gels cannot be dried and easily rehydrated, in thin-layer gel chromatography, the sample is spotted onto a wet layer equilibrated with the appropriate solvent.
2. Thin-layer chromatography is done with an ascending mobile phase (see Figure 8-7), whereas in thin-layer gel chromatography the descending method is used. The plate is put in a tight chamber and connected to a reservoir at both ends by filter-paper bridges (Figure 8-22). Liquid flows through the layer at a rate determined by the angle (usually 20°). As in thin-layer chromatography, the run terminates before the material of interest enters the lower reservoir. However, there is a continuous flow of liquid so that, unlike thin-layer chromatography, there is no solvent front. Therefore, there is no measurement of R_f values and positions must be measured with respect to added standards.

Applications of Thin-layer Gel Chromatography

Whereas thin-layer chromatography is primarily used for amino acids, sugars, oligosaccharides, lipids, steroids, and other small molecules, thin-layer gel chromatography is applied to proteins, peptides, nucleic acids, nucleotides, and other large hydrophilic substances.

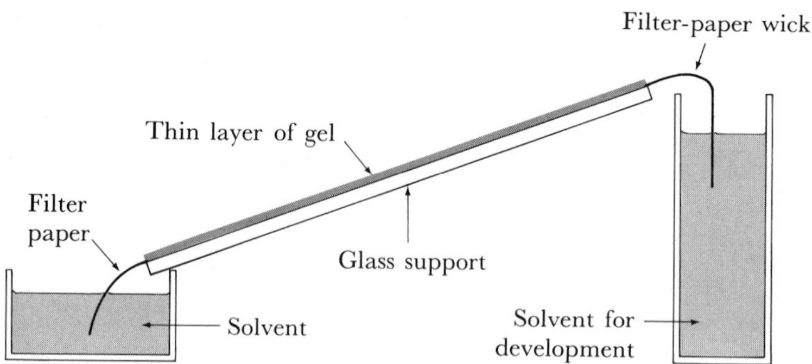

FIGURE 8-22
Experimental arrangement for thin-layer gel chromatography. The gel and the filter paper are usually enclosed in various ways to prevent evaporation.

The most important use of thin-layer gel chromatography is in determining molecular weights of proteins and peptides. Plots similar to those in Figure 8-19 can be obtained in which the y-axis is the *relative distance* migrated. This gives fairly accurate values, especially if several determinations are made.

Thin-layer gel chromatography is being used in clinical diagnosis to detect pathological proteins from blood serum, spinal fluid, and urine. It is, of course, of great value in assaying the purity of proteins, enzymes, and so forth. It is also being developed as a modification of immunodiffusion, immunoelectrophoresis, and isoelectric focusing.

The literature on thin-layer gel chromatography is limited, but the use of this method should certainly increase rapidly.

ION-EXCHANGE CHROMATOGRAPHY

An *ion exchanger* is a solid that has chemically bound charged groups to which ions are electrostatically bound; it can exchange these ions for ions in aqueous solution. Ion exchangers can be used in column chromatography to separate molecules according to charge.

The principle of ion-exchange chromatography is that charged molecules *adsorb* to ion exchangers reversibly so that molecules can be bound or eluted by changing the ionic environment. Separation on ion exchangers is usually accomplished in two stages: first, the substances to be separated are bound to the exchanger, using conditions that give stable and tight binding; then the column is eluted with buffers of different pH or different ionic strength, and the components of the buffer compete with the bound material for the binding sites.

Properties of Ion Exchangers

An ion exchanger is usually a three-dimensional network or matrix that contains covalently linked charged groups. If a group is negatively charged, it will exchange positive ions and is a *cation exchanger.* A typical group used in cation exchangers is the sulfonic group, SO_3^-. If an H^+ is bound to the group, the exchanger is said to be in the acid form; it can, for example, exchange one H^+ for one Na^+ or two H^+ for one Ca^{2+}. The sulfonic acid group is called a *strongly* acidic *cation* exchanger. Other commonly used groups are phenolic hydroxyl and carboxyl, both *weak*ly acidic cation exchangers. If the charged group is positive—for example, a quaternary amino group—it is a strongly basic *anion exchanger.* The most common weakly basic anion exchangers are aromatic or aliphatic amino groups.

The matrix can be made of various materials. Commonly used materials are dextran, cellulose, and copolymers of styrene and vinylbenzene in which the divinylbenzene both cross-links the polystyrene strands and contains the charged groups. Table 8-3 gives the composition of many common ion exchangers.

The *total capacity* of an ion exchanger measures its ability to take up exchangeable ions and is usually expressed as milliequivalents of exchangeable groups per milligram of dry weight. This number is supplied by the manufacturer and is important because, if the capacity is exceeded, ions will pass through the column without binding.

The *available capacity* is the capacity under particular experimental conditions (i.e., pH, ionic strength). For example, the extent to which an ion exchanger is charged depends on the pH (the effect of pH is smaller with strong ion exchangers). Another factor is ionic strength because small ions

TABLE 8-3
Properties of various ion exchangers.

Matrix	Exchanger*	Functional group	Trade name
Dextran	SC	Sulfopropyl	SP-Sephadex
	WC	Carboxymethyl	CM-Sephadex
	SA	Diethyl-(2-hydroxypropyl) aminoethyl	QAE-Sephadex
	WA	Diethylaminoethyl	DEAE-Sephadex
Cellulose	C	Carboxymethyl	CM-cellulose
	C	Phospho	P-cel
	A	Diethylaminoethyl	DEAE-cellulose
	A	Polyethyleneimine	PEI-cellulose
	A	Benzoylated-naphthoylated, diethylaminoethyl	DEAE(BND)-cellulose
	A	*p*-Aminobenzyl	PAB-cellulose
Styrene-divinylbenzene	SC	Sulfonic acid	AG 50
Acrylic	WC	Carboxylic	Bio-Rex 70
Phenolic	SC	Sulfonic acid	Bio-Rex 40
Epoxyamine	WA	Tertiary amino	AG-3

*C, cationic; A, anionic; S, strong; W, weak.

near the charged groups compete with the sample molecule for these groups. This competition is quite effective if the sample is a macromolecule because the higher diffusion coefficient of the small ion means a greater number of encounters. Clearly, as buffer concentration increases, competition becomes keener.

The *porosity* of the matrix is an important feature because the charged groups are both inside and outside the matrix. Large molecules may be unable to penetrate the pores; so the capacity will decrease with increasing molecular dimensions. The porosity of the polystyrene-based resins is determined by the amount of cross-linking by the divinylbenzene (porosity decreases with increasing amounts of divinylbenzene). With the Dowex series, the percentage of divinylbenzene is indicated by a number after an X—hence, Dowex-50-X8 is 8% divinylbenzene.

Ion exchangers come in a variety of particle sizes, called *mesh size.* Finer mesh means an increased surface-to-volume ratio and therefore increased capacity and decreased time for exchange to occur for a given volume of the exchanger. On the other hand, fine mesh means a slow flow rate, which can increase diffusional spreading.

Such a collection of exchangers having such different properties—charge, capacity, porosity, mesh—makes the selection of the appropriate one for accomplishing a particular separation difficult. How to decide on the type of column material and the conditions for binding and elution is described in the following sections.

Choice of Ion Exchanger

The first choice to be made is whether the exchanger is to be anionic or cationic. If the materials to be bound to the column have a single charge (i.e., either plus or minus), the choice is clear. However, many substances (e.g., proteins), carry both negative and positive charges and the net charge depends on the pH. In such cases, the primary factor is the stability of the substance at various pH values. Most proteins have a pH range of stability (i.e., in which they do not denature) in which they are either positively or negatively charged. Hence, if a protein is stable at pH values above the isoelectric point, an anion exchanger should be used; if stable at values below the isoelectric point, a cation exchanger is required. The choice between strong and weak exchangers is also based on the effect of pH on charge and stability. For example, if a weakly acidic substance that requires very low or very high pH for ionization is to be chromatographed, a strong ion exchanger is called for because it functions at extremes of pH. However, if the substance is labile, weak ion exchangers are preferable. Weak ion exchangers are also excellent for the separation of molecules with a high charge from those with a small charge, because the weakly charged ions usually fail to bind. Weak exchangers also show greater resolution of substances if charge differences are very small.

The Sephadex and Bio-gel exchangers offer a particular advantage for macromolecules that are unstable in low ionic strength. Because the cross-links in these materials maintain the insolubility of the matrix even if the matrix is highly polar, the density of ionizable groups can be made several times greater than is possible with cellulose ion exchangers. The increased charge density means increased affinity so that adsorption can be done at higher ionic strengths. On the other hand, these exchangers retain some of their molecular sieving properties so that sometimes molecular weight differences annul the distribution caused by the charge differences.

Choice of Porosity and Mesh Size

Small molecules are best separated on matrices with small pore size (high degree of cross-linking) because the available capacity is large, whereas macromolecules need large pore size. However, except for Sephadex, most ion exchangers do not afford the opportunity for matching the porosity with the molecular weight.

The cellulose ion exchangers have proved to be the best for purifying large molecules such as proteins and polynucleotides. This is because the matrix is fibrous, and hence all functional groups are on the surface and available to even the largest molecules.

Selecting a mesh size is always difficult. Small mesh size improves resolution but decreases flow rate, which increases zone spreading and decreases resolution. Hence, the appropriate mesh size is usually determined empirically.

Choice of pH, Buffer, and Ionic Conditions

Because buffers themselves consist of ions, they can also exchange, and the pH equilibrium can be affected. To avoid these problems, the *rule of buffers* is adopted: use *cationic buffers with anion exchangers* and *anionic buffers with cation exchangers.* Because ionic strength is a factor in binding, a buffer should be chosen that has a high buffering capacity so that its ionic strength need not be too high. Furthermore, for best resolution, it has been generally found that the ionic conditions used to apply the sample to the column (the so-called *starting conditions*) should be near those used for eluting the column.

Techniques of Ion-exchange Chromatography

The basic principle of ion-exchange chromatography is that the affinity of a substance for the exchanger depends on both the electrical properties of the material and the relative affinity of other charged substances in the solvent. Hence, bound material can be eluted by changing the pH, thus altering the charge of the material, or by adding competing materials, of which salts are but one example. Because different substances have different electrical properties, the conditions for release vary with each bound molecular species. In general, to get good separation, the methods of choice are either continuous ionic strength gradient elution or stepwise elution. (A gradient of pH only is not used because it is difficult to set up a pH gradient without simultaneously increasing ionic strength.) For an anion exchanger, either pH and ionic strength are gradually increased or ionic strength alone is increased. For a cation exchanger, both pH and ionic strength are increased. The actual choice of the elution procedure is usually a result of trial and error and of considerations of stability. For example, for unstable materials, it is best to maintain fairly constant pH. Figure 8-23 shows a typical ion-exchange chromatogram, using gradient elution.

For resolution of very similar materials, elution can be carried out without varying pH or ionic strength. For example, suppose that the ionic conditions are selected so that the binding of a molecule with many bind-

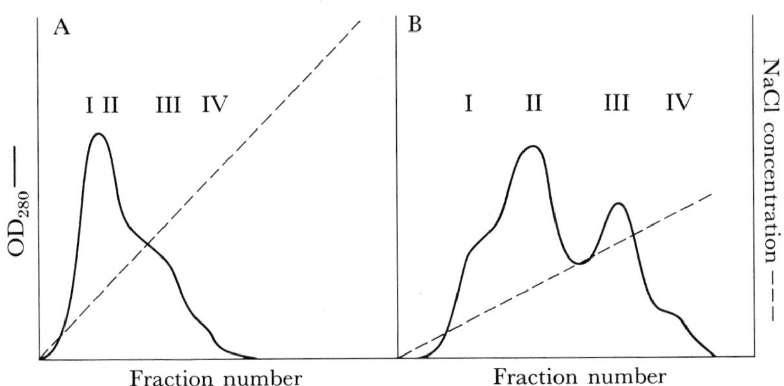

FIGURE 8-23
Separation of four proteins by ion-exchange chromatography on DEAE-cellulose. Note the strong effect of the steepness of the NaCl gradient. In part A, the steeper gradient allows only partial separation of two fractions. With the shallower gradient of part B, proteins II and III clearly separate and the existence of I and IV becomes obvious.

ing sites is so weak that there is a finite probability that simultaneous dissociation of all bonds will occur in a given time. Hence, the molecule is in equilibrium with the exchanger and will move down the column at a finite rate. The migration rate will then depend precisely on this probability so that molecules having slightly different affinities for the exchanger will migrate at different rates and will therefore separate. (This is, of course, formally equivalent to adsorption chromatography.) A simple variation of this method is called the *starting-condition procedure* in which the sample is adsorbed and eluted in the same solution. In general, resolution is improved if the flow rate is very slow; however, with decreasing flow rate, diffusional spreading becomes more of a problem and resolution can be reduced. The proper flow rate is usually determined empirically. Figure 8-24 shows a typical chromatogram developed by the starting-condition procedure.

In deciding on the elution conditions, it is important to consider how the eluting fluid will affect the assay for the material. For example, if spectral analysis is to be used, the buffer should not absorb in the required wavelength range. If the sample is to be assayed by its radioactivity using Geiger counting, volatile buffers help to prevent encasing the radioactivity in crystals resulting from drying down the buffer. If the radioactivity is to be detected by scintillation counting, it is important that the eluting liquid does not contain quenchers.

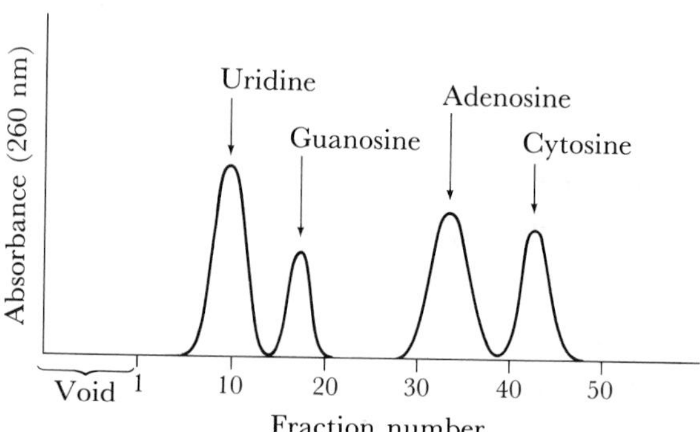

FIGURE 8-24
Separation of RNA nucleosides by chromatography on Dowex 50W4X. The sample, which is an RNA hydrolysate treated with alkaline phosphatase to remove terminal phosphates, was applied and eluted with 0.4 M NH_4 formate, pH 4. The absorbance can be used to calculate the amount of each substance because the relation between absorbance at 260 mμ and concentration is known.

Applications of Ion-exchange Chromatography

In principle, any substance that is charged can be chromatographed on an ion exchanger. Resin exchangers are most useful for small organic molecules and can even be used to separate metallic ions (e.g., Ca^{2+} from Mg^{2+}). Proteins and polysaccharides are best used with the cellulose, dextran, and polyacrylamide exchangers. The dextran and polyacrylamide exchangers have also been widely used for the separation of nucleotides, amino acids, and other biologically important small molecules.

SPECIAL TECHNIQUES IN THE CHROMATOGRAPHY OF NUCLEIC ACIDS AND OF PROTEINS THAT BIND NUCLEIC ACIDS

Complex molecules such as proteins and nucleic acids have such a variety of functional groups, binding sites, and stabilizing forces that their behavior on standard chromatographic materials is difficult to predict. Instead, ingenious chromatographic procedures based on specific molecular interactions have been developed.

DNA-Cellulose Chromatography

DNA-cellulose binds many proteins that bind to DNA and can therefore be used to separate these proteins from nonbinding proteins or other substances. The preparation of DNA-cellulose consists of mixing cellulose powder with a solution of either single- or double-stranded DNA and then drying the mixture in a vacuum. The powder is then washed to remove unbound DNA and redried for storage. The mechanism of binding the DNA to the cellulose is unknown. With prolonged use, columns made of this material tend to lose DNA; to avoid this, the dried DNA-cellulose powder is sometimes resuspended in ethanol and irradiated strongly with ultraviolet light. This chemically couples the DNA to the cellulose in an unknown way.

DNA-cellulose chromatography is used primarily to purify DNA-binding proteins. A mixture of proteins in a buffer having low ionic strength (in which DNA-binding proteins adsorb to the DNA) is passed through the column. The column is then washed to remove unbound proteins and eluted with a gradient of increasing ionic strength. The proteins are thereby removed, with those that bind more weakly eluting first.

This method was first used by Bruce Alberts to detect DNA-binding proteins in *E. coli* infected with phage T4. An example of this work is shown in Figure 8-25.

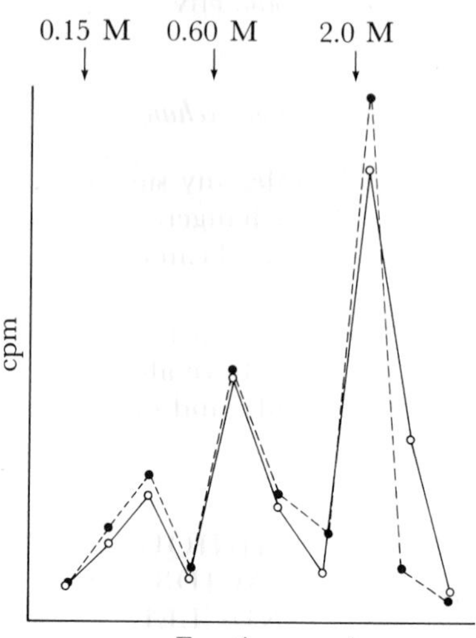

FIGURE 8-25
An example of DNA-cellulose chromatography. *E. coli* cells were infected with phage T4 in medium containing ^{14}C-leucine; another culture was infected with a T4 mutant in gene 32 in medium containing ^{3}H-leucine. Proteins were isolated, mixed, and loaded onto a column of single-stranded DNA-cellulose. The column was eluted stepwise with 0.15 M, 0.60 M, and 2.0 M NaCl. In the third step there was a deficiency of ^{3}H-leucine, indicating that this fraction contains ^{14}C-labelled gene-32 protein. This was the first use of DNA-cellulose chromatography to purify a DNA-binding protein. Open circles— ^{14}C; solid circles—^{3}H. [Redrawn from B. M. Alberts, *Fed. Proc.* 29 (1970): 1154–1175.]

The technique has now become a standard step in the purification of polymerases, nucleases, repressors, and so forth. Because both single- and double-stranded DNA can be used, proteins can also be separated according to the relative binding to each.

Methylated Albumin-Kieselgur (MAK) Columns

Methylated serum albumin adsorbs to kieselgur (diatomaceous earth) and binds tightly. Joseph Mandell and Alfred Hershey showed that columns of this material bind single- and double-stranded DNA and RNA and can fractionate them by elution with a gradient of increasing ionic strength. Using various conditions, this method has been used to separate native DNAs according to molecular weights, base composition, and degree of glucosylation; to separate single- from double-stranded DNA; to separate double-stranded DNA with single-strand ends from those without such termini; to separate transfer RNA from ribosomal RNA; and to separate different transfer RNAs from one another. An example of the use of this column material is shown in Figure 8-26.

Histone-Kieselgur Columns

Histones also adsorb to kieselgur. Because histones form complexes with double- but not single-stranded DNA or RNA, double-stranded DNA can

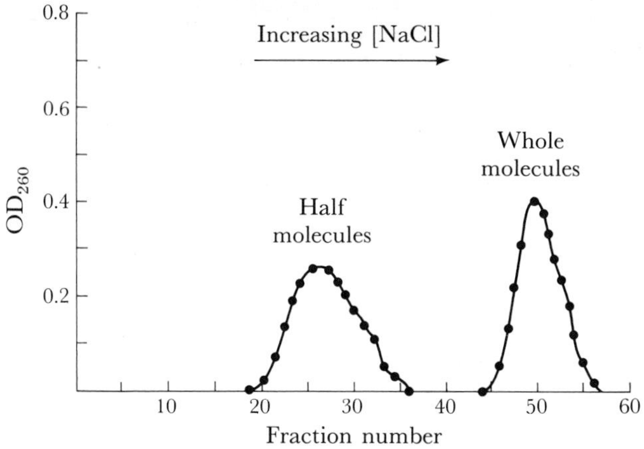

FIGURE 8-26
Separation of equal amounts of whole and half T2 DNA molecules by MAK chromatography. The half molecules were prepared by hydrodynamic shearing (see Chapter 18). Elution is by an increasing NaCl concentration gradient. Note that the larger molecules are bound more tightly and need a higher NaCl concentration for elution. Because all "halves" are not the same size, the zone for half molecules is broader than for whole ones.

be purified by this method. If eluted with increasing ionic strength, DNA separates according to nucleotide base composition, high guanine-cytosine DNA eluting first.

Hydroxyapatite Chromatography

Hydroxyapatite (HA) is a crystalline form of $Ca_{10}(PO_4)_6(OH)_2$ prepared from $CaHPO_4 \cdot 2H_2O$ crystals. Columns of hydroxyapatite bind substances that interact with calcium, including DNA, RNA, nucleohistone, poly- and oligonucleotides, and phosphoproteins. Binding is apparently by means of the phosphate groups in the binding material because binding requires a low phosphate concentration and elution is accomplished merely by increasing the concentration of various phosphate buffers. A particular advantage with nucleic acids is that the dependence of the elution behavior on molecular weight is very small. The following mixtures have been successfully fractionated on hydroxyapatite: double- from single-stranded DNA; glucosylated from nonglucosylated DNA; different nucleohistones from one another; ribosomal from viral RNA; and double- from single-stranded RNA. An example of the use of hydroxyapatite is shown in Figure 8-27.

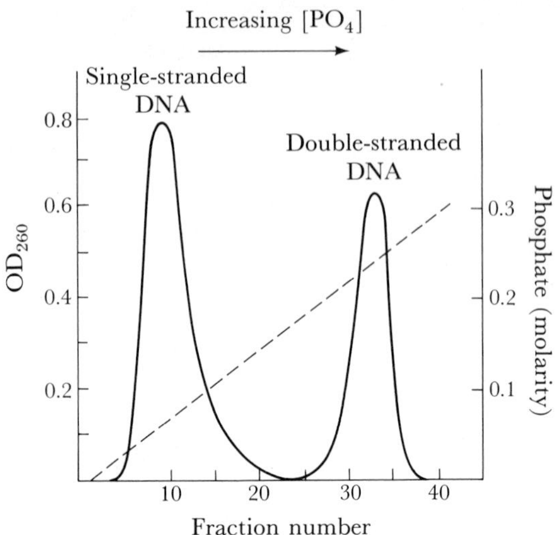

FIGURE 8-27
Separation of single and double-stranded DNA by chromatography on hydroxyapatite. Elution is by a gradient of increasing phosphate concentration (dashed line).

AFFINITY CHROMATOGRAPHY

Affinity chromatography is a type of chromatography that makes use of a specific affinity between a substance to be isolated and a molecule that it can specifically bind (a *ligand*). The column material is synthesized by covalently coupling a binding molecule (which may be a macromolecule or a small molecule) to an insoluble matrix. The column material is then specifically able to adsorb from the solution the substance to be isolated. Elution is accomplished by changing the conditions to those in which binding does not occur.

Several requirements must be met for success in affinity chromatography: (1) the matrix should be a substance that does not itself adsorb molecules to any significant extent; (2) the ligand must be coupled without altering its binding properties; (3) a ligand should be chosen whose binding is relatively tight because, although weak binding will enhance retardation, it may not be adequate for separation to result; (4) it should be possible to elute without destroying the sample. The most useful matrix material is agarose (Sepharose, Pharmacia Fine Chemicals, Inc.) because it exhibits minimal adsorption, maintains good flow properties after coupling, and tolerates the extremes of pH and ionic strength as well as

7.0 M guanidinium chloride and urea, which are often needed for successful elution. That it is possible to purchase agarose to which are covalently coupled either reagents for coupling proteins, membranes, and steroids, or concanavalin A, a ready-to-use adsorbent for polysaccharides and glycoproteins containing α-D-mannosyl and α-D-glucosyl residues (e.g., cell membranes and whole cells), is an additional benefit.

The major use of affinity chromatography to date has been the purification of proteins, membranes, and polysaccharides. Examples of its use follow.

Purification of Enzymes. Either the substrate, a tight-binding inhibitor, or a cofactor can be coupled to the matrix. If a mixture of proteins or even a crude cell extract is passed through the column, only materials that bind remain on the column. In some cases, enzymes can be substantially purified directly from very complex mixtures. However, because there often are many enzymes that can bind to the substrate or cofactor, only partial purification usually results.

Purification of Antibodies. This has been accomplished mainly with cyanogen bromide-Sepharose to which has been coupled various antigens such as proteins, viruses, or bovine serum albumin coupled with haptens; it is the method of choice for antibody purification.

Purification of Transport Proteins. This is usually done with Sepharose to which the substance that is being transported is coupled (e.g., thyroxin-binding globulins, estradiol-binding proteins, and hormone and drug receptors).

Purification of Membranes and Particles Containing Known Substances. Membranes to which a hormone binds can be purified using Sepharose coupled with that hormone; influenza virus, which contains neuraminidase on its surface, has also been purified using Sepharose to which inhibitors of neuraminidase are coupled.

Purification of Glycoproteins. This is efficiently done with concanavalin A-Sepharose.

Separation of Specific Animal Cells. This has been done using coupled agglutinins, such as concanavalin A, wheat germ agglutinin, or phytohemagglutinin. For example, some virus-induced tumor cells can be separated from normal cells because the tumor cells bind more tightly to concanavalin A-Sepharose.

Appendix

Purification of an Enzyme: An Example of the Use of Many Chromatographic Procedures

Justice can hardly be done to the subject of enzyme purification in a few paragraphs; nonetheless, it is instructive to examine a typical purification scheme to see how different chromatographic procedures are made use of. The example is the purification of *E. coli* DNA polymerase I. The method is that of Paul Englund, modified from the original procedure from Arthur Kornberg's laboratory.

STEP 1. BREAKAGE OF BACTERIA

The first step is almost always to break open the bacteria. This can be done by sonication (see Chapter 18) or by alternate freezing and thawing to weaken the cell wall, followed by stirring in a food blender with fine glass beads. After most of the cells are broken, the glass beads, unbroken cells and cell walls are removed by centrifugation.

STEP 2. PARTIAL REMOVAL OF THE NUCLEIC ACIDS

Nucleic acids are precipitated by the addition of streptomycin sulfate. This step reduces the viscosity of the material, thus allowing substantial concentration, and eliminates one of the major cell constituents. It is done at this point because the nucleic acids would bind to some of the chromatographic columns used later.

STEP 3. FINAL REMOVAL OF NUCLEIC ACIDS AND PRELIMINARY REMOVAL OF SOME PROTEINS: AUTOLYSIS

Magnesium chloride is added to the mixture to activate nucleases. After prolonged incubation, almost all nucleic acids are degraded to nucleotides and small oligonucleotides, which remain soluble when the proteins are precipitated in step 4.

STEP 4. AMMONIUM SULFATE FRACTIONATION

Proteins differ in their solubility in concentrated salt solutions and hence can be separated from one another by precipitation at high ionic strength. Many salts are possible, but ammonium sulfate is preferred because it does not significantly affect pH; it is inexpensive, very soluble, and does not destabilize proteins (see Chapter 18). In fact, many proteins are stabilized by the NH_4^+ ion.

Hence, the extract is treated with $(NH_4)_2SO_4$ at a concentration that precipitates proteins other than the polymerase. This precipitate is removed by centrifugation and more $(NH_4)_2SO_4$ is added to the supernatant to reach a concentration that precipitates the polymerase. This precipitate is collected and redissolved in an appropriate buffer in preparation for step 5.

STEP 5. DEAE-CELLULOSE ION-EXCHANGE CHROMATOGRAPHY

A DEAE-cellulose column is loaded with the material from step 5, using conditions of pH and ionic strength such that the column adsorbs all nucleic acids without binding the polymerase. This is a necessary step because small amounts of nucleic acids that remain after step 4 would interfere with adsorption to phosphocellulose in the next step. The void volume is then collected for step 6.

STEP 6. PHOSPHOCELLULOSE CHROMATOGRAPHY

Chromatography on a phosphocellulose column separates most of the proteins from polymerase. The few remaining proteins are eliminated in the next step.

STEP 7. GEL CHROMATOGRAPHY ON SEPHADEX

Fractions from phosphocellulose chromatography containing polymerase are chromatographed on Sephadex and the separation is based on molecular weight. This is a useful step because the polymerase has a very high molecular weight compared with that of the remaining proteins. All proteins in the effluent from step 6 separate on the Sephadex. The fractions containing the polymerase are therefore pure. They are then treated with $(NH_4)_2SO_4$ to precipitate the polymerase in order to concentrate the purified enzyme. The precipitate is then dissolved in a small volume of a suitable buffer.

Selected References

Bobbit, J. M., A. E. Schwarting, and R. J. Gutter. 1968. *Introduction to Chromatography.* Reinhold.

Chovin, P., 1962. "Gas Chromatography," in *Comprehensive Biochemistry,* vol. 4, edited by M. Florkin and E. H. Stotz. Elsevier.

Fisher, L. 1969. "An Introduction to Gel Chromatography," *Laboratory Techniques in Biochemistry and Molecular Biology,* edited by T. S. Work and E. Work. American Elsevier.

Helferich, F. 1962. *Ion Exchange.* McGraw-Hill.

Ingram, V. M. 1959. "Abnormal Human Haemoglobins. 1. The Comparison of Normal Human and Sickle-Cell Haemoglobins by 'Fingerprinting.'" *Biochim. Biophys. Acta* 28:539–545. The development of fingerprinting.

Mandell, J. D., and A. D. Hershey. 1960. "A Fractionating Column for Analysis of Nucleic Acids." *Analyt. Biochem.* 1:66–77. MAK chromatography was first described here.

Martin, A. J. P., and R. L. M. Synge. 1941. "A New Form of Chromatography Employing Two Liquid Phases." *Biochem. J.* 35:1358–1368. Partition chromatography was first described in this paper.

Peterson, E. A. 1970. "Cellulosic Ion Exchangers," in *Laboratory Techniques in Biochemistry and Molecular Biology,* vol. 2, edited by T. S. Work and E. Work, pp. 228–400. North-Holland.

Purnell, H. 1962. *Gas Chromatography.* Wiley.

Randerath, K. 1966. *Thin-layer Chromatography.* Academic Press.

The best information by far can be found in brochures available free of charge from Pharmacia Fine Chemicals, Inc.; for example:

Sephadex, Gel Filtration in Theory and Practice
Sephadex Ion Exchangers
Thin-layer Gel Filtration
CNBr-activated Sepharose 4B

and from Bio-Rad Laboratories; for example, Materials, Equipment and Systems for Chromatography, Electrophoresis and Membrane Techniques.

The latest information can be found in such journals as Journal of Chromatography and Journal of Gas Chromatography.

Problems

8-1. Suppose that an enzyme is dissociated into four identical subunits and that you want to test for the enzymatic activity of the individual subunits, but you must be sure that there are no tetramers remaining in the sample. What chromatographic system would you choose to free the monomers from the tetramer?

8-2. In attempting to purify a protein, it is found that the protein is tightly bound to DNA. When DNase is added to destroy DNA, it is found that the DNase binds to the complex without digesting the DNA. Therefore, DNase treatment cannot be used to eliminate the DNA. To dissociate the DNA-protein complex, 2 M NaCl is needed; at this concentration neither DNA nor protein are retarded by ion exchangers. How might the protein be purified?

8-3. In preparing hybrid DNA by renaturation of denatured DNA, there is usually some unrenatured DNA. How would you eliminate this?

8-4. In comparing a molecule of DNA that is circular with one that is linear, both of molecular weight 6×10^6, which form would elute first from an agarose column? What about native and denatured ribosomal RNA? (Ribosomal RNA has considerable secondary structure.)

8-5. A procedure has been devised for purifying a particular enzyme using gel chromatography. In an effort to increase the amount of material to be handled, you use a sample whose concentration of protein is ten times that normally used with this procedure. The enzyme activity now elutes principally in the void volume. Explain what happened.

In general, to increase the amount of material to be handled, should the diameter of the column be increased? Should the length be increased? or should both?

8-6. If chromatographed on paper, the substances A, B, C, and D have the following R_f values in the solvent systems indicated.

Substance	Butanol-H_2O	Isopropanol-HCl	Ethanol	Acetic acid
A	0.23	0.31	0.42	0.09
B	0.24	0.20	0.51	0.62
C	0.38	0.58	0.40	0.64
D	0.41	0.56	0.53	0.10

What would be the best way to separate A, B, C, and D?

8-7. Twenty milligrams of a protein mixture is applied to a DEAE-cellulose column. Thirty percent of the protein is known to be enzyme X. After eluting the column, the total amount of protein in all samples is 18.9 mg. There is no detectable enzyme X activity. Give a possible explanation for the lack of enzymatic activity.

8-8. A DNA molecule consisting of a 6×10^6 dalton double-strand piece and a single-strand extension equal in length to the double-strand segment is adsorbed to hydroxyapatite. Given the fact that there is very little dependence of elution behavior on molecular weight in this range, if the column were eluted with an increasing gradient of $[PO_4]^{3-}$, where would you expect this molecule to elute—in the single-strand or double-strand region or elsewhere?

8-9. An enzyme is known to require a high concentration of Mg^{2+} for activity. If the Mg^{2+} is removed, the protein is irreversibly denatured. Suppose that, in establishing a purification scheme, you try both ion-exchange and gel chromatography and that, in both cases, the enzyme loses activity. Explain why this might happen. In view of your explanation, what modifications could you make to improve the situation?

8-10. Certain drugs, such as morphine, bind to specific receptors in neural tissue. Design a procedure for the partial purification of the receptors.

8-11. Two proteins have the same molecular weights. At pH 5.5 both have considerable secondary structure and are at least 75% α helix. At pH 8.5, one of them loses all structure and is a random coil; when returned to pH 5.5, its structure is restored. Devise a procedure for separating these proteins.

CHAPTER 9

Electrophoresis

Most biological polymers are electrically charged and will therefore move in an electric field. The transport of particles through a solvent by an electric field is called electrophoresis.

A useful way to characterize macromolecules is by their rate of movement in an electric field. This property can be used to determine protein molecular weights, to distinguish molecules by virtue of their net charge or their shape, to detect amino acid changes from charged to uncharged residues or vice versa, and to separate different molecular species quantitatively. How this is done is described in this chapter.

THEORY OF ELECTROPHORESIS

The detailed theory of electrophoresis is highly complicated and at present incomplete; a simple description of the electrophoretic principle is sufficient to understand how the technique is used for most purposes.

In many ways, electrophoresis is like sedimentation (Chapter 11): a force is applied and countered by viscous drag. If a particle with charge q suspended in an insulating medium is in an electric field, E, the particle will move at a constant velocity, v, determined by the balance between the electrical force, Eq, and the viscous drag, fv, in which f is the frictional coefficient, that is,

$$Eq = fv \tag{1}$$

The mobility, u, is defined as the velocity per unit field, or

$$u = v/E = q/f \tag{2}$$

Because mobility depends on the frictional coefficient, which in turn is a function of some of the physical parameters of the molecules, in principle the value of u should give information about the size and shape of the molecule. However, the supporting medium is normally not an insulator but an electrolyte consisting of charged ions, and this introduces great complexity. The principal complication is that a charged particle suspended in an electrolyte attracts the ions and hence is surrounded by an ion atmosphere that shields the particle from the applied field. However, this ion atmosphere is partly disrupted both by the field and by the motion of the particle through the medium. To date, the theory of electrophoresis has failed to account adequately for these complications, as well as several others, so that electrophoresis has not turned out to be very useful in supplying *detailed* information about macromolecular structure. It is, however, enormously useful as both an analytical and a preparative tool* because of its ability to separate different molecules. Although the theory is correct in providing the working rule of electrophoresis—that is, that mobility increases with q, decreases with f, and is zero for uncharged molecules—in fact, the optimal conditions for separation are almost always determined empirically.

TYPES OF ELECTROPHORESIS

There are three types of electrophoresis: *moving boundary, zone,* and *continuous.*

In moving boundary electrophoresis, macromolecules are present throughout a solution and the position of the molecules (actually, the boundary separating the solution from the solvent) as a function of time is determined by schlieren optics (Chapter 11). This method, which is in many ways equivalent to boundary sedimentation, is an analytical method that has been used primarily for the determination of mobilities and isoelectric points of proteins. However, because there is not usually much

---

*Reference will often be made to preparative and analytical procedures. By a preparative procedure is meant a technique designed to provide relatively large quantities of pure (or nearly pure) material to be used in later experiments. An analytical procedure is used in determining purity, evaluating the number of components in a mixture and possibly the proportions of each, detecting changes in charge, and so forth—that is, obtaining information to find answers to particular questions.

to be gained from the quantitative determination of mobility, moving boundary electrophoresis is rarely used.

In zone electrophoresis, a solution is applied as a spot or a band, and particles migrate through a solvent that is almost always supported by an inert and homogeneous medium such as paper or in a gel. It is used to analyze mixtures, to determine purity, to assay for changes in mobility and/or conformation, and for purification. The quantitative determination of mobility is never done because of the virtual impossibility of accounting theoretically for various effects of the support material.

In continuous electrophoresis, the sample is also applied as a zone except that it is continuously added.

In the following sections, various electrophoretic techniques are described. Moving boundary electrophoresis is so rarely done at present that it is not included. Information about this method can be found in the Selected References near the end of the chapter.

Zone Electrophoresis

In zone electrophoresis, a spot or thin layer of solution is placed in contact with a semisolid or gelatinous medium, an electric field is applied, and molecules migrate on or through the supporting material. Because the molecules are applied in a zone, small samples are used and the separation of solutes can be complete. The function of the supporting medium is primarily to prevent mechanical disturbances and convection that arise both from temperature changes and from the high density of concentrated macromolecular solutions. However, the supporting medium sometimes adsorbs various molecular species or acts as a molecular sieve and therefore has a chromatographic effect (see Chapter 8) that can either aid or detract from the separation. Methods for using various supporting substances are described in the following sections.

PAPER ELECTROPHORESIS

Figure 9-1 shows a typical low-voltage (20 V/cm) paper-electrophoresis arrangement. The paper strip is first dipped in the buffer solution and then placed in the tank as indicated. The sample is then applied either as a spot or a line. The paper is enclosed in a tank to prevent evaporation and voltage is applied. When separation is completed, the paper is removed and dried. If the sample contains sufficient material, it can be located either by its color or fluorescence or by staining with various dyes. If quantitative measurement is required, the dye can be eluted and determined spectrophotometrically. Because dye uptake is rarely quantitative, the accuracy is about 20%. If the sample is radioactive, spots can be located

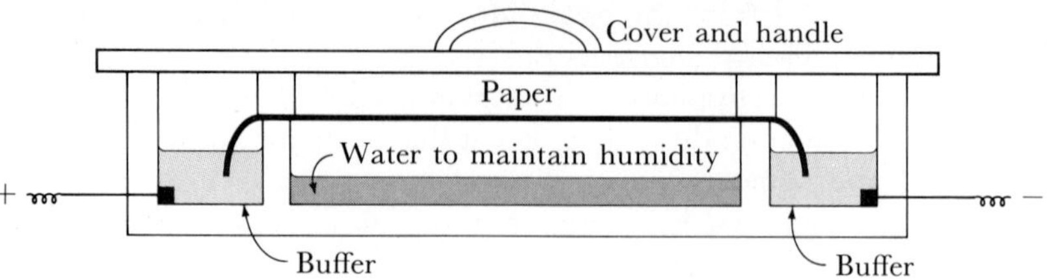

FIGURE 9-1
Experimental arrangement for low-voltage paper electrophoresis. A sample is spotted onto paper as in paper chromatography (see Figure 8-6) but usually near the middle. The paper is saturated with a buffer and both ends are placed in a buffer reservoir to make electrical contact with the power supply. The system is enclosed to prevent drying of the paper. At the end of a predetermined time, the paper is removed and dried.

by cutting up the paper and counting the radioactivity (see Chapter 5) or by autoradiography of the whole sheet using x-ray film (see Chapter 6). A particularly useful technique is to combine stains with radioactivity to determine which substances are radioactive. Low-voltage paper electrophoresis has been most useful for the analysis of protein mixtures that are poorly separated by chromatography. However, this technique has been superceded by gel electrophoresis (page 216).

Low-voltage electrophoresis is inefficient for small molecules (e.g., amino acids and nucleotides because their small charge results in low mobility (i.e., slow separation) and their small size allows considerable diffusional spreading. In *high-voltage electrophoresis,* the speed of separation is increased by the use of a potential gradient of 200 V/cm. The high voltage produces a high current, which heats the paper so that cooling is necessary. This can be accomplished by immersing the paper in a large volume of an immiscible and nonconducting liquid or by pressing it against cool paper or glass. The immersed-strip technique (see Figure 9-2 for the arrangement) is not often used because the required coolants (e.g., toluene, carbon tetrachloride, or various oils) are toxic and sometimes inflammable. The *enclosed-strip method,* shown in Figure 9-2, is by far more common. Note that the paper is not immersed in the buffer. After electrophoresis is complete, the paper is dried and spots are identified as they are in the low-voltage method.

High-voltage paper electrophoresis is of great value for resolving amino acids and peptides. Because complete resolution of mixtures is not always possible in a single high-voltage electrophoresis operation, it is frequently coupled with chromatography in the *two-dimensional separation* technique. The sample is first electrophoresed and then chromatographed at right angles or vice versa, resulting in the kind of separation shown in Figure 9-3.

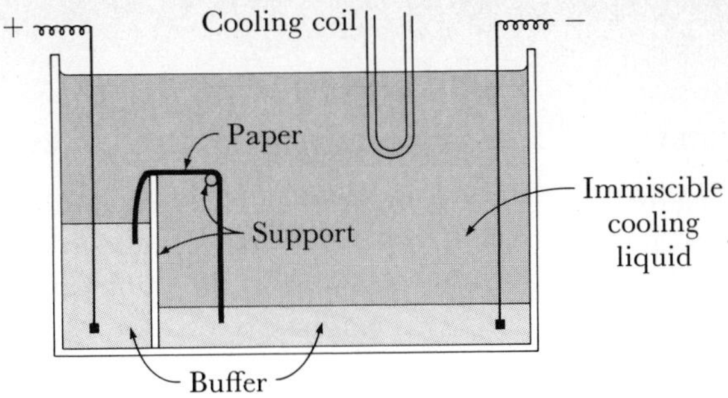

A. Immersed-strip method

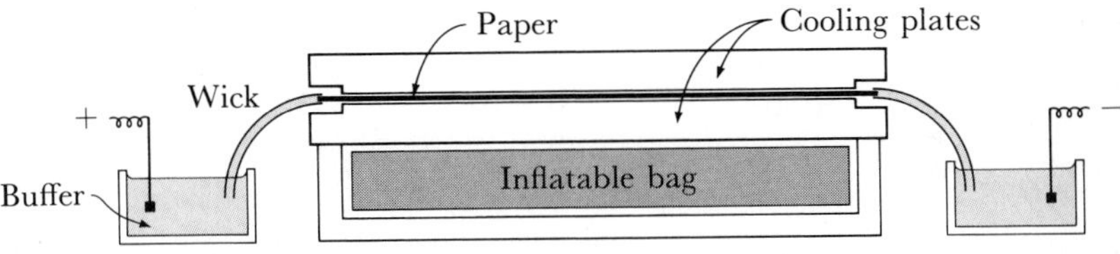

B. Enclosed-strip method

FIGURE 9-2
Two experimental arrangements for high-voltage electrophoresis. Because the higher current produces sufficient heat to destroy the sample and evaporate the solvent, the paper is either immersed in a cooling liquid or placed between two cooling plates. The inflatable bag supports the cooling plates, preventing bowing.

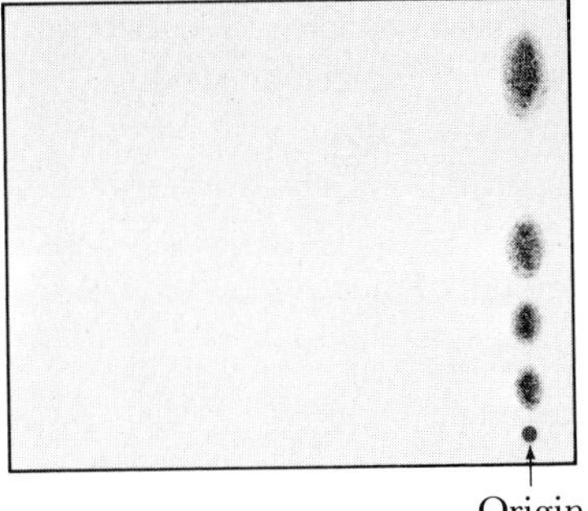

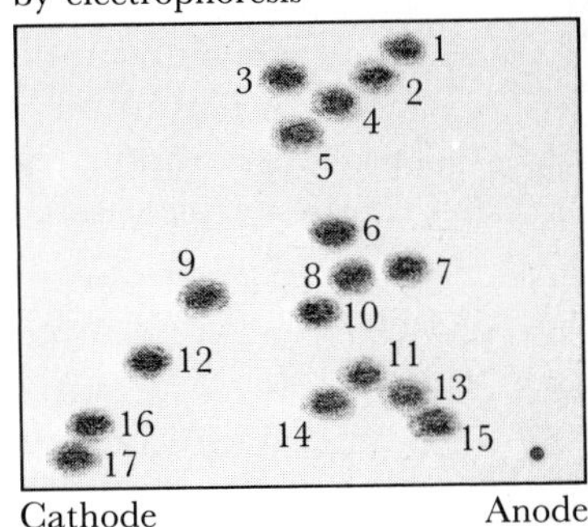

FIGURE 9-3
Two-dimensional chromatography combined with electrophoresis to separate amino acids: chromatography in lutidine and water in a ratio of 2 to 1; electrophoresis at pH 2.25. The numbered spots are: (1) tryptophan; (2) tyrosine; (3) leucine; (4) phenylalanine; (5) methionine; (6) threonine; (7) hydroxyproline; (8) proline; (9) alanine, (10) serine; (11) glutamine; (12) glycine; (13) glutamic acid; (14) asparagine; (15) aspartic acid; (16) arginine; (17) lysine.

CELLULOSE ACETATE STRIP ELECTROPHORESIS

Many biological macromolecules adsorb to cellulose (i.e., paper) by means of the cellulose hydroxyl groups. Adsorption impedes movement (see Chapter 8) and therefore causes the tailing of spots or bands, which reduces resolution. This can be avoided by using a cellulose acetate membrane instead of paper because most of the hydroxyls are converted into acetate groups, which are generally nonadsorbing. The elimination of this skewing improves resolution so that separations are more rapid at low voltage. The fact that spots are smaller also means that the material in the spot is more concentrated and easier to detect. The low adsorption of cellulose acetate also reduces background staining, thus improving the sensitivity of detection.

Two other advantages in using cellulose acetate are that the material is transparent, which aids in the spectrophotometric determination of material, and is easily dissolved in various solvents, thus allowing simple elution of material.

For simplicity of operation coupled with high resolution, cellulose acetate cannot be surpassed. However, for maximum resolution, gel electrophoresis is the method of choice.

GEL ELECTROPHORESIS

The use of gels such as starch, polyacrylamide, agarose, and agarose-acrylamide as supporting media provides enhanced resolution, particularly for proteins and nucleic acids. The reasons are not clear but certainly include a combination of reduced diffusion by the gel network and the separating action of gel chromatography ("molecular sieving").

The earliest work in gel electrophoresis was done with *starch gels.* Figure 9-4 shows a typical arrangement for starch-gel electrophoresis. The gel consists of a paste of potato starch whose grains have been burst by heating in the buffer. When the gel is mounted horizontally, as shown, the

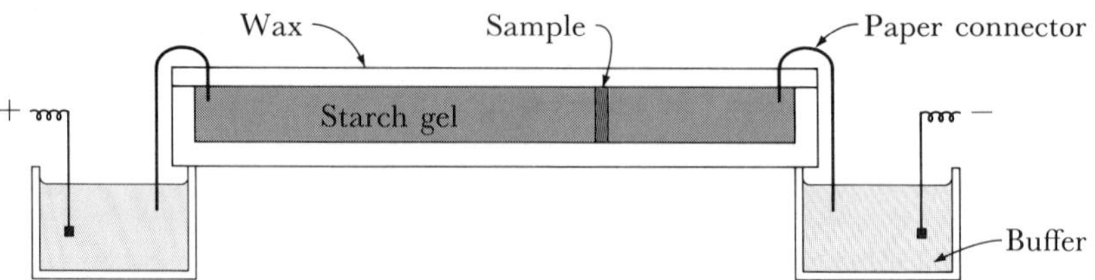

FIGURE 9-4
Arrangement for starch-gel electrophoresis. The gel is formed in place by allowing the starch grains to swell in a buffer. The gel is cut and the sample is placed in the cut. The system is covered with wax to prevent drying.

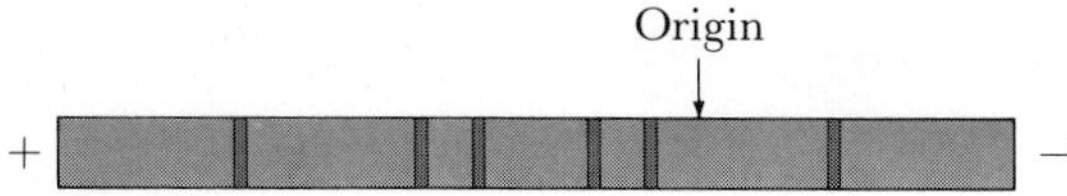

FIGURE 9-5
A starch-gel electrophoregram of separated proteins. When electrophoresis is finished, the gel is stained to make the proteins visible. Note that proteins have migrated both in positive and in negative directions, depending on their charge.

sample is applied to a slot cut with a razor blade either as a single solution or as a slurry with starch grains. The slot is sealed with wax or grease and the voltage is applied. After electrophoresis, the semirigid gel is removed and frequently sliced into two or three layers, each to be differentially stained. The various components appear as a series of bands in the gel (Figure 9-5). Starch gels are rarely used now, having been superceded by polyacrylamide gels.

Polyacrylamide gels have replaced starch gels because the amount of molecular sieving can be controlled by the concentration of the gel and the adsorption of proteins is negligible. Polyacrylamide is currently the most effective support medium in use for proteins and small RNA molecules. (For nucleic acids that are too large for the polyacrylamide pores, agarose and agarose-acrylamide are superior). Polyacrylamide gel is prepared by cross-linking acrylamide with *N,N′*-methylene-bis-acrylamide in the container in which the electrophoresis is to be carried out.

There are basically two experimental arrangements used for polyacrylamide-gel electrophoresis: column gels and slab gels. These arrangements are shown in Figures 9-6 and 9-7. Slab gels are rapidly replacing column gels because a large number of samples can be run simultaneously.

Polyacrylamide-gel electrophoresis is probably the most versatile and useful electrophoretic system for the analysis and separation of macromolecules. Two variations that have been developed for special purposes are described next.

SDS-Gel Electrophoresis. Klaus Weber and Mary Osborn showed that the molecular weights of most proteins could be determined by measuring the mobility in polyacrylamide gels containing the detergent sodium dodecyl sulfate (SDS). At neutral pH, in 1% SDS and 0.1 M mercaptoethanol, most multichain proteins bind SDS and dissociate, disulfide linkages are broken by the mercaptoethanol, secondary structure is lost, and the complexes consisting of protein subunits and SDS assume a random-coil configuration. Proteins treated in this way behave as though they have uniform shape and an identical charge-to-mass ratio. This is

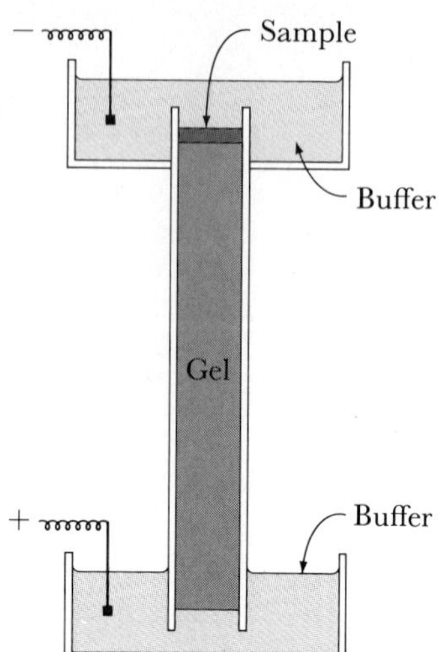

FIGURE 9-6
Experimental arrangement for gel electrophoresis in a column. The gel is polymerized in the column and then placed in the buffer reservoirs. The sample is suspended in a concentrated sucrose solution and layered on the gel with a fine pipette. After electrophoresis, the gel is pushed out of the column and then either stained to make the bands visible or, if a radioactive sample is used, sliced and the radioactivity in each slice measured.

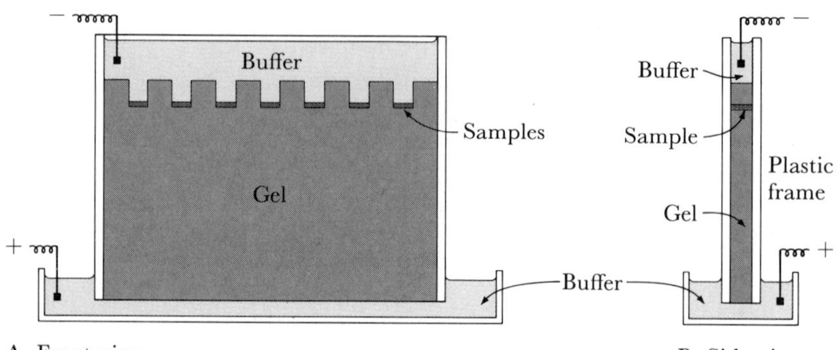

FIGURE 9-7
Apparatus for slab-gel electrophoresis capable of running seven samples simultaneously. The gel is polymerized in place. An appropriately shaped mold is on top of the gel during the polymerization to make notches for the samples. After electrophoresis, the slab is stained either by adding the stain to the upper buffer or by removing the plastic frame and immersing the gel in the stain. Excess stain is removed either by washing or by electrophoresis.

because the amount of SDS bound per unit weight of protein is constant—1.4 g of SDS/g of protein; the charge is in fact determined by the SDS rather than the intrinsic charge of the amino acids. This has the useful result that the effective mobility is related only to molecular weight because of the molecular-sieving property of the gel. If a series of proteins of known molecular weight are electrophoresed in a column gel, they will separate into a series of bands (Figure 9-8) and a plot of the distance migrated versus log M gives a straight line (Figure 9-9). Hence, if a protein

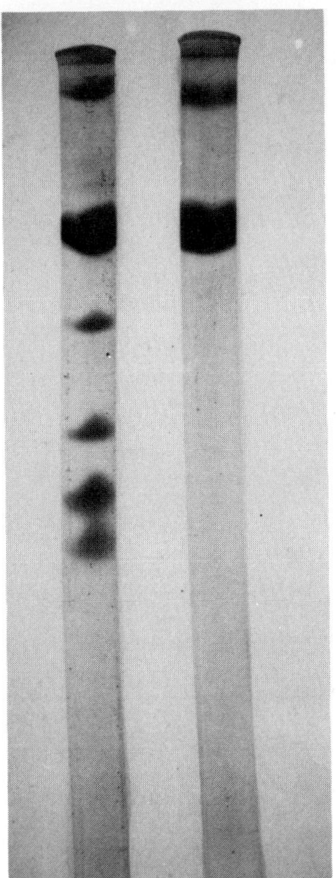

FIGURE 9-8
Result of the electrophoresis of proteins in SDS. Disrupted polyoma virus (left) and polyoma virus capsid protein (right) were treated with SDS and mercaptoethanol and separately electrophoresed in polyacrylamide containing SDS. The gels were stained to make the proteins visible. Bands identified as histones (the lower two bands) were found in the whole virus and, because the virus contains RNA, this has been used to prove that cellular histones are bound to viral RNA [Courtesy of William Murakami.]

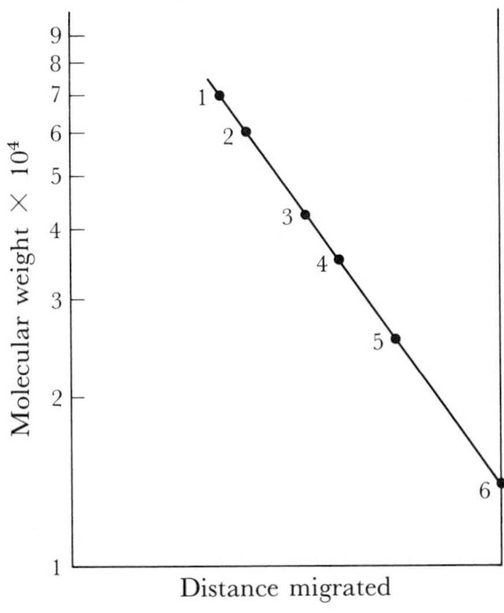

FIGURE 9-9
Typical semilogarithmic plot of M versus distance migrated for determining molecular weight by SDS–polyacrylamide-gel electrophoresis. Proteins were reduced to eliminate disulfide bonds. The proteins are: (1) bovine serum albumin; (2) catalase; (3) ovalbumin; (4) carboxypeptidase A; (5) chymotrypsinogen; (6) lysozyme.

of unknown molecular weight is electrophoresed with two of known molecular weight, that of the unknown can be calculated to an accuracy ranging between 5% and 10%. Greatest accuracy is obtained if the known values for M bracket the unknown. This is certainly the most common way of estimating the molecular weight of protein subunits in use today (although thin-layer gel chromatography—see page 193—will probably prove to be of equal value).

Note that the SDS-gel technique can also be used to determine whether S—S groups are present because, if they are, the mobility will depend on whether the protein is treated with the S—S cleaving agent mercapto-ethanol. This complements the viscosity measurements described on page 365.

In earlier work, 7.0 to 12.0 M urea was used to eliminate secondary and tertiary structure of proteins, but this has been supplanted by SDS.

Polyacrylamide-Gel and Polyacrylamide-Agarose-Gel Electrophoresis of Nucleic Acids. In electrophoresis of nucleic acids, the molecular-sieving effect is also the principal factor in separation because the charge-to-mass ratio is nearly the same for all polynucleotides. Hence, small molecules move faster than large ones. Because naturally occurring nucleic acid molecules are very large, the pore size of the gel must be large, that is, the gel must be dilute. To strengthen the gels, especially for large molecules ($M > 5 \times 10^6$), agarose (a highly porous polysaccharide) is added; sometimes agarose alone is used. Electrophoresis is done in slab gels or column gels. For RNA molecules (e.g., messenger and ribosomal RNA), resolution is far better than in zonal centrifugation (Figure 9-10) and electrophoresis is certainly the method of choice. Furthermore, the distance migrated is proportional to $1/\log M$ so that M is measurable if two standard samples are included.

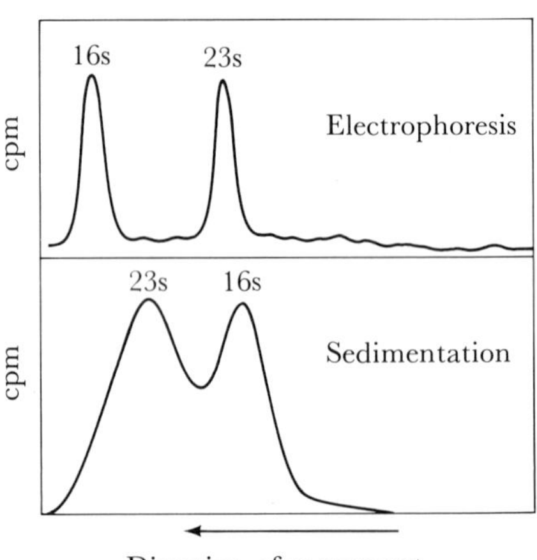

FIGURE 9-10
Separation of ribosomal RNA species (16s and 23s) by polyacrylamide-gel electrophoresis compared with sedimentation through sucrose gradients. Note that the electrophoretic bands are narrower than the zones in sedimentation; also the relative positions of the two classes of molecules are reversed.

Separation in polyacrylamide and agarose is also very good for single-stranded DNA up to $M = 50 \times 10^6$. However, because double-stranded DNA is highly extended, gels of sufficient pore size to allow the passage of DNA greater than $M = 8 \times 10^6$ cannot be prepared, although separation is highly effective below this value. Figure 9-11 shows a slab gel of DNA molecules separated in agarose.

For reasons that are unclear, in certain buffer systems single-stranded DNA molecules separate not only by size, but apparently on some other basis. For example, the two polynucleotide strands obtained by denaturing

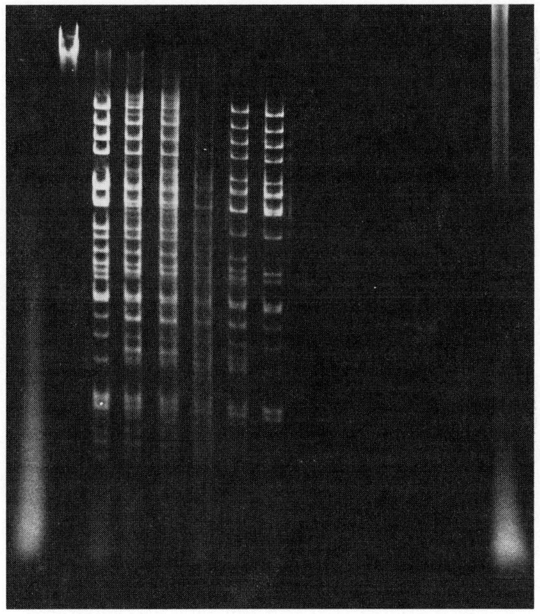

FIGURE 9-11
Slab-gel electrophoresis of DNA fragments. Electrophoresis of fragments of DNA obtained by the digestion of various λ *para* phage DNAs with Hin2 and Hin3 restriction enzymes. The gel was 1.4% agarose. After electrophoresis, the gel was soaked in a solution of ethidium bromide, illuminated with exciting light, and photographed to reveal the fluorescence produced by the dye bound to DNA. Because the fluorescence is strongly enhanced when bound, the gel itself does not show significant fluorescence. Each vertical set of bands represents a distinct DNA. [Courtesy of Robert Schleif.]

several *E. coli* phage DNAs are separable in 0.6% agarose using a Tris-phosphate buffer.

Another interesting use of polyacrylamide-gel electrophoresis is to separate circular from linear DNA molecules. It is possible to find a gel concentration that allows linear but not circular molecules (of a given molecular weight) to penetrate the gel. Hence, circular molecules remain at the origin. This probably works because the pore size is so small that only linear molecules *entering longitudinally* can penetrate the gel. This method has not yet been seriously exploited.

Nucleic acid bands can be detected in gels in a variety of ways. Radioactivity is usually measured by slicing the gel, solubilizing it in 0.5 M NaOH or H_2O_2, and using a scintillation counter. With material of high molecular weight, slicing can be difficult because the gel is necessarily soft to insure large pore size. Double-stranded DNA is most easily visualized by adding the fluorescent dye ethidium bromide (Figure 9-11), whose quantum yield (see Chapter 15) is considerably enhanced when bound to native DNA. The dye Stains-All has the interesting property of staining DNA, RNA, and protein and yielding a different color for each.

DISC ELECTROPHORESIS IN POLYACRYLAMIDE GELS

For maximum resolution, disc electrophoresis—an important refinement of zone electrophoresis—is used. (As will be seen, "disc" does not refer to the bands that the technique yields, but to the fact that the system uses a pH discontinuity that leads to a voltage *disc*ontinuity.) It consists of concentrating an initially thin (1–2 mm) sample layer into an ultrathin starting zone from 1 to 100 μ thick, as shown in Figure 9-12. Note that the gel system is prepared in a vertical column and consists of three separate regions: the uppermost or sample, the middle or spacer gel, and the lower separation gel. The sample and spacer gels are less concentrated (larger pore size) than the separation gel and are prepared in a buffer of lower ionic strength and different pH. The larger pore size in the upper gels means that molecules will be impeded less and will move faster than in the separation gel. Similarly, the lower ionic strength means higher electrical resistance so that the electric field (V/cm) is greater in the upper region and hence makes the molecules move faster than in the lower gel. The relative pH values have the same effect on mobility. This rapid movement through the upper gels results in an accumulation of material at the boundary between the spacer and separation gels. However, note that the components, though very close, are also arranged into stacks in order of mobility. (Hence, the spacer gel is sometimes called the *stacking* gel). As the molecules proceed downward through the separation gel, the various zones separate according to mobility. At the end of the operation, the gel is removed from the glass tube. The zones can be identified in many ways:

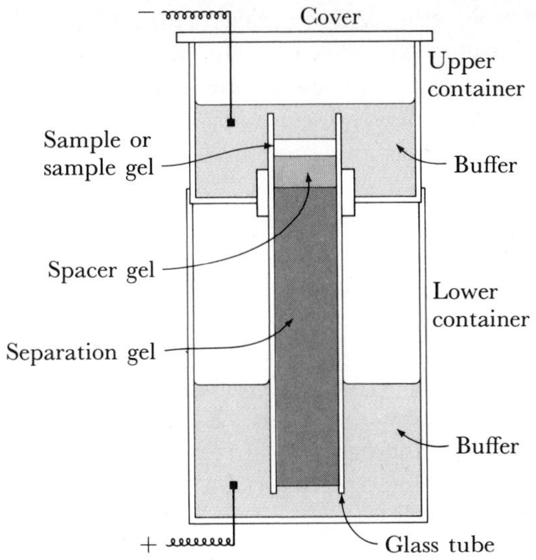

FIGURE 9-12
Apparatus for disc electrophoresis. The gels are separately polymerized in the column and then placed in the buffer reservoir. Sometimes a sample gel is not used and the sample is in a dense sucrose solution, which is layered with a fine pipette on the spacer gel but under the buffer. After electrophoresis, the gel is pushed out. To make the band visible, the sample is either stained or, if the sample is radioactive, sliced transversely, each slice being counted in a scintillation counter. In some cases (Figure 9-13), the gel is sliced vertically and placed against autoradiographic film. The slab-gel arrangement (Figure 9-7) can also be used for disc electrophoresis.

First, the gel can be stained by immersion in dye followed by extensive washing to remove unbound dye. (In some cases, destaining can be done electrophoretically. This is possible because, in the staining process, the dye becomes covalently coupled to the protein and the protein is coupled to the gel. Unbound dye is free to move out of the gel when the field is applied again.) The gel can be photographed if necessary or, if quantitative information is necessary, dye uptake can be measured by a recording densitometer. Second, the gel can be sliced transversely into many layers, each assayed for radioactivity, enzymatic activity, optical density at particular wavelengths, or dye uptake. Third, if radioactive, the gel can be sliced vertically and assayed by autoradiography.

The principal uses of disc electrophoresis are to determine the purity of a presumably pure protein and to analyze the components of a mixture with very high resolution (i.e., if there is a very large number of components).

Purity is, of course, a relative term and a single band may result because the impurities are at too low a concentration to be detectable. Therefore, in purity tests a large amount of protein must be used. Because the limit of detection by staining is about 1 μg, to see if something is 99% pure at least 100 μg should be used—and that quantity is sufficient only if the impurity is a single protein! Even at very high concentrations (0.5–1.0 mg) a single band is not a criterion of purity because the impurities might not have been resolved by the particular pH and/or buffer. Hence, to have some confidence about homogeneity, it is usually necessary to use several buffers and a variety of pH values.

The following example illustrates the use of disc electrophoresis to resolve a complex mixture.

Example 9-A. Analysis of proteins made during infection of the bacterium *E. coli* by phage T4.

If *E. coli* is infected with phage T4, a large number of phage-mediated proteins are synthesized. The relative proportions of each and the timing of their synthesis have been studied by disc electrophoresis, using a radioactive label to detect the protein. If a mixture of all the proteins in an infected cell were electrophoresed and then stained, the gel would be nearly continuously stained because there would be thousands of proteins present. However, infection with phage T4 shuts off host protein synthesis; therefore, if the infection is done in the presence of ^{3}H-leucine, no bacterial protein will be radioactive. Hence, to study the sequence of protein synthesis during infection with T4, such an infection was performed, ^{3}H-leucine was added at various times after infection, and the cells were harvested several minutes after each addition of ^{3}H-leucine. The proteins were isolated and then separated by disc electrophoresis, and the bands were detected autoradiographically. It was observed that different bands were formed at different times.

When a phage mutant that makes only a fragment of a particular protein was studied, one band was missing and a new one (the fragment) appeared (Figure 9-13). Hence, the missing band corresponds to the gene that has the mutation. By the use of many such mutations, each band could be identified; in this way the sequence of synthesis of different proteins was elucidated. It is worth emphasizing that this analysis was done in the presence of all the bacterial proteins but, because they were not radioactive, they were not seen.

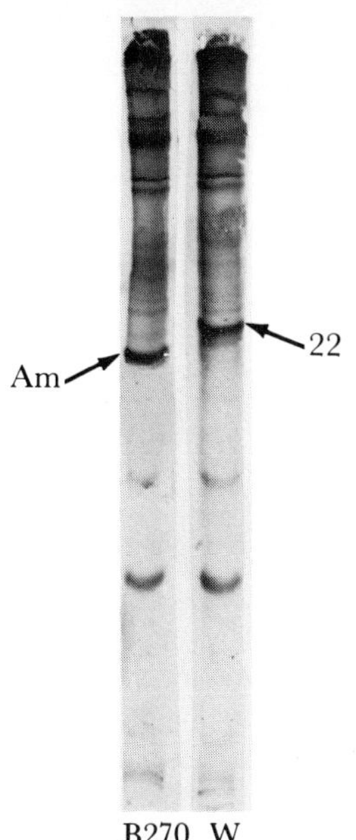

FIGURE 9-13
Electrophoregram of proteins synthesized in *E. coli* infected either with wild-type (W) T4 phage or with a mutant (B270) in the major head protein. Infected cells were labeled with ^{3}H-leucine from 10 to 12 minutes after infection. Proteins were extracted and electrophoresed in a polyacrylamide gel containing alkaline urea. After electrophoresis, the gels were sliced vertically and autoradiographed. Note that band 22 (the major head protein) is absent in the chain-terminating mutant B270, and a new band (Am) corresponding to the fragment appears. [From J. Hosoda and C. Levinthal, *Virology* 34(1968):709–716.]

Continuous (Curtain) Electrophoresis

Continuous electrophoresis, a preparative procedure, is achieved by simultaneously displacing a continuously added sample in one direction by solvent flow and perpendicularly by an electric field. This is frequently done using filter paper as the supporting medium but can be done in the absence of a supporting medium. The system is shown in Figure 9-14. A large sheet of paper is cut into the shape shown and placed between two sheets of Lucite. The upper edge of the paper is in a buffer reservoir and initially the paper is completely wet with buffer. As long as the reservoir is full, the buffer moves continually down the paper by capillary flow and drips from the serrated points into collecting tubes. At the top of the paper, a sample is continuously added, using a motor-driven syringe or a wick. The electric field causes a constant horizontal displacement, the extent of which depends on the value and sign of the mobility. The factors determining the separation are the rate of flow of the buffer, the rate of the addition of the sample, the applied voltage, and the pH and ionic composition of the buffer; optimal conditions are usually determined by trial. This system has been used successfully to separate proteins, peptides, amino acids, other small molecules, and inorganic ions.

With some materials, adsorption to the paper is a serious problem. To avoid this, Kurt Hannig developed a system that does not require

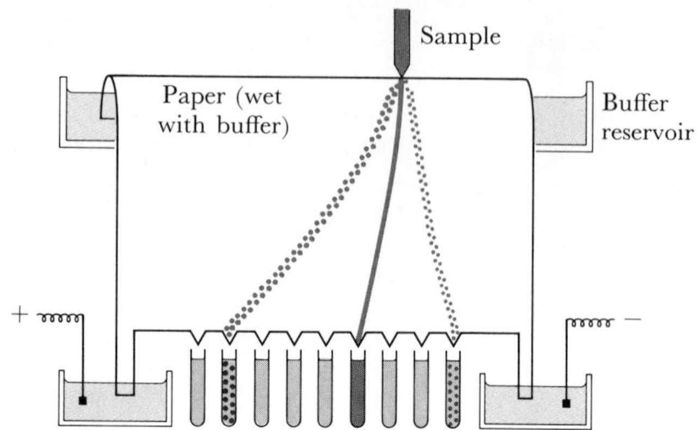

FIGURE 9-14
Apparatus for curtain electrophoresis. The sample is applied continuously, as indicated, while the buffer flows vertically down the paper and drips into the collection tubes. The potential is applied perpendicular to the direction of flow of the buffer.

supporting material. In this system, the way in which buffer is supplied from the reservoir produces a continuous flow between two glass plates very near one another. Precise temperature control and the capillarity resulting from the narrow space between the plates reduce convective flow. A series of conical outlets at the base allows various fractions to be collected. This expensive and demanding instrument has not had widespread use but has given superb separations of proteins, peptides, and different blood cells (e.g., antibody-producing human lymphocytes have been separated from leukocytes and macrophages with this instrument).

ISOELECTRIC FOCUSING

Proteins are ampholytes—that is, they contain both positively and negatively charged groups. All ampholytes have the property that their charge depends on pH; they are positively charged at low pH and negatively charged at high pH. For every ampholyte, there exists a pH at which they are uncharged, and this is called the *isoelectric point.* At the isoelectric point, the ampholyte will not move in an electric field. If a protein is placed in a pH gradient, the molecules will move until they reach a point in the gradient at which they are uncharged; then they will cease to move. With a mixture of proteins, each type will come to rest at a point in the pH gradient corresponding to its own isoelectric point. This method of separating proteins according to their isoelectric points in a pH gradient is called isoelectric focusing and is the electrophoretic analog of centrifugation to equilibrium in a density gradient (see Chapter 11). The process of the migration of two different proteins with different isoelectric points is shown schematically in Figure 9-15.

The pH gradient is established in an unusual way. If it were established by simply allowing two buffers at different pH to diffuse into one another or by mixing two buffers in the way that is standard for preparing concentration gradients (Chapter 8), the resulting gradient would not be stable in an electric field; the ions of the gradient would migrate in the field so that fractionation could not occur unless the macromolecules were to migrate faster than the pH gradient was being disrupted. The method used to produce a stable pH gradient consists of distributing synthetic, low-molecular weight (300–600) polyampholytes (multicharged structures) that cover a wide range of isoelectric points. These polyampholytes are usually mixed polymers of aliphatic amino and carboxylic acids (e.g.,

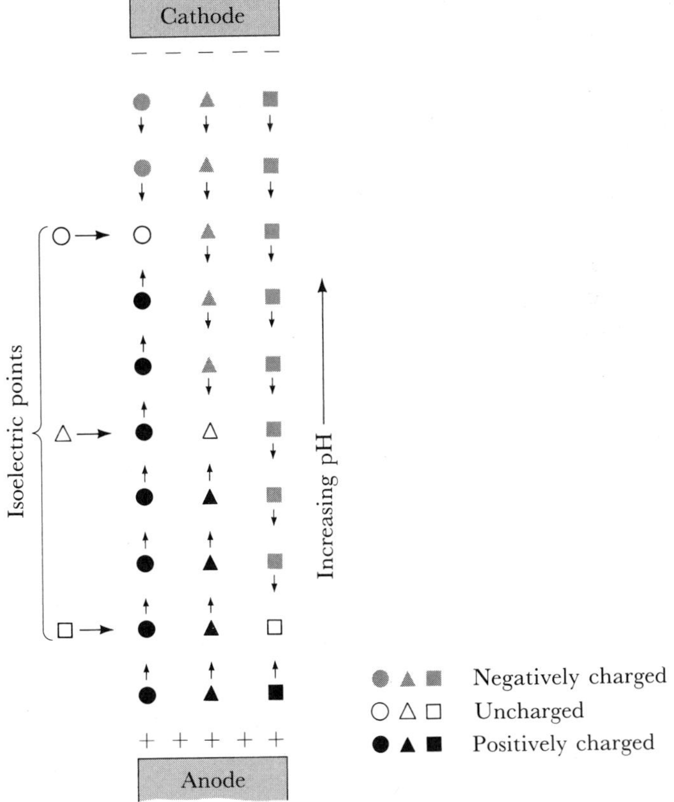

FIGURE 9-15
The process of migration in isoelectric focusing. Proteins move until they reach a position in the pH gradient at which they are uncharged.

Ampholine, LKB Industries, Inc.) A pH gradient is established by starting with a mixture of polyampholytes in distilled water whose isoelectric points cover a small range. Before the application of an electric field, the pH throughout the system is constant and is averaged from all the polyampholytes in the solution. When the field is applied, the polyampholytes start to migrate. Because of their own buffering capacities, a pH gradient is gradually established. Soon each particle will come to rest in this self-established gradient at the point corresponding to its own isoelectric point. If the mixture contains proteins of different isoelectric points, they will migrate to the positions corresponding to their isoelectric points as long as the concentration and buffering capacity of the protein (which is also a polyampholyte) is not so high that the pH gradient is disrupted.

The apparatus used is of the type shown in Figure 9-16. The pH gradient is formed in a water-cooled glass column containing a cathode tube and a cathode. The tube is filled with a uniform concentration of the poly-

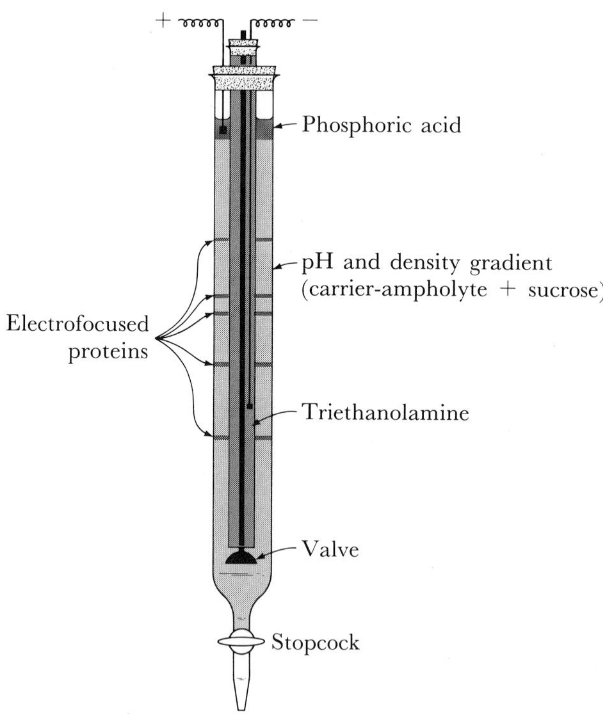

FIGURE 9-16
Apparatus for isoelectric focusing. A description of its operation is given in the text.

ampholyte prepared in a sucrose concentration gradient to eliminate convection. The sample material is also contained in the polyampholyte suspension. The cathode tube is filled with a strong base (typically triethanolamine) and the main column is overlaid with phosphoric acid; the anode is in this acid layer. The valve at the bottom of the cathode tube is opened, followed by the application of a few hundred volts between the electrodes. The polyampholyte near the cathode will have a negative charge and will move to the cathode. From one to three days later, the system will be at equilibrium and the proteins distributed throughout the pH gradient according to their own isoelectric points. The tube is then drained and fractionated through the stopcock at the bottom. The various proteins can be detected by spectrophotometry, enzyme activity, or radioactivity. Isoelectric focusing can be used both as an analytical and as a preparative procedure. Figure 9-17 shows the kind of separation usually obtained. The isoelectric focusing method has two great advantages: (1) separations that are not obtainable by any other method are possible, and (2) proteins are significantly concentrated.

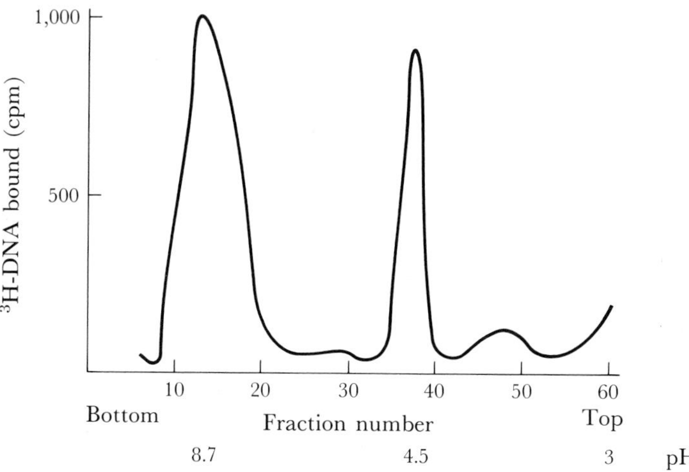

FIGURE 9-17
Isoelectric focusing. An enzyme, UV-endo, was purified to apparent homogeneity by various chromatographic procedures. However, when tested by isoelectric focusing, it appeared to consist of two proteins, each having enzymatic activity. Activity of the enzyme was determined by its ability to bind ultraviolet irradiated ^{3}H-DNA to a nitrocellulose filter. Note that the pH gradient is not linear. [Courtesy of Sheikh Riazuddin.]

LARGE-SCALE PREPARATIVE ELECTROPHORESIS

Block Electrophoresis

A slurry of starch grains (or glass powder, silica gel, cellulose, polyvinyl chloride, or agar) in a buffer is poured into a large rectangular trough, which is mounted so that there is an electrode at either end. The sample is applied by removing starch from a large groove near one electrode, mixing the protein solution with the starch, and refilling the groove. After electrophoresis, the block is cut up and each section eluted.

Column Electrophoresis with Polyacrylamide

This procedure employs the apparatus shown in Figure 9-18. Material continually flows into the elution buffer, which flows through as indicated. The elution buffer is collected by a fraction collector.

IMMUNOELECTROPHORESIS

As mentioned earlier in the chapter, substances with the same mobility can often be separated by subsequent chromatography in a perpendicular direction or by chromatography followed by electrophoresis (see Figure 9-3). This allows separation to be based on two properties. Immunoelectrophoresis also provides this dual system: first, molecules are separated by electrophoresis, and then they are detected by immunodiffusion (see Chapter 10).

Immunoelectrophoresis is usually done horizontally on a slab of agar or agarose as shown in Figure 9-19. A uniform layer is prepared and a well and trough are cut for antigen and antibody, respectively. The antigen mixture is placed in the well and the slab is submitted to electrophoresis. During this period, molecules migrate in both directions, the distances depending on their mobilities and the sign of the charge. When electrophoresis is complete, antiserum is added to the trough. Both antiserum and antigens diffuse through the gel. If an antigen-antibody reaction occurs where they meet, a visible, curved precipitin line (see Chapter 10) forms (Figure 9-19C). This technique allows the identification of the number of antigens producing antibody and the relative amounts of each, and it allows the assay of materials for which only immunoassay is possible (e.g., cross-reacting, inactive fragments of enzymes).

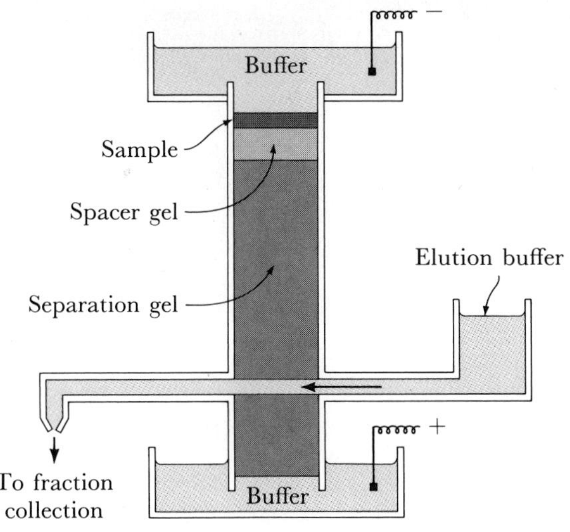

FIGURE 9-18
Apparatus for large-scale preparative disc electrophoresis. See Figure 9-12 for comparison.

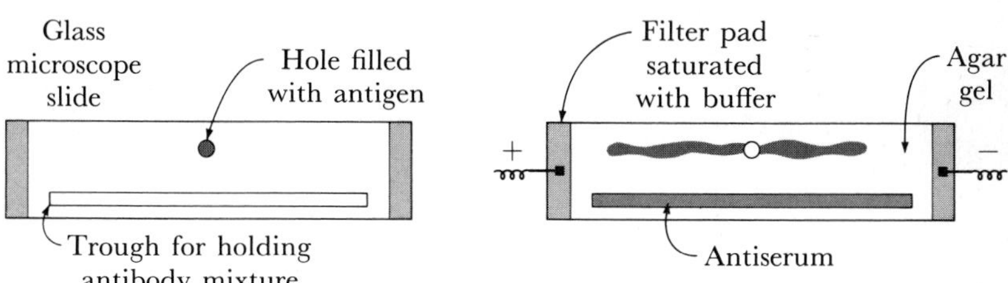

A. Hole is filled with protein.

B. Electrophoresis causes migration of proteins. Voltage removed and antiserum added to trough.

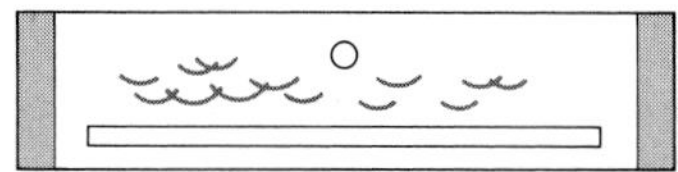

C. Immunodiffusion forms precipitin lines.

FIGURE 9-19
Arrangement for immunoelectrophoresis.

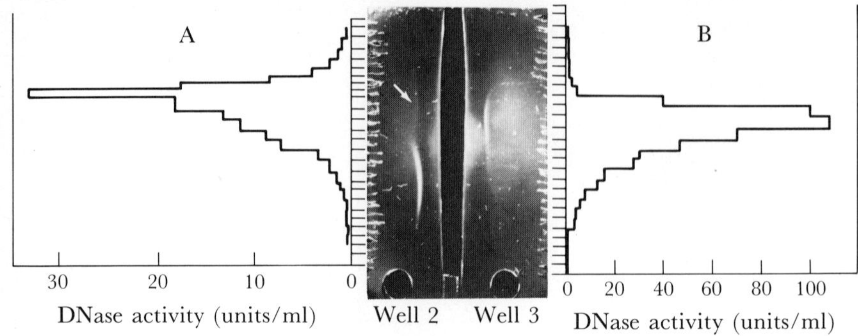

FIGURE 9-20
The use of immunoelectrophoresis to assay the purity of *E. coli* phage λ exonuclease. A fairly well purified fraction of λ exonuclease (A) and a more stringently purified sample (B) were assayed as follows: Four wells were made. Two on the left were filled with fraction A and two on the right with B. The system was electrophoresed so that movement was toward the top of the illustration. The outer strips corresponding to wells 1 and 4 were sliced perpendicular to the direction of migration. Each slice was eluted and assayed for exonuclease activity. The graphs show the activity from these slices as a function of position. The center trough was filled with antiserum directed against partially purified exonuclease and immunodiffusion (see Chapter 10) was allowed to occur. Note that in fraction A the major precipitin band is not in the same position as the activity; only a very weak band (arrow) is at that position. Hence, the enzyme has a major contaminant. In fraction B, more extensive purification has eliminated the band of the impurity and only a single band in the appropriate position is seen. Hence, the protein appears pure. [From C. M. Radding and D. C. Shreffler, *J. Mol. Biol.* 18(1966):251–261.]

Selected References

DeWachter, R., and W. Fiers. 1971. "Fractionation of RNA by Electrophoresis on Polyacrylamide Gel Slabs," in *Methods in Enzymology,* vol. 21, edited by L. Grossman and K. Moldave, pp. 167–178. Academic Press.

Efron, M. 1960. "High Voltage Paper Electrophoresis," in *Chromatographic and Electrophoretic Techniques,* vol. 2, edited by I. Smith, pp. 158–189. Interscience.

Gordon, A. H. 1968. "Electrophoresis of Proteins in Polyacrylamide and Starch Gels," in *Laboratory Techniques in Biochemistry and Molecular Biology,* vol. 1, edited by T. S. Work and E. Work, pp. 1–149. North-Holland.

Maures, H. R. 1971. *Disc Electrophoresis.* deGruyter.

Ornstein, L. 1964. "Disc Electrophoresis." *Ann. N.Y. Acad. Sci.* 121:321–349.

Raymond, S. 1964. "Acrylamide Gel Electrophoresis." *Ann. N.Y. Acad. Sci.* 121:350–365.

Smithies, O. 1953. "Zone Electrophoresis in Starch Gels: Group Variations in the Serum Proteins of Normal Human Adults." *Biochem. J.* 61:629–641. First use of starch gels.

Weber, K., and M. Osborn. 1969. "The Reliability of Molecular Weight Determined by Dodecyl Sulfate-Polyacrylamide Gel Electrophoresis." *J. Biol. Chem.* 244:4406–4412. The definitive paper on SDS-gel electrophoresis.

Zwaan, J. 1967. "Estimation of Molecular Weights by Polyacrylamide Gel Electrophoresis." *Analyt. Biochem.* 21:155–168.

Zweig, G., and J. R. Whitaker. 1967. *Paper Chromatography and Electrophoresis,* vol. 1: *Electrophoresis in Stabilizing Media.* Academic Press.

Problems

9-1. Why is electrophoresis done in solutions having low salt concentration?

9-2. In two-dimensional chromatography and electrophoresis, does it matter which is done first?

9-3. In determining the conditions for maximum separation of two components, what parameters (e.g., pH, ionic strength, temperature, etc.) should be varied? What effect might the variation of each have on the resolution?

9-4. A mixture of proteins is electrophoresed in a polyacrylamide gel at three different pH values. In each case five bands are seen. Can you reasonably conclude that there are only five proteins in the mixture? Explain.

9-5. An enzyme has been extensively purified. By a variety of criteria it is thought to be pure—that is, it shows a single peak when chromatographed, electrophoresed, or centrifuged in a variety of ways. When subjected to SDS-gel electrophoresis, two bands result, one twice the area of the other. What information does this give about the protein? Because purity is always difficult to prove, how could you prove that your hypothesis is correct? (*Hint:* Use gel chromatography.)

9-6. A virus contains 256 proteins, 64 having a molecular weight of 1,800 and 192 with a molecular weight of 26,000. If the virus were disrupted and analyzed by SDS gel electrophoresis, what would be the relative distances migrated and the relative areas of the bands?

9-7. Suppose that you have isolated a protein that seems to have two enzymatic activities. This makes you suspect that you may have two proteins that copurify. To check this, you electrophorese the preparation in a polyacrylamide gel at a variety of pH values. In each case, a single band results, but the band is sufficiently broad that you suspect that it is really two bands which do not resolve. In an SDS gel, a single, broad band is also found. Because a protein with two enzymatic activities is rare, it is necessary to try a little harder to see if the breadth is due to the presence of two bands. What parameters could you vary to improve resolution by electrophoresis? What other methods (nonelectrophoretic) might you try?

CHAPTER 10

Immunological Methods

The identification and assay of biological material is most commonly done by chemical tests, spectroscopic methods, and the determination of various physical parameters such as sedimentation coefficients, electrophoretic mobilities, chromatographic constants, and so forth. If very small quantities of material must be assayed, traditional chemical methods fail, and an approach has been to use radiochemical tracers and the technology described in Chapter 5. Unfortunately, in the assay of complex systems such as cells in culture and certainly of whole animals or plants, it is not always possible to add enough radioactive material to label detectable quantities of a particular biochemical—either because of dilution of the label or because of danger to the organism.

Immunological procedures provide the solution to these difficulties because they make it is possible to assay minute amounts of nonradioactive material in complex mixtures.

THE IMMUNOLOGICAL SYSTEM

In response to the injection of a foreign substance into a higher animal, an *antibody* (Ab) is produced that can react with the substance. Antibodies are proteins found in the bloodstream and are part of a class of serum proteins called *immunoglobulins.* Any substance that can elicit antibody production is called an *antigen* (Ag). An antibody produced by exposure to an antigen has the important property of reacting specifically with the antigen that stimulated its production and not with most other antigens. Similarly, the antigen fails to react with any antibody other than that which it elicited.

A *hapten* is a substance that cannot by itself stimulate antibody synthesis but can react with a hapten-specific antibody. Most haptens are molecules of low molecular weight (<1,000); an antibody to a hapten is usually prepared by chemically coupling the hapten (by covalent bonds) to an immunogenic substance such as the protein serum albumin and injecting this conjugated protein into an animal (e.g., a rabbit). An antibody prepared against a substance X (which may be either an antigen or a hapten) is commonly called *anti-X.* An antibody-antigen complex is often written Ab-Ag.

In addition to the original antigen, there are other substances that react with a specific antibody, though often with a somewhat lower efficiency. This weaker reaction is called a *cross-reaction.* One kind of cross-reaction is that which takes place when antigen A reacts partially with anti-B and antigen B partially with anti-A. Asymmetric cross-reactivity also occurs—that is, antigen A reacts with anti-B, but antigen B does not react with anti-A.

Antibodies, being proteins, are also antigenic in other animals. For example, an antibody obtained from rabbit serum can be injected into a goat and elicit an antibody. The goat antibody does not react with the antigen that stimulated the production of the rabbit antibody but to the rabbit antibody itself; it is usually called *goat antirabbit antibody.*

If the biological activity of a substance is destroyed by an antibody, the substance is said to be *neutralized.*

PREPARATION OF ANTIBODY

Because antibodies are produced in the bloodstream of an animal in response to the injection of a foreign substance, antibody can be obtained by bleeding an animal that has been repeatedly injected with the same antigen or conjugated hapten. Because of the specificity of the Ag-Ab reaction, it is rarely necessary to isolate the specific antibody, or even the immunoglobulin fraction. Hence, in most immunological work, blood serum from which all cells have been removed by centrifugation is used. Serum known to contain a particular antibody is called *antiserum.* There are some occasions when an antiserum must be partially purified. For example, when a hapten conjugated to serum albumin is injected into an animal to prepare antihapten, the animal also makes antiserum albumin. And if, for example, antibody to ultraviolet-irradiated DNA is made, an antibody will be made that is active only against unirradiated DNA. The secondary antibody may be unacceptable in a particular experiment and must be removed. This is done by *adsorption.* That is, the antigen to the unwanted antibody is added to the serum, and the antigen-antibody complex is allowed to form. Under appropriate conditions, this complex can be removed by centrifugation. The repeated addition of antigen usually results in eliminating the unwanted antibody. On rare occasions, a pure antibody is needed; it can be prepared by means of affinity chromatography (see Chapter 8).

The preparation of an antihapten requires special procedures. Because the hapten itself is not antigenic, it is coupled to a strongly antigenic molecule such as bovine serum albumin or certain synthetic polypeptides and then injected into a rabbit. The coupling is sometimes a problem because the hapten must contain at least one functional group that can be used to be coupled to the available functional groups (e.g., amino, carboxyl, phenolic, imidazolyl, sulfhydryl, indolyl, and guanidino) of the protein under chemical conditions that destroy neither the hapten nor the protein structure. Otherwise, a chemical derivative of the hapten must be synthesized to allow this coupling to be done. Table 10-1 lists haptens to which antibodies have been made. The ability to prepare antihaptens is essential to the radioimmunoassay that is described in a later section.

REACTION OF ANTIBODY AND ANTIGEN

After an antibody has been prepared, the antibody-antigen reaction is carried out merely by mixing the two substances in a buffer and incubating until the reaction is complete. The buffer is usually at neutral pH and

TABLE 10-1
A variety of haptens to which antibodies have been made.

Peptide hormones	Insecticides
Nonpeptide hormones	Carbohydrates (e.g., glucose, maltose, lactose)
Coenzymes	Catecholamines
Vitamins	Lipids
Drugs	Steroids
Toxins	Nucleic acid constituents
Carcinogens	Plant hormones

NOTE: For a complete description, see V. P. Butler and S. M. Beiser, *Advan. Immunol.* 17(1973):255–310.

is isotonic to prevent denaturation of the antibody protein, although a range of ± 2 pH units and a factor of 2 in ionic strength can be tolerated.

The reaction is remarkable in that it usually proceeds in complex mixtures containing other proteins, small molecules, polysaccharides, lipids, and so forth, with virtually no effect on the rate or extent of reaction. In quantitative studies, control experiments are always performed to insure that there are no interfering substances, because in isolated cases a reaction mixture may contain substances that degrade or metabolize either the antigen or the antibody.

IMMUNE REACTIONS USEFUL IN BIOASSAYS

Precipitin Reaction

Under certain conditions, if an antibody is added to a solution containing an antigen, a complex forms and precipitates. The amount of antibody precipitated by various amounts of antigen depends on the antibody–antigen ratio and is described by a *precipitin curve* (Figure 10-1). Precipitin refers to a particular type of *reversibly dissociating* antigen-antibody complex. The reversibility is made evident by the fact that, even after precipitin has formed, the amount present can be varied by changing the ratio of antigen to antibody. In the regions marked *antibody-excess* and *equivalence,* the antigen is totally precipitated; hence, precipitation with excess antibody can be used to remove particular substances from solution. For example, in a mixture of radioactive proteins, the amount of a particular protein (e.g., A) can be determined by adding anti-A and sedimenting the precipitate. The ratio of sedimentable-to-total radioactivity yields the weight fraction of the protein mixture that is A. That the precipitin reaction is an important part of the radioimmunoassay will be seen later.

Gel-Diffusion Precipitin Reactions and Immunodiffusion

The precipitin reaction can also occur in agar gels. Consider the arrangement (and method) shown in Figure 10-2A. A test tube is partly filled with agar containing antiserum, and this is overlaid with antigen. The antigen diffuses downward into the agar and a precipitate forms in the agar at the advancing, diffusing front. The precipitate is a band due to the reversibility of the precipitin reaction described in the preceding section. That is, in the upper part of the tube, there is little precipitin because of excess antigen (see Figure 10-1). As downward diffusion of the antigen

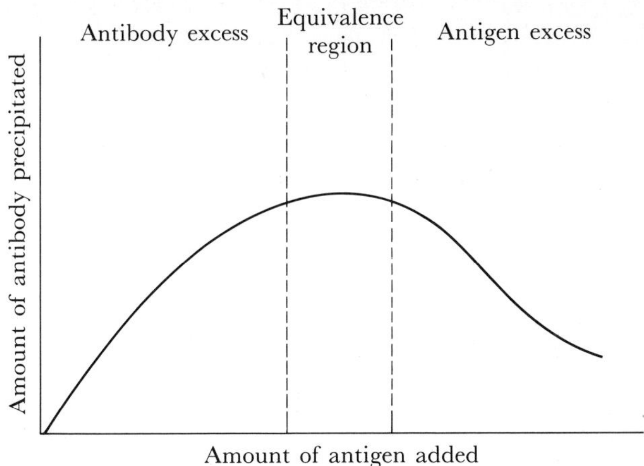

FIGURE 10-1
Precipitin curve. To a fixed amount of antibody various amounts of antigen are added. After a suitable period of incubation, the mixture is centrifuged and the amount of antibody precipitated is determined by measuring the amount of antibody remaining in the supernatant. In the antibody-excess region, all antigen is precipitated—that is, the supernatant contains the excess antibody. In the antigen-excess zone, much of the antibody does not form precipitin and the supernatant contains a mixture of antigen and antibody. At equivalence, all of the antigen and antibody are precipitated.

continues, the precipitate appears to move downward. This is because, at the leading side, more antibody is encountered, whereas, at the trailing side, more antigen is arriving; the excess antigen solubilizes the precipitate (Figure 10-1). If both antigen and antibody are mixtures of several reacting species, several bands will be present, presumably one for each antigen-antibody complex. The separation of these bands depends on the diffusion coefficients of the antigens and the relative concentrations of antigens and antibodies; hence, if two antigens have the same diffusion coefficients and the concentrations of antibodies and antigens are the same, only one band will appear. The number of observed bands is always a minimum value for the number of antigens present.

Additional information can be obtained by means of the *Ouchterlony double-diffusion method,* which is shown in Figure 10-2B–D. A Petri dish of agar contains three wells (holes in the agar), two of which are filled with antigen and the third with antiserum. Diffusion proceeds in all directions from all wells; when the equivalence point is reached in the space between

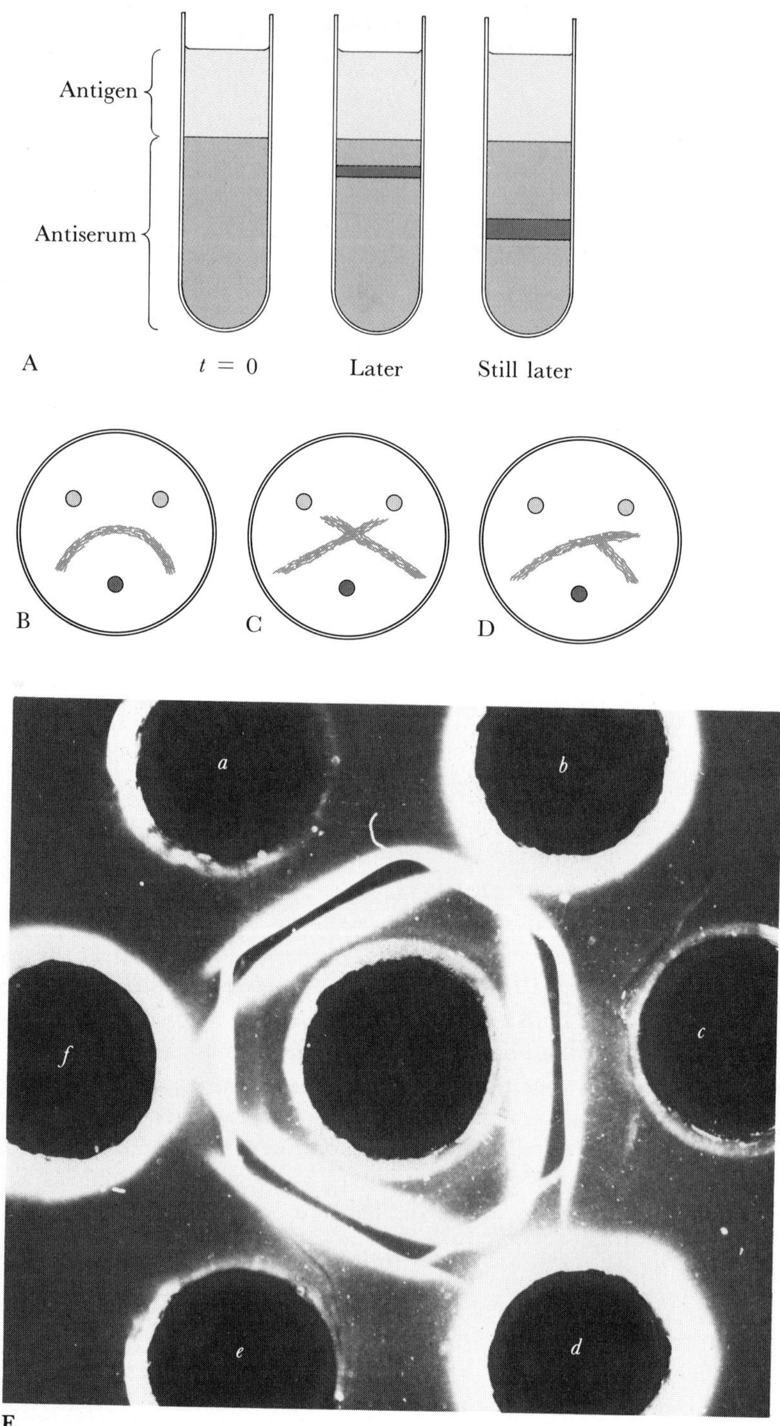
Antigen
Antiserum
A
t = 0
Later
Still later
B
C
D
a
b
c
d
e
f
E

an antibody and an antigen well, a precipitin line appears. If the two antigen wells contain the same antigen, a curved line such as that shown in Figure 10-2B is the result. If the antigens are different, the two lines are independent and will intersect as shown in Figure 10-2C. If there is cross-reaction between the two, the pattern of Figure 10-2D will result. The small region projecting above the curve in Figure 10-2D is called a *spur* and always points toward the less reactive antigen. Two different weakly reacting but cross-reacting antigens can appear to give intersecting bands as in Figure 10-2C and could be mistaken for independent substances. However, the crossovers are spurs only if they are less dense than the main line and show pronounced curvature. The Ouchterlony procedure can be used to determine whether an unknown material is the same as some known substance because, if it is, the pattern of Figure 10-2B results.

Complement Fixation Assay

Blood sera contain a class of proteins that are not immunoglobulins and are collectively called *complement,* or C′. Complement reacts not with antigen or antibody alone but with a variety of antigen-antibody complexes. For reasons that are not clear, complement does not react with hapten-antibody complexes. If haptens are to be measured, inhibition assays (see next section) are usually used.

The complement fixation assay for an antigen-antibody reaction makes use of the fact that complement is consumed ("fixed") by the antigen-antibody complex, making less complement available. The usefulness of this reaction derives from the fact that, if the antigen is to certain elements

FIGURE 10-2
Gel diffusion precipitin reaction. A. Single diffusion in one dimension (Oudin method). The test tube is partly filled with agar containing antiserum. After the agar has hardened, it is overlaid with agar containing antigen. This agar also hardens. Antigen then diffuses into the antiserum. B–D. Double diffusion in two dimensions (Ouchterlony method). A Petri dish is filled with agar in which three cylindrical wells are cut. One is filled with antiserum (solid spot) and the others with antigen. In part B, both upper wells are filled with the same antigen. In part C, each upper well is filled with a different antigen; the lower well contains antibodies to both antigens. In part D, a single antibody is in the lower well; the upper left is the antigen and the upper right is a cross-reacting antigen. E. Photograph of an Ouchterlony plate. Consider the outer bands only because the inner band is a contaminating protein. Wells *a, c,* and *e* contain *E. coli* phage λ exonuclease. Well *b* contains a protein that is immunologically indistinguishable because the bands are continuous. Wells *d* and *f* contain altered exonucleases that are cross-reacting (note the spurs). [From C. M. Radding, *Methods Enzymol.* 21(1971):458–462.]

on the surface of red blood cells, the addition of complement leads to cell *lysis* (bursting of the cells). This lysis is easily seen with red blood cells; if intact cells are centrifuged, the red hemoglobin pigment is carried to the bottom of the centrifuge tube; if lysis (hemolysis) has occurred, the hemoglobin remains in the supernatant and can be detected quantitatively by its absorbance at 541 mμ (see Chapter 14). The standard test uses guinea pig serum as a source of complement, sheep red cells, and rabbit antibodies to sheep red cells. The assay shown in Figure 10-3 is normally done in two steps:

1. Antibody, antigen, and a measured amount of complement are mixed and incubated until the binding of complement is complete. In this step, the complement is fixed by the antigen-antibody complex.

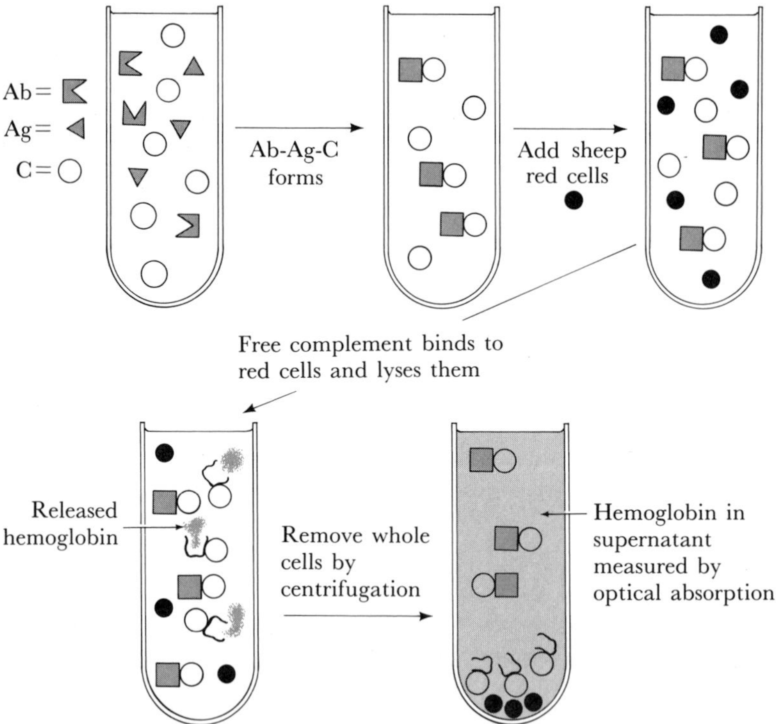

FIGURE 10-3
Complement fixation. Note that there are equal amounts of antibody and antigen and an excess amount of complement for the purpose of illustration. Normally, the amounts of antibody and complement are adjusted so that most of the complement is consumed at equivalence. The sheep red cells are precoated with rabbit antibodies to sheep red cells; this has been omitted from the illustration for clarity.

2. Sheep erythrocytes precoated with specific rabbit antibody are then added to the mixture. In this step, the erythrocyte-antibody complex binds that complement which was not consumed in Step 1. After a period of incubation, during which lysis occurs, the mixture is centrifuged and the optical density of the supernatant is recorded.

Because hemolysis requires free complement (i.e., that not consumed in the first antibody-antigen reaction), the amount of hemoglobin in the supernatant measures the amount of complement remaining. Note that an *increased* antigen-antibody reaction means *decreased* lysis.

Complement fixation can be used to measure the amount of a particular antigen present in a complex mixture with great sensitivity. Figure 10-4 shows a curve relating the amount of complement fixed (i.e., decrease in lysis) as a function of added antigen in the presence of a fixed amount of antibody. Typically, the amount of complement fixed increases with the amount of antigen in the antibody-excess region and decreases in the antigen-excess region. Hence, to measure the amount of a particular antigen, the following protocol is used. First, a standard curve such as that shown in Figure 10-4 is obtained, using known amounts of the antigen. Then the reaction is carried out with several dilutions of the antigen

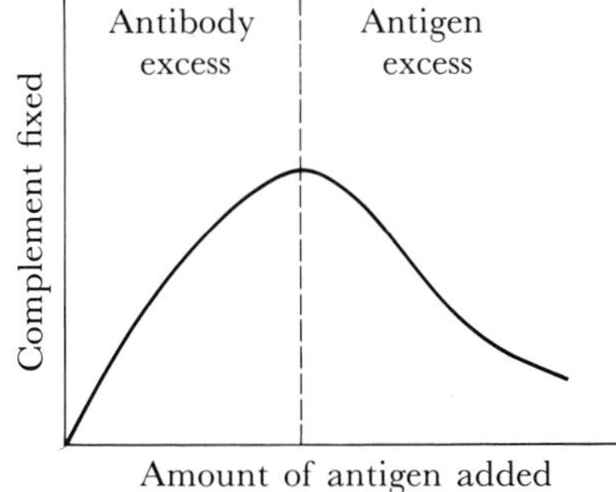

FIGURE 10-4
Standard complement fixation curve. To a fixed amount of antibody and complement are added various amounts of antigen. After a period of incubation, sheep erythrocytes and antisheep erythrocytes are added. After a second period of incubation, the mixture is centrifuged to remove unbroken erythrocytes. The absorbance of the supernatant is measured to determine the amount of hemoglobin released. If no complement is fixed, the maximum amount of hemoglobin is found in the supernatant. Note that two different amounts of antigen can fix the same amount of complement. However, to measure the amount of an antigen whose concentration is unknown, it is necessary to make measurements with dilutions of the sample. If diluting the antigen decreases the amount of complement fixed, the amount fixed corresponds to the antibody-excess part of the curve.

sample. This is necessary because, if a single dilution were used, it would not be known whether the amount of complement fixed corresponded to the antibody- or antigen-excess side of the curve. After this has been done, the amount of antigen can be read from the curve.

The complement-fixation assay is roughly from fifty to one hundred times as sensitive as the precipitin reaction; it is possible to detect as little as 0.01 μg of certain antigens. Its main disadvantage is that, in some crude mixtures being assayed for the presence of a particular antigen, substances may be present that inactivate complement (anticomplementary substances) or stimulate hemolysis.

INHIBITION ASSAYS USING COMPLEMENT FIXATION AND PASSIVE AGGLUTINATION

Many substances can be assayed by their ability to inhibit antigen-antibody reactions. This is especially useful in the assay of haptens, because haptens that react with an antibody usually do so *without fixing complement.* Therefore, haptens cannot be assayed directly by complement fixation. However, if an antibody-antigen reaction is set up so that all antibody is saturated with antigen and all complement is fixed, then any agent that can bind antibody without fixing complement can be detected as a reduction in the amount of complement fixed. Hence, if a hapten is added along with an antigen with which it cross-reacts, it can compete with the antigen for the antigen-binding site on the antibody and therefore reduce the amount of complement fixed. By adding various amounts of a given hapten to a given antigen-antibody mixture, a standard curve can then be constructed, which can be used to measure an unknown amount of the same hapten in a mixture. This procedure is called the *complement fixation inhibition test.* An example is shown in Figure 10-5.

It should be realized that this inhibition assay cannot be used for antigens or even cross-reacting antigens because in both cases complement would be fixed.

Antigens can be assayed by an inhibition test, using the phenomenon of *agglutination.* This refers to the fact that, in much the same way that precipitation of antigen-antibody complexes occurs, bacteria and cells in suspension will clump (agglutinate) if exposed to antibodies directed against the surface components of cells. Consider a modification of this reaction called *passive agglutination.* An antigen to which an antibody has previously been made is coupled to the surface of red blood cells or sometimes to polystyrene spheres. The coupling of the antigen may be by adsorption or by covalent bonds. The addition of an antibody directed against the surface-bound antigen will cause the cells or spheres to clump,

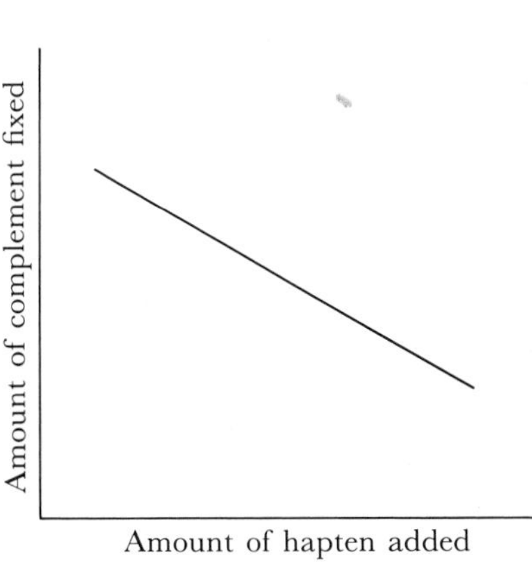

FIGURE 10-5
Complement fixation inhibition assay. Amounts of antigen, antibody, and complement are mixed so that the amount of complement corresponds to the antigen-excess region of the curve shown in Figure 10-4. Various amounts of hapten are added before the addition of the sheep erythrocytes and antisheep erythrocytes. The hapten (H) interferes with the antigen-antibody reaction and decreases the amount of complement fixed because: Ab + Ag + C′ → Ab · Ag · C′, but Ab + H + C′ → Ab · H + C′. If an unknown amount of hapten is added, the amount can be read from this standard curve.

whereas an unbound antigen or a hapten will compete with the surface-bound antigen and prevent clumping.

At one time these two inhibition assays were commonly used. However, they have been almost completely replaced by the enormously sensitive radioimmunoassay described in the following section.

RADIOIMMUNOASSAY

A highly sophisticated immunoassay, the radioimmunoassay (RIA), is capable of detecting extraordinarily small amounts of nonradioactive material and can do so in mixtures of huge numbers and amounts of extraneous materials. Its sensitivity equals or surpasses all known chromatographic and spectrophotometric assays. Its main use is to detect molecules (1) that cannot be radioactively labeled in vivo to a suitable specific activity or without also labeling other compounds; (2) that cannot fix complement when combining with specific antibody (e.g., haptens); and (3) whose identity is unknown but that can cross-react and thereby compete with known antigens.

Because the radioimmunoassay is such a powerful method, it is worthwhile examining more carefully how the inhibition procedure is used. The principle is shown schematically in Figure 10-6. Antibody (Ab)

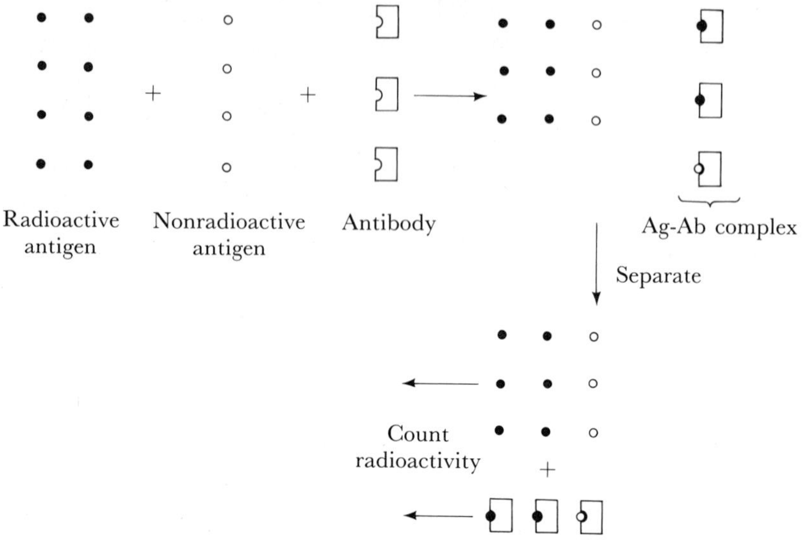

FIGURE 10-6
Principle of the radioimmunoassay. See text for details.

directed against the antigen (Ag) is saturated with radioactive Ag (i.e., Ag*), using excess Ag*. If nonradioactive Ag is added to Ab along with Ag*, less Ag* will be found in the antibody-antigen complex as the ratio Ag/Ag* increases (i.e., there will be an increasing amount of Ab-Ag and a decreasing amount of Ab-Ag*). For example, if Ag/Ag* = 1, only 1/2 of Ag* will be bound; if Ag/Ag* = 2, only 1/3 will be bound, and so forth. If the Ab-Ag* complex can be physically separated from Ag*, the amount of Ag can be determined.

To measure Ag, a standard curve must be constructed (Figure 10-7). This is done by mixing a fixed amount of Ab and Ag* and placing the mixture in a set of test tubes. Known amounts of Ag are added to each. When the reaction is complete, the Ab-Ag* is separated from Ag* (how this is done is described on page 247). A graph is then made that relates the radioactivity in the collected Ab-Ag* to the amount of added Ag (Figure 10-7). To determine the amount of Ag in an experimental sample, an aliquot of the sample is added to the same Ab-Ag* mixture used to get the standard curve, Ab-Ag* is collected and the radioactivity measured, and the amount of Ag is read from the standard curve. This is possible with any sample (no matter how complex) as long as nothing in the mixture interferes with the Ab-Ag* reaction.

The separation of Ab-Ag* from Ag* is in practice performed in two ways (although other methods are also possible):

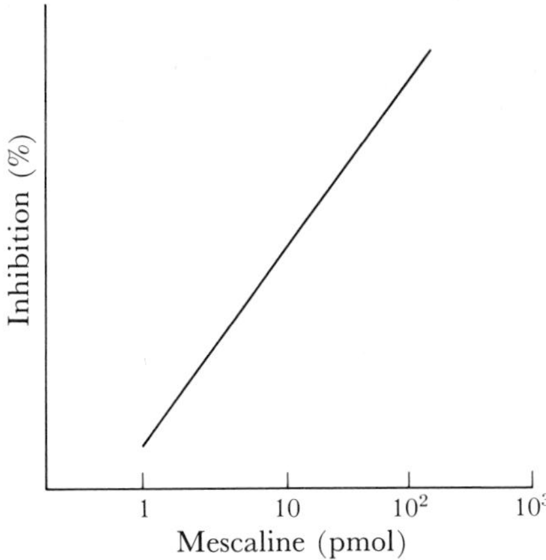

FIGURE 10-7
A typical standard curve for the radioimmunoassay. Antibody to mescaline and ^{125}I-mescaline were mixed with various quantities of unlabeled mescaline. The inhibition of precipitation of radioactivity by added mescaline provides the standard curve. Note that the x-axis is a logarithmic scale. [Courtesy of Helen Van Vunakis.]

Dextran-coated Activated Charcoal Method. Activated charcoal has the widely used property of adsorbing many small molecules virtually instantaneously. Proteins and the Ab-Ag* complex also adsorb, though more slowly and to a lesser degree. However, when charcoal is coated with the cross-linked polysaccharide dextran (a molecular sieve that cannot be penetrated by large molecules, see Chapter 8), both Ab and Ab-Ag* fail to adsorb. Hence, if dextran-coated charcoal is added to a mixture containing free Ag* and Ab-Ag*, immediate centrifugation results in pelleting of non-Ab-bound radioactivity (i.e., Ag*) and Ab-Ag* radioactivity remains in the supernatant. This is a convenient and rapid technique and is shown in Figure 10-8A.

Double-antibody Method. Antibody is normally prepared in a rabbit. Rabbit γ-globulin is antigenic in other animals and has been used to generate antirabbit γ-globulin in the goat, sheep, horse, and cow. The addition of goat antirabbit serum to rabbit γ-globulin results in the formation of precipitin, which can be collected by centrifugation (Figure 10-8B). Hence, the following procedure is used. To each of the tubes used to generate the standard curve and to the sample tubes (all of which have been previously incubated sufficiently long to form Ab-Ag*) is added goat antirabbit Ab at the equivalence point of the precipitin reaction (see Figure 10-1). The mixture is then incubated (typically, for a period ranging from 16 to 48 hours) and, after incubation, the sample is centrifuged. Ab-Ag* is found in the pellet; free Ag* is in the supernatant. This general method is highly satisfactory but requires maintaining large animals as a source of antiserum.

The radioimmunoassay differs from other immunoassays in that radioactive antigen is required. Occasionally, the appropriate ^{3}H- or ^{14}C-labeled compound can be purchased or synthesized from labeled precursors. More

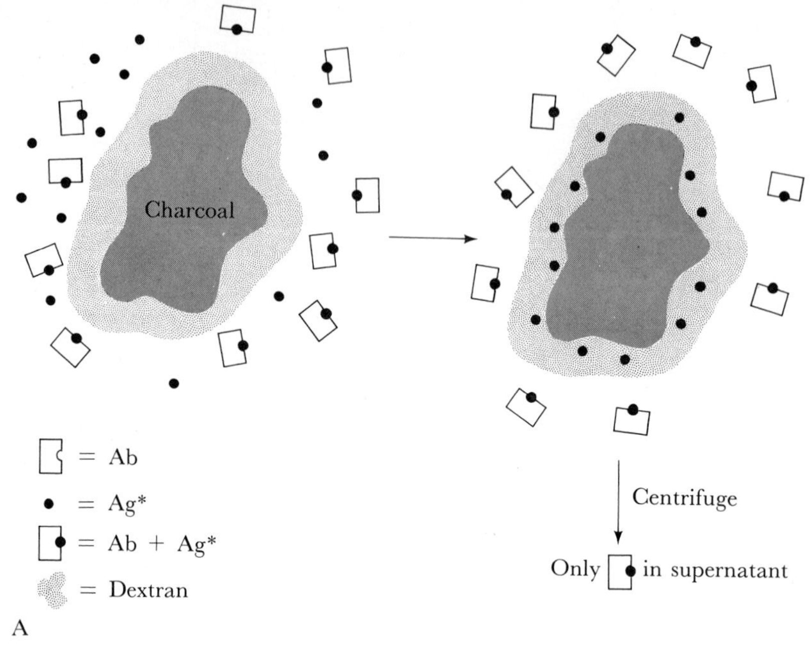

$Ab(g)$ = goat antirabbit γ-globulin
$Ab(r)$ = rabbit antiantigen

$$Ab(r) + Ag^* \longrightarrow Ab(r)Ag^* + Ag^*$$

Excess (Ag*) — Less (Ag*)

↓ Add Ab(g)

$$Ab(g)[Ab(r)Ag^*] + Ag^*$$

↓ Centrifuge

$Ab(g)[Ab(r)Ag^*]$ in pellet
Ag^* in supernatant

B

FIGURE 10-8
Methods for separating Ab-Ag* from Ag*: (A) dextran-coated charcoal method (because Ag* is a small molecule, it can penetrate the dextran matrix); (B) double-antibody method (the Ab-Ag* complex forms precipitin in box when goat antibody is added).

commonly, the antigen or hapten is iodinated with ^{125}I or ^{131}I; there is a special advantage in using ^{125}I because this isotope is a strong γ-ray emitter. As explained in Chapter 5, the β emitters have to be counted by liquid scintillation counting and because the sample in a radioimmunoassay frequently contains a substantial amount of protein, quenching in the scintillation system is a severe problem. With ^{3}H and ^{14}C, the quenching can be quite strong. With γ emitters, quenching rarely occurs because of the high energy of the radiation.

The preparation of iodinated antigens is not always simple, especially because it is necessary that the substance not suffer any damage that might affect its antigenicity. Iodination of proteins or an amino group, or of haptens containing an amino group, can be accomplished by reaction with ^{125}I-3-(4-hydroxyphenyl)-propionic acid *N*-hydroxysuccinimide ester. For proteins containing tyrosine or haptens with a phenolic group, the aromatic ring can be iodinated in a variety of ways. Nucleic acids can also be iodinated by a reaction with cytosine.

In some cases, there are no functional groups that can be directly iodinated; in these cases, the compound is first conjugated with tyrosine or histamine and the conjugate is iodinated.

THE IMMUNORADIOMETRIC ASSAY

In some cases, there is no satisfactory way to prepare a radioactive antigen —which precludes use of the radioimmunoassay. A variation of this assay, the immunoradiometric assay (IRA), allows the assay of such substances. It is best described by the diagram in Figure 10-9. Purified antigen is

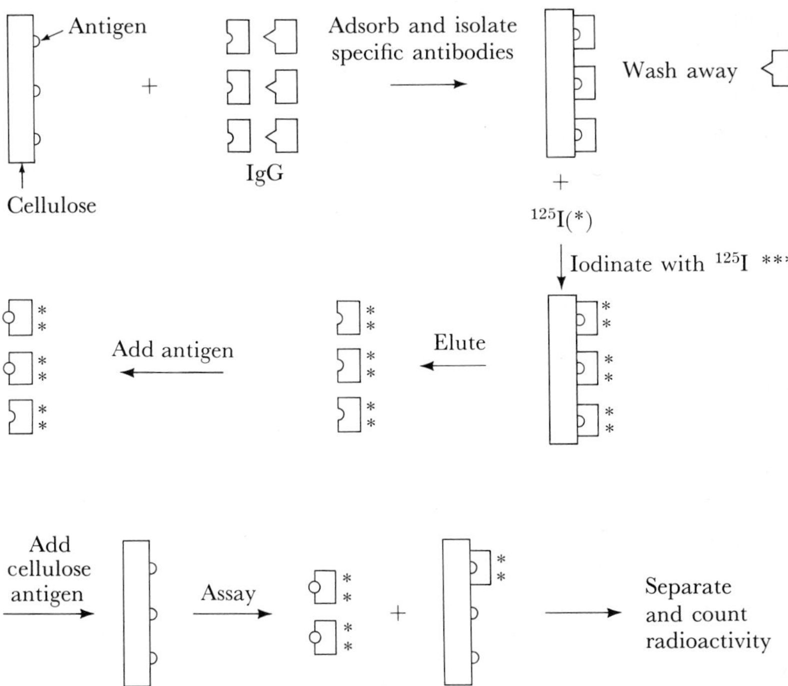

FIGURE 10-9
Principle of the immunoradiometric assay. See text for details.

adsorbed or coupled to a stable substance such as cellulose (or Sepharose treated with cyanogen bromide, see Chapter 8). Purified immunoglobulin (IgG) obtained from sera containing an antigen-specific antibody is added. Unadsorbed antibodies are removed by washing and the adsorbed antigen-specific antibody is iodinated with radioactive iodine (note that here, unlike RIA, the antibody is radioactive instead of the antigen). The radioactive antibody is then removed from the adsorbant and used in the assay. For assay, Ab* is mixed with the antigen to be determined, using antibody excess (in RIA, antigen excess is used). The mixture is then added to the cellulose, which contains Ag; only unreacted Ab* can bind to the cellulose. Reacted Ab* is washed off and counted and is a measure of the amount of Ag. The immunoradiometric assay is not yet widely used but certainly will be in the future.

EXAMPLES OF IMMUNOLOGICAL PROCEDURES USED IN BIOASSAYS

The following examples show the great variety of systems that can be examined by immunological procedures. Note that, in some cases, the immunoassay is used qualitatively and its value lies in its extraordinary sensitivity. In other cases, it is a quantitative and precise analytic tool.

Example 10-A. Identification and classification of various bacteria by agglutination.

Many bacteria isolated in nature or obtained from patients can be identified by precipitation or agglutination with specific antibodies directed against known strains (Figure 10-10). The reaction can be detected either by observation of clumping with the light microscope or by changes in turbidity of a suspension measured spectrophotometrically.

Example 10-B. Identification of viruses by inhibition of virus-induced hemagglutination.

It is often necessary both in the laboratory and clinically to identify viruses. This can be easily done with a simple inhibition assay. Certain classes of viruses have the property of causing the clumping of red cells (*hemagglutination*). If a particular virus is causing agglutination, the addition of an antibody directed against that virus will bind to the virus and thereby prevent hemagglutination, but other antibodies will not. Therefore, if one is armed with a collection of antibodies directed against many viruses, the particular virus can be identified.

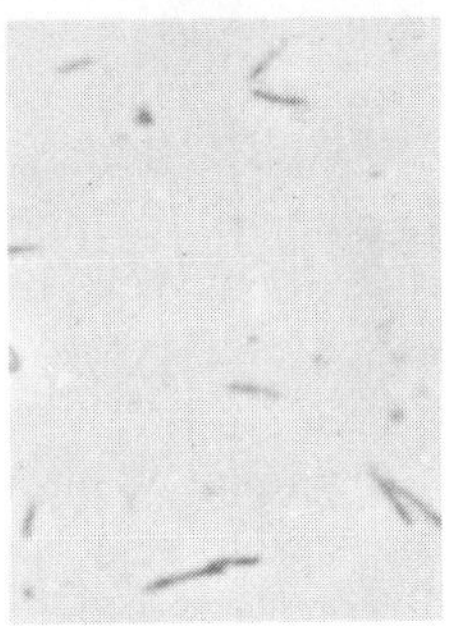

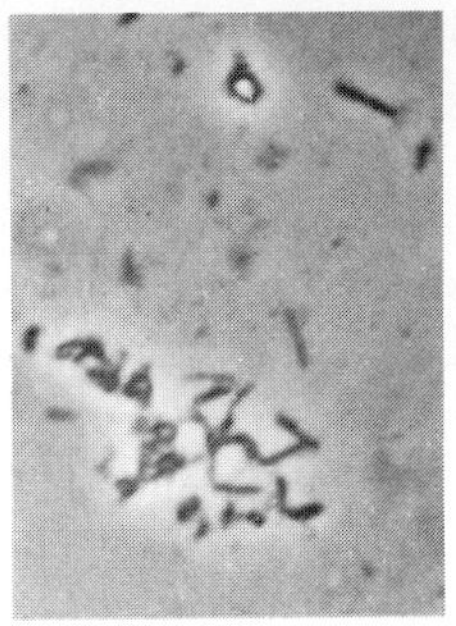

A B

FIGURE 10-10
E. coli bacteria agglutinated by the addition of specific antiserum: (A) no antiserum; (B) with antiserum.

Example 10-C. Detection of gonadotrophins in the urine of pregnant women by an inhibition assay and passive agglutination.

Small polystyrene particles can be coated with human gonadotrophin, a hormone found in the urine of pregnant women. The addition of antigonadotrophin causes these coated particles to clump (Figure 10-11); this clumping is easily seen with a light microscope. If urine from a pregnant woman is added to the coated particles before adding antibody, urinary gonadotrophin competes with the coated particles for sites on the antibody so that the aggregates do not form. This is a useful pregnancy test.

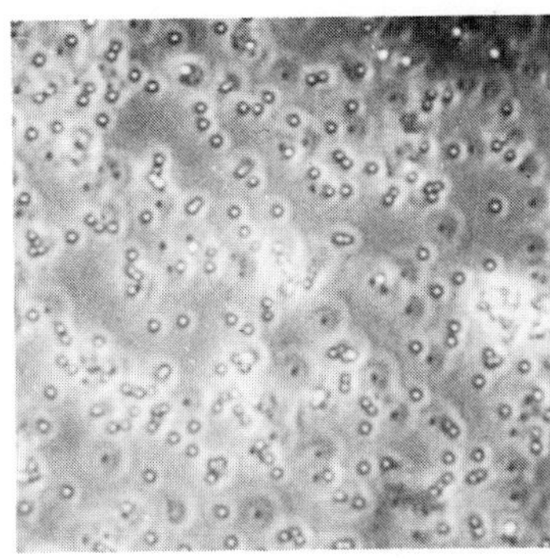

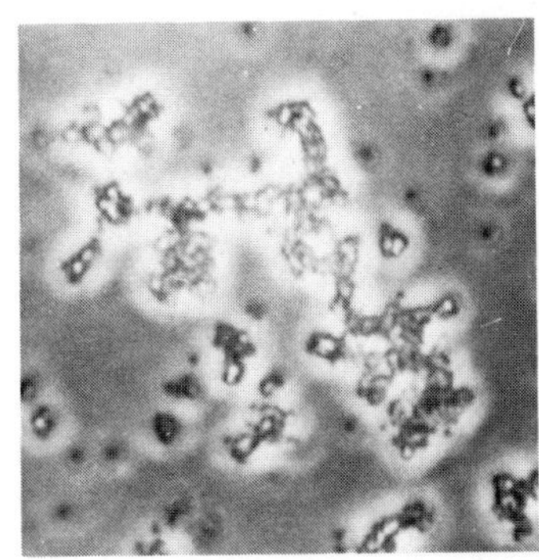

A B

FIGURE 10-11
Clumping of gonadotrophin-coated polystyrene particles by antigonadotrophin: (A) no antiserum; (B) with antiserum.

Example 10-D. Measurement of the synthesis of bacteriophage tail fibers in infected *E. coli* bacteria, using an inhibition assay.

Antibodies prepared against *E. coli* bacteriophage T2 are directed primarily against T2 tail fibers; hence, the addition of anti-T2 inactivates the phage by preventing adsorption to the host. This is an example of the neutralizing power of antibodies. The intracellular synthesis of tail fibers can be followed by preparing cell extracts at various times after infection, mixing aliquots with anti-T2, and then adding phage. If tail fibers are present, they will bind to the anti-T2, the anti-T2 will be titrated (i.e., inhibited), and the phage will survive. Hence, the number of viable phage remaining after such treatment measures the amount of tail fibers in the extract (Figure 10-12). This inhibition assay is known in phage research as the *serum blocking power assay*.

Example 10-E. Measurement of T4 DNA synthesis in infected *E. coli* by complement fixation.

Antibodies to denatured, glucosylated DNA can be prepared. *E. coli* bacteriophage T4 DNA contains glucosylated 5-hydroxymethylcytosine instead of cytosine and is therefore distinguishable from *E. coli* DNA in that T4 but not *E. coli* DNA will react with the antibodies. Using complement fixation, the course of synthesis of T4 DNA during infection has been followed by preparing cell extracts, denaturing the DNA, and assaying for glucosylated, denatured DNA (Figure 10-13).

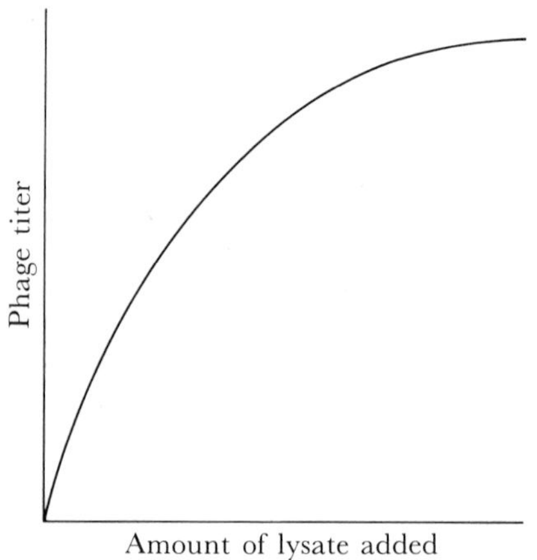

FIGURE 10-12
Serum blocking power assay. Anti-T2 is added to phage T2, along with various amounts of an extract of bacteria infected with a phage mutant that fails to make phage heads. In the absence of the extract, the antiserum inactivates the phage. If phage tails (to which the antiserum is directed) are present in the extract, they compete with the phage for the antibody so that phage survive. This can be used to determine the number of phage equivalents of tails present in the extract.

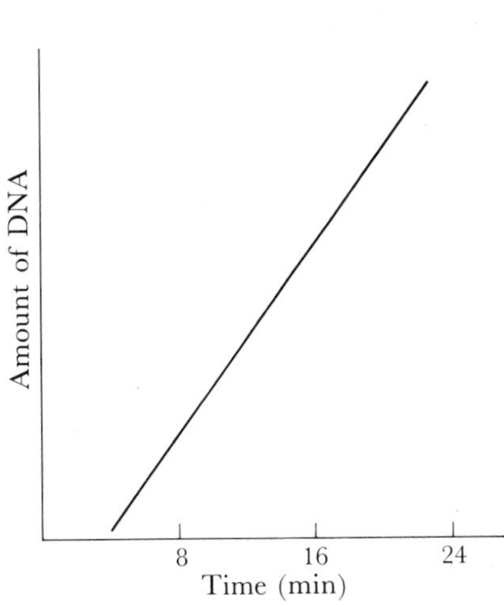

FIGURE 10-13
Measurement of *E. coli* phage T4 DNA synthesis by complement fixation. *E. coli* B was infected with five phages per bacterium. Samples were taken at various times and assayed for the presence of glucosylated DNA, using complement fixation with antibody to glucosylated DNA. The virtue of this method is that *E. coli* DNA is nonglucosylated and does not react with the antibody, whereas T4 DNA contains glucosylated 5-hydroxymethylcytosine. Therefore, the antibody detects the T4 DNA and ignores the huge amount of bacterial DNA. [From L. Levine, W. T. Murakami, H. Van Vunakis, and L. Grossman, *Proc. Nat. Acad. Sci.* 46(1960):1038–1043.]

Example 10-F. Effect of denaturants on DNA.

The denaturation of DNA is usually studied by measuring the increase in optical density at 260 mμ (see Chapter 14). If a potential denaturant also absorbs light of this wavelength, this simple assay cannot be carried out. However, denaturation can be detected immunologically, because several DNA antibodies react only with denatured DNA. In fact, the effects of a large number of destabilizing substances have been assayed by complement fixation, the extent of denaturation being determined as a function of denaturant concentration. An example of some of the data obtained is given in Table 10-2.

Example 10-G. Identification of the *E. coli* phage λ beta protein.

If *E. coli* is infected with phage λ, a large amount of an exonuclease is synthesized. This nuclease was purified and anti-λ exonuclease was prepared. When tested for homogeneity by Ouchterlony immunodiffusion (Figure 10-14), two bands were found, indicating that at least two proteins were present in the "pure" sample. One band resulted from the product of a recombination gene called *exo.* The other proved to be the product of another gene called *beta.* The beta protein complexes

TABLE 10-2
Reagent concentration giving 50% denaturation at 73°C and ionic strength of 0.043 of *E. coli* bacteriophage T4 DNA, assayed by complement fixation.

Reagents	Concentration (molarity)
Methanol	3.50
Ethylene glycol	2.20
Glycerol	1.80
1,4-Dioxane	0.64
Phenol	0.08
Formamide	1.90
Urea	1.00
Thiourea	0.41

SOURCE: Data from L. Levine, J. A. Gordon, and W. P. Jencks, *Biochem.* 2(1963):168–175.

NOTE: DNA was placed in solutions containing various concentrations of the reagents and incubated at 73°C. The solutions were then diluted and assayed by complement fixation, using a DNA antibody that detects only single-stranded DNA. Fifty-four different reagents were tested.

with the exonuclease but has no detectable enzymatic activity in vitro. Its synthesis and genetic control has been studied, using immunodiffusion as an assay because no other assay for beta protein exists.

Example 10-H. Detecting *E. coli* DNA polymerase III in the presence of polymerase I by neutralization.

If two enzymes participate in similar reactions, it is often difficult to assay one in the presence of the other. However, in assaying DNA polymerases, polymerase I activity can be eliminated from a cell extract by the addition of antipolymerase I before adding the extract to the reaction mixture. The neutralizing power of an antibody is employed as a biochemical tool, thus allowing polymerase III to be assayed unambiguously.

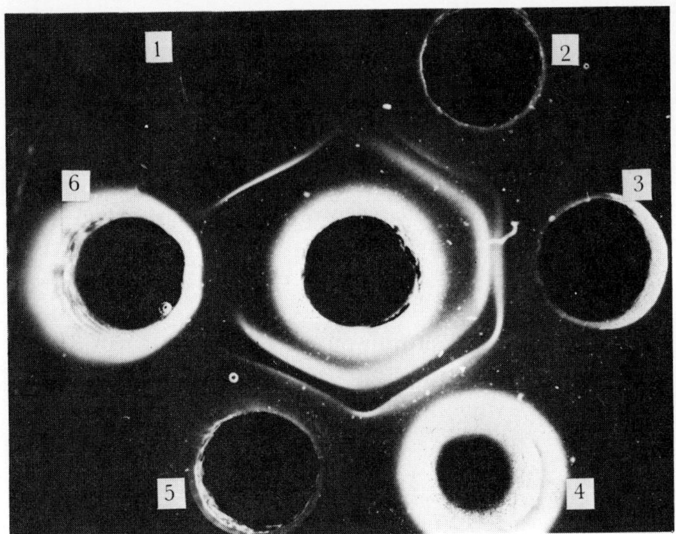

FIGURE 10-14
Ouchterlony analysis of the *E. coli* phage λ exonuclease and β protein. The center well contains antibody directed against presumably pure λ exonuclease. Well 1 contains highly purified λ exonuclease and a single precipitin line is seen. Well 3 is the same as Well 1 but less pure. Well 2 contains a partially purified extract of a cell infected with wild-type λ. The exonuclease (Exo) band and the inner band (β protein) is present. Well 4 is the same as Well 2 but less pure. Well 5 contains the result obtained with a λ mutant called *t11,* which overproduces Exo and fails to make all replicative and structural phage proteins. Beta is present, indicating that it is made in the part of the genome containing the *exo* gene but is not in the replicative or structural region. Well 6 contains an extract from a λ N-mutant. Both bands are absent indicating that Exo and β are both under the control of gene *N.* [From C. M. Radding and D. C. Shreffler, *J. Mol. Biol.* 18(1966):251–261.]

Example 10-I. Detection of mutant proteins or of protein fragments by cross-reaction.

Proteins made from mutant genes contain either a new amino acid replacing the one found in the wild-type protein or, with chain-termination mutations, fragments of the normal protein. In both cases, the activity of the protein (e.g., as an enzyme) is lost and there is no longer a bioassay. If the altered protein or the fragment still contains

TABLE 10-3
Immunological cross-reaction between different mutants of *E. coli* β-galactosidase.

Region of genetic map in which mutant occurs	Amount of protein (in μg) for 50% complement fixation*
Wild-type	0.06
1	NF†
2	NF
3	NF
4	NF
5	4.0
6	1.0
7	2.0
8	NF
9	NF
10	1.4
11	0.5
12	1.4
13	0.6
14	1.3
15	0.5
16	0.5

Conclusions

1. The enzyme β-galactosidase can be detected immunologically even if no enzymatic activity is detectable.

2. Regions 1–4 and 8–9 either are the antigenic sites or more effectively alter the three-dimensional conformation of the antigenic site, because mutations in those regions eliminate all antigenic activity.

SOURCE: Data from A. V. Fowler and I. Zabin, *J. Mol. Biol.* 33(1968):35–47.

*The nearer the value is to the wild-type figure of 0.06, the more closely the antigenic determinant of the mutant protein resembles that of the wild type.

†No fixation detectable.

the same antigenic determinant (e.g., an amino acid sequence) as that of the wild-type protein, it can be detected immunologically. Complement fixation, immunodiffusion, or the radioimmunoassay would be the methods of choice. An example of this approach is shown in Table 10-3.

Example 10-J. Detecting relatedness between proteins by cross-reaction. Certain proteins—for example, those found in the glial cells of the brain—are widespread in nature. Antibody to a particular rabbit glial protein has been found by means of complement fixation to react with proteins found in nerve tissues of species ranging throughout the evolutionary scale from frogs to humans. Representative data, which can be taken as evidence for related structure, are given in Table 10-4.

TABLE 10-4
Cross-reaction between the brain S-100 proteins from various organisms and antibody to cattle S-100 protein.

Animal	Relative antiserum concentration required for 50% fixation of complement*
Cow	1.0
Sheep	1.0
Rat	1.0
Guinea pig	1.2
Mouse	1.3
Pig	1.8
Rabbit	1.9
Human	2.0
Chicken	2.2
Pigeon	2.3
Bullfrog	2.5

SOURCE: Data from D. Kessler, L. Levine, and G. Fasman, *Biochemistry* 7(1968):758–764.

*The nearer the value is to 1, the more closely related is the protein.

EXAMPLES OF THE USE OF THE RADIOIMMUNOASSAY

Table 10-5 gives a variety of substances that have been assayed using the radioimmunoassay. The catalog of such substances increases daily.

Example 10-K. Tumor diagnosis.

Some tumors produce tumor-associated antigens that can be found in blood or urine. In some cases—for example, human chorionic gonadotrophin, human placental lactogen, and placental alkaline phosphatase—these antigens are not present in normal blood but are found in serum from a certain percentage of patients with particular kinds of carcinoma.

If the tumor is in tissue that normally controls or makes a hormone, often associated with the tumor is an increased level of the particular hormone—for example, high insulin with pancreatic cancer. Each of these substances can be detected by the radioimmunoassay, with far greater sensitivity than with any known chemical or bioassay.

Example 10-L. Assay of clinically important substances in children or in general if only small amounts are available.

Many clinically important substances can be assayed in large blood samples by conventional chemical tests. However, for babies, small children, seriously ill patients, and small animals, the required sample size may be greater than that which can be safely withdrawn. The great sensitivity of the radioimmunoassay avoids this problem.

Example 10-M. Measurement of pyrimidine dimers in ultraviolet irradiated DNA.

In the study of mechanisms of repair of irradiated DNA in vivo, it is almost always necessary to measure the number of pyrimidine dimers (the principal product of ultraviolet irradiation of DNA). This can be done by paper chromatography or electrophoresis if the intracellular DNA can be labeled to high specific activity with ^{3}H. However, this cannot be done with animal or plant cells. Furthermore, it is difficult to detect a very small number of dimers—for example, 1 per 10^8 base pairs—by conventional means, whereas it can be done by the radioimmunoassay.

Example 10-N. The role of prostaglandins in humans.

Human venous blood contains from 10 to 100 picograms per milliliter of various polyunsaturated fatty acids called prostaglandins. Understanding their significance has been hampered by the inability to assay these substances quantitatively and with high sensitivity; the radioimmunoassay provides this opportunity as well as the ability

TABLE 10-5
Substances that have been used in the radioimmunoassay.

Hormones	
Gastrointestinal (glucagon, gastrin, enteroglucagon, secretin, pancreozymin, vasoactive intestinal peptide, gastric inhibitory peptide, motilin, insulin)	Prolactin Thyrocalcitonin Parathyroid hormone Human chorionic gonadotrophin
Corticotrophin Follicle-stimulating hormone	Posterior pituitary peptides (oxytocin, vasopressin, neurophysin)
Antidiuretic hormone	Bradykinin
Thyroid-stimulating hormone	Thyroid hormones
Pharmacologic agents	
Morphine and opiate alkaloids	Tetrahydrocannabinol
Cardiac glycosides	Barbiturates
Prostaglandins	Nicotine and metabolic products
Lysergic acid and derivatives	Phenothiazines
Amphetamines	
Vitamins and cofactors	
D, B12, folic acid, cyclic AMP	
Hematological substances	
Fibrinogen, fibrin, and fibrinopeptides	Prothrombin
Plasminogen and plasmin	Transferrin and ferritin
Antihemophilic factor	Erythropoietin
Virus antigens	
Hepatitis antigen	Polio
Herpes simplex	Rabies
Vaccinia	Q fever
Several Group A arboviruses	Psittacosis group
Nucleic acids and nucleotides	
DNA, RNA, cytosine derivatives	

FIGURE 10-15
Detection of thymine dimers ($\widehat{TT}$) in DNA by the radioimmune assay, using an antibody directed against ultraviolet-irradiated DNA, that is, anti-$\widehat{TT}$. A mixture is made of anti-$\widehat{TT}$ and iodinated UV-irradiated DNA and the amount of radioiodine precipitated by goat antirabbit serum is measured. The decrease in precipitated radioactivity as a function of added UV-irradiated DNA is measured. From a standard curve, the percentage of inhibition can be related to the number of thymine dimers. Each curve represents DNA obtained from a different bacterium. *Micrococcus luteus* DNA has a lower thymine content than *Proteus vulgaris* DNA and therefore requires a higher dose to yield the same number of dimers. [From E. Seaman, H. Van Vunakis, and L. Levine, *J. Biol. Chem.* 247(1972):5709–5715.]

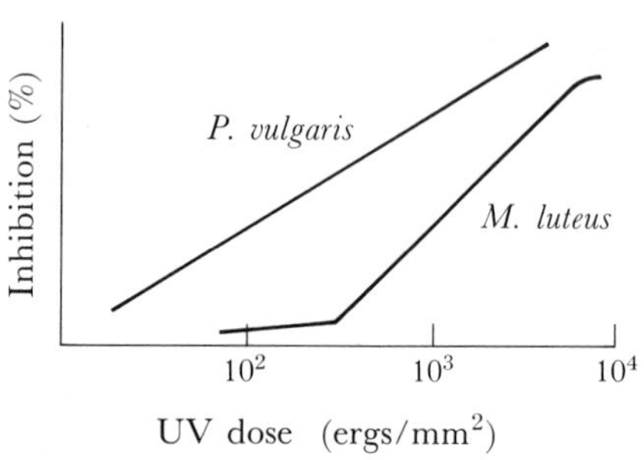

to analyze a large number of samples in a short time. This becomes especially important (i.e., to monitor blood levels) because many prostaglandins have recently been shown to have pharmacological activity—but with a narrow therapeutic margin of safety.

IMMUNOLOGICAL TECHNIQUES FOR LOCALIZING SUBSTANCES IN CELLS, TISSUES, AND MOLECULES; THE FLUORESCENT ANTIBODY AND THE FERRITIN-CONJUGATED ANTIBODY

The fluorescent dye fluorescein can be chemically coupled to most antibody molecules without loss of antibody activity. This material can be used to localize structures in cells and tissues by fluorescence microscopy (see Chapter 2). An example is the identification of cells to which a particular virus is bound at the surface, using fluorescein-conjugated antivirus. The presence of the virus is indicated by the fluorescence even though the size of the virus is beyond the limit of resolution of the microscope (see Chapter 2, page 24).

The iron-containing protein ferritin can also be coupled to antibody. If specific ferritin-conjugated antibody is reacted with cells and the cells are examined by electron microscopy, the antibody molecules can be localized by the position of the dark dots corresponding to the ferritin. This allows localization of particular molecules at high resolution. An example is shown in Figure 10-16.

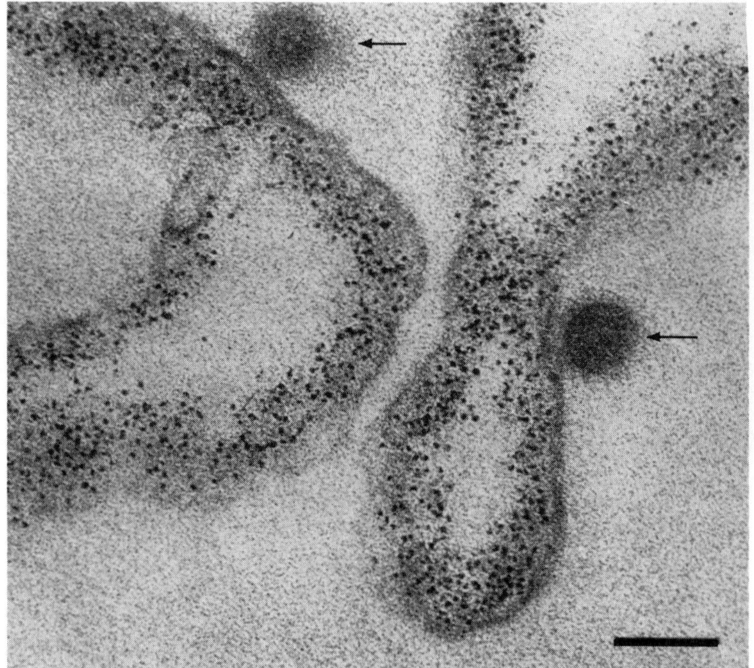

FIGURE 10-16
An example of the use of ferritin labeling. A human red cell has been broken open and incubated with a mixture of influenza virus (arrow) and ferritin-conjugated antibody directed against the protein spectrin. The influenza virus adsorbs to the outer membrane of the red cell. The ferritin is seen as dark dots and is observed only on the opposite side of the membrane from the virus. This shows that spectrin is primarily, if not solely, on the inner surface of the red cell membrane. Bar = 0.01 μ. [From G. L. Nicolson, V. T. Marchesi, and S. J. Singer, *J. Cell Biol.* 51(1971):265–272.]

Selected References

Berson, S. A., and R. S. Yalow. 1971. "Radioimmunoassay: A Status Report," in *Immunobiology,* edited by R. A. Good and O. W. Fisher, pp. 287–293. Sinauer Associates. An excellent description of the radioimmunoassay.

Bremer, J. M., A. J. Pesce, and R. B. Ashworth. 1974. *Experimental Techniques in Biochemistry,* ch. 4. Prentice-Hall.

Clausena, J. 1969. "Immunochemical Techniques for the Identification and Estimation of Macromolecules," in *Laboratory Techniques in Biochemistry and Molecular Biology,* vol. 1, edited by T. S. Work and E. Work, pp. 405–556. North-Holland.

Kabat, E. A. 1968. *Structural Concepts in Immunology and Immunochemistry.* Holt, Rinehart and Winston.

Kwapinski, J. B. 1972. *Methodology of Immunochemistry and Immunological Research.* Wiley-Interscience.

Wasserman, E., and L. Levine. 1961. "Quantitative Micro-complement Fixation and Its Use in the Study of Antigenic Structure by Specific Antigen-Antibody Inhibition." *J. Immunol.* 87:290–295. The definitive paper on micro-complement fixation.

Williams, C. A. 1967. *Methods in Immunology and Immunochemistry.* Academic Press.

Problems

10-1. A complement fixation assay is performed, using an antigen X and anti-X. A standard curve is constructed by adding various amounts of X. The data are:

X added (μg)	OD	X added (μg)	OD
0.00	1.15	0.05	0.17
0.01	0.82	0.06	0.19
0.02	0.50	0.07	0.23
0.03	0.22	0.08	0.50
0.04	0.18	0.09	0.75

You wish to determine the amount of X in a cell extract. This is done by adding 1 μl of the extract and then 1:1 and 1:2 dilutions of the extract to the mixture (without adding any X); the OD readings are 0.20, 0.66, and 0.85. What is the concentration of X in the extract?

10-2. Suppose that you have a mixture of two proteins and antiserum prepared against the mixture. If the two proteins have the same molecular weight and shape, they will have nearly the same diffusion coefficient (see Chapter 12). If this mixture is subjected to the Ouchterlony diffusion test, the probability is very high that two bands will result. Why? Under what circumstances would only one band appear? Suppose that the two proteins had very different diffusion coefficients; under what circumstances might the bands fail to resolve?

10-3. Would you expect the reaction between specific antibody and a protein antigen (assayed in any way) to depend on whether the protein is native or denatured? Explain.

10-4. Why should single amino acid changes in a protein (a) eliminate all antigenicity, (b) have no effect, or (c) produce cross-reacting material?

10-5. Suppose that you wanted to measure the level of a particular drug in the bloodstream of an animal. The standard chemical assay for the drug cannot be carried out in the presence of the amount of protein in blood. How would you go about setting up an assay using immunological methods? Describe all necessary steps.

PART IV

Hydrodynamic Methods

CHAPTER 11

Sedimentation

One of the more common techniques in use today for the characterization of macromolecules is sedimentation.* By using the appropriate variant of the technique, the molecular weight, density, and shape of a macromolecule can be obtained, changes in these parameters can be detected, and any of them can be used as the basis of the separation of the components of a mixture for preparative or analytic purposes. Furthermore, the ease with which measurements can be made with modern automatic instruments makes ultracentrifugation especially useful.

Basically, only one thing is done with an ultracentrifuge: particles are made to move by centrifugal force and the distribution in concentration of the particles along the length of the centrifuge tube is determined at one or more times. A measurement made while the molecules are moving along the centrifugal axis is called a *sedimentation velocity determination* and the result is a *sedimentation coefficient,* a number that gives information about the molecular weight and the shape of the particle. When the concentration distribution is measured under conditions such that the distribution no longer changes with time, the particles are said to have reached *sedimentation equilibrium;* this type of measurement yields data about molecular weights, density, and composition.

---

*Sedimentation is a general term for motion in a centrifugal field. If sedimentation analysis is performed using a high speed centrifuge called an ultracentrifuge, the word ultracentrifugation is commonly used.

In principle, sedimentation is very simple. However, in practice, certain complications must be taken into account to obtain meaningful information.

In this chapter, both the principles and the complexities of many ultracentrifugal procedures are described, and the kinds of measurements that are made are shown by example.

SIMPLE THEORY OF VELOCITY SEDIMENTATION

If a particle is in a centrifugal field generated by a spinning rotor with angular velocity, ω, it will experience a centrifugal force, $F_c = m\omega^2 r$, in which m is the mass of the particle and r the distance from the center of rotation. If this particle is not in vacuum but in a solvent,* the solvent molecules will, of course, be displaced by the motion of the particle. Their resistance to this displacement constitutes a buoyant force, opposed to the centrifugal force. This *buoyancy* reduces the net force on the macromolecule by $\omega^2 r$ times the mass of the displaced solution; this mass is simply the volume of the particle multiplied by the density, ρ, of the solvent. The particle volume is $m\bar{v}$, in which $\bar{v}$ is the partial specific volume† of the particle, so that the buoyant force is $\omega^2 r m\bar{v}\rho$.

Clearly, the particles and the solvent molecules cannot slip by one another without experiencing friction. This frictional force—which opposes the motion of both—is proportional to the difference between the velocities of the particle and of the solvent molecules and is expressed as fv, in which f is the frictional coefficient and v is the velocity relative to the centrifuge cell, which holds the solvent. Because the velocity is constant when the net force is zero, the velocity of the particle is

$$v = \frac{\omega^2 rm(1 - \bar{v}\rho)}{f} \tag{1}$$

Equation (1) says that (all other things being equal):

1. A more massive particle (or molecule) tends to move faster than a less massive one.
2. A denser particle (i.e., small partial specific volume) moves faster than a less dense one.

---

*Solvent is used in the general sense of the suspending medium. That is, if the particle is in water, the solvent is water; if it is in 0.1 M NaCl, the NaCl solution is considered to be the solvent.

†A detailed discussion of partial specific volume is the subject of Chapter 12.

3. The denser the solution, the more slowly the particle will move.
4. The greater the frictional coefficient, the more slowly the particle will move.

These four statements constitute the basic rules of sedimentation and apply to all particles whether they are large structures, macromolecules, or small molecules.

Because the velocity of a molecule is proportional to the magnitude of the centrifugal field (i.e., $\omega^2 r$), it is common to discuss sedimentation properties in terms of the velocity per unit field, or

$$s = \frac{v}{\omega^2 r} = \frac{m(1 - \bar{v}\rho)}{f} \tag{2}$$

in which s is the *sedimentation coefficient;* the determination of s is the immediate goal of most sedimentation velocity experiments (see page 334 for method).

Ideally, at a single temperature s would be a constant for a particle in a given solvent. However, this is not the case for macromolecules, owing to their great size, and the theory must be expanded. This is an appropriate place to do so but, because it will be more understandable through reference to real sedimentation data, the instrumentation of ultracentrifugation and the kinds of data obtained will be explained first.

INSTRUMENTATION OF ULTRACENTRIFUGATION

Centrifugation experiments require instruments that operate at accurately known speeds with small variation and without temperature fluctuations.* Modern ultracentrifuges operate at forces as great as 420,000 ± 100 g and with temperature control within approximately 0.1°C. Two types of instruments exist—*analytical* and *preparative.* Analytical centrifuges are equipped with *optical systems* designed to determine concentration distributions at any time during the measurement, whereas preparative centrifuges require fractionation of the contents of the centrifuge cell and measurement of the concentration in each fraction to obtain a concentration distribution. Preparative centrifuge is certainly a misnomer because, in addition to being used in the preparation and purification of various macromolecules and cell organelles, it is used at least as frequently to

*The field of ultracentrifugation was born when the first ultracentrifuge was designed and built by The Svedberg in 1923 (see the Selected References near the end of the chapter).

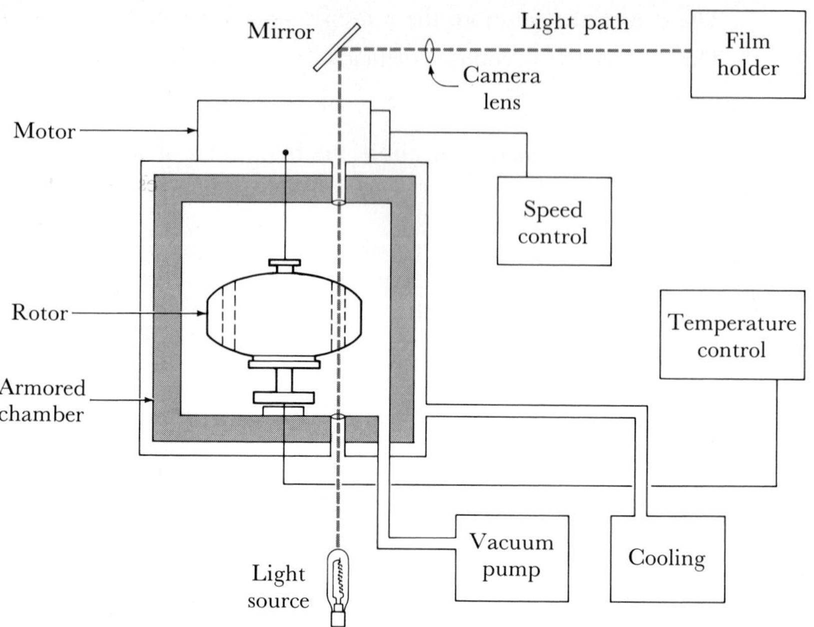

FIGURE 11-1
A Beckman analytic ultracentrifuge. The rotor is in an evacuated and cooled chamber and is suspended on a wire coming from the drive shaft of the motor. The tip of the rotor contains a thermistor for measuring temperature. Electrical contact of the thermistor to the control circuit is by means of a pool of mercury, which the rotor tip touches. The rotor chamber contains an upper and a lower lens. The lower lens collimates the light so that the sample cell is illuminated by parallel light. The upper lens and the camera lens focus the light on the film.

analyze mixtures in a quantitative way (i.e., as an analytical instrument).

Figure 11-1 is a schematic diagram of a Beckman analytical ultracentrifuge. This instrument consists basically of a motor; a centrifuge rotor, which is contained in a protective armored chamber; and a photographic system for recording the distribution in concentration of the sample in the centrifuge cell (see figure legend for details).

Figure 11-2 shows a standard rotor for the Beckman analytical centrifuge. The rotor is suspended in the center from the drive motor by a wire. One of the holes in the rotor contains the centrifuge cell; the other contains a counterbalance and reference hole for determining distances from the center of rotation. Figure 11-3 shows a diagram of one of the simplest sample holders—a single-sector cell. The walls of the cell are designed

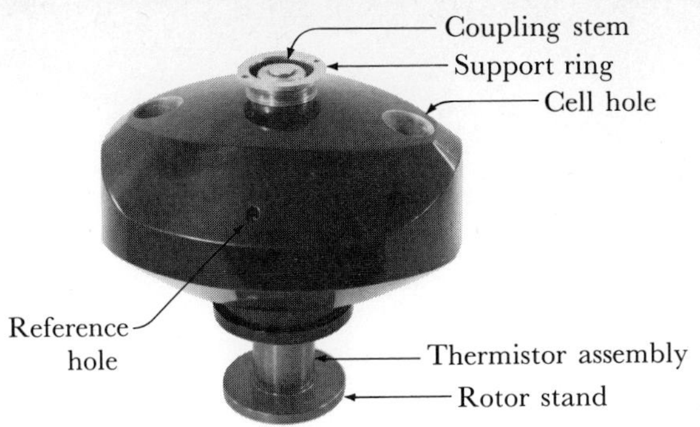

FIGURE 11-2
An analytical rotor. [Courtesy of Beckman Instruments.]

so that, if the cell is carefully oriented in the rotor, the walls will be parallel to the lines of centrifugal force. This insures that there is no pile-up of material against the walls. The cell consists of a centerpiece, which contains the liquid sample; two windows; a housing to hold the centerpiece and the windows; and a filling port. The windows are usually made of quartz, although sapphire is sometimes used for very high speed operation because it deforms less at high forces. The centerpiece may be made of metal (usually aluminum) or plastic (either Kel-F or an epoxy resin). Metal centerpieces have the advantage that thermal equilibrium is rapidly achieved, thus minimizing convection produced by thermal gradients. On the other hand, the metal frequently interacts with the solvent or solute, in which case the inert plastic centerpieces must be used. Centerpieces are also made of titanium and have the advantages of both, but they are limited to lower speeds because of the high density of the metal.

For some purposes, a double-sector cell is used; one of the sectors contains the solution and the other contains the solvent, which provides a baseline for the optical system.

Concentration distributions within the cell are determined by passing light through the moving cell and recording the intensity of the transmitted light either on photographic plates or by means of an electronic scanner. The electronic scanner, though exceedingly expensive, provides accuracy not achievable with the photographic system.

Figure 11-4 is a schematic of a Beckman preparative ultracentrifuge. This instrument is much simpler than the analytical instrument.

Figure 11-5 shows two types of rotors for the preparative instrument. The angle rotors are primarily used for pelleting materials.* They are very efficient for this purpose because the material moves only a short distance to the wall of the centrifuge tube; the resulting accumulated material then

*Pelleting means sedimenting particles until they are tightly packed at the bottom of a centrifuge tube.

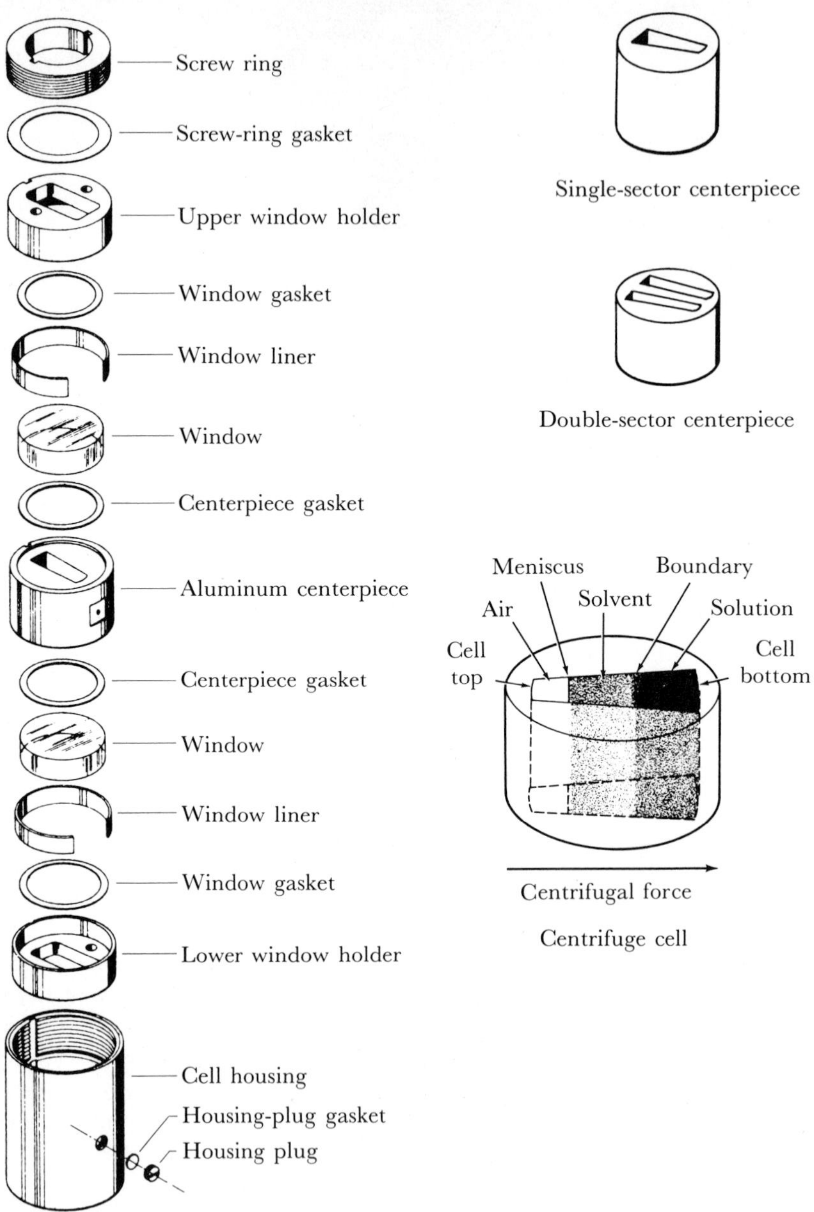

FIGURE 11-3
A centrifuge cell and the single- and double-sector centerpieces. [Courtesy of Beckman Instruments.]

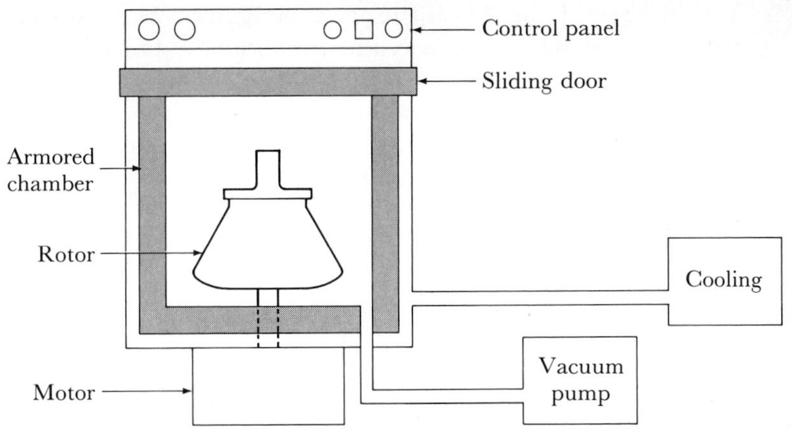

FIGURE 11-4
A Beckman preparative ultracentrifuge.

FIGURE 11-5
Beckman swinging-bucket rotor (left) and angle-head rotor (right). These rotors are designed to generate forces of 420,000 and 368,000 gravity, respectively, at their maximum speed of 65,000 rpm.

rapidly slides down the wall to the tube bottom. The swinging bucket rotors are used mostly for analytical equilibrium and zonal centrifugation (to be described shortly).

If the purpose of centrifugation is to pellet material on the tube bottom, the procedure is merely to pour off the supernatant liquid after the rotor has stopped spinning and the tube has been removed. If the centrifuge operation is for analytical purposes, the contents of the tube can be fractionated by puncturing a hole in the tube bottom and collecting the material by drops. If the rate of dripping is sufficiently slow that turbulence is not produced, each drop represents a single lamella from the tube as shown in Figure 11-6.

It should be noticed that the walls of the tubes used with the preparative rotors are parallel, in contrast with the sector-shaped cell of Figure 11-3. This produces convection and will be discussed later in the section on zonal centrifugation. To date, no one has designed a satisfactory preparative rotor tube that is sector-shaped.

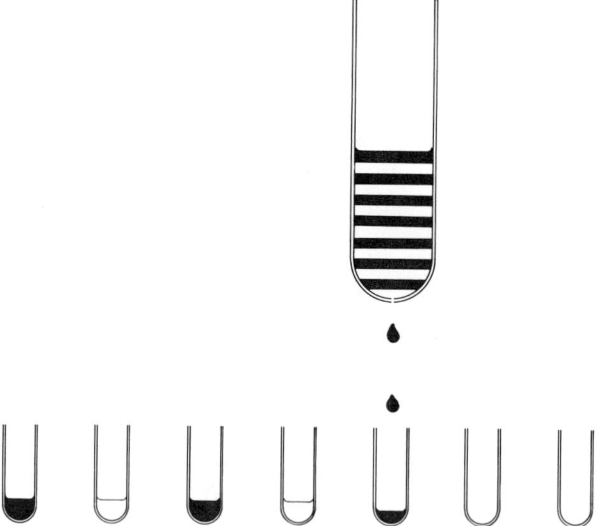

FIGURE 11-6
Fractionating the contents of a centrifuge tube by drop collection. The bottom of the tube is pierced with a needle. As long as the system is stabilized against convection (e.g., by a concentration gradient), the drops represent successive layers of liquid. These layers are shown schematically as alternating black and white.

DISTRIBUTION OF CONCENTRATION IN A BOUNDARY SEDIMENTATION VELOCITY EXPERIMENT PERFORMED IN A SECTOR-SHAPED CELL

Two types of initial conditions are used in sedimentation experiments. In one, the particles are distributed uniformly throughout the solution. As sedimentation proceeds, the particles move through the solution, leaving behind a region of pure solvent. This is called *boundary sedimentation* because all information is obtained from the parameters of the boundary between solvent and solution. The other type is *zonal sedimentation,* in which the sample is layered on top of a denser solvent. This section will confine itself to a discussion of boundary sedimentation.

Figure 11-7 shows the concentration distribution in sector-shaped centrifuge cells filled with different solutions of macromolecules after a period of centrifugation sufficient to move the material roughly halfway down the cell. The cell in series A contains material with a single sedimentation coefficient (it is common to use the phrase "homogeneous with respect to sedimentation coefficient" or the term "monodisperse"). It also has a low diffusion coefficient so that it gives a sharp boundary. The cell in series B contains a mixture of equal amounts of two components, each having a unique sedimentation coefficient and sharp boundaries, as in series A. The cell in series C contains a mixture of components whose sedimentation coefficients are slightly different from one another. The second row consists of plots of the concentration of the material in each cell that sediments. The region in which concentration changes is called the *boundary;* the upper flat portion is the *plateau.* The third row consists of plots of the derivative of the concentration curve, that is, the *concentration gradient.*

The particular points to be noticed are that (1) a homogeneous material gives a very sharp boundary, (2) a mixture of two components gives two sharp boundaries that look like steps, (3) a mixture of many components gives a boundary consisting of many steps that produces a broad, smooth curve, and (4) in each case, particles are pelleting on the tube bottom.

Let us consider the plot of concentration versus distance for series B in a little more detail. The curve has two measurable features—the distance sedimented (from which s is determined) and the relative heights of the two boundaries. Because concentrations are additive, the heights a and b tell the ratio of the two components. Clearly, if the two components are present in equal concentrations and if they sediment independently, $a = b$. In general, if the slower material a is a fraction, f, of the mass of the total material, then $f = a/(a + b)$. Let us now consider a system with many components. If there were three components, the boundary shown in Figure 11-8A would result. In general, if a mixture consists of a

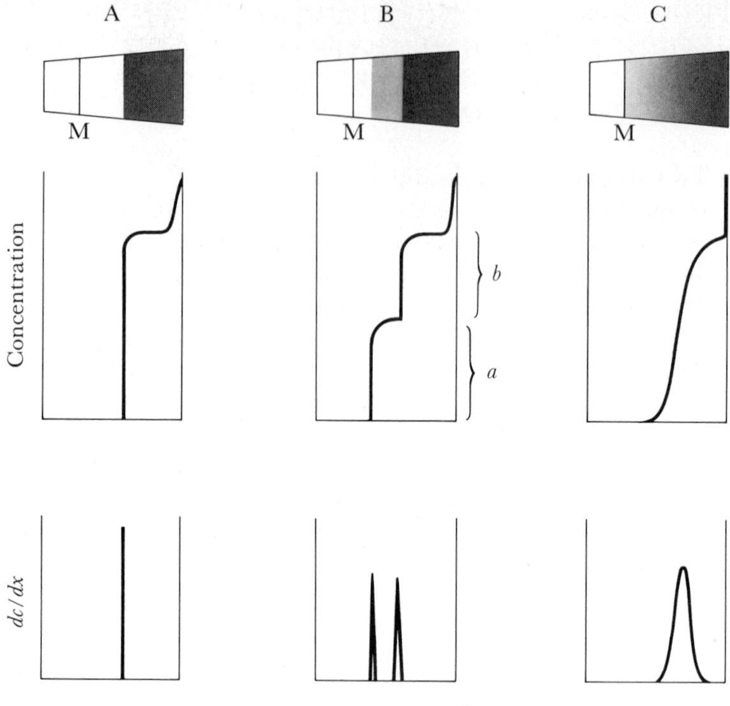

FIGURE 11-7
Concentration distribution after a given period for cells filled with three different solutions: (A) a single type of solute molecule having only one sedimentation coefficient; (B) two components, each having a distinct sedimentation coefficient; and (C) a heterogeneous mixture having a range of s values. Note the material accumulating on the cell bottom. The bottom row shows the concentration gradient; this is equivalent to a schlieren photograph. The letter M indicates the meniscus.

weight fraction f of a heterogeneous mixture and $1 - f$ of a fast component, the pattern of Figure 11-8B will result with $a/(a + b) = f$. Hence, because a plot of concentration versus distance can easily be converted into a plot of concentration versus sedimentation coefficient, the weight fraction of total material with s between s_1 and s_2 is $(c_2 - c_1)/c$, as shown in Figure 11-8C.

This same line of reasoning applies to plotting the concentration gradients, as in Figure 11-7. However, in this case, the measurement is of areas rather than distances.

How these simple observations can yield significant information about a real molecule can be seen by the following example.

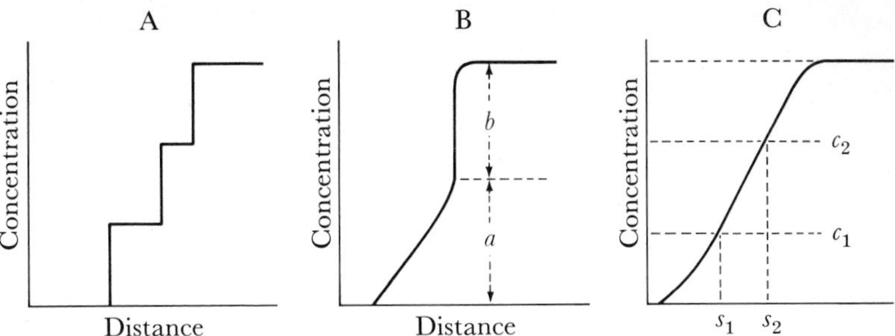

FIGURE 11-8
Determination of the composition of a solution by sedimentation velocity: (A) the solution contains three components, each at the same concentration; (B) the solution contains a major fraction having a single s value, the remainder being a slower heterogeneous mixture, and $a/(a + b)$ is the fraction of the material that is heterogeneous, whereas $b/(a + b)$ is the homogeneous fraction; (C) the solution is a heterogeneous mixture. c_1 and c_2 are the concentrations that have s less than or equal to s_1 and s_2, respectively, and $c_2 - c_1$ is the concentration having $s_1 < s < s_2$.

Example 11-A. Structure of *E. coli* phage T7 DNA.
E. coli bacteriophage T7 DNA gives the boundary shown in Figure 11-9A. This DNA has a single, sharp boundary at all concentrations so that the DNA can be considered to be homogeneous with respect to molecular weight. If the DNA is sedimented instead at pH 12.5, at which pH the two strands of the DNA separate, the resulting boundary is that shown in Figure 11-9B. This boundary consists of two parts—one sharp (homogeneous) and the other broad (heterogeneous). The distance $a/(a + b)$ is 0.5, which means that 50% of the single strands sediments more slowly than the homogeneous component. It can be concluded, then (because in this case it is known that a low sedimentation coefficient means a lower molecular weight), that 50% of the DNA single strands must be broken. Because the slowly moving material sediments quite heterogeneously, the broken strands are of a large range of sizes and therefore the breaks must be at a large number of places in the molecule—perhaps randomly distributed. Because half of the single strands are broken, either all the molecules in the population of double-stranded DNA molecules contain one single-strand break, or half the molecules contain no breaks and half have breaks in both strands, or there is a mixture of those having no broken strands, one broken strand, or two (see Figure 11-9C). These models are not distinguishable from these data alone.

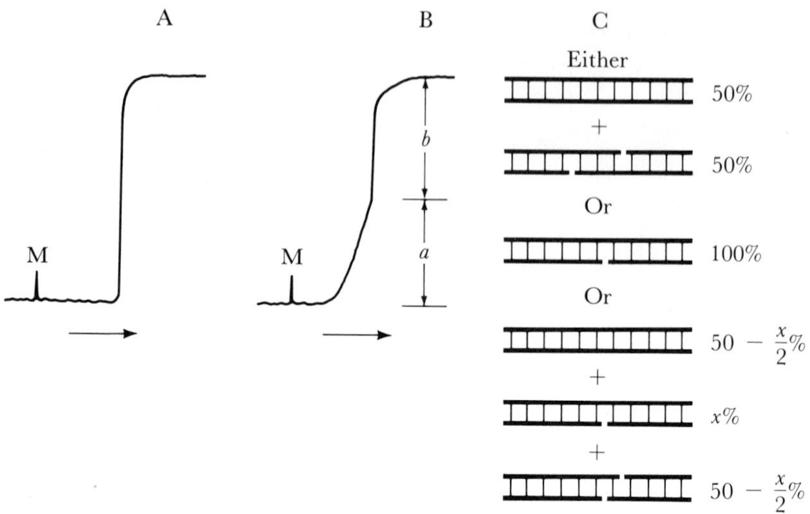

FIGURE 11-9
Sedimentation analysis of *E. coli* phage T7 DNA; M is the meniscus and the arrow indicates the direction of sedimentation. A. Native DNA has a single, sharp boundary, indicating that no molecules are broken. B. Sedimentation in alkali to separate the strands, which shows that some single strands are intact (b) and some are broken (a). Because $a = b$, the concentration of intact molecules equals the concentration of broken molecules. The number of broken molecules is, of course, greater than the number intact. C. Three possible interpretations of the data.

MEASUREMENT OF THE CONCENTRATION DISTRIBUTION IN AN ANALYTICAL CELL

In all the methods of analyzing distributions of molecules in the ultracentrifuge, it is necessary to measure the concentration of the components. Existing analytical ultracentrifuges utilize three different optical methods: *schlieren, interference,* and *absorption* (usually ultraviolet absorption). It is not of great importance to understand in detail how these optical systems form the image from which the concentration is obtained (this can be found in the manufacturer's manual), but it is worthwhile understanding how the image is related to the concentration distribution in the centrifuge cell.

The schlieren system depends on the fact that light passing through the regions of uniform concentration is undeviated but that passing through the regions of varying concentration is deviated because of the change in index of refraction (which varies with concentration). The optical system

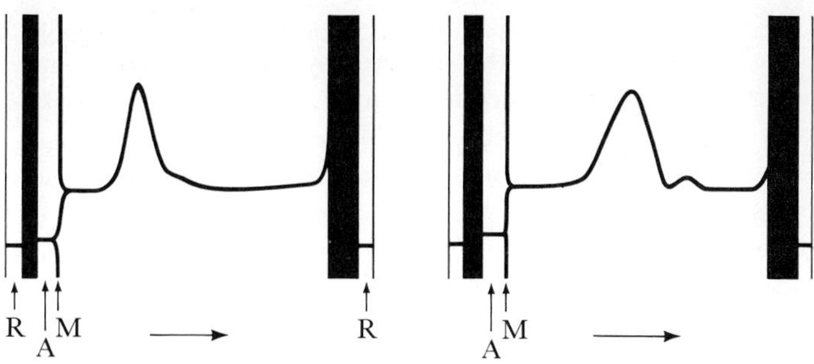

FIGURE 11-10
Schlieren patterns of a mixture of two proteins. The centrifugation period is different for each illustration, that on the right being the longer. Note that, with the longer time, a small amount of a fast-moving material is resolved. The blackened areas are the top and the bottom of the cell; A is the air space above the solution but in the cell; R stands for reference holes in the rotor that are fixed distances from the axis of rotation; and M is the meniscus.

of modern centrifuges converts this deviation into a curve that shows the concentration gradient, as in Figure 11-10.* The optical system can be adjusted so that the curve is more or less sharp; however, in making this adjustment the area bounded by the schlieren curve and the baseline must be constant and proportional to concentration. By measuring this area and by making use of certain optical constants of the instrument, the concentration can be determined. As the concentration of material in the cell decreases, the area decreases until the peak is barely resolved above the baseline; in practice, the lower limit for the schlieren system is a concentration of a few milligrams per milliliter. The value of the schlieren system is its great ability to examine boundary shape and to detect the presence of inhomogeneity in *s*. It has been widely used especially for the determination of the sedimentation coefficient of proteins.†

---

*Devising this ingenious optical system, which actually plots a graph of concentration gradient versus distance was a major accomplishment in the development of ultracentrifugal technique, without which the method might never have become useful. The system was designed by J. St. L. Philpot and modified by H. Svenson.

†As will be seen later, the single problem with the schlieren system is that the required concentrations are so high that there is often strong concentration dependence. This does not affect work on proteins, but it severely affects that on nucleic acids (Figure 11-13). Almost no information could be gained about the sedimentation properties of DNA until the more sensitive, UV-absorption optical system was developed.

For a variety of techniques (especially equilibrium and approach-to-equilibrium methods) the schlieren system is not sufficiently sensitive to detect small changes in concentration. For this reason, the *interference system* was developed, in which a double-sector cell (Figure 11-3) is used, one sector containing solvent only and the other the solution. The optical interference pattern is produced by light passing through both of the sectors. (The mechanism by which the interference pattern is produced is described in the texts listed in the Selected References near the end of this chapter. For the present purpose, suffice it to say that the optical system is based on the Rayleigh interferometer and that the fringe pattern is a function of the index of refraction at each point in the cell. Whereas the schlieren system plots refractive index *gradient* versus distance along the cell, with interference optics, each fringe traces a curve of the index of refraction versus distance.)

Figure 11-11 shows the interference patterns obtained for various situations. In part A, both sectors are filled with solvent only and the fringes are undeflected across the cell. In part B, one sector contains solute and the fringes are actually deflected a distance proportional to concentration, although this is not easily detectable because all the individual fringes look the same. If the material under study is homogeneous and has a small diffusion coefficient so that the boundary is narrow (as in Figure 11-7), the pattern shown in part C results, in which the fringes have shifted upward. In a real situation in which the boundary has finite width, the pattern shown in part D is obtained. Part E shows how concentration can be determined from the fringe shift. This is accomplished by extending a line from a fringe near the center of the pattern across the pattern along the centrifugal axis and counting the number of fringes crossed in going from left to right from an arbitrary position, A (toward the tube bottom). When this line has been extended so that no other fringes will be crossed, a point B is chosen and a perpendicular is erected as shown. The distances p and q are measured. The fractional fringe is then p/q. The change in index of refraction (Δn) in moving along the cell axis from the left side of the boundary where concentration is zero to point B is

$$\Delta n = \Delta \mathscr{F} \lambda / d \tag{3}$$

in which $\Delta \mathscr{F}$ is the number of fringes crossed, λ is the wavelength of the light used to form the fringes, and d is the length of the cell along the optical path (i.e., not the distance in the centrifugal direction). Because the relation between n and protein concentration is accurately known (and can be easily measured for any substance with a refractometer), concentration can be plotted against distance along the cell (or change in concentration in the boundary region) with accuracy. (This yields the boundary shape from which s can be calculated.) The interference system

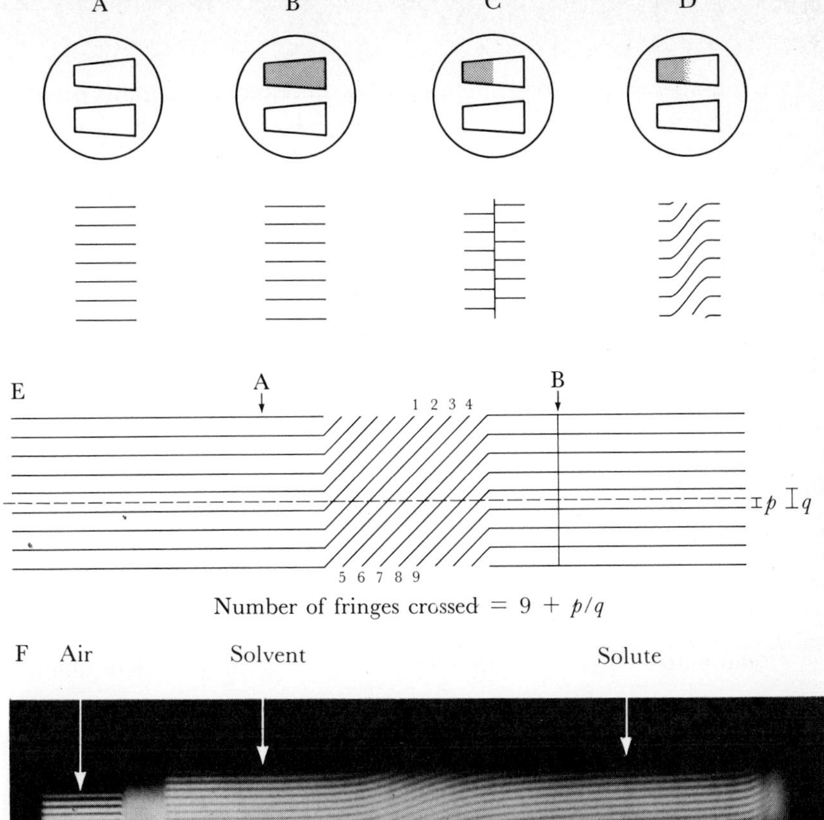

FIGURE 11-11
Generation and analysis of interference patterns from double sector cells: (A) interference pattern resulting if both sectors contain the same solution; (B) one sector contains solvent, the other solution (but before sedimentation has begun); (C) same as part B, but after a period of centrifugation (the sample is homogeneous and has a very low diffusion coefficient); (D) same as part B, but the solute has a very high diffusion coefficient; (E) method for counting the number of fringes crossed (see text); (F) interference optical photograph of bovine serum albumin in H_2O sedimenting at 60,000 rpm for one hour. In part F, sedimentation is from left to right. With interference optics, a double-sector cell is used, one sector filled with solvent, the other with solution. In this photograph, the two sectors were not filled with the same volume; hence, two menisci (M) can be seen. The boundary (B) is indicated by the curvature of the fringes. Note that the fringes curve upward at the bottom of the cell because material is pelleting. Because the position of the fringes is determined by solute concentration, the concentration of solute along the cell can be plotted.

is usable to a concentration of a few hundred micrograms per milliliter, approximately one order of magnitude better than the schlieren system.

For nucleic acids, the dependence of s on concentration is so great that a method was sought so that concentrations of $< 20\ \mu g/ml$ could be used. Fortunately, because the molar extinction coefficient (see Chapter 14) of nucleic acid is very high at 2537 Å (a strong emission line from a Hg arc light source), an absorption optical system was easily developed. To understand, refer to Figure 11-7, which shows a plot of concentration along the cell. If the cell is photographed with ultraviolet light, the blackening of the film will decrease as nucleic acid concentration increases. Figure 11-12A shows a typical photograph of sedimenting DNA. If the film is traced with a photometer, a plot of concentration versus distance is easily obtainable, as shown in Figure 11-12B. Many instruments in current use eliminate the photographic intermediate by measuring the transmitted light directly and converting this by photocells and appropriate circuitry into a plot of concentration versus distance, which appears on a chart recorder. Such instruments are called scanners.

Some analytical ultracentrifuges are now equipped with a monochromator for wavelength selection, which permits the detection of molecules by absorption optics, using a wavelength corresponding to their absorption maximum. This allows much lower concentrations of proteins to be used if the protein has strong absorption (e.g., cytochrome c or hemoglobin).

FACTORS AFFECTING SEDIMENTATION VELOCITY

Before discussing the complexities of sedimentation velocity, it is worthwhile to have an intuitive understanding of the problems. Imagine yourself walking across a room. Then imagine walking through a swimming pool with water up to your neck; then through a pool of molasses. The effect of solvent viscosity on your speed should be clear. Now consider walking through the pool with your arms extended. Your frictional coefficient has increased and it will take longer to walk through the pool as equation (1) states. If there are one hundred people in the pool, it will be harder to get across and even worse if your arms are extended; you should now understand why s is lower at higher concentrations and that the concentration dependence is greater with very asymmetric molecules. Suppose that you tried to run through the water with your arms partly extended. With twice the energy expended, would you get across in half the time? You should now appreciate speed dependence. Later you will see that, although intuition tells the direction of the effects, the causes at the molecular level may differ from that in the pool.

The sedimentation coefficient of macromolecules depends on a variety

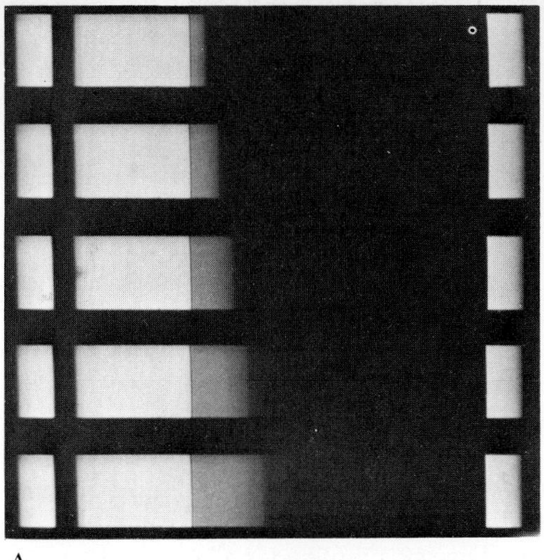

A

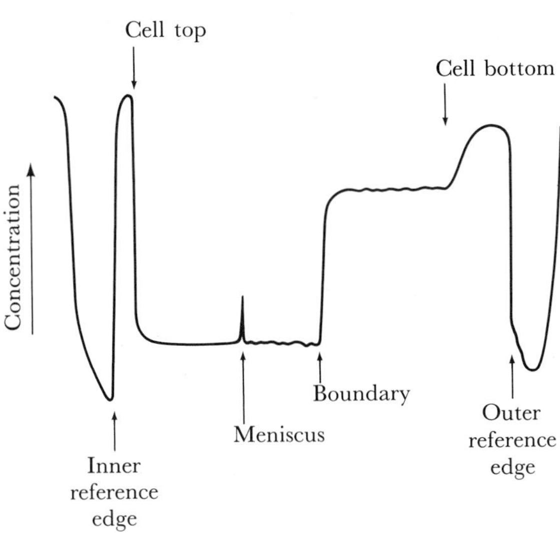

B

FIGURE 11-12
A. A UV-absorption pattern of sedimenting *E. coli* phage DNA. Sedimentation from left to right. Note that the picture shown is a positive print rather than the negative referred to in the text. B. A photometric trace of the blackening of the second frame of part A.

of factors because (1) the shape of the molecules depends on solvent composition, (2) many of the molecules may be charged, (3) they are very large compared with solvent molecules, and (4) they become deformed as they move. The manifestation of these effects is a dependence of s on the concentration, the centrifugal speed, and the ionic strength of the solvent.

Concentration Dependence of s

Consider a solution of macromolecules of such a size that they frequently collide with one another. This is characteristic of very large or extended molecules (such as proteins and nucleic acids) in solution because, as they rotate, they effectively occupy a relatively large volume of the solution. When they approach one another, it is more difficult for solvent molecules to move in the opposite direction—that is, the solvent viscosity is effectively increased in the vicinity of the macromolecule. This reduces the forward velocity of the particle in a given centrifugal field and therefore reduces s. Because the probability of collision (or close approach) increases both with molecular volume and with the degree of extension of the molecule, the magnitude of the concentration dependence also increases with these parameters. This is exemplified in Figure 11-13, which shows several representative curves for s as a function of the concentration of DNA or protein. It can be seen that the greatest concentration dependence is with the large, extended molecules (i.e., the DNAs) and that it is very slight with small spheres (phage ϕX174) and globular proteins.*

There is no detailed theory to describe concentration dependence.† However, it has been found empirically that, for most systems, the concentration dependence is best described by the equation

$$s(c) = s^\circ/(1 + kc)$$

or

$$\frac{1}{s(c)} = \frac{1}{s^\circ} + \frac{kc}{s^\circ} \tag{4}$$

in which s° is the s at zero concentration, $s(c)$ is the s at concentration c, and k is a constant for the particular molecule. The value of s° is obtained experimentally by measuring s at various concentrations, making a plot of $1/s$ versus c and determining $1/s^\circ$ by extrapolation to $c = 0$. Clearly, s° is a more useful parameter to describe a molecule and should always

*Very slight does not mean inconsequential. For example, proteins are usually studied at such high concentrations (such as 0.5–5.0 mg/ml) that the *total* decrease in s may be substantial.

†An attempt has been made to improve on equation (4). For example, it has been calculated that $s(c) = s^\circ/(1 + k[\eta]c)$, in which $[\eta]$ is the intrinsic viscosity described in Chapter 13 and k is a constant having a value of 1.2 for rodlike molecules and 1.6 for spheres. However, this theory must still be considered semiempirical because $[\eta]$ cannot be readily calculated for most proteins and polynucleotides.

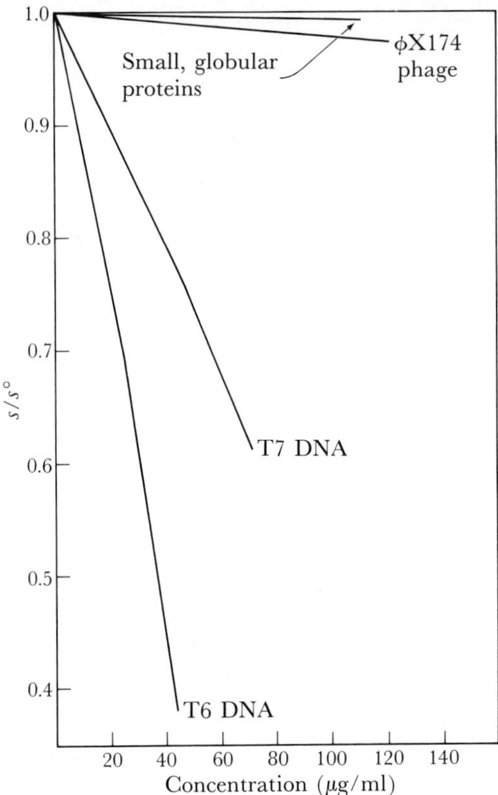

FIGURE 11-13
Dependence of s on concentration plotted as $s/s°$ where $s°$ is the value of s extrapolated to zero concentration. Note that the concentration dependence becomes more severe as the molecular weight increases and as the molecule is more extended.

be determined. This has not always been done in published work and it is important to realize that there may be substantial differences between $s°$ and s measured at standard concentrations, as is clear from Figure 11-13.

A particular problem arises if the macromolecules in a mixture have different s values and if the concentration is relatively high. In this case, the molecules interfere not only with the sedimentation of their own species, but also with that of the other species. Furthermore, the molecule with the higher s and greater concentration dependence must sediment through a solution of the slower-moving molecule. At high concentration, the faster molecules are so impeded that they sediment near or at the same rate as the slower molecules, which can disguise the fact that there is really a mixture and gives the misinformation that the material is homogeneous. This is known as the Johnston-Ogston effect. In fact, with large nucleic acids it is often necessary to use very low concentrations before the two species can be resolved; even at these concentrations, the data frequently indicate that there is less of the faster component than is really present. An example of this is given in Figure 11-14. As will be seen in a later section (page 297), this problem is minimized in the technique of zonal centrifugation.

Speed-dependent Sedimentation

Speed-dependent sedimentation refers to two independent phenomena—a speed-dependent aggregation of molecules that occurs at high concentration and an actual reduction of $s°$ at high speed. In neither case is the phenomenon completely understood, but, owing to the work of Bruno Zimm and his co-workers, we have some ideas about the causes.

Speed-dependent aggregation refers to an apparent loss of material from the bulk of the solution, which decreases the effective concentration of the macromolecule (and therefore increases the apparent s). The explanation currently given for this phenomenon is that at high velocity a molecule leaves behind it a wake, which enhances the speed of the molecule just behind it. This results in the formation of molecular clusters that have very high s values and rapidly form a pellet at the bottom of the centrifuge tube. This process continues until the concentration is too low for clusters to occur. Figure 11-15 shows data in which this speed-dependent aggregation is occurring. The concentration of DNA in a centrifuge cell is plotted for two different times and three different speeds of centrifugation. At the higher speed, the DNA concentration decreases with time because of the pelleting. At the lower speed, it remains fairly

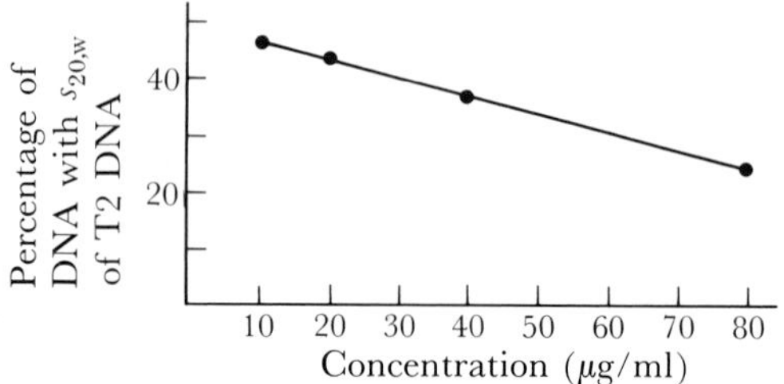

FIGURE 11-14 Amount of T2 DNA seen in a 50:50 mixture of T2 and T7 DNA at various DNA concentrations—the Johnston-Ogston effect.

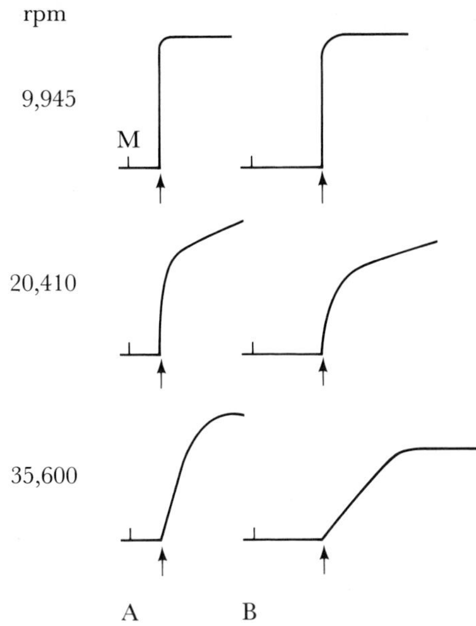

FIGURE 11-15
Photometric traces of UV-absorption patterns of *E. coli* phage T4 DNA sedimented at three speeds. In column A, all the arrows at the base of the boundaries are roughly equidistant from the meniscus, M. This is true of column B also, but the distance is greater. Note that, as the speed increases, a greater fraction of the DNA seems to sediment more rapidly than the *s* value calculated from the position of the arrows. This *s* value is that expected for the molecular weight of the DNA. The rapid sedimentation is due to the speed-dependent aggregation described in the text. Sedimentation is from left to right.

constant. Note that the measured *s* values are 50 and 37 at the higher and lower speeds, respectively. Such an apparent loss of material is a strong indication of speed dependence and demands that measurements be done at lower speeds.

The second type of speed dependence occurs with very large molecules at low concentrations. With T4 DNA (molecular weight = 106×10^6), the increase in *s* is 9% in going from 65,000 to 10,000 rpm. With larger DNA molecules (e.g., molecular weight greater than 500×10^6), the speed dependence becomes significant and can lead to three to eightfold errors in estimates of molecular weight. This is an important phenomenon because the chromosomes of bacteria and eukaryotic cells possess DNA molecules having molecular weights ranging from 2×10^9 to 10×10^9. This phenomenon is indicated in Figure 11-16, which shows that above a certain speed the dependence of *s* on molecular weight is lost. Zimm has explained this surprising result as being due to a change in shape (an extension) of the DNA molecule at high speed because of enhanced frictional drag at the ends of the molecule. (Presumably, there is no speed dependence of this sort for circular DNA molecules; this has not yet been tested though.) Clearly, to distinguish molecular-weight differences for very large DNAs, it is necessary to use very low centrifuge speeds (or to make measurements of its visco-elastic properties—see Chapter 13).

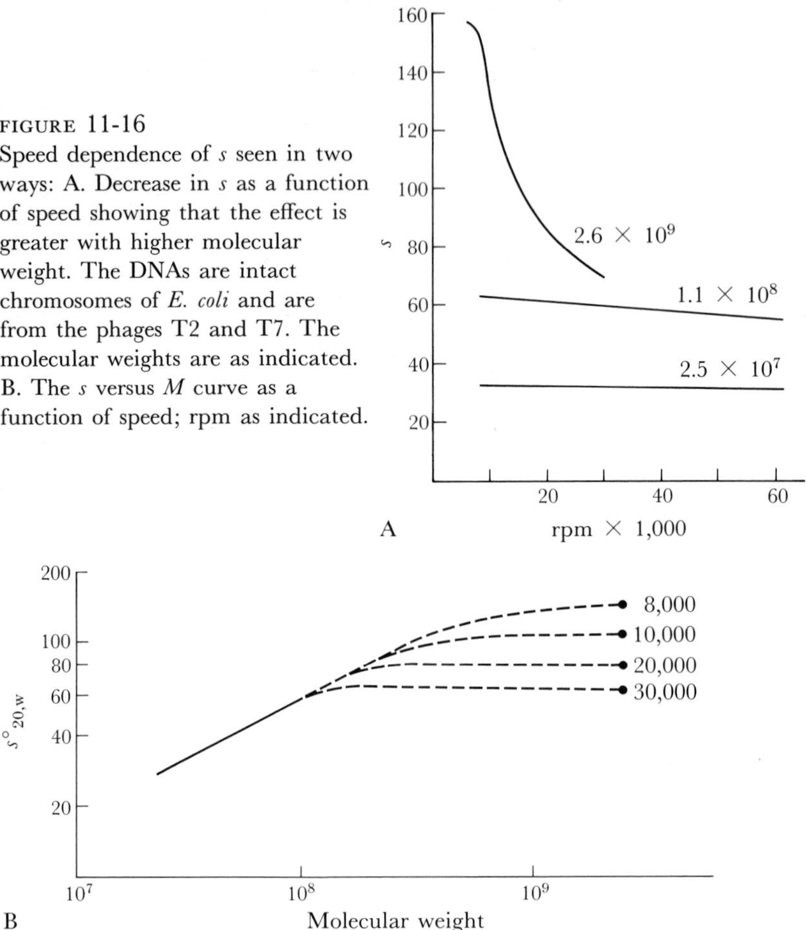

FIGURE 11-16
Speed dependence of s seen in two ways: A. Decrease in s as a function of speed showing that the effect is greater with higher molecular weight. The DNAs are intact chromosomes of *E. coli* and are from the phages T2 and T7. The molecular weights are as indicated. B. The s versus M curve as a function of speed; rpm as indicated.

Effect of Charge

A strong solvent effect that does not involve a change in the shape of a macromolecule arises if the molecule is charged (which is usually the case with biological macromolecules) and sedimentation is carried out in solutions of low (<0.01) ionic strength. Because of the large s of the macromolecule compared with that of the neutralizing ions (the "counterions" such as Na^+ or Mg^{2+}), the ionized macromolecule sediments more rapidly than the counterions. The distribution of charge results in a potential gradient against the direction of sedimentation, which tends to reduce s. This complication is simply avoided by the use of excess counterions—for example, in practice, by always using ionic strength in excess of 0.05.

STANDARD SEDIMENTATION COEFFICIENTS

If sedimentation velocity studies are carried out in different solvent-solute systems, the measured s is affected by the solution density (i.e., by the $1 - \bar{v}\rho$ term), by the solution viscosity (i.e., a molecule will move more slowly in a viscous medium), and by the temperature (primarily, by its effect on density and viscosity). To be able to compare experimental values obtained in different solvent systems and at different temperatures, observed values are converted into a standard solvent and temperature. In so doing, it is assumed that the partial specific volume and the friction factor are unaffected by the change in the system. This is probably a reasonable assumption (although not always justified) as long as the counterion concentration is not very high and solute molecules that bind to the macromolecule are not present.*

By agreement, sedimentation coefficients are corrected to a reference solvent having the viscosity and density of water at 20°C by means of the equation

$$s_{20,\mathrm{w}} = s_{\mathrm{observed}} \cdot \frac{1 - \bar{v}\rho_{20,\mathrm{w}}}{1 - \bar{v}\rho_T} \cdot \frac{\eta_T}{\eta_{20}} \cdot \frac{\eta}{\eta_0} \tag{5}$$

in which $s_{20,\mathrm{w}}$ is the standard s, $\bar{v}$ is the partial specific volume, $\rho_{20,\mathrm{w}}$ and ρ_T are the densities of water at 20°C and of the solvent, respectively, η/η_0 is the relative viscosity of the solvent to that of water, and η_T/η_{20} is the relative viscosity of water at temperature T compared with that at 20°C.

The appropriate values of η for many solvent-solute systems can be found in various references; others can be measured directly by viscometry (see Chapter 13).

Note that, for this correction as well as for calculations of molecular weight to be described in later sections of this chapter, the value of $\bar{v}$ is necessary. How this value is obtained is described in Chapter 12.

---

*The problem is that almost all equations used in centrifugation have been derived by making the assumption that the system has only two components, the solvent molecules and the molecules whose sedimentation properties are being studied. However, in practice this is rarely the case, because of the addition of counterions. In fact, the concentration of the counterions usually ranges from 0.01 to 1.0 M, whereas the molecule of interest is between 1.0×10^{-5} and 1.0×10^{-3} M. In practice, this complexity is usually ignored because, if the salt concentration is < 0.05 M, nothing anomalous is detectable. However, in 1 M NaCl, a common solvent for DNA, some corrections have to be made. This is rarely done because in fact the hydrodynamic theory of DNA sedimentation has not yet achieved the degree of sophistication such that these corrections would make much of a difference. Subtle problems do arise, though, if proteins are sedimented in 5-molar guanidinium chloride or 7-molar urea.

FACTORS AFFECTING STANDARD SEDIMENTATION COEFFICIENTS

Friction: The Shape Factor

Let us consider by a simple example the factors affecting the frictional coefficient. If a ball and a stick of the same mass and density are centrifuged through a viscous medium, it is intuitively obvious that the ball, which is more compact, will move faster. (If the rod could be oriented so that it moved only along its axis, this need not be true. However, Brownian motion keeps the rod—and a particle—rotating so that the statement is correct.) The general rule is that the more extended an object is, the greater will be its resistance to motion and therefore the greater its frictional coefficient. This simple consideration enables one to understand why the sedimentation coefficients of extended semirigid molecules increase markedly if the intramolecular structure giving rise to the rigidity is destroyed. A striking example of this can be seen in studies with DNA.

Example 11-B. Detection of DNA denaturation by ultracentrifugation. The rigidity of DNA is produced primarily by its double-stranded helical structure in which the nucleotide bases tend to stack one on top of the next (see Chapter 16) and in which bases on opposite strands are held in contact by hydrogen bonds. As the temperature of a DNA solution is increased, hydrogen bonds break and the stacking tendency of the bases decreases. When a region of bases loses its rigidity because of a breakdown of hydrogen bonds and of stacking, the DNA molecule will then have points at which bending can occur. This decreases the

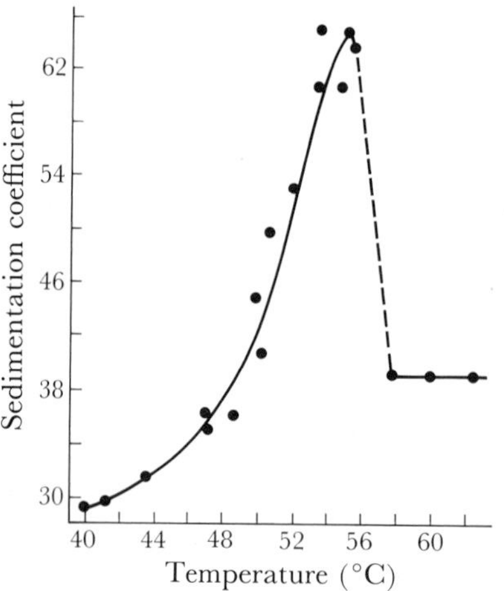

FIGURE 11-17
Sedimentation coefficient of the sharp boundary of T7 DNA fully or partly denatured by heating at the indicated temperature in 0.1 M PO_4, pH 7.8, containing formaldehyde. The formaldehyde prevents the reformation of hydrogen bonds when the solution is cooled to 20°C for centrifugation. The dashed line represents separation of the strands. [From D. Freifelder and P. F. Davison, *Biophys. J.* 3(1963):49–63.]

extendedness of the molecule and therefore increases s. As more and more intramolecular disruption occurs, the molecule becomes more and more flexible, less rigid, and less extended, and it has a higher s. This continues until the last hydrogen bond is broken. At that point, the two DNA strands separate, thus producing two units having half the molecular weight of the original structure. This phenomenon can be clearly seen in Figure 11-17, which plots the s of a DNA having a molecular weight of 25 million as a function of the temperature to which the DNA has been heated. The s increases by 220% during the period of disruption and then drops precipitously at the point of strand separation (because the molecular weight is halved). It is interesting that this simple measurement was one of the earliest indications that the strands of DNA could be separated.

Molecular Weight

Equation (2) shows that s increases with M if the partial specific volume ($\bar{v}$) and the frictional coefficient (f) are constant. Values of $\bar{v}$ do not usually change significantly with M, although they vary with the amino acid composition of proteins. The frictional coefficient, on the other hand, is a strong function of M. Many attempts have been made in the past twenty years to calculate the relation between s and M (which in fact means calculating the f-M relation). For solid spheres and rigid rods, this is a relatively simple task and, insofar as proteins can be treated as globular structures, the theory is also adequate. For flexible rods such as DNA, these attempts have not been very successful, although recently some headway has been made.

For the more common purpose of using s as an assay of M or of changes in M, the approach to this problem is to obtain empirical relations between s and M. This approach has been successful for DNA, although the equations in common use have been continually modified as the measured values of M for various DNAs have become more precise.* The best equation to date for double-stranded Na-DNA at neutral pH in 1 M NaCl sedimented in a sector-shaped cell (Figure 11-3) is

$$s^{\circ}_{20,\mathrm{w}} = 2.8 + 0.00834\, M^{0.497} \tag{6}$$

This equation is, unfortunately, not based on a very large number of values of M and may be changed slightly in the future. An important point

*Many other equations can be found in the literature. Unfortunately, they have been based on values of M that are considerably in error. The source of these errors is discussed in several of the references listed near the end of the chapter.

to be noticed is that s varies more slowly than M so that s alone is not a good measure of M unless a significant error can be tolerated. For example, a 5% error in s yields a 10% error in M. The value of s for DNA can be measured to 2% accuracy, but this is rarely done. Because the s-M relation is strictly empirical, it is important that M values are not evaluated from these equations if the s lies outside of the s values (10–60) used to obtain the equation.

For globular proteins, s varies roughly as $M^{2/3}$ so that a 5% error in s yields a 7% error in M. This is fairly reliable because s for proteins can be measured to 1% accuracy. However, the empirical equation for proteins suffers from the fact that $\bar{v}$ varies with amino acid composition.

As will be seen in a later section, there are better ways to determine precise values of M and s is usually measured to assay *changes* in M or in shape.

PROCEDURE FOR THE ACCURATE MEASUREMENT OF SEDIMENTATION COEFFICIENT

Now that some constraints have been placed on the means of measuring s, it is worthwhile defining a procedure that will result in meaningful s values. First, to avoid the effects of charge, ionic strength should be maintained in the range of 0.05 to 1.0 and pH should be controlled by a suitable buffer. Measurements should be performed at several speeds, each differing by about 50%, to ascertain that the effects of speed are absent. At a given speed, at least four concentrations should be used (including the lowest possible concentration) to ascertain that either a single species is present or that the true ratio of several components is being observed and to obtain $s°$. Finally, $s^{\circ}_{20,\mathrm{w}}$ should be calculated. A sample calculation of $s^{\circ}_{20,\mathrm{w}}$ is given in the appendix to this chapter.

At this point, it must be remembered that $s^{\circ}_{20,\mathrm{w}}$ has been determined not for the bare macromolecule but for the molecule *with bound counterions and other ligands.* This fact cannot be emphasized too strongly. Unless there is a strong effect of counterion concentration on shape or on $\bar{v}$ (as described in an earlier section), it is usually the case that $s^{\circ}_{20,\mathrm{w}}$ is the same in, for example, 0.1 M and 1.0 M NaCl and the same in 0.1 M and 1.0 M CsCl. However, for DNA, for example, it is certainly not the same for identical concentrations of NaCl and CsCl because in one case Na-DNA is produced and in the other Cs-DNA, which have average nucleotide molecular weights of 336 + 23 = 359 and 336 + 137 = 453, respectively. (The average molecular weight for a nucleotide is 336; Na and Cs have atomic weights of 23 and 137, respectively.) In both solute systems, s will probably reflect the actual molecular weight, but this conclusion must be made cautiously because it must be realized that the empirical equation (6) was obtained for Na-DNA only. This is also the case for proteins, which bind either cations or anions or both.

EXAMPLES OF THE USE OF BOUNDARY SEDIMENTATION

In the following examples, boundary sedimentation is used to determine the relative amounts of material having different values of s. It is interesting to note that strong conclusions are made without the necessity of knowing the actual s values precisely.

Example 11-C. Mechanism of inactivation of bacteriophages by x irradiation.

If bacteriophages are irradiated with x rays, they lose viability. A possible explanation is that the DNA within the phages is broken by the radiation. To investigate this, phages were irradiated and samples were taken after various doses for determination of viability and for analysis of the DNA by ultracentrifugation. Figure 11-18A shows sedimentation diagrams for the DNA of phages given two x-ray doses. As the dose increased, more of the DNA sedimented more slowly; this represents broken molecules. The fraction of unbroken molecules was determined by the procedure shown in Figure 11-8. When this fraction and the fraction of surviving phages is plotted against dose (Figure 11-18B), it can be seen that phage killing is more rapid than DNA breakage and that DNA breakage occurs at about half the rate of killing. If the irradiation is carried out in the absence of oxygen (i.e., in a nitrogen atmosphere), the phages become more resistant to radiation, although the rate of DNA breakage is unchanged. If a plot such as that shown in Figure 11-18B is made, it is found that the rate of killing is the same as the rate of breakage. Hence, from this simple analysis, the conclusion can be drawn that there are two modes of killing: an O_2-independent process, which is DNA breakage, and an O_2-dependent process, which does not include DNA breakage.

Example 11-D. Binding of a cofactor to an enzyme.

The cofactor reduced diphosphopyridine nucleotide (DPNH) binds to the enzyme chicken heart lactic dehydrogenase (LDH). To understand the mechanism of action of this enzyme, the number of binding sites and the dissociation constants characterizing the various equilibria must be known. Hence, an assay for binding must be done. This can be done with the ultracentrifuge. Using absorption optics and light at 340 nm (which is absorbed by DPNH but not LDH), the sedimentation

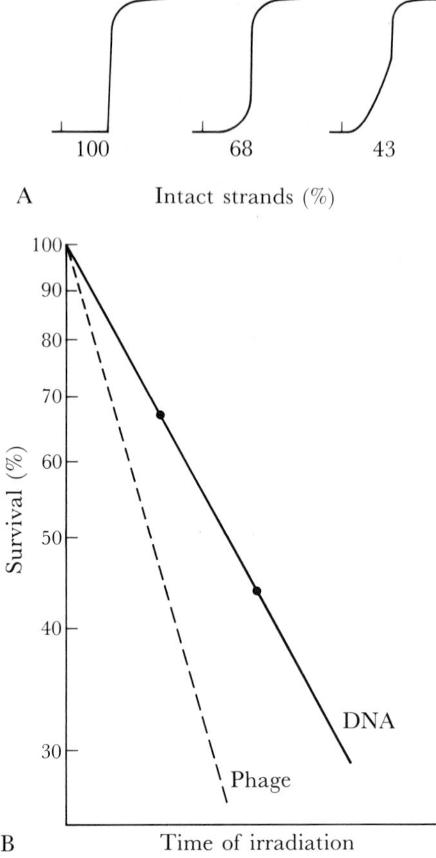

FIGURE 11-18
A. Sedimentation diagram of *Pseudomonas aeroginosa* phage B3 DNA after two doses of x rays. B. Plot of the percentage of intact double strands as a function of time of x irradiation compared with the phage survival. A comparison of the inactivation and strand-breakage rates requires measuring the relative doses that give the same survival level.

of DPNH can be detected (Figure 11-19). When only DPNH is present, a boundary having $s = 0.2$ is seen. When LDH is added, a boundary appears at $s = 7$, the value for pure LDH. Hence, DPNH is sedimenting with the LDH and must be bound to it. As the amount of LDH increases, the relative amount of DPNH at $s = 0.2$ and $s = 7$ can be measured, and this yields the fraction bound. Using this simple assay, the ratio of bound to unbound DPNH can be determined as a function of DPNH and LDH concentration, and of pH, ionic strength, and so forth.

Example 11-E. An indication of the strength of base pairs of DNA. When DNA is treated with an endonuclease or with x rays, single-strand breaks are produced. As these breaks accumulate, ultimately two single-strand breaks in separate strands become near enough that

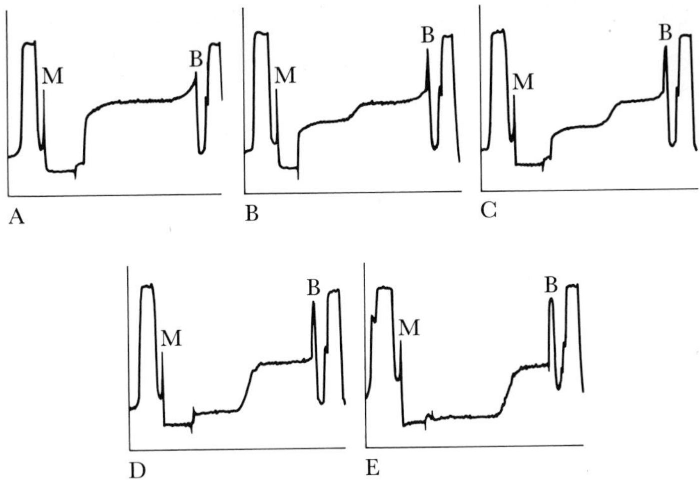

FIGURE 11-19
Sedimentation velocity patterns of various mixtures of lactic dehydrogenase (LDH) and the coenzyme reduced diphosphopyridine nucleotide (DPNH). The cell was photographed with light of wavelength 314 nm and the absorbance was plotted by a scanner. (A) DPNH with no LDH; (B–E) DPNH/LDH = 12, 8, 4 and 2, respectively. Sedimentation is from left to right; M and B are the meniscus and bottom of the cell, respectively. [From H. K. Schachman and S. I. Edelstein, *Methods Enzymol.* 27(1973):3–58.]

they constitute a double-strand break, and the double-stranded DNA is fragmented. Because the strength of hydrogen bonds depends on temperature and ionic strength, it might be expected that the minimum separation of two single-strand breaks that would not cause double-strand breakage would also depend on these agents. To measure single-strand breaks requires only the sedimentation of DNA in alkali because, at high pH, the double helix unwinds to form two single polynucleotide strands. To investigate this, DNA was x-irradiated and single- and double-strand breaks were determined by sedimentation in alkali and at neutral pH (Figure 11-20). The fraction of unbroken single- and double-stranded DNA molecules was measured for each dose, using the procedure shown in Figure 11-8. From a relatively simple equation,

$$h = [L(1 - F^{1/b})/2b] - 1/2$$

in which h is the minimum number of base pairs required to maintain the structure, L is the number of nucleotides per single strand, F is

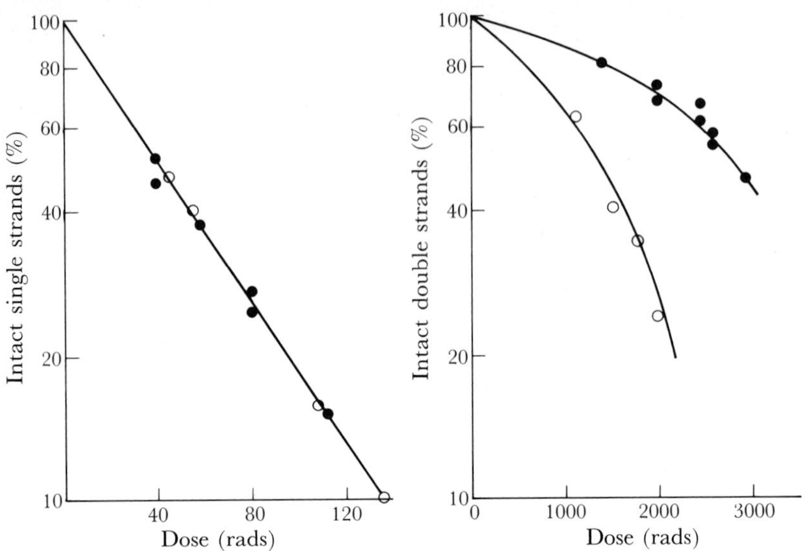

FIGURE 11-20
The percentage of intact single and double strands of *Pseudomonas aeroginosa* phage B3 DNA following introduction of single-strand breaks by x irradiation. Irradiated DNA was sedimented at alkaline pH (left) and at neutral pH (right) to measure single- and double-strand breakage. Solid circles, high ionic strength; open circles, low ionic strength. [From D. Freifelder and B. Trumbo, *Biopolymers* 7(1969):681–693.]

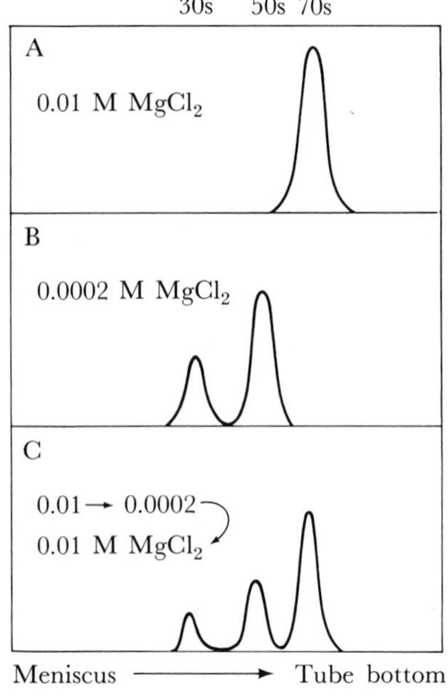

FIGURE 11-21
A sedimentation velocity study of dissociation and association of *E. coli* ribosomes. Drawings of schlieren patterns of ribosomes sedimented in different Mg^{2+} concentrations: (A) in 0.01 M $MgCl_2$, the 70s ribosome is stable; (B) in 0.0002 M $MgCl_2$, it dissociates to form one 50s and one 30s ribosome (the areas are in a 2:1 ratio because the 50s has about twice the molecular weight of the 30s); (C) partial reassociation by restoring the sample in part B to 0.01 M $MgCl_2$. At higher Mg^{2+} concentration, a 100s peak appears.

the fraction of double-stranded molecules that are unbroken, and b is half the average number of breaks per single strand, the minimum number of base pairs required to keep the double-stranded molecule intact was calculated. It was found to be 2.6 and 15.8 in 0.01 M and 1.0 M NaCl at 25°C, respectively.

Example 11-F. Structure of ribosomes.
In isolating ribosomes from the bacterium *E. coli,* it is found that there are at least four species having s = 30, 50, 70, and 100 (Fig. 11-21). It can be shown by sedimentation analysis that these species constitute an associating system dependent on Mg^{2+} concentration in which a 30s and a 50s particle combine to form a 70s particle and two 70s particles combine to form 100s. This can be seen in Figure 11-21, in which peaks disappear and reappear at new positions and in which the relative areas reflect the concentration of material having each s value.

ZONAL CENTRIFUGATION THROUGH PREFORMED DENSITY GRADIENTS

In the centrifugation procedures described in the preceding sections, the starting solution consists of macromolecules uniformly distributed throughout the cell and all information is obtained from observations of the trailing boundaries of the molecules as they sediment down the centrifuge tube. Whereas this method has many advantages and is ideally suited to the analytical ultracentrifuge, it has the disadvantage that molecules having different s values never separate from one another so that faster molecules are always sedimenting through a solution of slower molecules. This has two effects: (1) concentration dependence is always at work to decrease resolution and (2) if a fast component is to be physically separated from a slow one, the fast one is always contaminated by the slower; the level of contamination can only be reduced by repeated sedimentation. An ideal technique would allow each molecular species to sediment only through the solvent. This situation can be approximated by *zonal,* or *band,* centrifugation. In the description given here, it is carried out in the preparative centrifuge, using swinging-bucket rotors (Figure 11-5). A later section will show how it can be performed in special cells in an analytical centrifuge.

The methods used in zonal centrifugation are shown in Figure 11-22. A small volume of a solution containing the molecules to be characterized is layered on top of a preformed concentration gradient contained in a centrifuge tube. The solution being layered always has a density less than that at the top of the gradient (otherwise it would sink into the gradient). The tube is then centrifuged and the molecules in the starting layer sedi-

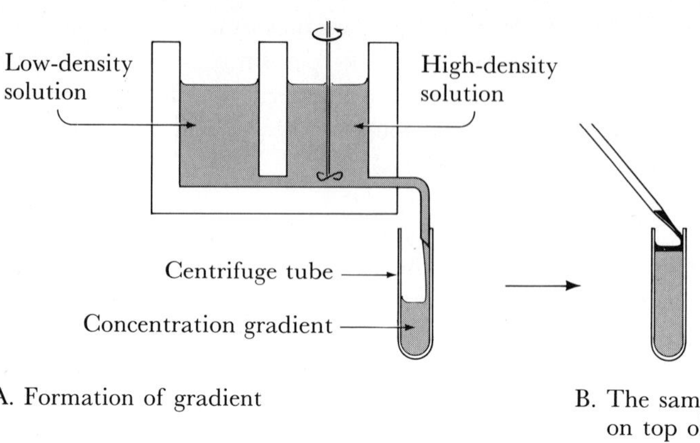

A. Formation of gradient

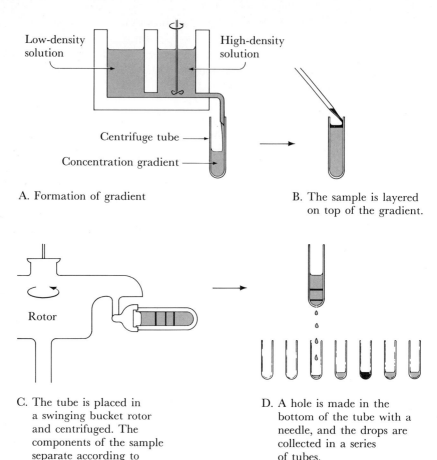

B. The sample is layered on top of the gradient.

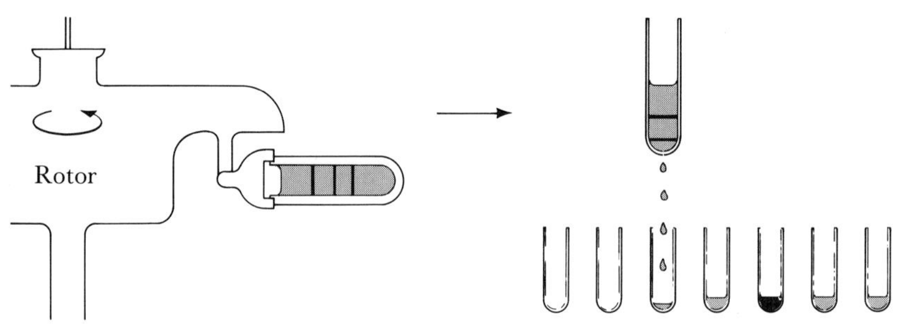

C. The tube is placed in a swinging bucket rotor and centrifuged. The components of the sample separate according to their *s* values.

D. A hole is made in the bottom of the tube with a needle, and the drops are collected in a series of tubes.

FIGURE 11-22
Operations in zonal centrifugation.

ment through the gradient. If the molecules have the same sedimentation coefficient, they will sediment within a narrow zone as shown in Figure 11-23. If there are molecules having several sedimentation coefficients, they will separate from one another as sedimentation proceeds. The different components will then be resolved into a series of zones or bands, which then sediment only through solvent and independently of one another. After centrifugation is complete, the contents of the tube are fractionated, most commonly by drop collection from the tube bottom (Figure 11-6). If the dripping rate is sufficiently slow so that there is no turbulence, each drop represents a single lamella in the tube. The fractions then consist of one or more drops each. A great variety of techniques can be used to assay

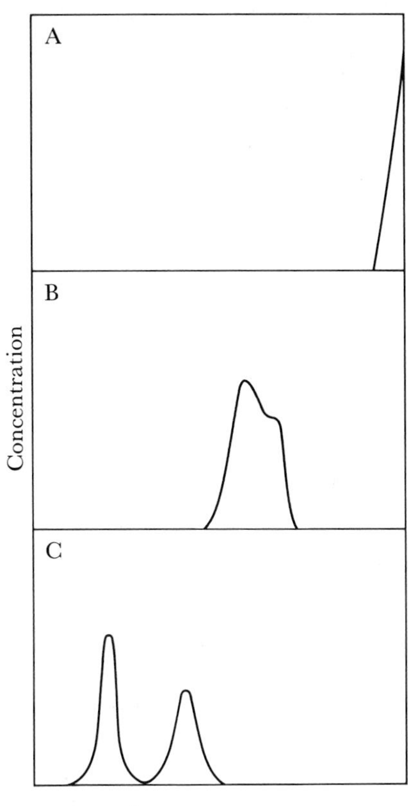

FIGURE 11-23
Separation of two components by zonal centrifugation in a sucrose density gradient—sedimentation is from right to left: (A) before centrifugation; (B and C) at successively longer periods of centrifugation.

the materials and thereby determine the concentration distribution. In analytical centrifugation, optical techniques (UV-absorption, schlieren, or Rayleigh interference) are relied on entirely to detect the molecules. However, the fractions obtained from a drop-fractionated tube can be assayed by radioactivity, chemical tests, enzymatic activity, absorption, fluorescence, or combinations of these. Several examples of these methods are shown in Figure 11-24.

Why is a concentration gradient needed for zonal centrifugation? Let us consider the consequence of layering a low-density solution on top of a higher-density solution in which there is no concentration gradient. Initially, the system is stable because of the density difference at the boundary—that is, the upper solution is simply floating on the lower. Suppose now that there are small fluctuations in temperature in the solution (as there certainly will be). Because the density of most solutions decreases with increasing temperature, these fluctuations will create local density inversions (i.e., higher-density regions above lower), which will result in a local flow of the liquid called *convection.* This will have little effect on the

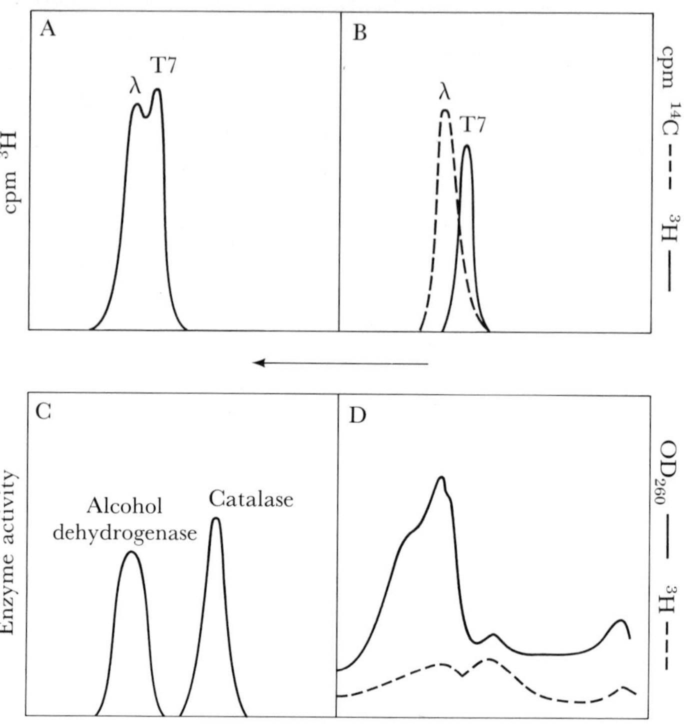

FIGURE 11-24
Several ways to assay a zonal gradient to get the concentration distribution: (A) ^{3}H-T7 DNA and ^{3}H-λ DNA; (B) ^{3}H-T7 DNA and ^{14}C-λ DNA; (C) a mixture of two enzymes; and (D) the incorporation of ^{3}H-uracil into RNA (OD_{260} indicates the amount of total RNA).

initial boundary because the density difference at the boundary is usually great enough that mixing across the boundary does not occur (unless the temperature difference is huge—10°–20°). However, after sedimentation, the molecules to be studied will have moved into the denser lower layer, and such convection can destroy any band that might exist. The introduction of a steep density gradient ensures that temperature changes have to be very great to create a density difference sufficient to cause flow within the gradient. A second important function of the density gradient is to prevent mixing due to mechanical disturbances; any perturbation would be counteracted by the attempt to return to the situation in which low density is above high density. The gradient also serves a third purpose. Consider a system without a gradient and free of all temperature fluctuations and mechanical disturbances in which the sedimenting molecules have entered the lower layer and formed a zone. In this zone, the presence of the molecules increases the density of the solution by their own contribution to the density (normally a very small effect but, when high concentrations are used, it is significant). Hence, the density of the zone is also greater than the solution just below it, and this results in the convective flow of the zone toward the bottom of the cell. If instead sedimentation were through a preformed concentration gradient, the sedimenting molecules would continually be entering a region of higher density. This would still increase the density of that region but, if the gradient were sufficiently steep, the contribution of the molecules to the density would be insufficient to cause a density inversion and the system would remain stable. The material most commonly used to form density gradients is sucrose because of its purity, low cost, and lack of interference with most chemical, enzymatic, and optical assays. If the macromolecule to be studied is an enzyme or an unstable protein, glycerol is frequently used because many proteins are more stable and less easily denatured in the presence of glycerol. H_2O-D_2O gradients have also been used successfully; this is the method of choice if the particle under study requires constant osmotic pressure. Examples of the use of these systems will be described in a later section.

DETERMINATION OF THE SEDIMENTATION COEFFICIENT BY ZONAL CENTRIFUGATION IN PREFORMED DENSITY GRADIENTS

Zonal centrifugation is often used to determine s by calibration with markers for which s values have been measured in an analytical ultracentrifuge. However, even if the concentration is low enough to eliminate concentration-dependent sedimentation (which is of great value if the material is not pure), the determination of s by means of zonal centrifugation has pitfalls not encountered with the analytical ultracentrifuge. There are three reasons for this: First, the density and viscosity of the solution increase with distance along the tube so that the net force on the molecules is not a simple function of distance. Second, the walls of the preparative tubes are parallel and therefore not aligned with the centrifugal direction, and this results in sedimentation against the walls and accumulation of material. This accumulation causes a local density inversion and some convection. Third, s is calculated from only two points, the position at zero time and at one later time (when the centrifuge has stopped) so that the rate of movement is not really known. The first reason can be dealt with theoretically in a complicated way; the second has been considered theoretically but the effect on s is not clearly understood. To avoid the problems of reason 1, the common practice is to use isokinetic gradients—

that is, the concentration and viscosity gradient is selected so that the molecules move at constant velocity at all distances from the center of rotation. The currently standard 5% to 20% sucrose gradient was empirically selected to have this property. It is important to realize that the isokinetic property of a concentration gradient depends on the change in concentration as a function of *distance* along the tube; therefore, a 5% to 20% sucrose gradient is isokinetic in a particular centrifuge tube (e.g., the tubes for the Beckman SW39, SW50, and SW65 rotors) but not for longer tubes (e.g., SW25, SW41). Thus, gradients of different composition must be selected for tubes of different length if isokinetic conditions are needed. The student is cautioned to keep this in mind when reading the literature because this has rarely been appreciated and is one of the reasons that different values of s are reported from different laboratories.

Because of the second and third reasons, absolute determination of s is difficult, and the usual practice is to determine relative s by cosedimentation with molecules of known s, for which s has been determined in the analytical ultracentrifuge. However, this practice is not valid for the following reason: $s^{\circ}_{20,\mathrm{w}}$ for double-stranded DNA varies with $M^{0.497}$ [equation (6)] if measured in a sector-shaped cell. This equation would also be valid if the cell contained an isokinetic gradient in which case the distance traveled (d) per unit time would also vary as $M^{0.497}$. However, if, in a 5% to 20% sucrose isokinetic gradient, the relation between d and M is measured in a preparative centrifuge, it is found to be

$$\frac{d_1}{d_2} = \left(\frac{M_1}{M_2}\right)^{0.38} \tag{7}$$

in which the subscripts 1 and 2 refer to two independently sedimenting components. Because s is proportional to distance in an isokinetic gradient, the s of DNA in a 5% to 20% sucrose gradient in a preparative cell (with parallel walls) will not be the same as that obtained in a sector-shaped cell because the exponents of M are not the same. (It is thought that this is due to the convection described in reason 2.) Therefore, s cannot be determined from the relative distance traveled. There is no reason to believe this situation is any better for proteins. However, the problem is not really serious because normally s is only wanted for an estimation of M. Because the s-M relation is already an empirical one for most substances, an empirical d-M relation can be obtained just as well, using relative distance traveled at a given time. Equation (7) describes such a relation for DNA. However, equation (7) is valid *only* for an isokinetic gradient and, because of reason 3, it can never be known whether it is isokinetic. This problem can be avoided by the rarely used procedure of cosedimenting two or three substances of known molecular weights that bracket that of the unknown and measuring relative d for two different periods of centrifugation. A simple plot of relative d versus log M will give an accurate M.

EXAMPLES OF THE USE OF ZONAL CENTRIFUGATION IN THE PREPARATIVE ULTRACENTRIFUGE

In the following examples of zonal centrifugation, very small amounts of material that cannot be detected by optical methods are studied. Note the variety of the assays and the strength of the conclusions.

Example 11-G. Demonstration that the *E. coli* ρ factor terminates RNA synthesis.

When RNA polymerase synthesizes RNA from a DNA template, synthesis begins at base sequences known as promoters and ends at terminator sequences. However, in vitro there is no evidence that termination sequences are recognized. Jeffrey Roberts isolated a protein called ρ, which decreased the amount of RNA synthesized in vitro. Evidence that ρ supplies the information for the recognition of termination sites came from analyzing the synthesized RNA by zonal centrifugation in sucrose gradients. One of the relevant experiments is shown in Figure 11-25. If ρ decreased either the rate of initiation of synthesis or of polymerization, after long periods of synthesis (when many molecules had been synthesized), the size distribution of the RNA molecules would be the same whether ρ is present or not. If, however, it induced termination, the molecules would be smaller. Because the overall shape of all RNA molecules is the same, this would mean that, when ρ is present, the synthesized RNA would sediment more slowly as shown in Figure 11-25.

Example 11-H. Measurement of the change in size of a mutant enzyme. Many of the methods used to determine molecular weights of enzymes either require highly purified samples (e.g., sedimentation equilibrium, as will be described) or, if impure samples can be used, yield subunit weights (e.g., SDS-gel electrophoresis, Chapter 9). The size of an active enzyme can be determined from a crude lysate, using zonal sedimentation and enzyme activity as an assay. For example, *E. coli* DNA polymerase I sediments at $s = 5.4$. The polymerizing and 5′–3′ exonuclease activities associated with this protein cosediment so that either activity can be used to localize the enzyme in a sucrose gradient. A mutant form of this enzyme lacks polymerase activity but retains 5′–3′ exonuclease

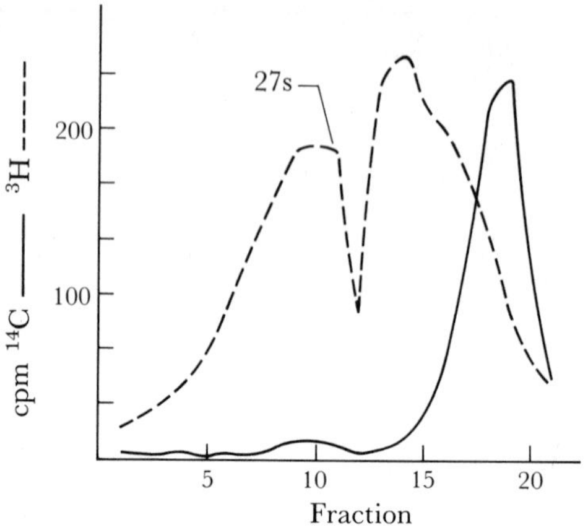

FIGURE 11-25
Zonal sedimentation (5%–20% sucrose gradient) of RNA synthesized from *E. coli* phage λ DNA template. Two reaction mixtures were prepared containing λ DNA, *E. coli* RNA polymerase, and appropriate buffers. The dashed line represents RNA synthesized when ^{14}C-uridine triphosphate (UTP) is incorporated. The solid line represents RNA made with ^{3}H-UTP in the presence of the rho factor. Rho decreases the size of the RNA. Sedimentation is from right to left. [From J. W. Roberts, *Nature* 224(1969):1168–1174.]

activity. As shown in Figure 11-26, the s of the mutant enzyme is 2.8. Because the particular mutation is a chain-termination mutation, the reduced s indicates that the enzyme is a small fragment of the wild-type protein, which has 0.4% activity.

Example 11-I. Determination of the molecular weight of T7 DNA by end-group labeling.

The enzyme T4 polynucleotide kinase can be used to label the 5′ ends of a double-stranded DNA molecule (by transferring ^{32}P from γ-^{32}P-ATP to 5′-OH groups). This can be used to determine molecular weights if the amount of DNA and the number of ^{32}P atoms coupled are known. However, if the DNA sample contained broken DNA molecules, a low value of M would result because too much ^{32}P would be bound—for example, if 1% of the molecules were broken into one hundred fragments, the observed M would be half the true M. To determine the number of ^{32}P atoms per intact DNA molecule, the reaction

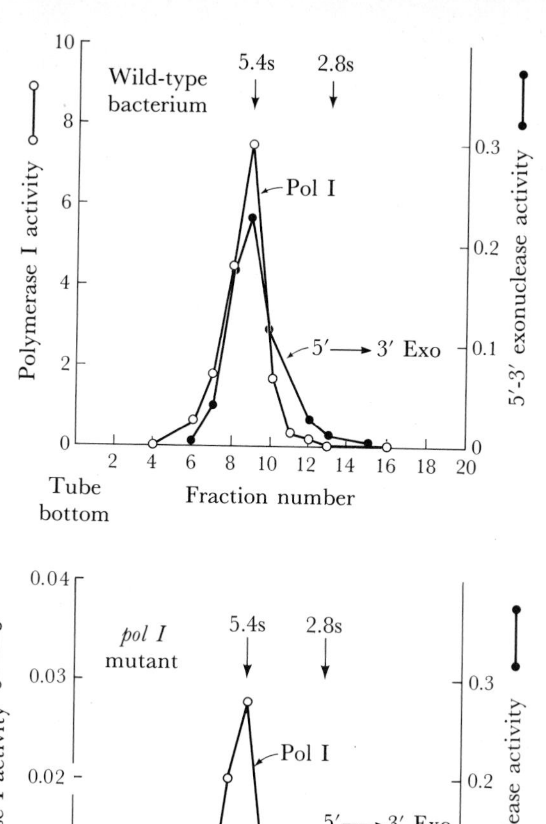

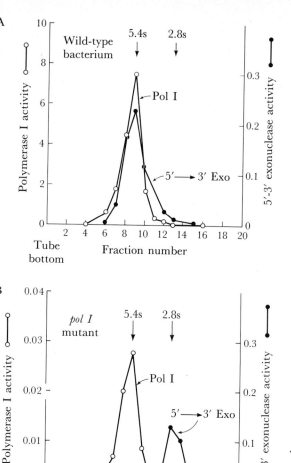

FIGURE 11-26
Zonal sedimentation through a 5% to 20% sucrose gradient of *E. coli* DNA polymerase I isolated from (A) a wild-type strain and (B) a strain making only a fragment of the enzyme. The wild-type enzyme possesses two activities: the polymerase activity is measured by the incorporation of a ^{32}P-containing nucleotide into acid-insoluble material (left axis); the 5′–3′ exonuclease is assayed as the removal of ^{32}P from a terminally ^{32}P-labeled polynucleotide (right axis). In part A, the two activities cosediment because they are possessed by a single protein. In part B, the mutant enzyme has 0.4% of the polymerizing activity, but the more slowly moving fragment retains the 5′–3′ exonuclease activity. [From I. R. Lehman and J. Chien, *J. Biol. Chem.* 248(1973):7717–7723.]

mixture can be sedimented through a sucrose gradient (Figure 11-27). Figure 11-27 shows that, as is often the case, a significant fraction of the radioactivity barely sediments and remains at the top of the gradient. However, the main peak represents intact molecules so that, from the ratio of ^{32}P to DNA concentration in the peak, M can be calculated.

Example 11-J. Partial purification of an osmotically fragile structure containing a hormone.

A hormone called substance P can be isolated from the hypothalamus of a rat. Dorothy Freifelder and Susan Leeman found that the

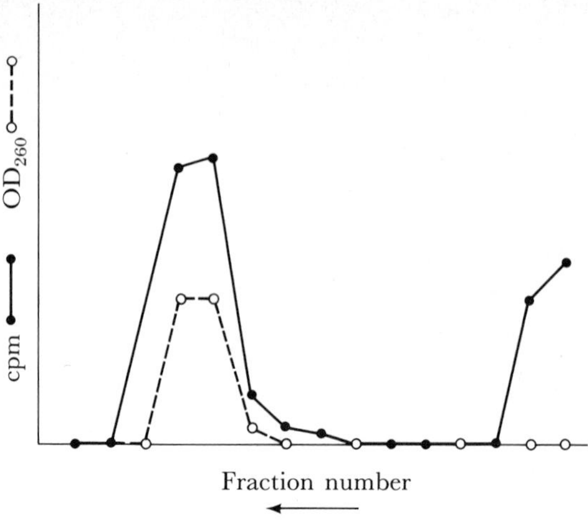

FIGURE 11-27
Sedimentation pattern of T7 DNA labeled at its termini with ^{32}P, using T4 polynucleotide kinase: OD_{260} is a measure of the total DNA concentration; ^{32}P cpm measures number of termini. Note that, at the top of the gradient, no OD_{260} is detectable. Hence, the DNA concentration is very low and the cpm represents the labeling of tiny fragments.

hormone can be pelleted from macerated tissue at very low centrifugal force if the tissue were suspended in 20% sucrose before disrupting the cells; in the absence of sucrose, it was not sedimentable even at very high forces. This suggests that the hormone is contained in osmotically fragile particles. It is not possible to sediment the particles through a 5% to 20% sucrose gradient because the sample layer would immediately sink to the bottom. A 25% to 70% gradient is not useful because the particles are apparently destroyed at very high osmotic strength. To avoid these problems a gradient was prepared from 20% sucrose in H_2O and 20% sucrose in D_2O. This provides a stable gradient because D_2O is 11% denser than H_2O, and the constant sucrose concentration maintains the necessary osmotic strength. Sedimentation through this gradient yielded partly purified particles.

BAND SEDIMENTATION IN SELF-GENERATED DENSITY GRADIENTS

It is often desirable to carry out zonal sedimentation with the analytical ultracentrifuge and ultraviolet-absorption optics. This method is preferable if only very small samples are available that cannot be labeled with radioisotopes or assayed by chemical or enzymatic means. Because of difficulties in preparing density gradients in cells of the analytical ultracentrifuge, Jerome Vinograd developed a procedure for zonal centrifugation without using a preformed gradient.

As indicated in an earlier section, if a sample is layered onto a denser solution without a preformed density gradient, convection will ultimately disrupt or destroy the boundary. Vinograd realized that, if the sample is in a solution of very low density, because of the diffusion of the solute used to increase the density of the denser solution, in time the density discontinuity generated at the interface will slowly be propagated down the centrifuge tube, forming a shallow gradient. Below the boundary generated by the propagating gradient, there is no gradient and convection can occur. However, at the propagating gradient and above it, there is stabilization against convection. Clearly, as long as the material being sedimented does not move faster than the gradient being generated, the material will be in a density gradient and sedimentation will be normal. Furthermore, if the solute of the denser solution has a fairly high density (which is the case for CsCl, the material most frequently used), the salt will also sediment slightly and form an additional stabilizing gradient. This technique has been termed *band sedimentation* in self-generated density gradients.

Figure 11-28 shows the type of centrifuge cell that is used with this method. The cell is prefilled with a solution of high density (usually 2 M CsCl) and the sample hole is filled with the solution of the material to be studied. Shortly after acceleration, material leaks across the scratch on the face of the cell and is layered on top of the denser solution. As the

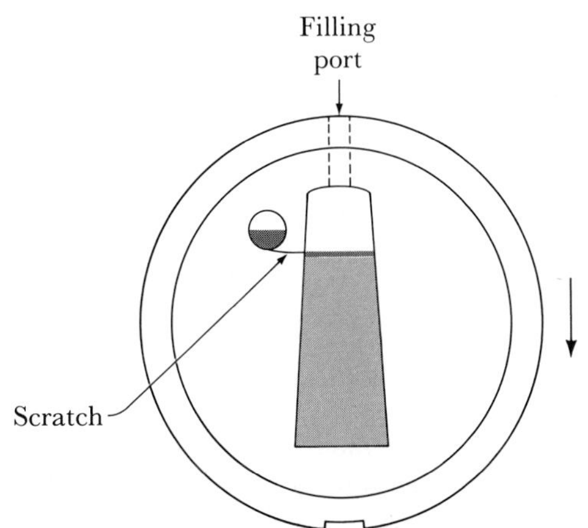

FIGURE 11-28
Band-forming centerpiece viewed from the top. The solvent (shaded area) is placed in the sector through the filling port. The sample is placed in the circular well. During acceleration, the sample flows along the scratch and forms a layer on the solvent. The sample in this figure is in the process of layering. Note that the sample flows only through the scratch because the entire surface of the centerpiece is in close contact with an end window. (Centerpiece designed by Jerome Vinograd, California Institute of Technology.)

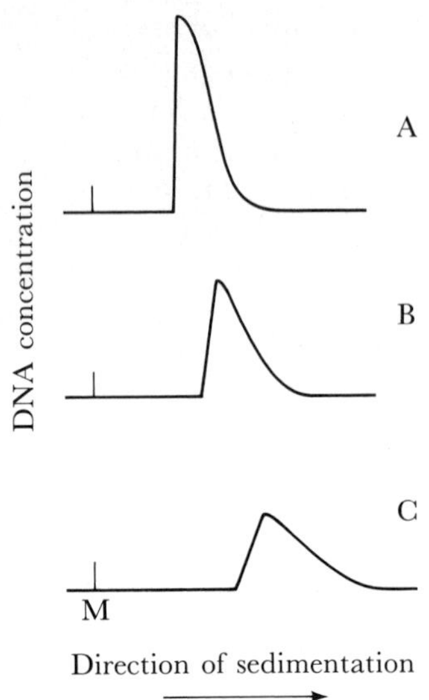

FIGURE 11-29
Sedimentation of double-stranded T7 DNA in a band-forming cell like that shown in Figure 11-28. The time of centrifugation increases in going from part A to part C and M is the meniscus.

speed of rotation increases, sedimentation occurs. As long as the speed is not too high, the zone is undisturbed. Figure 11-29 shows representative concentration distributions (i.e., ultraviolet-absorption patterns) of DNA sedimenting through concentrated CsCl solutions. Note that the boundary is asymmetric with material moving faster in the leading direction. This is clearly understood from concentration dependence; material on the leading side of the boundary is at the lowest concentration and therefore moves fastest. The trailing side of the boundary is sharp because any molecule that diffuses in the centripetal direction finds itself in a region of zero concentration and therefore sediments at maximum rate. As a matter of fact, the complete relation between s and concentration for a given molecule can be derived from the shape of the boundary.

Band centrifugation is the procedure to use for determining whether a sample contains a very small amount of *rapidly* sedimenting material. In standard boundary sedimentation, the rapidly sedimenting material moves through the slower material and is slowed down by the Johnston-Ogston effect; therefore, it is frequently not observable. In band centrifugation, this small fraction is sedimenting in a region of zero concentration and is easily observable.

Band sedimentation is rapidly becoming the method of choice for analytical centrifugation of DNA; it has not yet become popular for proteins.

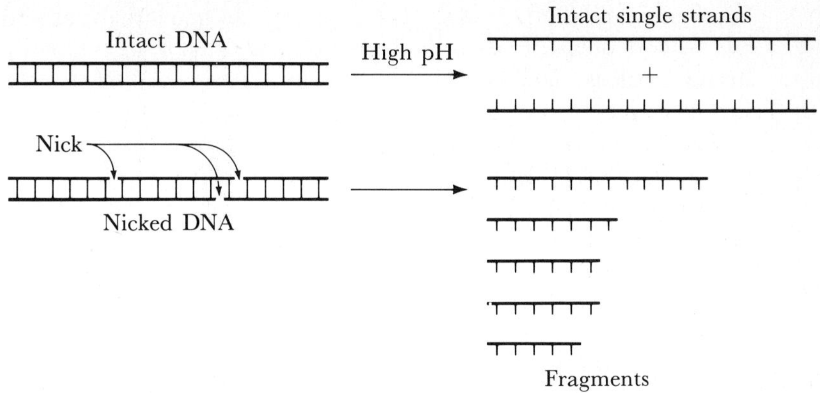

FIGURE 11-30
Result of the denaturation of intact and nicked double-stranded DNA.

SEDIMENTATION OF DNA AT ALKALINE pH

DNA is denatured at high pH. Hence, by preparing alkaline sucrose gradients (i.e., 5%–20% sucrose in 0.1–0.3 M NaOH), denatured DNA can be studied.*

Sedimentation through alkali has two important uses: to identify single-strand breaks in double-stranded DNA and to distinguish various conformations of DNA. The identification of single-strand breaks is based on the following considerations. If the strands of a double-stranded molecule that contains no interruptions are separated in alkali, the two strands that result have the same size and therefore the same *s* (Figure 11-30). However, if there is a break—let us assume it is in a unique position in one strand—the result is one intact single strand and two fragments of unique size, that is, three *s* values. If the break is randomly situated, the

*The only complication with this technique is that sucrose at high concentration is a reasonably good buffer in the pH range of 10 to 11. Furthermore, its buffering capacity increases substantially as the temperature is lowered. Hence, alkaline sucrose gradients cannot be prepared by making a dilute pH 12.5 buffer and then adding sucrose. To insure denaturation, the pH must be adjusted after sucrose is added, and the value must be checked at the temperature at which it will be used. The use of 0.3 M NaOH eliminates this problem. All alkaline sucrose solutions should contain a chelating agent such as EDTA (ethylene diamine tetracetate) to avoid a metal-ion–catalyzed alkaline hydrolysis of phosphodiester bonds.

fragments can be of all sizes. This reasoning can of course be extended to one or more breaks in one or both strands. The main point is that a single-strand break is made evident in alkali by the appearance of DNA that sediments more slowly than intact strands.

The distinction between various conformations of DNA is based on the following considerations. As mentioned in Chapter 1, some types of DNA may be either linear, an open circle (with at least one single-strand break), or a covalent circle (containing no interruptions). Combinations of these are also possible—for example, two circles may be linked as in a chain; such linked circles are called *catenanes.* Each of these structures has a distinct s value as shown in Figure 11-31, although the differences are not large. Consider, however, the effect of placing each of these structures in alkali (Fig. 11-32). A linear molecule is dissociated into two single strands; a circle with one single-strand break separates into one linear molecule and one single-strand circle. However, the two single-strand circles of a covalent circle cannot separate because they are intertwined. The result is that, in alkali, a denatured covalent circle collapses to a tangled, compact mass. Because the molecular weight is unchanged and the molecule becomes compact, its s becomes very large. Figure 11-33 shows two hypothetical sedimentation patterns for a mixture of the three forms centrifuged in neutral and alkaline sucrose. The extraordinary separation of covalent circles is a useful property that is exploited in some of the examples that follow.

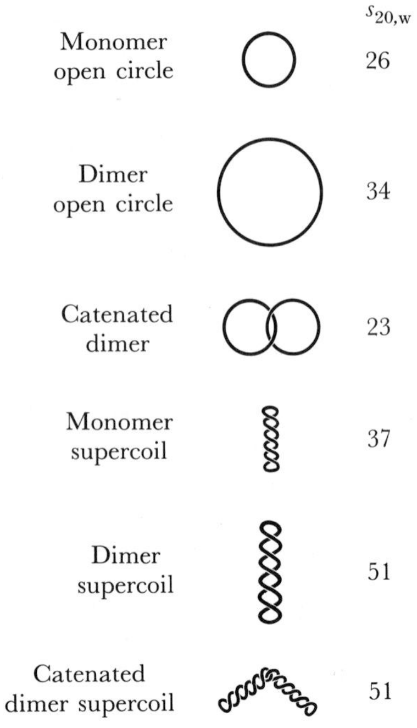

FIGURE 11-31
Sedimentation coefficients of different circular structures of the DNA isolated from the mitochondria of human leucocytes. [From B. Hudson, D. A. Clayton, and J. I. Vinograd, *Cold Spring Harbor Symp. Quant. Biol.* 33(1968):435–442.]

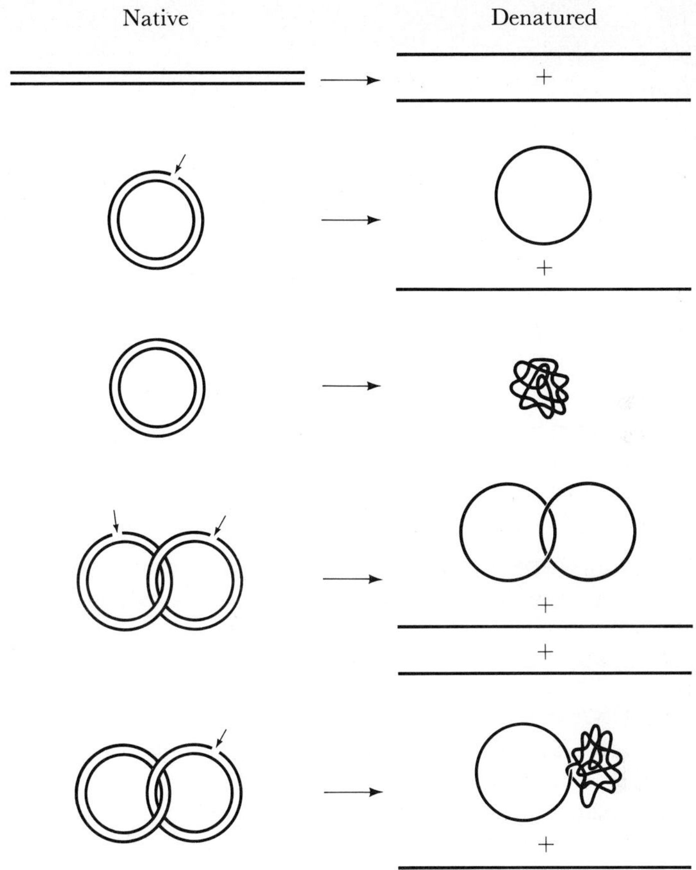

FIGURE 11-32
Products of the denaturation of different forms of DNA.

Example 11-K. Identification of small fragments of newly synthesized DNA: the Okazaki fragments.

If the bacterium *E. coli* is grown for many generations in a growth medium containing ^{3}H-thymidine, the synthesized DNA is labeled with ^{3}H. If this DNA is isolated and sedimented at neutral pH, its s is characteristic of DNA whose molecular weight is very high. However, if sedimented in an alkaline sucrose gradient, a small fraction of the DNA sediments very slowly (Figure 11-34A). This implies that somewhere in the double-stranded DNA there are closely spaced single-strand breaks. If nonradioactive cells are grown for only 5 seconds (0.2% of a generation) in radioactive medium and then collected and analyzed in an alkaline sucrose gradient, most of the radioactivity (i.e.,

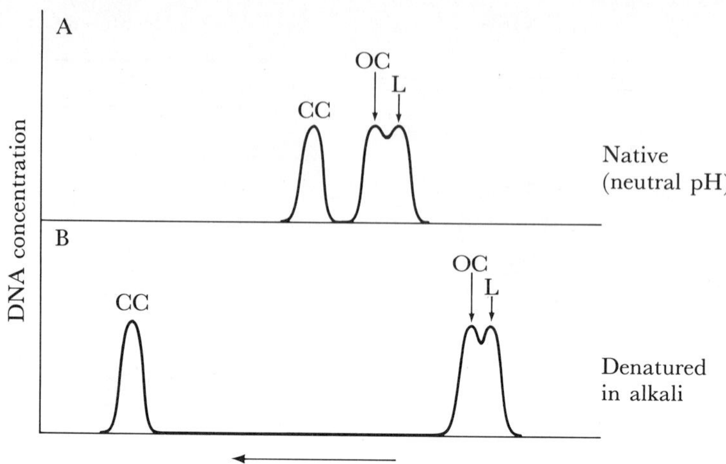

FIGURE 11-33
Separation of covalent circles (CC), open circles (OC), and linear (L) molecules by sedimentation at neutral pH (native DNA) or in alkali (denatured DNA). The time of centrifugation of part B is half that of part A. Note the improved separation of covalent circles in alkali.

all DNA made during the 5-second period) has a very low *s* (Figure 11-34B). Hence, recently made DNA must contain many closely spaced single-strand breaks. If the cells labeled for 5 seconds are transferred to nonradioactive medium and allowed to grow for 10 minutes (1/4 generation), sedimentation analysis in alkaline sucrose shows that all radioactivity has a high *s* (Figure 11-34C); that is, "old" DNA does not contain single-strand breaks. These simple results lead to the conclusion that DNA is synthesized in small fragments, which are then joined to form a larger unit.

Example 11-L. Detection of radiation damage in bacterial DNA. If bacteria labeled with ^{3}H-thymidine are treated with the enzyme lysozyme, which removes part of the bacterial cell wall, and then layered directly on an alkaline sucrose gradient, the alkali will complete the lysis of the bacteria and denature the DNA. If this is then centrifuged, the sedimentation pattern indicates that most of the DNA has a very high *s* (Figure 11-35). If the cells are x-irradiated before lysis, the *s* is much lower, indicating that the x rays cause single-strand breaks. From an empirical relation between *s* and *M*, the number of single-strand breaks can be calculated and compared with the x-ray dose. In that way, the relation between strand breakage and x-ray induced killing can be investigated.

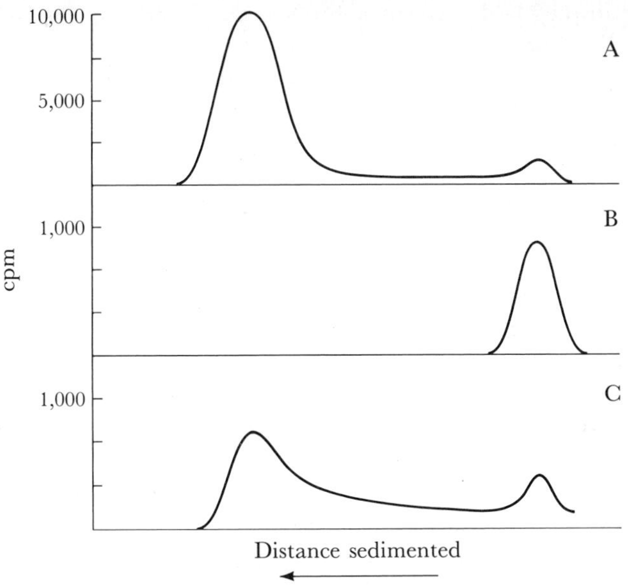

FIGURE 11-34
Detection of short single-strand DNA fragments in *E. coli* by sedimentation in alkaline sucrose: (A) the bacteria were grown in ^{3}H-thymidine for many generations; (B) the bacteria were pulse-labeled; (C) pulse-labeling followed by growth in the absence of label (see text for details). Note the change in scale (cpm) between part A and parts B and C. [Data from the laboratory at Brandeis University.]

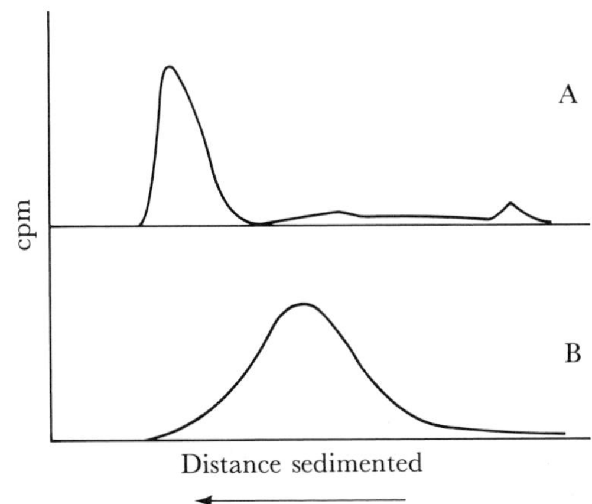

FIGURE 11-35
Detection of single-strand breaks in *E. coli* DNA by the technique of Richard McGrath and R. W. Williams [*Nature* 212(1966): 534–535.] (A) unirradiated bacteria; (B) x-irradiated bacteria. The cells were treated with lysozyme and layered on an alkaline sucrose gradient. Lysis and the release of DNA occurred in the layer. [Data from the laboratory at Brandeis University.]

Example 11-M. Determination of the structure of the DNA of the *E. coli* F factor.

The *E. coli* F factor is responsible for the ability of a male strain to transfer its DNA to a female strain. Physical analysis has shown that it is a piece of DNA distinct from that of the chromosome and only about 1% as large. If a male strain of *E. coli* is grown for many generations in a medium containing ^{3}H-thymidine, all DNA is labeled. If these cells are lysed and the DNA is sedimented through alkaline sucrose, a rapidly sedimenting fraction (approximately 1% of the total DNA) is observed (Figure 11-36) whose s is greater than that of the intact *E. coli* chromosomal DNA. Such rapid sedimentation is usually indicative of a covalent circle. A covalent circle has the property, described earlier, that one single-strand break—converting it into an open circle—allows the strands to separate in alkali and thereby sediment much more slowly. Indeed, several agents that are known to produce single-strand breaks result in the disappearance of this rapid peak. The kinetics of disappearance are first order, indicating that loss of the peak is a result of one single-strand break. Hence, it can be concluded that F is a covalent circle. From an empirical s versus M relation (for covalent circles in alkali), M can also be calculated.

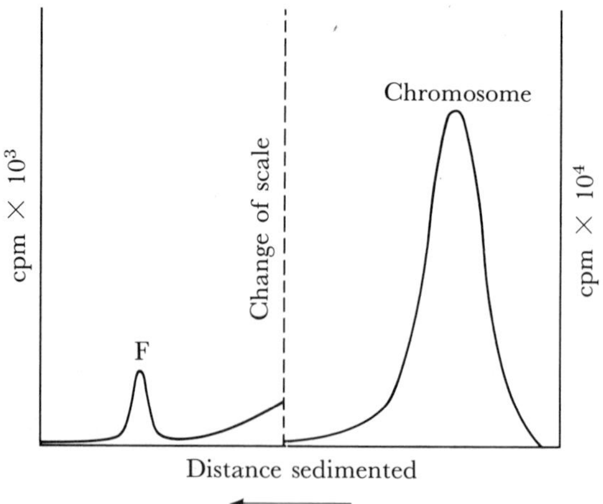

FIGURE 11-36
Detection of covalent circles of *E. coli* F factor DNA. Bacteria were grown in ^{3}H-thymidine, lysed, and sedimented through an alkaline sucrose gradient. The F supercoils account for approximately 1% of the total DNA but are easily resolved because of their high s in alkali.

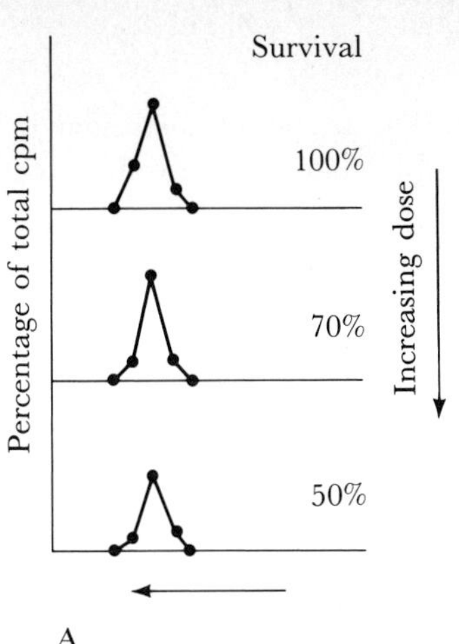

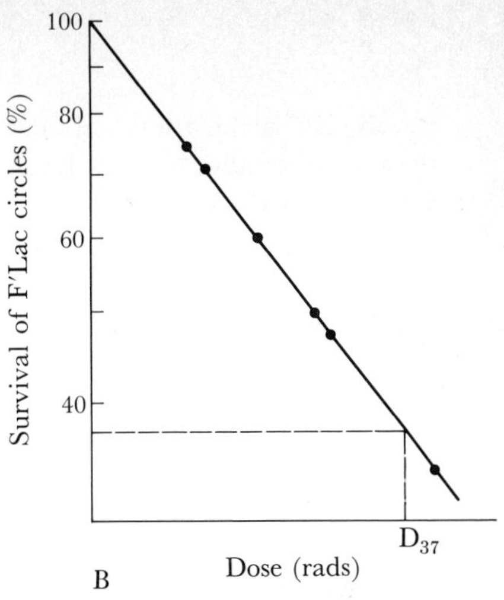

FIGURE 11-37
Effect of x irradiating *E. coli* that contain F′Lac. Part A shows the portion of the sucrose gradient containing the F′Lac covalent circles, for three different x-ray doses. Part B shows a semilogarithmic plot of the survival of F′Lac covalent circles; survival was determined from gradient analysis, as in part A. D_{37} is the dose giving a survival of 1/e, or 37%. At this dose, there is an average of one single-strand break per F′Lac molecule.

Example 11-N. Measurement of the rate of production of single-strand breaks by the measurement of the survival of covalent circles.

In Example 11-L, single-strand breaks are detected by breakage of linear DNA molecules. This can be done more accurately by analyzing covalent circles. Bacteria containing the *E. coli* sex factor F′Lac were labeled, x-irradiated, and sedimented, as in Example 11-M. The survival of F′Lac covalent circles as a function of x-ray dose was measured by comparing the amount of radioactivity in the rapidly sedimenting fraction. The curve of log(survival) versus dose is shown in Figure 11-37. Because the curve is linear, the dose yielding a survival of 1/e represents that producing an average of one single-strand break per covalent circle. Because M for F′Lac is known, a value of the number of breaks per dose per unit of molecular weight can be calculated.

DETERMINATION OF MOLECULAR WEIGHT BY THE SEDIMENTATION-DIFFUSION METHOD

Earlier in this chapter, it was shown that

$$s = m(1 - \bar{v}\rho)/f \tag{8}$$

in which f is the frictional coefficient. The frictional coefficient is very difficult to predict but can be shown to be related to the diffusion coefficient, D, by the equation

$$D = RT/f \tag{9}$$

in which R is the gas constant and T is the absolute temperature. These equations can be rearranged to yield

$$m = sRT/D(1 - \bar{v}\rho) \tag{10}$$

Hence, by measuring s and D (both extrapolated to zero concentration), m can be calculated. The coefficients s and D are either measured in the same solvent and at the same temperature (the better procedure) or corrected to water at 20°C. This is a reliable method.

The diffusion coefficient is not difficult to measure with modern instruments and is discussed in Chapter 12. However, for very large and extended molecules such as DNA, D is too small to be measured accurately.

MEASUREMENT OF MOLECULAR WEIGHT BY SEDIMENTATION EQUILIBRIUM

The sedimentation equilibrium method allows a direct determination of m because it eliminates the need to determine D. With this method, centrifugation is carried out at relatively low speeds so that sedimentation of the molecule is slow enough to be counterbalanced by diffusion—that is, the tendency of the centrifugal force to cause a decrease in concentration at the meniscus and an increase at the bottom of the cell is antagonized by diffusion, which tends to maintain the same concentration everywhere in the cell. At equilibrium the concentration is indeed lower at the meniscus and greater at the bottom but is unchanging. A description of the state of affairs at equilibrium is provided by the well-known Boltzmann distribution, which in terms of a system in a centrifugal field is

$$\frac{c_1}{c_2} = e^{-(E_1 - E_2)/kT} \tag{11}$$

in which c_1 and c_2 are the concentrations of solute molecules at distances r_1 and r_2 from the axis of rotation at which points the molecules have potential energies E_1 and E_2 and k is the Boltzmann constant. The potential energy difference, $E_1 - E_2$, is the work necessary to move a molecule of mass m from r_2 to r_1, that is, *against* the force field, or

$$E_1 - E_2 = -\int_{r^2}^{r_1} m(1 - \bar{v}\rho)\omega^2 r\,dr = \tfrac{1}{2}m(1 - \bar{v}\rho)\omega^2(r_2^2 - r_1^2) \qquad (12)$$

Combining equations (11) and (12) and replacing k by the gas constant R so that the units of m (which are now given as M) will be molecular weight units (daltons) instead of grams produces the following equation, which can be used to calculate M directly from centrifugation data:

$$M = \frac{2RT}{(1 - \bar{v}\rho)\omega^2} \ln\frac{c(r)}{c(a)} \cdot \frac{1}{r^2 - a^2} \qquad (13)$$

in which $c(r)$ is the concentration of the solute at a distance r from the axis of rotation and a is the distance of the meniscus from the axis of rotation. Hence, a plot of ln $c(r)$ versus r^2 is a straight line from whose slope M can be calculated (Figure 11-38).

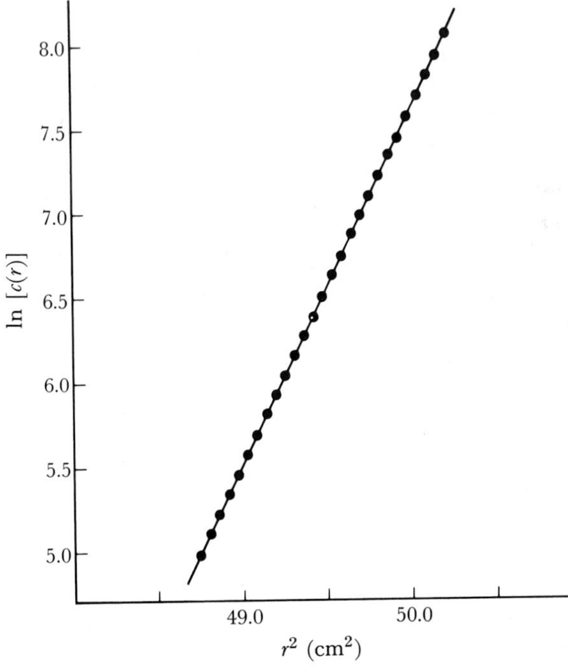

FIGURE 11-38
Sedimentation equilibrium data for bovine serum albumin in 6 M guanidinium chloride at 27,690 rpm and 20°C. Initial protein concentration was 650 μg/ml.

In the derivation, the system was assumed to consist of only two components and the variation in density due to changes in c and pressure* was assumed to be negligible. Pressure effects can in fact be ignored as long as the speed of centrifugation is low; the variation in ρ will be small as long as the concentrations used are low. However, the counterion concentration must be kept low enough to satisfy the two-component assumption. One might ask why counterions should even be included. They are necessary because usually the solute molecule is charged so that an electrical potential gradient would be produced by the distribution of charged molecules. The counterion effectively neutralizes this gradient and eliminates the enormous complexities that would result if this were not possible.

The advantages of the equilibrium method are that (1) it has a firm theoretical basis; (2) it requires only a small amount of material; and (3) the precision of measurement is very high (error $< 1\%$). Its only minor disadvantage is the length of time required to reach equilibrium (16–24 hours).

MOLECULAR WEIGHTS BY APPROACH TO EQUILIBRIUM (ARCHIBALD METHOD)

In the approach-to-equilibrium method (often called the "Archibald" method, after its inventor), as in the equilibrium method, molecular weight is obtained directly from the concentration distribution but in considerably less time. This is often an advantage if unstable molecules are being studied.

The equilibrium method was originally developed because the mathematical equations describing the transport of molecules in a centrifugal field become solvable for the condition of equilibrium in which there is no net flow at any point in the centrifuge cells. The Archibald method instead makes use of the fact that at all times (not only at equilibrium) there is no net flow of material at either the meniscus or the cell bottom. Although it is clear that, before equilibrium is reached, the macromolecule being studied is certainly moving away from the meniscus and accumulating at the cell bottom, this statement simply means that material neither leaves nor enters the cell. The mathematical consequences of this are that M can be determined by measuring the depletion of the macromolecules from the meniscus and the accumulation at the cell bottom *at any time.* Using either of the equations

*Pressure is a factor caused by the action of the centrifugal field.

$$\frac{(\partial c/\partial r)_{\mathrm{a}}}{\omega^2 r_{\mathrm{a}}} = -\frac{M_{\mathrm{a}}(1-\bar{v}\rho)(c_0-c_{\mathrm{a}})}{RT} + \frac{M_{\mathrm{a}}(1-\bar{v}\rho)c_0}{RT} \tag{14}$$

and

$$\frac{(\partial c/\partial r)_{\mathrm{b}}}{\omega^2 r_{\mathrm{b}}} = -\frac{M_{\mathrm{b}}(1-\bar{v}\rho)(c_0-c_{\mathrm{b}})}{RT} + \frac{M_{\mathrm{b}}(1-\bar{v}\rho)c_0}{RT} \tag{15}$$

in which c is the concentration of the macromolecule at a distance r, c_0 is the initial concentration, r is the distance from the axis of rotation at which c is measured, and the subscripts a and b refer to the meniscus and bottom, respectively.

The advantage of the Archibald method is that only very short periods of sedimentation are required. A major drawback of the method is that measurements are made very near the meniscus and cell bottom where the precision of the data is poorest.

SEDIMENTATION EQUILIBRIUM IN A DENSITY GRADIENT

So far, the molecules under consideration have had a density that is much more than that of the solvent. In this section, the important technique of sedimentation equilibrium in a density gradient (developed by Matthew Meselson, Franklin Stahl, and Jerome Vinograd) in which the density of the solvent is almost precisely that of the molecule under study is examined. This technique requires a third component (typically a cesium salt) of low molecular weight and high density. When the solution is centrifuged, the salt redistributes [according to equation (13)] and reaches equilibrium, thus forming a concentration gradient that is less dense at the top and more dense at the tube bottom—that is, a density gradient. The macromolecules also sediment, but the material at the top of the cell moves centrifugally through the gradient and that at the bottom moves centripetally (see Figure 11-39). This continues until the macromolecules form a band at a position in the concentration gradient at which the density of the macromolecule equals the density of the solution (Figure 11-40). (In this case, the density of the macromolecule refers to the density of the molecular unit including water and the cesium salt.) The band width is determined by a balance between the centrifugal force (which tends to make the band narrow), diffusion (which tends to broaden the band), and the density gradient (a steeper gradient will give a narrower band).

To date, this technique has been used almost exclusively for DNA and

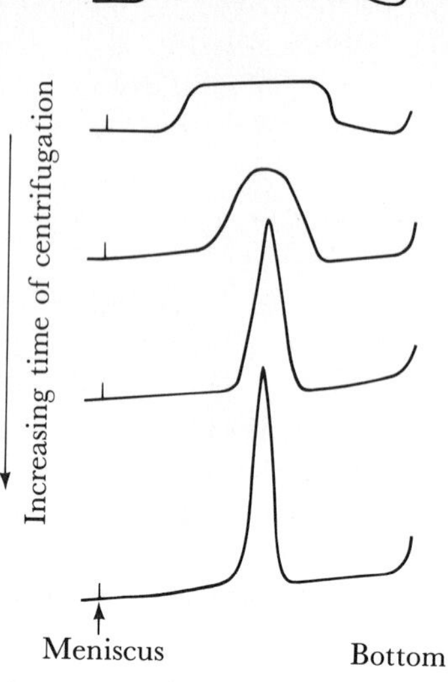

FIGURE 11-39
Movement of DNA molecules to the equilibrium position during centrifugation in a CsCl density gradient. The height of the curve is the concentration of DNA at each point in the cell. The area of the curve remains constant.

bacteriophages; in a few cases, it has been used for RNA and ribosomes, but for proteins its use is rare. The discussion herein will pertain to DNA and phages.

MEASUREMENT OF MOLECULAR WEIGHT BY SEDIMENTATION EQUILIBRIUM IN A DENSITY GRADIENT

The theory of sedimentation equilibrium in a density gradient is complicated because (1) the salt used to generate the gradient is usually at a very

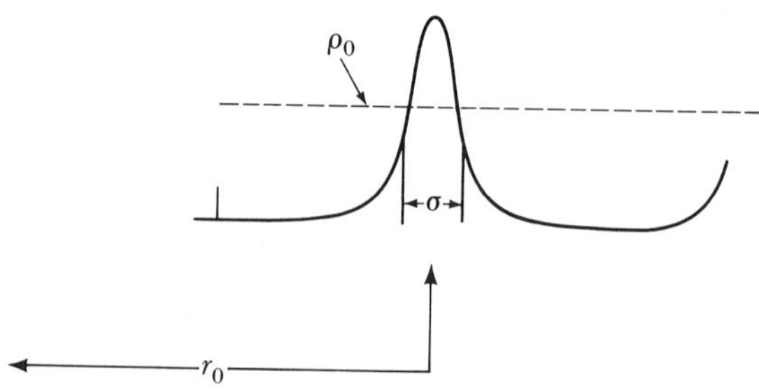

FIGURE 11-40
Parameters of a band in a CsCl gradient: the solid and dashed lines show the DNA concentration and the solution density, respectively; r_0 is the distance from the center of the band to the axis of rotation; ρ_0 is the density at the center of the band; σ is the band width at 1/e the height.

high concentration (5–7 molar), so that the system cannot be treated as a two-component system; (2) the positive and negative ions of the salt, because of differences in molecular weight and diffusion coefficient, do not distribute in the same way so that an electrical potential gradient is set up, and (3) the centrifugal speeds are so high that there are pressure effects. For these reasons, it is not common to use this method to determine molecular weights, but, because it is possible if the data are handled appropriately, the method will be described briefly.

If the fact that there is a three-component system is ignored, the following equation can be derived:

$$M = \frac{8RT\rho_0}{\sigma^2(d\rho/dr)\omega^2 r} \tag{16}$$

in which ρ_0 is the density at the center of the band, σ is the width of the band at $1/e$ the height, $d\rho/dr$ is the density gradient, ω is angular velocity in radians per second, and r_0 is the distance from the center of the band to the center of rotation. It has been repeatedly observed that this equation gives values of M that are too low (usually about twofold), the error increasing with increasing M. This is apparently a result of points 1 and 2 stated above. However, Carl Schmid and John Hearst showed that this is a result of inaccuracies in measuring $d\rho/dr$ and of the effects of the concentrations of the macromolecules in the band. The problem of measuring $d\rho/dr$ is solved by determining, for a given macromolecule whose density is known, the separation of two bands, which differ in density by virtue of isotopic substitution (e.g., ^{14}N for ^{15}N or ^{13}C for ^{12}C). Then equation 16 can be used to obtain accurate values of M by extrapolating to zero concentration, using the equation

$$\ln M = \ln M(c_0) + kc_0 \tag{17}$$

in which $M(c_0)$ is the value of M measured at c_0 (the concentration at the center of the band), k is a constant, and c_0 is calculated from

$$c_0 = \frac{0.20c_i L}{\sigma} \tag{18}$$

in which c_i is the initial concentration (before centrifugation) and L is the length of the liquid column, and σ is defined in equation (16).

The reader is cautioned to discount all measurements of the molecular weight of DNA made by density gradient equilibrium centrifugation before 1969 when Schmid and Hearst presented their work. At the present time, however, this method and the direct measurement of length by electron microscopy are probably the most reliable means of determining M for DNA.

All of these considerations are valid only for materials that are homogenous with respect to density. Density heterogeneity will clearly result in a broader band (owing to a superposition of many nearby bands). This will give an artificially large value of σ, which will produce lower values of M.

SEDIMENTATION IN DENSITY GRADIENTS TO DETERMINE DENSITY

The greatest value of equilibrium centrifugation in density gradients is to resolve material according to density. For example, DNA containing the isotope ^{15}N can be separated from ^{14}N-DNA. If sedimented to equilibrium in a dense CsCl solution, the density difference of 0.014 g/cc results in a separation of ^{14}N- from ^{15}N-DNA of about 0.5 mm in a standard centrifuge cell at 40,000 rpm. Other isotopes produce even greater separation, as indicated in Table 11-1.

The value of this ability to separate DNA molecules by density differences can best be seen in the first example of its use by Matthew Meselson and Franklin Stahl (the method was actually devised to do this experiment).

TABLE 11-1
Density increments in CsCl for DNA containing various isotopes or a substitute base.

Isotope*		g/cc
^{15}N	(100%)	0.016
^{13}C	(100%)	0.049
^{13}C	(60%)	0.030
^{2}H	(95%)	0.056
$^{15}N^{2}H$	(95%, 95%)	0.072
$^{15}N^{13}C$	(95%, 60%)	0.046
Substitute base:		
5-bromouracil	(100%)	0.119

*Number in parentheses refers to percent substitution for normal isotope.

Example 11-O. The Meselson-Stahl experiment demonstrating semiconservative replication of DNA.

Bacteria grown in a medium containing ^{15}N for many generations were shifted to one containing ^{14}N for two generations. DNA samples were isolated at various times and analyzed in CsCl density gradients. As the bacteria grew, the density of the isolated DNA shifted first to that of the average of ^{14}N- and ^{15}N-DNA (this is called the hybrid density), and then a band at the density of ^{14}N-DNA appeared (Figure 11-41). The shift in density as a function of the number of generations of growth gave clear indication that DNA had an even number of subunits and was replicated semiconservatively (i.e., the single strands but not the double strands remain intact).

Another use of isotopic labeling to study replication follows.

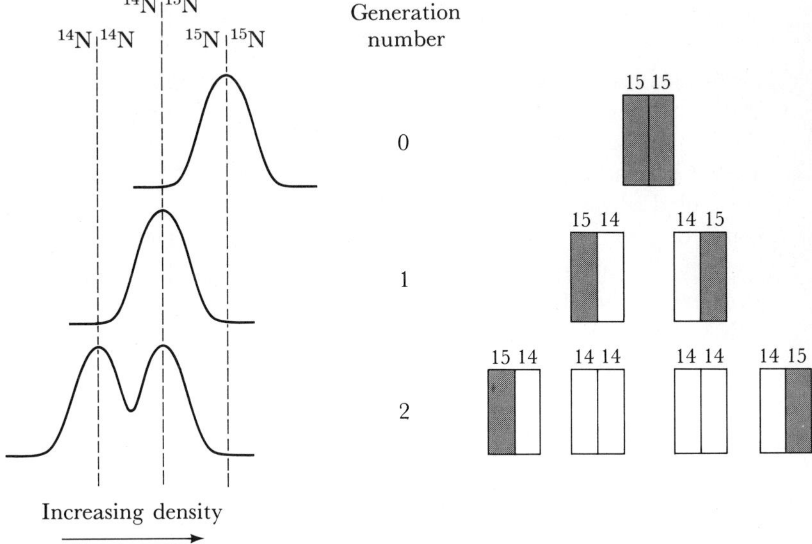

FIGURE 11-41
The type of data obtained in the Meselson-Stahl experiment [*Proc. Nat. Acad. Sci.* 44(1958):671–682]. *E. coli* were grown for many generations in ^{15}N medium and then at zero time transferred to ^{14}N medium. After one generation, all DNA had hybrid density. DNA of less than hybrid density is not seen before the first generation because the chromosome is fragmented into about 250 pieces during isolation. At the right, the state of the isolated DNA at various times is shown.

Example 11-P. Isolation of replicating DNA molecules.

An elegant procedure was used by Jun-ichi Tomizawa and Hideyuki Ogawa to isolate intact, replicating DNA molecules. *E. coli* phage λ containing the naturally occurring isotopes ^1H and ^{14}N were used to infect bacteria in a growth medium containing ^2H and ^{15}N. The infected cells remained in the $^2H^{15}N$ medium for a short time. All bacterial DNA had the $^2H^{15}N$ density; however, if the λ DNA had replicated once, it would have had a density corresponding to one strand of $^2H^{15}N$ and one strand of $^1H^{14}N$-DNA, as in Example 11-O. If, however, the λ had not replicated once but had partially replicated, its density would have shifted to a slightly higher value. By centrifugation of the DNA isolated from the infected cells and selection of fractions containing DNA with a small density shift from the position of unreplicated λ DNA, partially replicated molecules were isolated. These were then examined by electron microscopy and the first photographs of replicating DNA were obtained.

Density differences in CsCl solutions also arise from differences in the percentage of guanine-cytosine (GC) pairs in the DNA (Figure 11-42). This has been used to identify minor species of DNA. For example, the mean GC content of the DNA of the crab *Cancer borealis* is 57%. However, if the isolated DNA is centrifuged in CsCl, it forms two bands (Figure 11-43), one corresponding to the major fraction of chromosomal DNA (42% GC) and the other to the minor species (3% GC). Such minor bands are called *satellite bands* and are widespread in nature. Table 11-2 contains data for a variety of organisms containing satellite DNA.

In CsCl solutions, single-stranded DNA has a density that is from approximately 0.015 to 0.020 g/cc greater than that of double-stranded DNA. This fact can be used either as an analytical tool or to purify either single- or double-stranded DNA.

Example 11-Q. Demonstration of an early step in the infection of *E. coli* by phage ϕX174.

The phage ϕX174 contains single-stranded DNA. On infecting the bacterium *E. coli* with ϕX174 whose DNA is radioactive and analyzing intracellular DNA by centrifugation in CsCl, it was found in the laboratory of Robert Sinsheimer that the DNA decreases in density to the position of double-stranded DNA. This provided evidence that ϕX174 DNA serves as a template for the synthesis of a double-stranded replicating form.

The densities of DNA and protein are quite different in CsCl solutions—roughly 1.7 and 1.3 g/cc. Hence, banding in CsCl can be used to separate these molecules—for example, to prepare protein-free DNA. Another important consequence of this density difference is that nucleoproteins

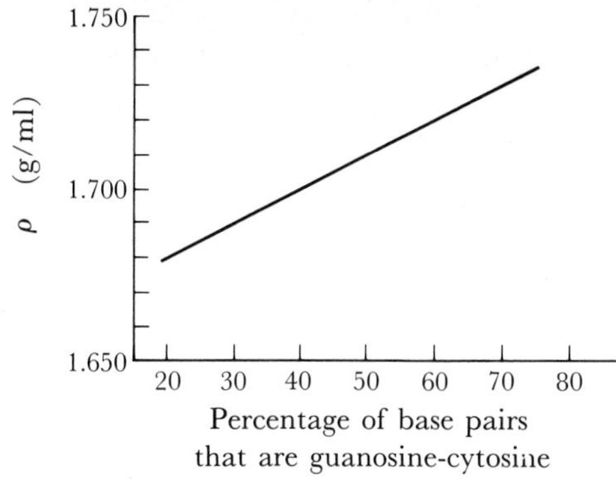

FIGURE 11-42
Density (ρ) of DNA in CsCl as a function of guanine-cytosine content. The equation that describes the curve is $\rho = (.098)(GC) + 1.660$ g/ml, in which (GC) is the mole fraction of G + C in native DNA. [From C. L. Schildkraut, J. Marmur, and P. Doty, *J. Mol. Biol.* 4(1962):430–443.]

such as phages and viruses have a density that reflects their ratio of nucleic acids to protein. For example, a phage that is half protein and half DNA has a density of $1/2\ (1.3) + 1/2\ (1.7) = 1.5$. On this basis, bacteriophage λ mutants that contain less DNA by virtue of a genetic deletion have been isolated by CsCl banding by Jean Weigle and Grete Kellenberger. Because the protein content does not decrease proportionally with DNA content, a deletion phage has a lower density than the original phage. Phage particles containing excess DNA have also been isolated (by Scott Emmons) because they have a higher density. This not only provides a tool for isolating interesting mutants, but allows different phages to be distinguished in physical experiments—that is, relative DNA content becomes a density label. This can be seen in the following example.

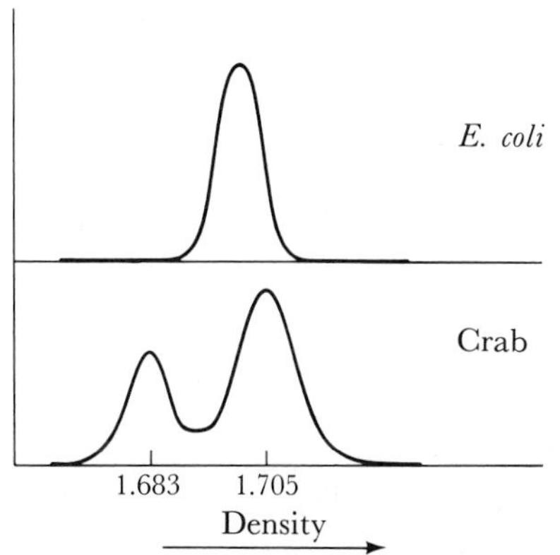

FIGURE 11-43
Photometric trace of the result of equilibrium centrifugation in CsCl of the DNA of the bacterium *E. coli* and the crab *Cancer borealis.* As is common with bacteria, the DNA has a narrow range of densities. The crab DNA consists of two discrete fractions, one of very low density. This minor band is called *satellite DNA.* [From N. Sueoka, *J. Mol. Biol.* 3(1961):31–40.]

TABLE 11-2
Several satellite DNAs detected by CsCl density gradient centrifugation.

Organism	Main band density	Satellite band density	Relative amount of satellite (%)
Crab (testis)	1.700	1.681	30
Mouse	1.701	1.690	8
Wheat (root)	1.707	1.716	Variable
Tomato	1.693	1.694 (chloroplast) 1.705 1.723	Variable
Spinach	1.697	1.694 (chloroplast) 1.703 1.719	Variable
Cricket (ovary)	1.699	1.716	Not known
Chicken embryo	1.698	1.718	Not known
Calf (thymus)	1.699	1.713 1.719	Not known
Bacillus subtilis	1.701	1.714	Not known
Escherichia coli 15	1.710	1.723	Variable

Example 11-R. Recombination studies with *E. coli* phage λ.

Phage λ contains a site in its DNA called *att* at which a recombinational exchange occurs, catalyzed by an enzyme called integrase. It was believed that this exchange involves the breakage of two DNA molecules and the joining of the fragments. This hypothesis was tested by performing a cross between a wild-type phage and a mutant that contains two deletions (b2 and b5) as shown in Figure 11-44. These deletions are large enough to have a significant effect on the phage density in CsCl. In the experiment to be described, all other recombination systems were eliminated by mutation so that the integrase system alone was active. As shown in the figure, λb2b5 labeled with ^{3}H-thymidine was crossed with wild-type nonradioactive phage in *E. coli*. After the infected cells lysed, the released phages were analyzed in CsCl. Four phage types can result from such an infection: b2b5 (parent), b2$^+$b5$^+$ (parent), and the recombinants, b2b5$^+$ and b2$^+$b5. Furthermore, they either will contain all parental DNA (i.e., unreplicated) or will be replicas. Although a great deal of information can be obtained from this experiment, the question here is simply whether parental DNA is ever found in the recombinants. Because all four phage types have different densities, they are readily distinguished and it is necessary to note only

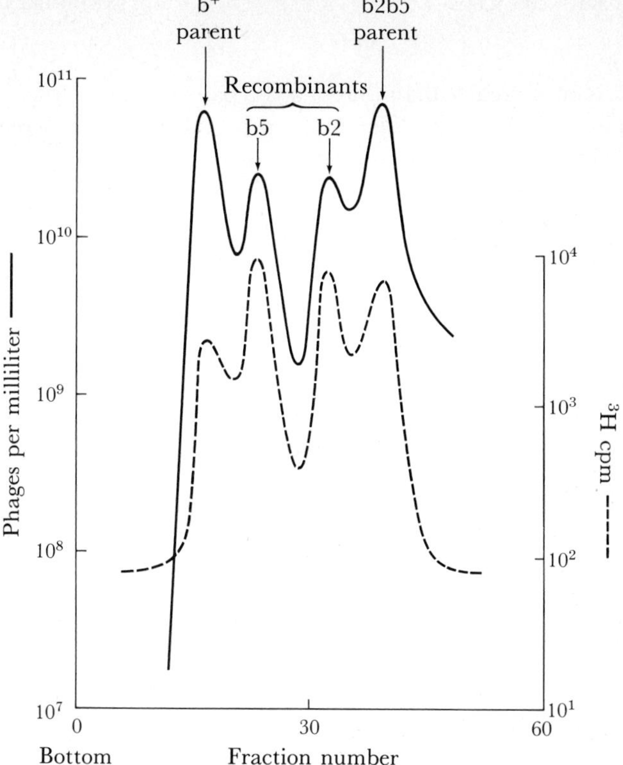

FIGURE 11-44
Resolution of parental and recombined λ phage by centrifugation in CsCl. Phage λb2b5 labeled with ^3H was crossed with unlabeled λb$^+$. The results show that there is ^3H associated with the recombinant DNA, indicating that parental DNA can get into the recombinants. [From G. Kellenberger-Gujer and R. A. Weisberg, in *The Bacteriophage Lambda*, edited by A. D. Hershey, pp. 407–415, Cold Spring Harbor Laboratory, 1971.]

whether ^3H is present in the bands corresponding to the recombinants. Figure 11-44 verifies that it is by showing that the DNA from the b2b5 parent appears in the recombinants. If the experiment is repeated with the ^3H in the b2$^+$b5$^+$ parent, ^3H is again found in the recombinant. This shows that both parents contribute DNA to the recombinants.

Density differences can be enhanced in several ways:

Separation of DNA Molecules in Solutions of Cs_2SO_4 Containing Hg^{2+} or Ag^+ salts. The Hg^{2+} ion binds strongly to the adenine-thymine (AT) base pair. Because this ion is very dense, the density of the Hg^{2+}-AT complex is even greater than that of a GC pair. With the proper concentration of Hg^{2+}, the density difference can be substantially increased. The Ag^+ ion binds to GC pairs and can be used in a similar way.

Example 11-S. Isolation of selected regions of phage λ DNA by centrifugation in Cs_2SO_4 containing Hg^{2+}.

If *E. coli* phage λ DNA is broken into small pieces by shearing or sonication (see Chapter 18), the fragments can be separated into different density classes by centrifugation in Cs_2SO_4 containing Hg^{2+}. This fractionation occurs because different regions of λ DNA have different ratios of AT to GC pairs. The extraordinary separation produced by this method is shown in Figure 11-45.

Separation of the Individual Polynucleotide Strands of DNA by Complexing with Polyribonucleotides and Centrifuging in Cs_2SO_4. The density difference between individual polynucleotide strands is rarely great enough to allow

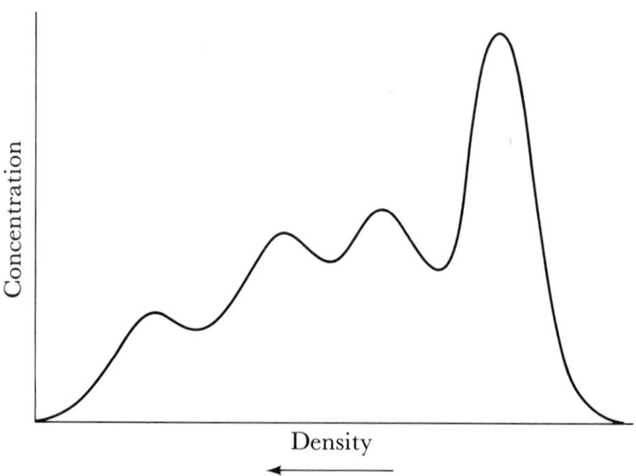

FIGURE 11-45
Equilibrium sedimentation of λ DNA fragments in Cs_2SO_4 containing Hg^{2+}: (A) various bands are seen that correspond to different regions of the DNA of *E. coli* phage λ. Note that high GC DNA has a lower density than low GC DNA in Cs_2SO_4-Hg^{2+} in contrast with CsCl. By similar studies of fragments derived from purified left and right halves and of quarters of λ DNA, the various density classes can be assigned to specific regions of the λ DNA. [From A. M. Skalka, A. D. Hershey, and E. A. Burgi, *J. Mol. Biol.* 34(1968):1–16.]

separation in density gradients. However, these strands frequently contain purine or pyrimidine tracts (i.e., base sequences consisting of only purines or only pyrimidines in a single strand). If a DNA molecule contains predominantly purine tracts on one strand (and therefore pyrimidine tracts on the other), the addition of a polyribopyrimidine will result in binding to one of the strands and not to the other. Polyribonucleotides have a very high density in Cs_2SO_4 so that the density of one of the DNA strands is markedly increased compared with the other. After separation, the polyribonucleotide can be easily removed by an enzyme such as RNase or by digestion of the polyribonucleotide with alkali. In this way, pure, separated strands of DNA can be obtained.

Example 11-T. Fractionation of the two complementary strands of phage λ DNA.

If λ DNA is heated and denatured to separate the strands, mixed with poly IG (a mixed polymer of inosinic and guanylic acids), and then sedimented to equilibrium in Cs_2SO_4, the two complementary strands separate, as shown in Figure 11-46.

Separation of Density Labeled DNAs Having Different AT Content with Cs_2SO_4. Occasionally, it is necessary to distinguish DNA molecules that not only have a density label but also differ in base composition. However, if a DNA whose AT content produces a density in CsCl of 1.700 and another whose density is 1.715 were labeled with ^{15}N and ^{14}N, respectively, their densities would be 1.714 and 1.715, which would be inseparable. However, in Cs_2SO_4 solution, the density is (for unknown reasons) insensitive to base composition. Hence, in Cs_2SO_4, only the isotopic differences will be detected.

Separation of Glucosylated from Nonglucosylated DNA in Cs_2SO_4 *Gradients.* Some phage DNA molecules contain glucosylated bases—for example, the *E. coli* phages T2, T4, and T6 contain 5-hydroxymethylcytosine instead of cytosine, and this base is glucosylated. The glucose decreases the density in CsCl slightly (0.002 g/cc), but in Cs_2SO_4 gradients (for reasons that are unclear) the difference is considerably greater (0.020 g/cc). This difference has been used to follow the synthesis of glucosylated phage DNA in hosts whose DNA is nonglucosylated.

Separation of Covalently Closed, Circular DNA Molecules from Linear DNA or DNA Containing Single-strand Breaks by Centrifugation in CsCl Containing Ethidium Bromide or Propidium Iodide. Ethidium bromide (Figure 11-47) binds very tightly to DNA in concentrated salt solutions and, in so doing, decreases the density of the DNA by approximately 0.15 g/cc. It binds by intercala-

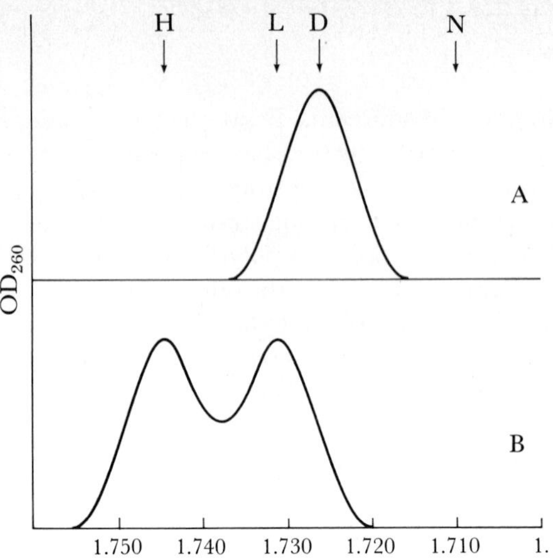

FIGURE 11-46
Separation of the complementary strands of phage λ DNA by complexing with poly IG (mixed copolymer of inosinic acid and guanylic acid). Phage λ DNA was denatured by heating to 95°C. In part A, this DNA is sedimented to equilibrium in CsCl and forms a single band at position D (denatured); N is the position for native DNA. In part B, the denatured DNA is mixed with equimolar amounts of poly IG before centrifugation in CsCl. The density increases because poly IG has a higher density than DNA. The two strands, H (heavy) and L (light), bind different amounts of poly IG and therefore band at different densities. The two bands can be collected by drop collection and freed from poly IG by enzymatic digestion with pancreatic RNase to yield pure H and L strands. In the early literature, the H and L strands are labeled C (Crick) and W (Watson). This widely used technique was developed by Waclaw Szybalski.

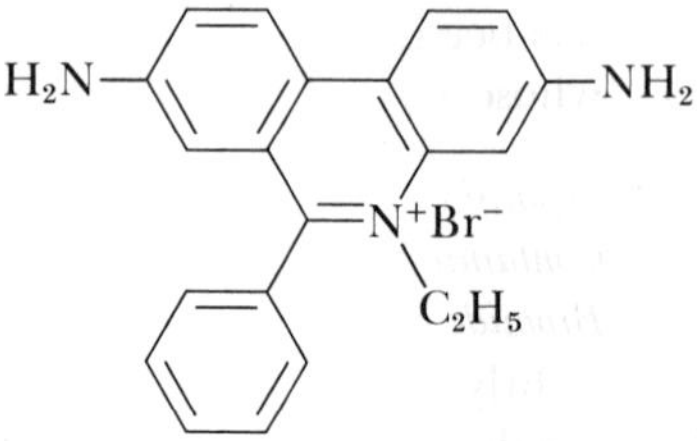

H_2N — NH_2
N^+Br^-
C_2H_5
Ethidium bromide

FIGURE 11-47
Chemical formula for ethidium bromide.

tion between the DNA base pairs and thereby causes the DNA molecule to unwind as more of the ethidium bromide is bound. A covalently closed, circular DNA molecule has no free ends so that, as it unwinds, the entire molecule twists in the opposite direction to compensate. For example, an O-shaped molecule that has bound enough ethidium bromide to produce one clockwise turn will twist in the counterclockwise direction to produce a molecule shaped like the figure eight. As more and more of the molecules intercalate, the 8-shaped molecule will become more twisted. Ultimately, the DNA molecule is unable to twist any more so that no more unwinding is possible. Therefore, no more ethidium bromide molecules can be bound. On the contrary, a linear DNA molecule or a circular DNA with one or more single-strand breaks does not have the topological constraint of reverse twisting and can therefore bind more of the ethidium bromide molecules. Because the density of the DNA and ethidium bromide complex decreases as more is bound and because more ethidium bromide can be bound to a linear molecule or an open circle than to a covalent circle, the covalent circle has a higher density at saturating concentrations of ethidium bromide. Therefore, covalent circles can be separated from the other forms in a density gradient as shown in Figure 11-48. Propidium iodide introduces a greater density difference.

This method, developed in the laboratory of Jerome Vinograd, has become one of the most widely used tools in biochemistry and molecular biology both as a means of purifying covalent circles and as an analytic tool.

Example 11-U. Demonstration that resistance transfer factors are plasmids.

Many strains of *E. coli* are simultaneously resistant to several antibiotics. This capability is transmitted from a resistant cell to a sensitive one, which suggests that the resistant cells contain a transferable DNA plasmid. If DNA is isolated from both a resistant and a sensitive cell and analyzed in CsCl containing ethidium bromide, it is found that the resistant cell contains a small amount of denser material not present in the sensitive cell. If a sensitive strain becomes resistant, it acquires this extra band. Therefore, the information for drug resistance is carried on a covalent circle of DNA. From the relative amount of material in the denser band, a minimum value can be calculated for the fraction of total DNA that is plasmid DNA. It is a minimum value because, in the course of DNA isolation, single-strand breaks might have been introduced in some covalent circles, which would result in the movement of the plasmid DNA to the main band. If the plasmid band is removed from the gradient, the DNA molecules can be visualized by electron microscopy and the molecular weight determined. Because the molecular weight of the *E. coli* chromosome is known, it is then

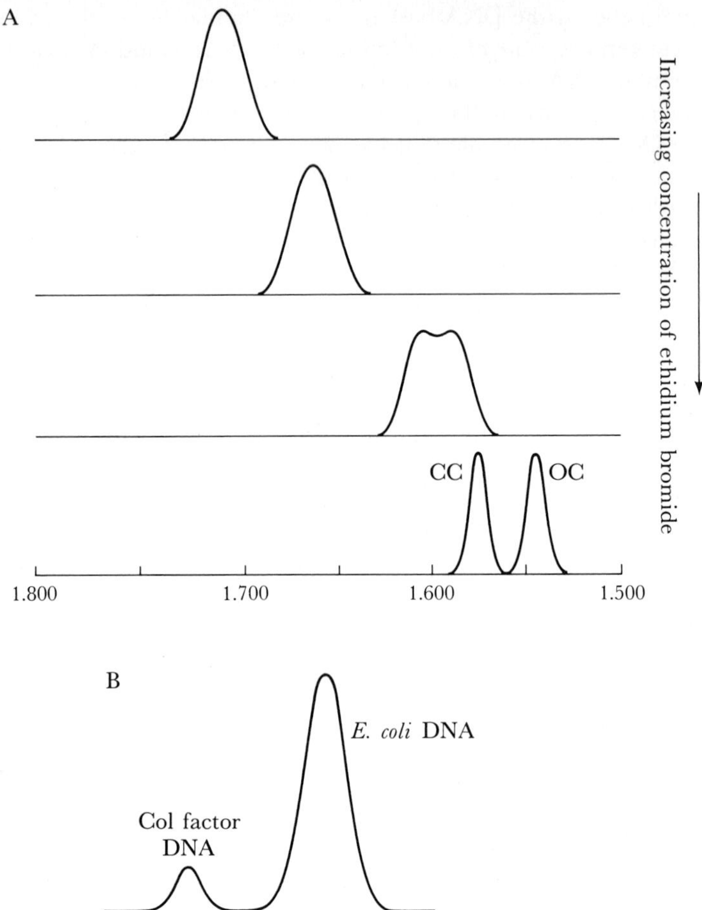

FIGURE 11-48
Effect of ethidium bromide on the density of DNA in CsCl. In part A, a mixture of equal amounts of open circles (OC) and covalent circles (CC) is centrifuged in CsCl containing various amounts of ethidium bromide. The density decreases until, at saturation, the two components separate. The covalent circles bind less ethidium bromide and are therefore at a higher density. Part B shows a photometric tracing of the DNA extracted from an *E. coli* strain containing a colicinogenic factor, centrifuged in CsCl containing ethidium bromide. Note that this differs from Figure 11-43, in which satellite DNA is separated from crab DNA, in that the Col DNA and the *E. coli* have the same density and separate here only because the Col DNA is a covalent circle.

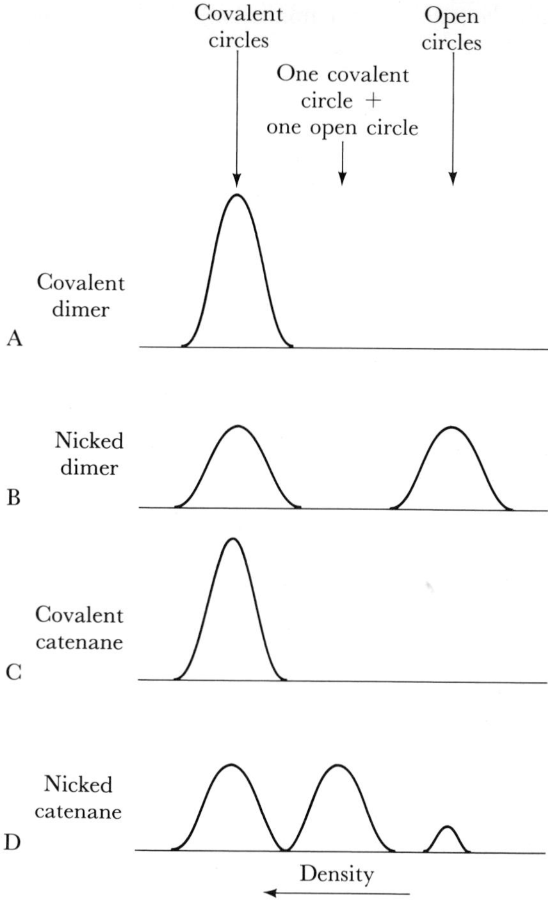

FIGURE 11-49
Method for distinguishing a covalent dimer from two linked covalent monomers (a covalent catenane) by centrifugation in CsCl containing ethidium bromide. The covalent dimer and the catenane band at the same position (A, C). However, if a single-strand break is introduced into half of the molecules (by x rays or, enzymatically, with DNase), the covalent dimer becomes an open circular dimer (B), but the covalent catenane will be a mixture of, mainly, one covalent circle linked to one open circle and a few linked open circles (D).

possible to calculate the minimum number of plasmid DNA molecules per cell.

Example 11-V. Distinction between DNA dimers and catenanes. DNA molecules sometimes interact to form circular structures having twice the molecular weight of the monomer. Such a dimeric structure might be a true circular dimer or it might be two linked monomers (a catenane). Both will band at the same position in CsCl containing ethidium bromide. However, consider the effect of introducing one single-strand break into either structure. If the molecule is a dimeric covalent circle, it will shift in the gradient to the position of a nicked circle (Figure 11-49). However, if it is a catenane, one single-strand break will convert one of the units into an open circle; hence, the density will be an average of the densities of a covalent and a nicked circle—that is, half-way between the two positions. If the catenanes consist of two linked units of unequal molecular weight, the introduction of single-strand breaks will produce two bands of intermediate density, corresponding to the two combinations of covalent and open circles.

Matthew Meselson and I used this technique to show that prophage DNA is inserted into bacterial DNA. A strain of *E. coli* that contains a circular sex factor with the attachment site for phage λ was lysogenized. The sex factor with the attached prophage was isolated and by a combination of centrifugation in alkaline sucrose gradients and in CsCl containing ethidium bromide was shown to be a larger circular structure. However, these tests did not reveal whether the lysogenic form was a larger circle or a catenane consisting of a linked sex factor and a λ circle. When single-strand breaks were introduced by x rays, the circles shifted in density to that of linear DNA—no intermediate density bands were found. Hence, it was concluded that the lysogenized form is a continuous circle.

Appendix

Determination of s

The definition of s [from equation (2)] is

$$s = \frac{v}{\omega^2 r} = \frac{1}{\omega^2 r} \cdot \frac{dr}{dt}$$

in which dr/dt is the rate of movement of the particle. If the boundary is at r_0 at $t = t_0$ and r_1 at some time t_1, then

$$\int_{t_0}^{t_1} s\,dt = \frac{1}{\omega^2}\int_{r_0}^{r_1} \frac{dr}{r}$$

or

$$s(t_1 - t_0) = \frac{1}{\omega^2}[\ln(r_1) - \ln(r_0)]$$

or

$$s = \frac{1}{\omega^2}\text{[slope of a plot of } \ln(r) \text{ versus time]}$$

The following discussion shows how the slope is determined.

The photometric traces from a sedimentation velocity experiment are shown in Figure 11-50. Note that the diagram at the left shows the reference mark (in the centrifuge rotor), which is 5.7 cm from the axis of rotation, and the meniscus, and each subsequent diagram also shows the meniscus. The combined magnification of the photographic and photometric systems is known so that distance as a function of time can be measured. This is done by determining the distance r_x from the meniscus to the boundary as indicated and then adding the distance d_0 from the reference edge to the meniscus. Dividing this sum by the magnification factor and adding the result to 5.7 cm yields the following data:

Time after reaching 33,000 rpm (min)	Distance from axis of rotation	ln r
2	6.225	1.8284
10	6.310	1.8421
14	6.358	1.8493
18	6.400	1.8561
22	6.445	1.8632
26	6.491	1.8705

These numbers are then plotted, as shown in Figure 11-51. The line is straight and the slope is $.0521/(24 \times 60)\ \text{sec}^{-1} = 3.62 \times 10^{-5}$. The centri-

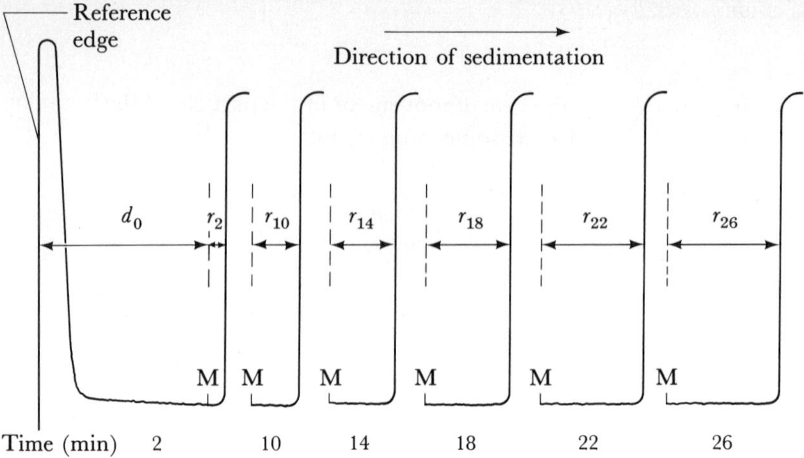

FIGURE 11-50
Photometric traces of sedimenting *E. coli* phage T7 DNA photographed by the ultraviolet-absorption system at six successive times of centrifugation (increasing from left to right). M is the meniscus and d_0 is the distance from the reference edge to the meniscus. Note that the distance between the meniscus and the boundary increases with time. See page 335 for calculation of r from d_0 and r_x.

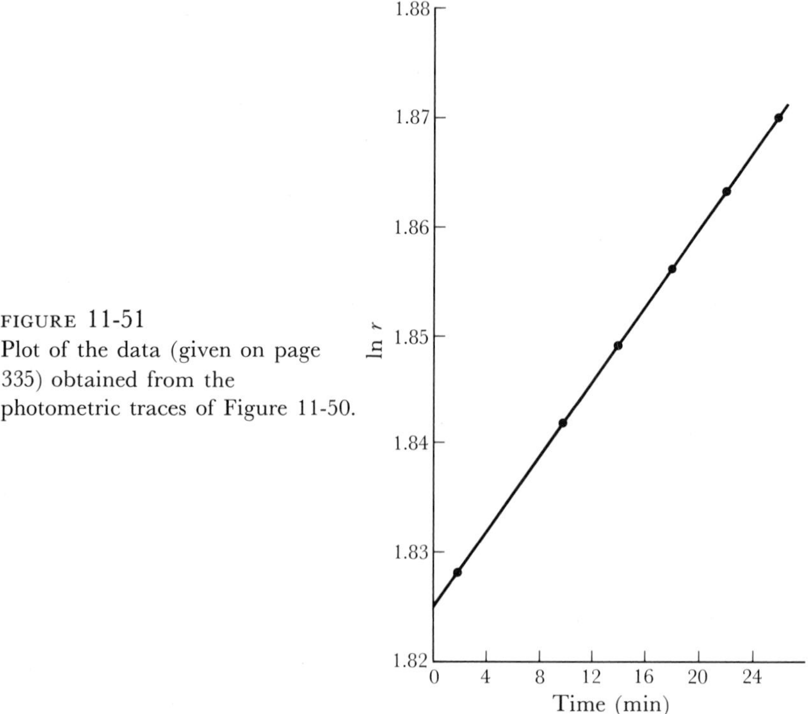

FIGURE 11-51
Plot of the data (given on page 335) obtained from the photometric traces of Figure 11-50.

fuge speed was 33,000 rpm = $(33{,}000 \times 2\pi)/60 = 3.45 \times 10^3$ radians per second. Therefore, ω^2 is 1.19×10^7.

$$s = 30.1 \times 10^{-13} \text{ sec}$$

Because one Svedberg unit = 10^{-13} sec,

$$s = 30.1 \text{ Svedberg units}$$

This can be converted into $s_{20,w}$ by the use of equation (5).

Selected References

Baldwin, R. L. 1957. "Boundary Spreading in Sedimentation-Velocity Experiments. 5. Measurement of the Diffusion Coefficient of Bovine Albumin by Fujita's Equation." *Biochem. J.* 65:503–512. Details of the measurement of *s*.

Brewer, J. M., A. J. Pesce, and R. B. Ashworth. 1974. *Experimental Techniques in Biochemistry,* ch. 6. Prentice-Hall.

Freifelder, D. 1970. "Molecular Weights of Coliphages and Coliphage DNA. 4. Molecular Weights of DNA from Bacteriophages T4, T5, and T7 and the General Problem of Determination of M." *J. Mol. Biol.* 54:567–577.

Freifelder, D. 1973. "Zonal Centrifugation," in *Methods in Enzymology,* vol. 27, edited by C. H. W. Hirs and S. N. Timasheff, pp. 140–150. Academic Press.

Hearst, J. E., and C. W. Schmid. 1973. "Density Gradient Sedimentation Equilibrium," in *Methods in Enzymology,* vol. 27, edited by C. H. W. Hirs and S. N. Timasheff, pp. 111–127. Academic Press.

Meselson, M., and F. W. Stahl. 1958. "The Replication of DNA in *Escherichia coli.*" *Proc. Nat. Acad. Sci.* 44:671–682.

Meselson, M., F. W. Stahl, and J. Vinograd. 1957. "Equilibrium Sedimentation of Macromolecules in Density Gradients." *Proc. Nat. Acad. Sci.* 43:581–588. The development of equilibrium centrifugation in CsCl.

Radloff, R., W. Bauer, and J. Vinograd. 1967. "A Dye-Buoyant-Density Method for the Detection and Isolation of Closed Circular Duplex DNA: The Closed Circular DNA in HeLa Cells." *Proc. Nat. Acad. Sci.* 57:1514–1521. The ethidium bromide method.

Schachman, H. K. 1959. *Ultracentrifugation in Biochemistry.* Academic Press.

Schumaker, V., and B. H. Zimm. 1973. "Anomalies in Sedimentation. 3. A Model for the Inherent Instability of Solutions of Very Large Particles in High Centrifugal Fields." *Biopolymers* 12:877–894. Speed-dependent aggregation.

Svedberg, T., and K. O. Pedersen. 1940. *The Ultracentrifuge.* Oxford University Press.

Szybalski, W., H. Kubinski, Z. Hradečna, and W. C. Summers. 1971. "Analytical and Preparative Separation of Complementary DNA Strands," in *Methods in Enzymology,* vol. 21, edited by L. Grossman and K. Moldave, pp. 383–413. Academic Press.

Szybalski, W. 1967. "Use of Cs_2SO_4 for Equilibrium Density Gradient Centrifugation," in *Methods in Enzymology,* vol. 12B, edited by L. Grossman and K. Moldave, pp. 330–360. Academic Press.

Van Holde, K. E., and R. L. Baldwin. 1958. "Rapid Attainment of Sedimentation Equilibrium." *J. Phys. Chem.* 62:734–743. The low-speed equilibrium method.

Vinograd, J., R. Bruner, R. Kent, and J. Weigle. 1963. "Band-centrifugation of Macromolecules and Viruses in Self-generating Density Gradients." *Proc. Nat. Acad. Sci.* 49:902–910. Development of the band-centrifugation method.

Yphantis, D. 1964. "Equilibrium Ultracentrifugation of Dilute Solutions." *Biochemistry* 3:297–317. The high-speed equilibrium method.

Zimm, B. 1974. "Anomalies in Sedimentation. 4. Decrease in Sedimentation Coefficients of Chains at High Fields." *Biophys. Chem.* 1:279–291. Theory of speed-dependent sedimentation.

The manual for the Beckman Model E ultracentrifuge contains a wealth of information.

Problems

11-1 Which would have the higher s, a rigid rod or a flexible rod, both having the same molecular weight, thickness, and length? A solid or a hollow sphere, both having the same radius and the same mass?

11-2. If two substances having different s values are sedimenting in the same centrifuge tube, their boundaries will continually separate with time. Write the equation describing this separation as a function of time.

11-3. Suppose that a dissociating and reassociating dimer was layered on a sucrose gradient and sedimented. Draw the concentration distribution throughout the cell at various times of sedimentation. Show roughly how this distribution would look with the extremes of rapid, slow, and medium dissociation rates.

11-4. A reagent is suspected of producing interstrand cross-links between the two polynucleotide strands of DNA. How can you determine by sedimentation whether this is the case?

11-5. A highly compact protein is treated with a reagent that causes the s to decrease by approximately 40%. A single boundary results. What do you think the reagent has done if the boundary has become (a) narrower or (b) broader?

11-6. Most proteins have a density of 1.3 in CsCl, although variations of up to 0.050 g/cc exist. Would it be feasible to separate two proteins whose molecular weight is 10,000 having densities of 1.300 and 1.340 by equilibrium centrifugation in buoyant CsCl? (The speed of the centrifuge usable for CsCl centrifugation ranges from 20,000 to 50,000 rpm. Although the density gradient is speed-dependent, for a rough calculation, 0.020 g/cc/mm can be used.)

11-7. Draw the concentration distribution in the centrifuge cell for a mixture of four components having s = 18, 26, 45, and 52 and being 15%, 25%, 50%, and 10% of the total concentration, respectively.

11-8. A protein having a molecular weight of 50,000 and a single, sharp sedimentary boundary is sedimented in 6 M guanidinium chloride. Two schlieren peaks are seen corresponding to molecular weights of approximately 5,000 and 15,000. The area of the slower peak is two-thirds that of the faster. What is the subunit structure of the protein?

11-9. *E. coli* ^{14}N- and ^{15}N-DNA (50% GC) is separated in a CsCl density gradient by 1.32 mm at a certain speed and temperature. What would be the distance from *E. coli* ^{14}N-DNA of DNAs that are 30% and 70% GC, centrifuged in the same cell?

11-10. A sample of DNA gives one band in CsCl but two in CsCl containing ethidium bromide. The ratio of the areas of the denser to the lighter band is 2:1. The molecular weight of the DNA is 30×10^6. Suppose that the DNA were treated with an enzyme that produces on the average one single-strand break in each molecule. What would the ratio of the band areas be after such a treatment? (Remember to use the Poisson distribution to determine the fraction of molecules receiving no breaks.)

11-11. The sedimentation velocity properties of a supercoiled DNA are being studied as a function of the concentration of added ethidium bromide. It is found that s decreases, reaches a minimum, and then increases. Explain.

11-12. A lipoprotein is suspended in 1 M NaCl and centrifuged to determine its sedimentation coefficient. When the schlieren patterns are looked at, it is found that a peak appears at the tube bottom and moves upward. Explain what has happened.

11-13. The sedimentation of a DNA solution is being studied with ultraviolet optics. The DNA solution has been divided into three parts. To part A, nothing is added. To parts B and C, two different materials are added. Then each is centrifuged. In part A, typical boundaries are observed. In parts B and C, no boundaries are seen, neither at the time of reaching speed nor after several hours. In part B, the cell appears uniformly transparent to UV, both above and below the meniscus. In part C, the cell remains nearly opaque. What have the reagents probably done to the DNA?

11-14. If DNA were labeled with ^{3}H-thymidine at a specific activity of 25 C/mmol, what shift in density in CsCl would result? Repeat for ^{14}C at 1 mC/mmol and ^{32}P for 1 mC/μmol.

11-15. The $s_{20,w}$ of a pure protein differs by 10% when sedimented in 0.25 M NaCl versus NaI. What are possible explanations for this? How could the possibilities be distinguished by sedimentation equilibrium?

11-16. Why is more protein than DNA required to obtain a value of s? (Consider how the boundaries must be detected.) For what kind of protein might this not be true?

11-17. You are doing a sedimentation velocity experiment. When the schlieren peak has moved half way down the cell, you remember that you forgot to add something to the solution. You stop the machine, add the necessary material, shake up the contents, and recentrifuge. You notice that the area of the schlieren peak is smaller than in the first centrifugation. Can this be due to the material you added? Is the s you obtain an accurate representation of what is in the cell?

11-18. Double-stranded DNA can be converted into single-stranded DNA either by heating in the presence of formaldehyde or by adjusting the pH to 12.5. The sedimentation patterns of DNA denatured by either procedure are normally identical. However, if DNA is mixed with acridine orange and then the mixture is irradiated with low doses of light absorbed by acridine orange, denaturation by the two procedures gives different patterns; that is, by the heat-formaldehyde treatment the sedimentation pattern of the DNA is that of unirradiated DNA—a single, sharp boundary with $s_{20,w}$ roughly 30% greater than that of native DNA—whereas by the alkaline treatment there is a small, sharp boundary with a great deal of material sedimenting more slowly. Explain this difference.

11-19. Describe how you would determine experimentally whether a particular density gradient would produce isokinetic sedimentation. Would a gradient that is isokinetic for DNA also be isokinetic for a protein?

11-20. Would you expect a circular DNA molecule to show speed-dependent sedimentation? Explain.

11-21. What effect would ethidium bromide have on $s_{20,w}$ of *linear* DNA?

11-22. If proteins are suspended in buffers containing mercaptoethanol, disulfide bonds are broken. Would this be detectable by sedimentation? Consider native and denatured protein.

11-23. The density of DNA in 7 M CsCl containing 0.1 M $MgCl_2$ is less than that in 7 M CsCl alone. Explain.

CHAPTER 12

Partial Specific Volume and the Diffusion Coefficient

In the preceding chapter, it was mentioned that measurements of the partial specific volume and the diffusion coefficient are frequently necessary in using hydrodynamic methods to characterize macromolecules. How this is done is the subject of this chapter.

MEASUREMENT OF PARTIAL SPECIFIC VOLUME

In the theory of ultracentrifugation, a term that is often encountered is $1 - \bar{v}\rho$, in which ρ is the solution density and $\bar{v}$ is the partial specific volume. The volume increment produced in a solution when unit mass of solute is added is $\partial v/\partial m = \bar{v}$. (The partial specific volume is sometimes approximated as the reciprocal of the density of the solute, which is not exactly true but often a good approximation.) Common experience indicates that the volume increment in solution differs from the volume of the solid—for example, a cup of sugar dissolved in a cup of water makes much less than two cups of solution (nearer one cup). It is also important to know that $\bar{v}$ is not an invariant parameter of a particular macromolecule but varies with the solvent composition—that is, the salt concentration, pH, presence of other dissolved substances, and so forth.

An evaluation of $\bar{v}$ is essential in determining both molecular weight and sedimentation coefficients. Furthermore, $\bar{v}$ must be measured with great precision because, due to the range of values for biological macromolecules (0.6–0.75), a 1% error in $\bar{v}$ gives about a 3% error in M or $s_{20,w}$. In fact, the measurement of $\bar{v}$ is frequently the limiting factor in determining molecular weight. The three major methods are the summation of $\bar{v}$ of the residues of a macromolecule, those that use density measurement, and parallel sedimentation equilibrium measurements in isotopically labeled solvents.

Summation of $\bar{v}$ of the Residues of a Macromolecule

If the amino acid composition of a protein is accurately known, $\bar{v}$ can be calculated from the $\bar{v}$ values of the individual amino acids as $n_i m_i \bar{v}_i / \Sigma n_i m_i$, in which n_i is the number of residues per mole of the ith amino acid in the protein, m_i is the residue molecular weight (the molecular weight of the amino acid minus the weight of one mole of water, because one mole of water is removed in the formation of a peptide bond), and $\bar{v}_i$ is the partial specific volume of the ith residue. Values of $\bar{v}$ for the residues can be found in various reference tables. If other groups such as lipids, carbohydrates, flavins, and so forth, are present, $\bar{v}$ for the group must be added in. This is an accurate method for proteins but has not been tested for nucleic acids.

Methods Using Density Measurement

Because $\bar{v} = \partial v / \partial m$, $\bar{v}$ can be determined from the variation of the density of a solution with solute concentration. Four methods for accurately measuring density are described first, and then the surprisingly formidable problem of precisely measuring concentration is discussed.

PYCNOMETRY

A pycnometer is simply a container whose volume can be accurately measured and which can be filled with great precision (Figure 12-1). The volume is measured by filling with water and weighing, since the density of water is accurately known. It is then filled with the *solution* and reweighed. Because of the temperature dependence of volume, the temperature of both the water and the solution must be accurately controlled and known. Pycnometry is the most direct way to determine $\bar{v}$ but for precision requires large volumes of solution of high concentration (50 mg/ml); frequently, it is very difficult to obtain so much material. Pyc-

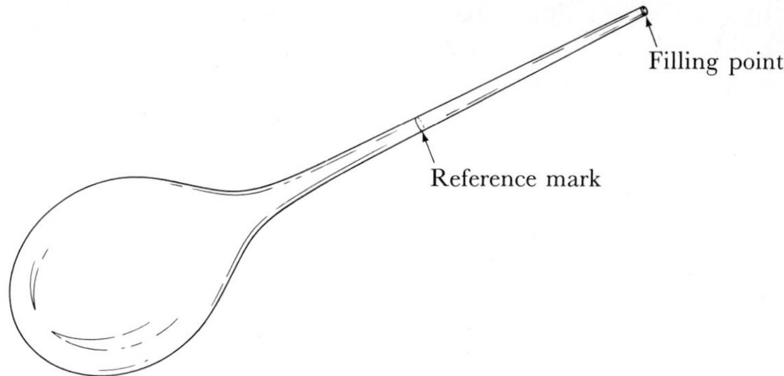

FIGURE 12-1
A pycnometer. The pycnometer is first weighed empty and then filled to the mark and reweighed. The volume up to the mark is accurately known.

nometry usually fails with highly extended molecules such as high-molecular-weight DNA because at 10 mg/ml the solution is a semisolid gel and the pycnometer cannot be filled.

LINDERSTRØM-LANG DENSITY GRADIENT COLUMN

A density gradient column of bromobenzene and kerosene, both of which are impermeable to water, is prepared (Figure 12-2). If a small droplet (usually 1 μl) of an aqueous solution is placed on the surface, it will fall through the column and come to rest at a point at which its density equals

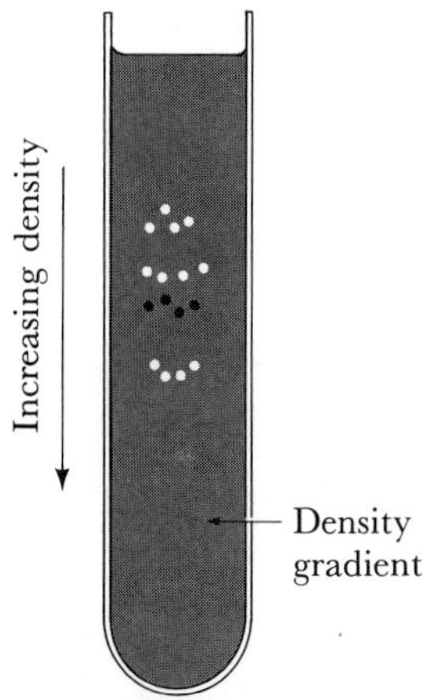

FIGURE 12-2
Linderstrøm-Lang density gradient column. A linear density gradient of two organic liquids (typically kerosene and bromobenzene, which are miscible with one another but immiscible with H_2O) is prepared using apparatus of the type shown in Figure 8-5. Droplets of solutions of known density (open circles) are introduced into the column for calibrating the density gradient. The position of the sample droplets (solid circles) is measured with respect to the reference drops.

that of the column. If the column is calibrated with solutions of known concentration (usually of KCl), the density of the sample can be determined from its position relative to the standards. A large number of standards and sample drops are needed to define the density gradient and to determine the position of the sample with precision. To avoid thermal convection, which would result in the movement of the drops, the temperature of the gradient must be controlled to $\pm 0.01\,^\circ C$. This method has the great advantage of using tiny amounts of material and is generally reliable although there have been a few instances of error caused by interaction with the solvents in the gradient.

CAHN ELECTROBALANCE

Like a pycnometer, this instrument accurately weighs a solution of known volume but uses small (1 ml) volumes at relatively low (~ 10 mg/ml) concentration. This is possible because its accuracy is $\pm 0.1\ \mu g$—roughly, 1,000 times as sensitive as standard laboratory balances. However, its cost ($5,000) prohibits its widespread use.

MECHANICAL OSCILLATOR TECHNIQUE

This method uses a commercially available mechanical oscillator that can be filled with fluid (Figure 12-3); its resonance frequency is related to the

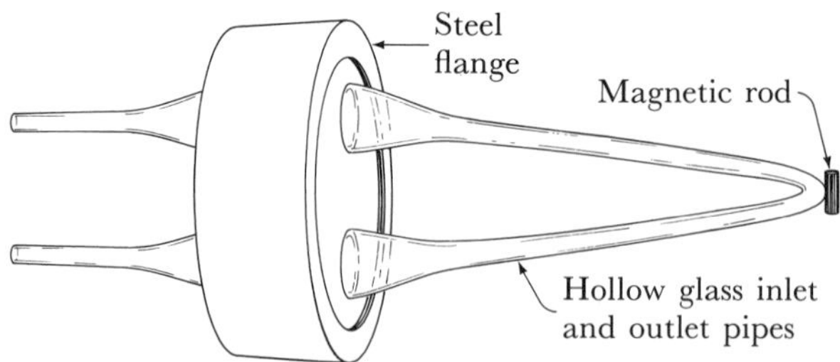

FIGURE 12-3
A mechanical oscillator for density measurement. An oscillating magnet elsewhere in the system causes the magnetic rod and therefore the entire V-shaped tube to vibrate. The natural frequency of vibration is determined by the geometry and mass of the tube. The tube is filled with the sample whose density is to be measured, and the natural frequency changes. From the measured frequency, the weight of added liquid is calculated. Because the volume of the V-shaped tube is accurately known, the density can be determined.

density of the liquid. The advantage of the instrument is high precision with a sample volume of less than 1 ml. The instrument is gaining widespread use.

With each of the methods requiring density measurement, it is necessary to know concentration precisely. This seems trivial because to make a known volume of solution necessitates only weighing a sample and dissolving it. However, the weight must be the *anhydrous weight* and, unfortunately, proteins and nucleic acids invariably contain bound water. With inorganic materials, an anhydrous sample can be obtained by heating to a high temperature, but proteins and nucleic acids are degraded at temperatures higher than 100°C. Hence, a standard method is to dry the protein or nucleic acid sample at 60° to 80°C in a vacuum until the weight becomes constant and assume that constant weight indicates that all water has been removed. However, to remove all water from proteins requires a temperature so high that degradation occurs. An alternative approach, which has been used only a few times, is to determine the dry weight from the elemental composition. For example, because the chemical formulas of all of the amino acids and nucleotides are known, the weight of amino acids or nucleotides can be calculated from the weight of nitrogen and phosphorus in a sample, both of which can be obtained to 1% accuracy. Therefore, the solution used for density determination can be analyzed for nitrogen and phosphorus content and dry weight can be calculated from the amino acid or nucleotide composition of the protein or nucleic acid (which must, of course, be determined if not already known).

Parallel Sedimentation Equilibrium Measurement in H_2O and D_2O Solutions

At equilibrium, the distribution of materials in a centrifuge cell is described by

$$M_{H_2O}(1 - \bar{v}\rho_{H_2O}) = \frac{2RT}{\omega^2}\left(\frac{d \ln c}{dr^2}\right)_{H_2O} \tag{1}$$

in which M_{H_2O} is the molecular weight in H_2O, R is the gas constant, T is the absolute temperature, ω is the angular velocity in radians per second, c is the concentration, and r is the distance from the axis of rotation in centimeters. A similar equation can be written for a D_2O solution. Howard Schachman and his colleagues have shown that these two equations can be solved to yield $\bar{v}$ because $\bar{v}$ is virtually the same in H_2O and D_2O. Because M_{H_2O} is not the same as M_{D_2O} (in D_2O the amide hydrogens exchange with deuterium), simultaneous equilibrium centrifugation analyses, one in

H_2O and the other in D_2O, yield $\bar{v}$. This method has the great advantages that only a tiny amount of material is needed and $\bar{v}$ is measured at the same time that M is being measured. It is the only method available if material is limiting; yet, in principle, it is not as accurate as the methods using the measurement of density. However, because dry weight determination is generally inaccurate, this method probably gives better precision in practice than the others. A considerable increase in accuracy can be achieved by the use of $D_2{}^{18}O$, which has recently become available.

MEASUREMENT OF THE DIFFUSION COEFFICIENT

As discussed in Chapter 11, molecular weight, M, can be determined from a measurement of the sedimentation coefficient, s, and the frictional coefficient, f. The direct determination of f is very difficult; fortunately, this can be bypassed by measuring the diffusion coefficient, D.

Diffusion is the net flow of molecules from a region of high concentration to one of low concentration *if there is no driving force*—that is, the result of random movement (Figure 12-4). The diffusion coefficient, D, of a molecule can be simply defined by Fick's law, which states that the number of molecules, dn, passing through an area, A, in time dt is related to the concentration gradient, dc/dt, by this equation:

$$\frac{dn}{dt} = -DA\left(\frac{dc}{dt}\right) \tag{2}$$

From the fact that molecules will move more slowly if the frictional coefficient, f, is large, it can be shown that

$$D = \frac{kT}{f} \tag{3}$$

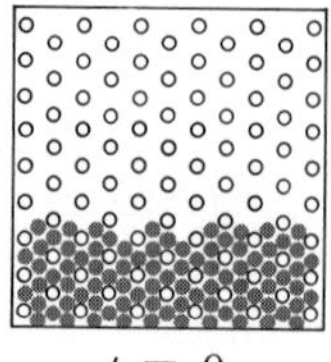

$t = 0$

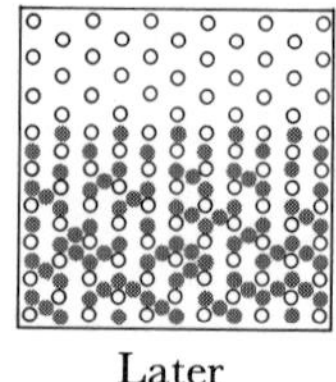

Later

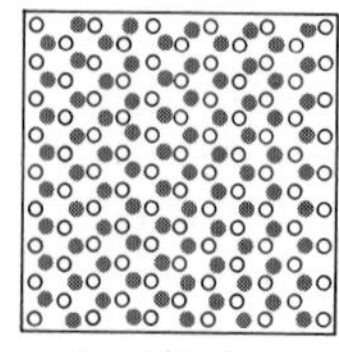

Equilibrium

FIGURE 12-4
The mechanics of diffusion. The solid circles are originally located at the bottom of the box. They diffuse upward until they are distributed uniformly throughout the system.

in which k is the Boltzmann constant and T is the absolute temperature. Hence, the Svedberg equation [Chapter 11, equation (1)] can be written with D instead of f as

$$M = \frac{RTs}{D(1 - \bar{v}\rho)} \tag{4}$$

in which R is the gas constant, T is the absolute temperature, $\bar{v}$ is the partial specific volume of the macromolecule, and ρ is the solution density, and M can be calculated if s and D are known.

The diffusion coefficient for macromolecules is usually measured by creating a boundary between a buffer and a solution of macromolecules of known concentration and observing the spreading of the boundary with time. The theory behind the measurement is simple, but in practice the measurement is filled with potential error. For example, D for a macromolecule is so small, from 10^{-8} to 10^{-5}, that it normally takes a day for measurable spreading to occur. During this time, the system must be free of all mechanical disturbances, and thermal convection must be avoided by accurate temperature control. Furthermore, if a molecule is charged (e.g., a net positive charge) the extra OH ions in solution, which diffuse more rapidly than the macromolecule, create an electric potential gradient that drives the charged molecules to the region of low concentration. Hence, it is necessary to conduct diffusion experiments at or near the isoelectric point and in the presence of sufficiently high ionic strength to neutralize or eliminate the effect of the developed electric field. In addition, because in theory the molecules must move independently of one another (i.e., they must not collide), it is necessary to extrapolate to infinite dilution.

Each of the methods that make use of boundary spreading start with a concentration distribution such as that shown in Figure 12-5A, and the change in time is measured as shown in Figure 12-5B. The measurement is facilitated if the concentration gradient is observed by schlieren optics (Chapter 11). If A and H are the area and height of the schlieren curve,

$$4\pi Dt = \left(\frac{A}{H}\right)^2 \tag{5}$$

and a plot of $(A/H)^2$ versus t gives a straight line of slope $4\pi D$ (Figure 12-6).

The initial boundary can be prepared in several ways. In a standard diffusion cell, the solvent is layered onto the solution of macromolecules. The initial boundary can also be formed in a special analytical centrifuge cell—a synthetic boundary cell as shown in Figure 12-7—in which, at a critical speed, solvent passes through a fine capillary into the solution.

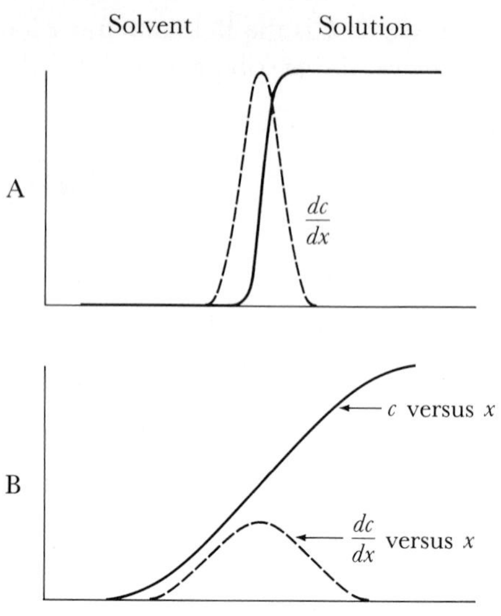

FIGURE 12-5
Measurement of diffusion: (A) initial concentration distribution in which the solvent and the solution are in contact; the curve c versus x shows the concentration of the solution across the cell and dc/dx versus x is the concentration gradient or the schlieren pattern; (B) at a later time.

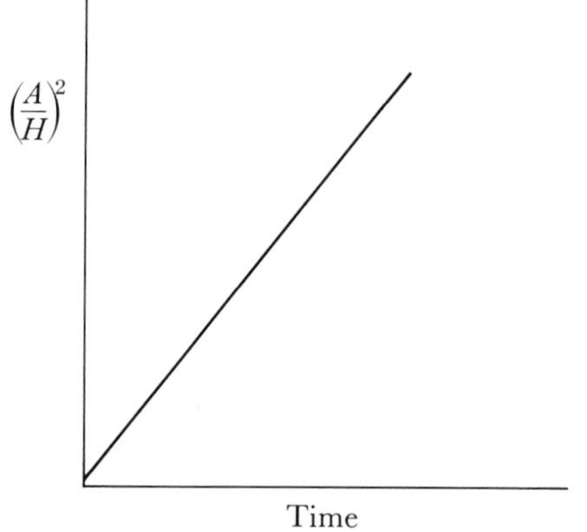

FIGURE 12-6
Typical curve for determining the diffusion coefficient; A and H are the area and height of the dc/dx curve of Figure 12-5. For small proteins, the time scale is of the order of several hours; for large viruses, it can be several days. Note that the curve does not pass through the origin. This is because the boundary is rarely perfect at zero time. The slope of the curve is $4\pi D$.

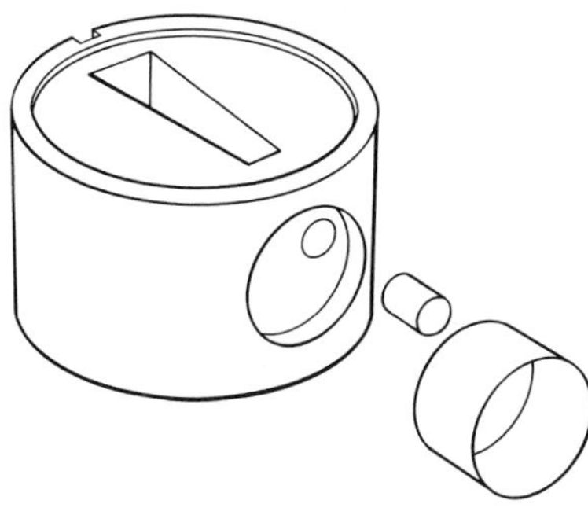

FIGURE 12-7
One type of synthetic boundary cell (valve type) for the Beckman ultracentrifuge: an aluminum centerpiece with a flat-bottomed cavity in place of the usual filling hole (see aluminum centerpiece in Figure 11-3). There is a small, round hole, which has a groove on its perimeter (not shown), on the bottom of the cavity in which a rubber plug is placed. A large cup containing a hole in which the plug fits snugly is placed in the cavity. This cup contains the solvent. The sector of the centerpiece contains the solution. At rest, the plug prevents entry of the solvent, but, during centrifugation, the plug is compressed and solvent leaks from the cup along the groove into the sector. [Courtesy of Beckman Instruments.]

The centrifuge is then operated at a speed sufficiently low that sedimentation does not occur appreciably. This method is useful if samples of only small size are available. In a variation of the centrifugal method, the cell is filled with the solution only and centrifuged: a boundary forms as a result of the sedimentation and the spreading of the boundary as sedimentation proceeds is measured, but a different equation is needed to analyze the data.

A remarkable method—optical mixing spectroscopy—involving the scattering of a laser light beam, has been developed. Because the molecules in solution undergo translational motion (i.e., diffusion), there is a Doppler shift in the scattered light so that the scattered light has a slightly different frequency. This frequency shift is measured by mixing the scattered and unscattered light (i.e., beating the scattered light against the incident light) and measuring the beat frequency (i.e., the frequency difference). This frequency can be simply related to D. This method gives 1% accuracy, uses small samples (a few hundred μg), is very rapid, and is not affected by convection or electrical effects. Its only disadvantages are that commercial instruments are not yet available and it often fails with very highly extended molecules.

To date, there has been no satisfactory measurement of D for large nonspherical molecules such as DNA, because D is so small for such molecules.

Selected References

Bancroft, F. C., and D. Freifelder. 1970. "Molecular Weights of Coliphages and Coliphage DNA. 1. Measurements of the Molecular Weights of Phage Particles by High Speed Equilibrium Centrifugation." *J. Mol. Biol.* 54:537–546. An example of $\bar{v}$ measurement by pycnometry and measurement of nitrogen and phosphorus.

Bauer, N. 1949. "Determination of Density," in *Physical Methods of Organic Chemistry,* vol. 1, part 1, edited by A. Weissberger, pp. 253–296. Wiley. A treatise on pycnometry.

Dubin, S. B., G. B. Benedeck, F. C. Bancroft, and D. Freifelder. 1970. "Molecular Weights of Coliphages and Coliphage DNA. 2. Determination of Diffusion Coefficients Using Optical Mixing Spectroscopy and Measurement of Sedimentation Coefficients." *J. Mol. Biol.* 54:547–566.

Dubin, S. B., J. H. Lunacek, and G. B. Benedeck. 1967. "Observation of the Spectrum of Light Scattered by Solutions of Biological Macromolecules." *Proc. Nat. Acad. Sci.* 57:1164–1171. The use of optical mixing spectroscopy to determine D is described in this paper and in the preceding one.

Edelstein, S. J., and H. K. Schachman. 1973. "Measurement of Partial Specific Volume by Sedimentation Equilibrium in H_2O–D_2O Solutions," in *Methods in Enzymology,* vol. 27, edited by L. Grossman and K. Moldave, pp. 83–98. Academic Press.

Einstein, A. 1956. *Investigations on the Theory of Brownian Movement.* Dover. The classic on diffusion.

Gosting, L. J. 1956. "Measurement and Interpretation of Diffusion Coefficients of Proteins." *Advan. Protein Chem.* 11:429–554. A good review of the technology of diffusion measurement.

Kratky, O., H. Leopold, and H. Stabinger. 1973. "The Determination of the Partial Specific Volume of Proteins by the Mechanical Oscillator Technique," in *Methods in Enzymology,* vol. 27, edited by L. Grossman and K. Moldave, pp. 98–110. Academic Press.

Kupke, D. W. 1973. "Density and Volume Change Measurements," in *Physical Principles and Techniques of Protein Chemistry,* part C, edited by S. J. Leach, pp. 1–75. Academic Press.

Problems

12-1. The partial specific volumes of amino acids have different values if the amino acids are in LiCl and KCl. Would you expect this to be true of proteins? Explain and estimate the magnitude of the effect.

12-2. Would $\overline{v}$ of RNA differ if the RNA were in NaCl rather than $MgCl_2$?

12-3. A pycnometer is being used to measure $\overline{v}$ of a solute. The pycnometer weighs 14.2056 g if empty and 24.1305 g if filled with H_2O at 20°C. A solution is prepared by dissolving 3.5921 g of the solute in 9.9413 g of H_2O. The pycnometer is filled with this solution at 20°C and weighs 25.5307 g. What is $\overline{v}$ for the solute? (The density of H_2O at 20°C is 0.9982 g/cc.)

12-4. If in determining D from measurements of A and H, $(A/H)^2$ plotted against t produces a curve rather than a straight line, what conclusion might you draw? Suppose that the curve has two distinct components, each asymptotic to a straight line. What conclusion might you draw?

12-5. A macromolecule with $\overline{v} = 0.74$ is sedimented in H_2O at 20°C; $s^{\circ}_{20,\mathrm{w}}$ is 14.2; $D^{\circ}_{20} = 5.82 \times 10^{-6}$ cm^2/sec. What is the molecular weight?

12-6. In the absence of convection, the width of a sedimentation boundary, observed in an analytical ultracentrifuge, is determined almost entirely by diffusion. Indeed, the rate of boundary spreading can be used to determine D. Would you expect the measurement of D to be more accurate at high or at low centrifugal speed? Explain. Would it ever be reasonable to perform such a measurement with a mixture of two components? When?

12-7. Will D increase or decrease as axial ratio increases? Which has the greater D, a rigid rod or a flexible rod, both having the same length and cross section and made of the same material? How are the diffusion coefficients of two spheres that have identical radii but different densities related?

CHAPTER 13

Viscosity

Solutions containing macromolecules have greater viscosities than does the solvent alone. The viscosity increment over the solvent alone is a function of several parameters of the molecule, each of which increases this increment. These parameters are the volume of the solution that is occupied, the ratio of length to width of the molecule (the *axial ratio* or the ratio of the axes of the smallest ellipsoid of revolution in which the molecules could fit), and the rigidity of the molecule. For globular molecules such as many of the proteins, the principal effect is through molecular volume and this is simply related to molecular weight. For very rigid, thin molecules, such as DNA, the major effect is due to the axial ratio and this is also a function of molecular weight. Hence, viscometry can be used for the determination of M; on the other hand, if M is known even approximately, information about the overall shape of the molecule can be obtained. These are the two main uses of viscometry.

SIMPLE THEORY OF VISCOSITY

When any substance moves across a surface, the motion is impeded by friction. If the substance is a liquid, this friction generates the effect called viscosity.

Consider a liquid between two large parallel plates (Figure 13-1A), one of which is stationary and the other moving in the x direction with velocity v. The infinitesimal layer of liquid next to each plate encounters

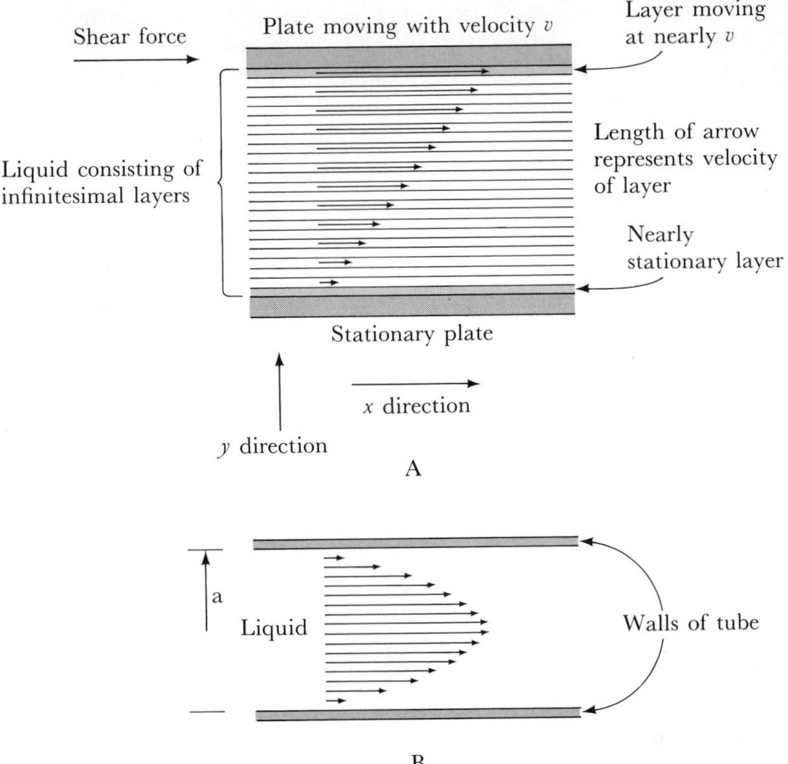

FIGURE 13-1
A. Shearing of a liquid between two parallel plates, one stationary and one moving at velocity v. The position of the tips of the arrows defines the velocity gradient. B. Velocity profile of a liquid flowing through a cylinder with radius "a."

frictional resistance to the motion. Hence, the moving plate carries liquid in the x direction at a velocity nearly equal to v and the layer next to the stationary plate moves very slowly. If the liquid is thought of as consisting of a large number of layers, each layer will slide along the adjacent one, and the frictional resistance between the adjacent layers generates a velocity gradient (Figure 13-1A). The kind of deformation of a liquid produced by a velocity gradient is called *shear.* Newton showed that the frictional force f between the layers is proportional to the area A of the layers and to the velocity gradient dv/dy between them; that is,

$$f = \eta A\left(\frac{dv}{dy}\right) \tag{1}$$

One usually calls η the *coefficient of viscosity,* or simply the *viscosity;* $f/A = F$, the *shear stress;* and $dv/dy = G$, the *shear gradient,* or *shear rate.* If η is a constant, the fluid is called Newtonian; if η is a function of F or G, the solution is called non-Newtonian.

If, instead of being between plates, the liquid is flowing through a cylindrical tube, the friction is encountered at the walls of the tube; in this case, the velocity is maximum along the axis of the tube and minimum adjacent to the walls (Figure 13-1B) so that the velocity gradient is parabolic instead of linear.

All viscometers in use in physical biochemistry employ either the parallel plate or the tube configuration.

The conditions of flow described in Figure 13-1A and B are called *laminar* flow and persist as long as the shear gradient is not too great; at very high shear gradients, *turbulence* sets in and the situation becomes difficult to treat both theoretically and experimentally. Turbulence will not be discussed here.

EFFECT OF MACROMOLECULES ON THE VISCOSITY OF A SOLUTION

The addition of macromolecules to a solvent* with viscosity η_0, yields a solution of higher viscosity, η. This can be thought to result from increased friction between adjacent unimolecular liquid planes (see Figure 13-1) caused by the fact that the macromolecules are larger than the solvent molecules and hence extend through several of these hypothetical planes. The change in viscosity is usually expressed as a ratio, η/η_0, called the relative viscosity, η_r. Einstein showed that η_r is a function of both the size and the shape of the macromolecules and derived the equation

$$\eta/\eta_0 = 1 + a\phi + b\phi^2 + \cdots \tag{2}$$

in which a is a shape-dependent constant ($a = 5/2$ for spheres), ϕ is the fraction of the solution volume occupied by the molecules, and b is a second shape-dependent constant. This equation can be rewritten in terms of the concentration, c, of the macromolecules by defining V as the specific volume of one molecule, so that $\phi = Vc$, to give

$$\eta/\eta_0 = 1 + aVc + bV^2c^2 + \cdots \tag{3}$$

**Solvent* as used here refers to either a pure solvent or a dilute solution of small molecules such as salts.

Viscosity is frequently expressed as the *specific viscosity*, η_{sp}, which is the fractional change in viscosity produced by adding the solute, that is,

$$\eta_{sp} = \frac{\eta - \eta_0}{\eta_0} = \frac{\eta}{\eta_0} - 1 = \eta_r - 1 = aVc + bV^2c^2 + \cdots \tag{4}$$

Neither η_r nor η_{sp} can be simply related to molecular parameters (i.e., shape and volume) because of intermolecular interactions (e.g., collision, entanglement). To avoid this problem, one must consider the situation at very low (i.e., zero) concentration. To do this, the *intrinsic viscosity* $[\eta]$ is defined as

$$[\eta] = \lim_{c \to 0} \frac{\eta_{sp}}{c} = \lim_{c \to 0} aV + bV^2c + \cdots \simeq aV \tag{5}$$

which depends only on the shape-dependent constant, a, and the specific volume, V. Operationally, this means that $[\eta]$ is determined by measuring η_{sp} at several concentrations, plotting η_{sp}/c versus c and extrapolating to $c = 0$. An example of the kind of data usually obtained is shown in Figure 13-2.

As defined above, $[\eta]$ is still not always a useful parameter because of the possible non-Newtonian dependence of $[\eta]$ on G. When a solution of macromolecules having a large axial ratio (i.e., long and thin, such as DNA) is sheared, the molecules tend to become oriented so that their long axes are parallel to the direction of flow of the solution. This orientation decreases η_r because the resistance contributed by the macromolecules to the sliding of the layers past one another decreases as the molecules are confined to fewer layers. The strength of this dependence of η_r on the

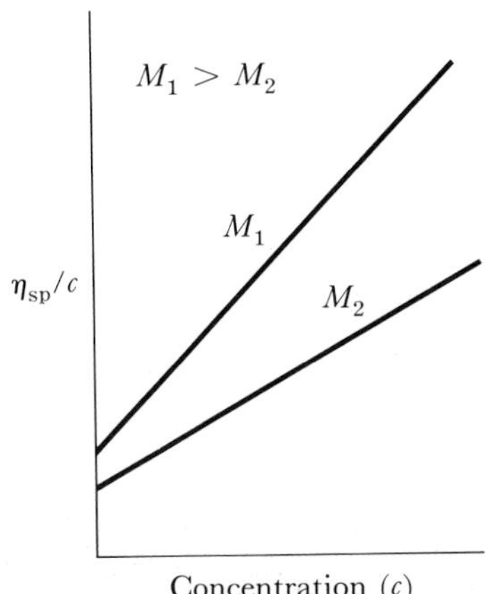

FIGURE 13-2
Plot of η_{sp}/c versus c for two different DNA molecules. Typically, the curves are linear and less steep as the molecular weight (M) decreases. Although not shown, the curve is much flatter for spherically symmetric molecules than for rods.

shear gradient increases with axial ratio because a long, thin molecule is more easily oriented than a short, thick one. This phenomenon can be seen in Figure 13-3B, which shows $[\eta]$ versus G for DNA molecules of different molecular weight—clearly, the greater M is (which for DNA means a greater axial ratio), the lower $[\eta]$ is for a given value of G.

If shape is dependent on the shear gradient (i.e., if the molecule is deformed by the shear), $[\eta]$, which depends on shape [see equation (5)], will also decrease with increasing G. Therefore, to determine molecular parameters from $[\eta]$, it is necessary to measure $[\eta]$ as a function of G and extrapolate to $G = 0$. Unfortunately, there is not a special notation for $[\eta]$ at $G = 0$ (although $[\eta]_0$ seems reasonable); however, in the scientific literature, $[\eta]$ invariably means the value extrapolated to $G = 0$ if there is dependence on G. Figure 13-3A shows why $[\eta]$ must be measured at low shear rather than simply extrapolated from high values.

These considerations lead to two qualitative rules-of-thumb, which can be used to estimate whether a given macromolecule has a large or a small axial ratio: (1) if η_r is large at low concentration, the macromolecule must have a high axial ratio and, similarly, if η_r is small at high con-

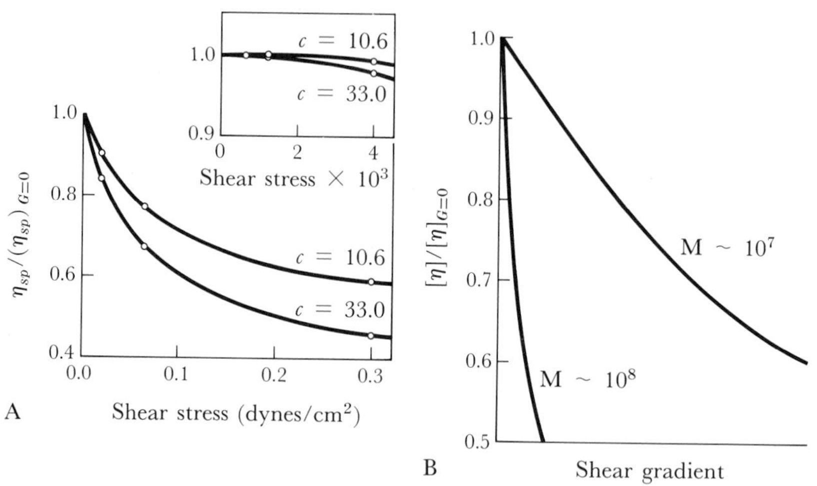

FIGURE 13-3
A. Dependence of η_{sp} of T2 DNA ($M = 106 \times 10^6$) on shear stress, expressed as a percentage of the value at zero shear. Curves for two different concentrations are shown (in μg/ml). The inset shows the change in shape at very low shear. The zero slope shows why low shear viscometers are important in the study of native DNA because the extrapolation to zero shear from large values can cause great errors. [From D. M. Crothers and B. H. Zimm, *J. Mol. Biol.* 12(1965):525–536.] B. Curves showing the effect of molecular weight on the dependence of $[\eta]$ on G. (Data only approximate; collected from a variety of unrelated experiments.)

centrations, the molecule must be somewhat compact; (2) if η_r decreases significantly with increasing shear gradient, the axial ratio must be high.

Two other effects of shear on η_r are worth mentioning: *degradation* and *rheopexy*.

Degradation refers to the fact that at high shear stress, long, thin molecules are broken. Figure 13-4 shows the state of such a molecule in a velocity gradient (illustrated by Figure 13-1B). Note that, because the velocity of flow is not constant across the tube, the ends of any molecule at an angle to the streamlines will move at two different velocities. This will tend to rotate the molecule until it is aligned with a streamline, in which case it experiences no force. Hence, when a molecule is at an angle to the streamlines, the forces at either ends of the molecule are not the same. These forces can be resolved into a perpendicular force (which produces rotation) and a parallel force, which produces stretching (Figure 13-4). It can be shown that the stretching force is maximal (1) in the center of the molecule and (2) when the molecule is at an angle of 45° to the streamlines. Because the force is greatest at the midpoint, the molecule will have the greatest probability of breaking in half. Several interesting

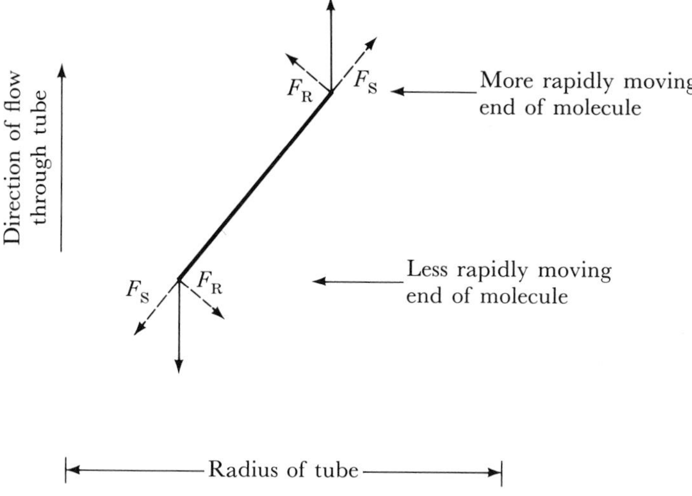

FIGURE 13-4
A rodlike molecule in a velocity gradient generated by flow through a tube. The streamlines move more rapidly near the center of the tube than near the walls. Hence, the molecule is subjected to a driving force at one of its ends and a retarding force at the other. These can be resolved into the forces F_R, which tends to rotate the molecule, and the stretching force, F_S, which causes the molecule to break near the center.

studies of this halving phenomenon for DNA have been made; some of the results are shown in Figure 13-5.

Experimentally, degradation is recognized by either a sharp drop in η_r at some value of G in an η_r versus G curve such as that shown in Figure 13-3B or a decrease in η_r in repeated measurements of η_r at a given value of G. Shear degradation is of great concern with high-molecular-weight DNA (see Chapter 18).

Rheopexy refers to the fact that at relatively high shear stress and concentration, η_r increases with time. This is observed only with material of very high molecular weight, such as the DNA of eukaryotic chromosomes, and the phenomenon is not clearly understood. Because η_r returns to the normal value after shearing is stopped, rheopexy is normally seen only if η_r is measured by repeated determinations under conditions of continuous shear, as with the Zimm-Crothers viscometer (see next section).

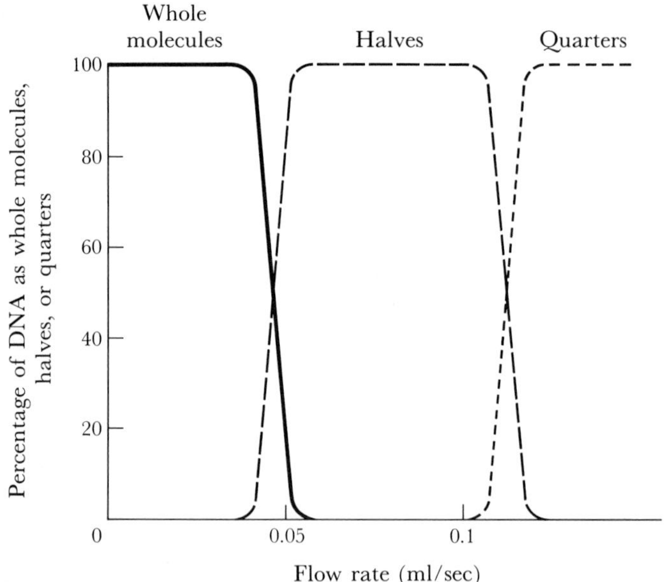

FIGURE 13-5
Breakage of T2 DNA by passage through a fine tube of 0.25-mm diameter at 0.1 μg/ml and at various flow rates. The size of the DNA was determined by zonal centrifugation in an H_2O-D_2O concentration gradient. At a critical flow rate, T2 DNA molecules are broken in half but no further. At a second critical rate, the halves are further broken into quarters. [Somewhat idealized curves are extrapolated from the data of C. Levinthal and P. F. Davison, *J. Mol. Biol.* 3(1961): 674–683.]

MEASUREMENT OF VISCOSITY

Many instruments have been developed for measuring viscosity. A description of those used with biological polymers follows. Viscosity varies by about 2% per °C at 20°C. Thus, precise temperature control (± 0.01°C) is needed and is done with a thermostat water bath.

Ostwald Capillary Viscometer

This viscometer consists of a capillary tube of radius r and length L through which a volume V passes (Figure 13-6). The instrument is used in the following way. The solution is added at opening 1 until the liquid level at rest is at scratch C. Suction is then applied at opening 2 until the liquid level is above scratch A. The suction is removed and the liquid falls owing to the difference in height between the two arms (i.e., the hydrostatic head). The time t required for the meniscus to move between

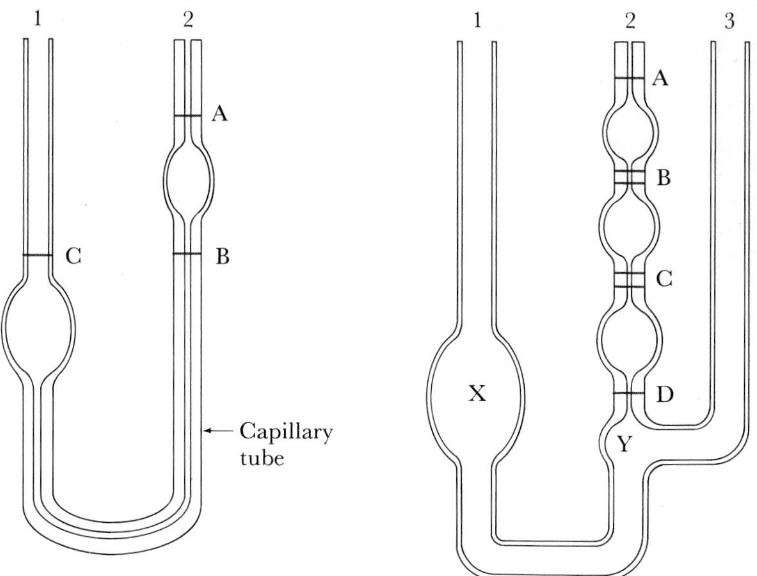

FIGURE 13-6
Capillary viscometers: (left) Ostwald type; (right) Ubbelohde type. See text for details of operation. In the Ubbelohde type, B and C are pairs of scratches. The upper scratch is for timing the movement of the meniscus from A to B and the lower from B to C. The double scratch is to allow time to restart a stopwatch. The relative shear gradients are usually calculated from constants provided by the manufacturer.

scratches A and B is measured. Because of the change in the relative liquid heights, the flow rate is not constant. The viscosity η and *average* G are

$$\eta = \frac{\pi hg\rho r^4 t}{8LV} \text{ and } G = \frac{8V}{3\pi r^2 t} \tag{6}$$

in which h is the average liquid height, g the gravitational constant, and ρ the density. Precise evaluation of h, r, and L is avoided by measuring

$$\eta_r = \frac{\eta}{\eta_0} = \frac{t}{t_0} \frac{\rho}{\rho_0} \tag{7}$$

This instrument has the advantage of low cost but the disadvantage that the shear gradient cannot be varied, and solutions must be relatively dust-free to avoid clogging the fine capillary.

Ubbelohde Capillary Viscometer

As indicated earlier, to measure $[\eta]$ at $G = 0$, one must be able to vary both concentration and shear. The Ubbelohde viscometer is designed with this in mind. As shown in Figure 13-6, this viscometer has several bulbs so that, with reduced pressure head, the average shear gradient decreases. The instrument is used in the following way. Liquid is added through opening 1 to fill bulb X. Opening 3 is closed and suction is applied at opening 2, until the liquid has been drawn above A. Opening 3 is then opened with opening 2 closed, which allows bulb Y to drain. Opening 2 is then opened and the times for the meniscus to pass scratches A, B, C, and D are determined. This gives the viscosity at three different values of G. For varying concentration, the liquid in bulb X can be diluted because the amount of liquid in X does not determine the volume of liquid contained between A and the bottom of the capillary. This instrument is fairly breakable and easily clogs but is highly useful as long as very low values of G are not required.* The Ubbelohde viscometer is no longer in common use, having been replaced by the Zimm-Crothers viscometer.

Couette Viscometer

This instrument, designed for use at relatively low shear gradients, consists of two concentric cylinders separated by a narrow annulus, which is

*The limitation on G is a result of the fact that G is primarily determined by the length and diameter of the capillary. As the diameter is increased, the flow rate would increase to the point that it would be too great to measure—that is, the meniscus would move too rapidly. To compensate for this, the length of the capillary would have to be increased. For very low shear gradients, the length would be unmanageably great.

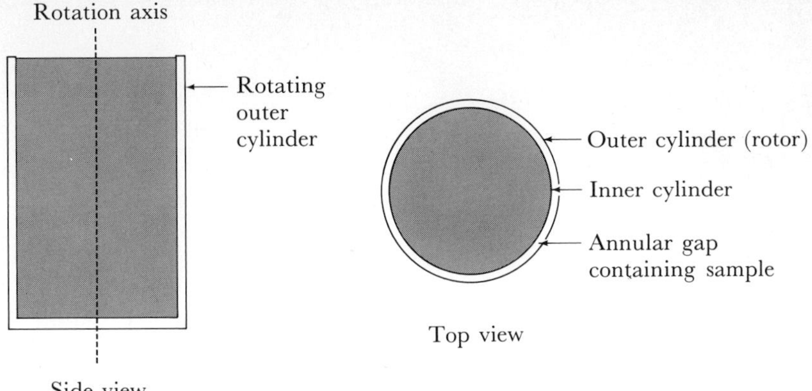

FIGURE 13-7
A Couette viscometer: the outer cylinder usually rotates; the inner cylinder is stationary.

filled with the sample (Figure 13-7). One cylinder is fixed and the other rotates. The situation is equivalent to the two parallel plates shown in Figure 13-1A. From the angular velocity and the dimensions of the cylinder, the terms in equation (1) can be calculated as follows. The velocity of the liquid layer next to the stationary cylinder is nearly zero; for the rotating cylinder, it is the speed of rotation. At intermediate layers, the velocities are proportional to the radial distance from the stationary cylinder. Therefore, the shear gradient is constant and is

$$G = \frac{\pi RS}{30d} \tag{8}$$

in which R is the average radius of the cylinders, S is the rotor speed in rpm, and d is the annular distance. The shear stress F is $T/2\pi R^2 h$, in which h is the height of the cylinder and T is the torque necessary to maintain the speed S. Hence, this instrument operates at a low, controllable, and easily measured value of G, which is of great value with non-Newtonian liquids for which an extrapolation of η to $G = 0$ is necessary. A complication of the Couette viscometer arises from the forces at the ends of the viscometer (the "end effects"). Note that the top surface of the liquid experiences the force of capillarity or surface tension; at the bottom, the liquid is between a stationary and rotating surface. In practice, a Couette viscometer is made sufficiently long that end effects become proportionally small.

Zimm-Crothers Floating-Rotor Viscometer and the Cartesian-Diver Modification

This modification of the Couette viscometer allows high precision work at very low shear gradients (Figure 13-8). It differs in principle from the

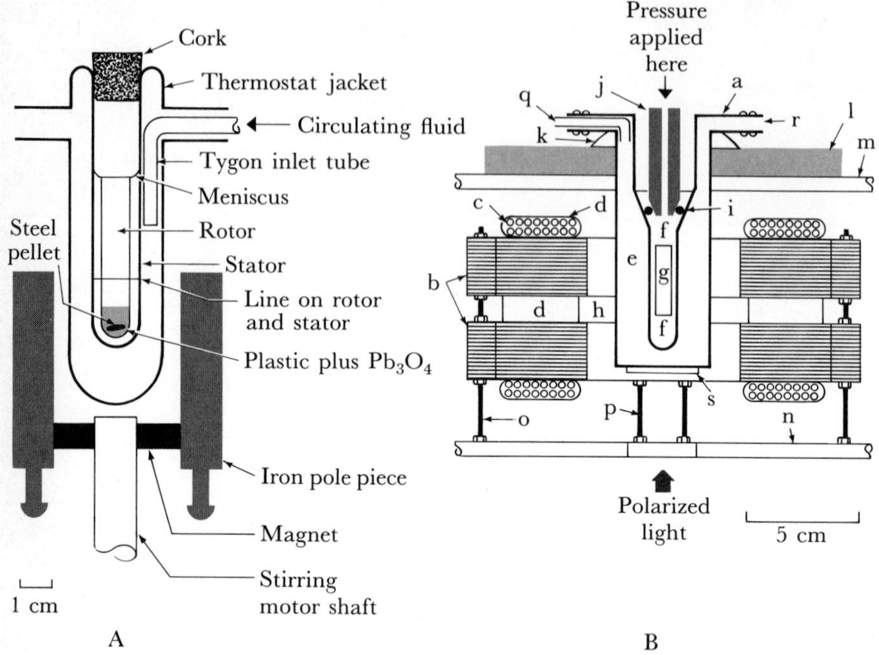

FIGURE 13-8
A. The Zimm-Crothers viscometer. [From B. H. Zimm and D. M. Crothers, *Proc. Nat. Acad. Sci.* 48(1962):905–911.] B. The Cartesian-diver modification: (a) the glass thermostated sample chamber (black section lining), (b) electromagnet laminations (horizontal section lining), (c) magnet wire, (d) insulating tape around outside of magnet wire, (e) circulating water, (f) DNA solution, (g) Cartesian-diver rotor (the rotor is shown in outline only), (h) air gap in magnets for observing rotor, (i) rubber O-ring used to form pressure seal, (j) "pressure bar," (k) plaster, (l) Bakelite block, (m) upper aluminum support plate, (n) middle aluminum support plate (o,p) nuts and bolts for attaching magnet poles to frame, (q,r) circulating water inlet and outlet, respectively, (s) frosted glass plate. All parts are drawn to the scale at the bottom of the figure except the magnet laminations (b) and the magnet wire (c). [From L. C. Klotz and B. H. Zimm, *Macromolecules* 5(1972):471–481. Copyright by The American Chemical Society.]

Couette type in that, with a Couette, the shear gradient is fixed and the shear stress is measured, whereas, with this viscometer, the shear stress is fixed and the shear gradient is measured. The solution is placed between a stationary outer cylinder and a potentially rotating inner cylinder, which floats in the liquid. The inner cylinder contains a steel pellet. External to the outer cylinder are rotating magnets, which cause the steel pellet, and therefore the inner cylinder, to rotate. The speed of this rotation depends on the amount of steel in the inner cylinder and the viscosity of the fluid and is independent of the speed of the rotating magnet and the density of the sample. The shear rate is varied by changing the amount of steel in the inner cylinder (by adding more pellets). Because the torque is proportional to the amount of metal in the rotating magnetic field, the rotor will turn more rapidly and the shear gradient will increase. The shear gradient is equal to the tangential velocity divided by the radial distance between the cylinders. This instrument simply measures the viscosity relative to a standard, such as water, by determining the relative amounts of time required for the inner cylinder to make a given number of rotations.

A problem with all rotating-cylinder viscometers is the effect of surface films at the meniscus (the end effects referred to in the preceding section). The Cartesian-diver rotating-cylinder viscometer avoids this problem. In this modification of the Zimm-Crothers instrument, the inner rotor contains a trapped air bubble, which is compressed by exerting pressure on the solution from above. This converts the inner cylinder into a Cartesian diver, which can be adjusted to be totally immersed in the liquid. This instrument is by far the best available for precise measurement at low shear rate. A modification of the Cartesian-diver viscometer called a viscoelastometer (used for another purpose) is described in a later section.

RELATION BETWEEN INTRINSIC VISCOSITY AND MOLECULAR WEIGHT

For polymers whose configuration is near that of a random coil, the relation between $[\eta]$ and M is of the form

$$[\eta] = KM^a \tag{9}$$

in which K and a are constants depending on the solvent. These constants are usually determined empirically for each solvent-solute system, using molecules of known M, because the theory is not yet adequate for calculating them. Equations of this type have not been derived for very compact molecules such as viruses because $[\eta]$ depends in a complicated way on the ratio of molecular volume to molecular weight. In general, for a given value of M, $[\eta]$ is smaller for the compact form than for a random coil of the same M. Actually, viscometry is rarely used to characterize highly compact molecules because other methods such as centrifugation are of greater value.

For proteins denatured in solutions of 6 M guanidinium chloride (a substance that breaks all hydrogen bonds so that a protein is a random coil if there are no intrastrand disulfide bridges), a and K are well known from extensive data collection. The relation is usually written as

$$[\eta] = 0.716n^{0.66} \tag{10}$$

in which n is the number of amino acid residues in the protein. The average molecular weight per residue can be determined from the amino acid composition so that M can be calculated from η. This relation can be used to determine other properties of proteins, as will be seen in Examples 13-A and 13-B.

For double-stranded linear DNA molecules, the relation between $[\eta]$ and M has been found to be

$$0.665 \log M = 2.863 + \log([\eta] + 5) \tag{11}$$

This strictly empirical equation can be used to calculate M if the DNA sample is homogeneous with respect to molecular weight. This precaution must be borne in mind because of the great sensitivity of DNA to degradation by shearing induced by handling and isolation procedures (see Chapter 18).

EXAMPLES OF THE USE OF VISCOMETRY

Viscometry can be used in a quantitative way to determine molecular weight or semiquantitatively to estimate shape or to detect changes in molecular weight or in shape. Molecular weight is calculated using data of the type shown in Figures 13-2 and 13-3; that is, η_r is measured at several values of c and G, and $[\eta]$ is determined for each value of G by extrapolating η_{sp}/c to $c = 0$ and then again extrapolating these values to $G = 0$. This is straightforward and will not be explained further. The examples that follow show the semiquantitative uses of viscometry. The basic rule used is that stated on page 356—an increase or a decrease in viscosity indicates an increase or a decrease in axial ratio, respectively. It should be noted in the examples that this type of information can often be obtained by measuring η_r only.

Example 13-A. Estimation of the overall shape of proteins.

As discussed in Chapter 1, a protein can be roughly categorized as being highly compact, a random coil, helical, or semirigid (or a combination of the last three). A statement about the overall shape can be made by comparing the viscosity of native and denatured proteins. (Denaturation can be accomplished by acid, high temperature, or the addition of denaturants such as guanidinium chloride.) Because the denatured form is usually a random or near-random coil (see Example 13-B for this distinction), one can determine whether the native form is more or less compact than a random coil by noting whether the viscosity increases or decreases on denaturation. For example, η_r of ribonuclease increases markedly on thermal denaturation in acid, indicating that it has a compact native structure (which is the case for most globular proteins). On the other hand, the viscosity of poly-γ-benzyl-L-glutamate in the rigid rod form decreases fourfold if placed in conditions in which it is random coil. Similarly, the viscosity of the protein collagen and of DNA, highly rigid triple- and double-stranded helices, respectively, decreases markedly on denaturation. (Note, however, the implicit assumption that the conditions of denaturation do not introduce any degree of depolymerization and that the decrease in viscosity is due solely to a change in configuration. This is not always the case—for example, low pH, which denatures DNA, also produces single-strand breaks and therefore is to be avoided if simple denaturation is desired.)

Example 13-B. Detection of intrastrand disulfide bonds in proteins. The cysteine moieties of proteins are frequently coupled by means of disulfide bonds. These disulfide bonds prevent a protein from assuming a completely random coil configuration when denatured, especially if the cysteines are separated by a large number of amino acids.

Such a molecule might be described as a nearly random coil because it is more compact than a true random coil. Disulfide bonds are broken by reduction with 2-mercaptoethanol and reformation of these bonds from the resulting sulfhydryl (SH) groups can be prevented by *S*-carboxymethylation with iodoacetamide. Hence, if disulfide bonds are present in the native structure, the viscosity in 6 M guanidinium chloride will be greater after treatment with mercaptoethanol, because the molecule will be less compact. For example, calf brain tubulin has $[\eta] = 36.0$ ml/g before reduction and 44.0 ml/g after reduction, indicating that disulfide bridges are present.

From the values of $[\eta]$ for the reduced and nonreduced form, information can be obtained about the distance between the cysteines participating in the S–S bond. It has been calculated that the ratio of $[\eta]$ for a single polymer in a straight-chain random-coil configuration to that of a ring structure is 1.6. Hence, in the tubulin example, because the value of 44.0 represents the straight chain (i.e., no S–S bonds), a circle (i.e., when the S–S bonds are between the two terminal amino acids) would have $[\eta] = 44.0/1.6 = 27.5$. Because the observed value is 36.0, the S–S bonds are clearly not between terminal amino acids. If the bonds were between adjacent amino acids, they would have little effect and $[\eta]$ should be very near (and probably indistinguishable from) 44.0. Hence, the S–S bonds are not between amino acids that are located close to one another. Because $[\eta]$ is about halfway between 27.5 and 44.0, it may be concluded that, if there is one or a small number of

S–S bonds, the cysteines must be reasonably far apart along the polypeptide chain but not too near the termini.

Example 13-C. Circularity versus linearity in DNA molecules.
In 1961, it was shown that the genetic map of *E. coli* phage T2 is circular. Because it was not possible to observe DNA with the electron microscope at that time, to test for circularity, the change in η_r of a single sample of T2 DNA, which was continuously being digested by pancreatic DNase, was studied. This enzyme produces single-strand phosphoester breaks that accumulate and ultimately match to form double-strand breaks (Figure 13-9A). The single-strand breaks alone have no effect on DNA viscosity (Figure 13-9B) because of the rigidity of the molecule conferred by base stacking (see Chapter 16). Hence, for a linear molecule, η_r is constant during the period in which there are only single-strand breaks and decreases when matching occurs, owing to decreasing molecular weight. However, when the first double-strand

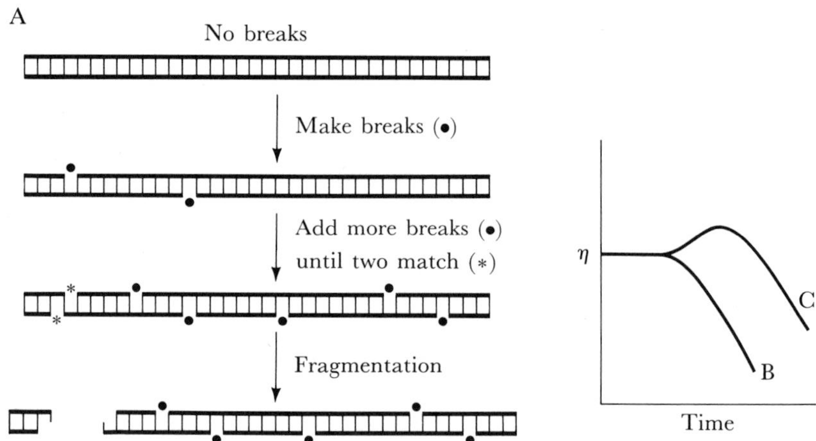

FIGURE 13-9
A. Process of the matching of single-strand breaks in DNA to make double-strand breaks. The dots indicate the positions of the breaks. The asterisks indicate a pair of breaks that result in a double-strand break: they are separated by only one base pair, which is insufficient to maintain the integrity of the molecule. B. Viscosity of T2 DNA as a function of time of digestion with pancreatic DNase. The DNase produces single-strand breaks, which ultimately match to produce double-strand breaks, resulting in a decrease in the viscosity. The curve is flat at first because single-strand breaks do not affect the viscosity. C. Expected curve (as in curve B) if the DNA were circular. At the time that double-strand breaks occurred, the circle would become linear and the viscosity would increase. Viscosity would not decrease until a second double-strand break had formed.

break occurs in a circular DNA molecule, the molecule becomes linear and η (Figure 13-9C) increases; because the axial ratio is increased, η_r does not begin to decrease until two double-strand breaks occur. For T2 DNA, η_r is constant for several hours of DNase treatment and then decreases, indicating that it is a linear molecule.

Example 13-D. Viscometric evidence that certain substances can intercalate between nucleotide bases of DNA.

The dye acridine orange binds tightly to double-stranded DNA. The sedimentation coefficient (Chapter 11) of the complex *decreases* compared with that of the free DNA; this decrease could result either from depolymerization or from an increase in the frictional coefficient due to an increase in the axial ratio. If the decrease is due to depolymerization, $[\eta]$ should decrease; if due to an increase in the frictional coefficient, it will increase. Experimentally, $[\eta]$ of the complex is greater than that of free DNA, so that the axial ratio of the complex must be greater than that of the free DNA. Hence, the DNA must increase in length when the dye is bound. Fluorescence polarization studies (Chapter 15) have shown that the dye is immobilized and is in the same plane as the base pairs. Hence, it has been inferred that the dye intercalates between the DNA bases, thus lengthening the molecule (Figure 13-10).

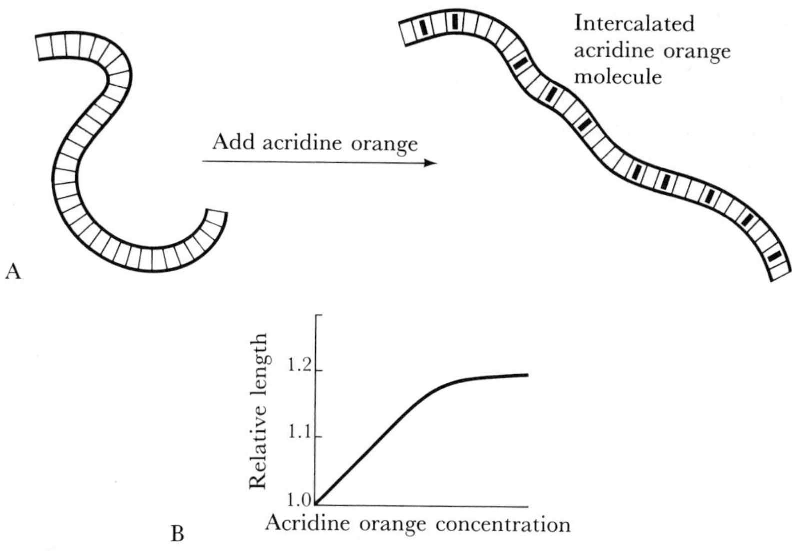

FIGURE 13-10
A. Acridine orange molecules intercalating between the base pairs of DNA. The molecule is extended and becomes more rigid. B. The length of *E. coli* phage λ DNA as a function of acridine orange concentration, as determined by electron microscopy. [From the laboratory at Brandeis University.]

This inference has been confirmed by electron microscopy of the complex; DNA becomes longer and less flexible when the dye is bound.

Example 13-E. Detection of the enzymatic polymerization of DNA from mononucleotides.

If a mixture of DNA and radioactive nucleotide triphosphate is incubated with the enzyme *E. coli* DNA polymerase I, some radioactive material becomes acid precipitable (see Chapter 5, Example 5-A). This could result from net synthesis leading to *an increase* in the amount of high-molecular-weight DNA or to nucleotide *exchange* into the original molecules. If there is net synthesis, η_r should increase because of an increase in the concentration of the DNA; if there is only exchange, the amount of DNA will be unchanged and η_r will be constant. At the time the work with polymerase I was beginning, DNA could be distinguished from nucleotides by light scattering (using purified protein-free samples with DNA at high concentrations), by ultracentrifugation or ultraviolet absorbance (at low concentration and in the absence of excessive UV-absorbing material), or by viscometry. The enzyme reaction contains DNA at low concentration and protein and UV-absorbing nucleotides at high concentrations, thus eliminating the first three possibilities. By viscometry, it was seen that η increased with time of incubation, indicating either that more DNA was present or that molecular weight was increased; either case implies DNA synthesis.

Example 13-F. Conformation of DNA-histone complexes.

Histones are proteins contained in chromosomes and bound to DNA. To understand the role played by the histones in chromosomal structure and in the regulation of DNA expression, the structure of the DNA-histone complex has been studied.

As histone is added to DNA, the sedimentation coefficient of the DNA increases. An increase in s means either an increase in M or a decrease in the frictional coefficient (or axial ratio). The intrinsic viscosity also increases, indicating either an increase in molecular weight or an increase in axial ratio. Qualitatively, these two results would suggest that M is increasing. However, as more histone is added, the ratio $s/[\eta]$ increases markedly, which indicates that the axial ratio also decreases with increasing bound histone. Hence, the increase in M by histone binding is probably accompanied by some folding of the DNA.

Example 13-G. Detection of the injection of DNA by phages.

Phages are highly compact and therefore have very low $[\eta]$ for their M. Treatment with certain reagents (e.g., alkaline buffers, $NaClO_4$, and guanidinium chloride) lead to extraordinary increases in $[\eta]$ due to the release of DNA. This can be used to measure the extent of the injection of DNA from the particles. It should be noticed though that, because viscometry measures the average property of a solution, the following cases cannot be readily distinguished: 100% of the phages injecting half of their DNA and half of the phages injecting all of their DNA. Hence, if a set of conditions results in a value of η less than that obtained when the phages are totally disrupted, the observed η cannot be readily interpreted. These cases could, of course, be distinguished by sedimentation experiments because the sedimentation coefficient would depend on the amount injected. Hence, in the first case, a single sedimentation coefficient whose value would be less than that of the intact phage and greater than that of free DNA would be observed; in the second case, two sedimenting species would be seen—the phage and the free DNA.

MEASUREMENT OF THE VISCOELASTICITY OF DNA SOLUTIONS

The DNA molecules of bacteria have a molecular weight greater than 2×10^9; that of animal cells approaches 10^{11}. In this range of M, hydrodynamic methods suffer from the complex phenomena indicated in Table 13-1. Note that the viscoelasticity method described herein is insensitive to these artifacts.

If solutions of DNA are subjected to a shear stress, the DNA molecules are extended as shown in Figure 13-11. After the stress has been removed, the molecule returns to a relaxed configuration in which the molecule, if it is very long, approximates a random coil. A solution of an extendable molecule shows viscoelasticity when sheared in a Couette-type viscometer. With the Zimm-Crothers Cartesian diver viscometer (Figure 13-8), the application of a shear stress causes the inner cylinder to rotate. When the shear stress is removed, if the viscometer contains a pure solvent, the angu-

TABLE 13-1
Problematic phenomena occurring with various hydrodynamic techniques.

Phenomenon	Sedimentation	Viscometry	Visco-elasticity
Speed-dependent sedimentation	+	−	−
Stress-dependent aggregation	+	+	−
Shear degradation	+	+	−
Variation with shear stress	−	+	−

NOTE: A plus indicates that a problem exists; a minus, that it does not.

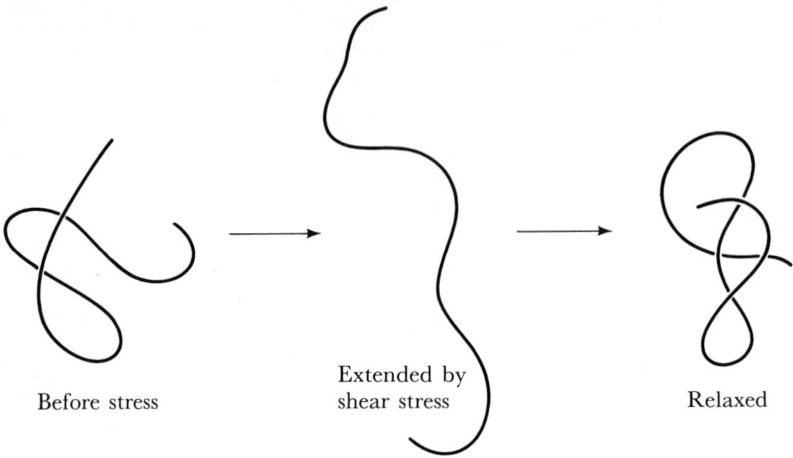

FIGURE 13-11
Extension of a DNA molecule followed by viscoelastic relaxation after the stress has been removed.

lar velocity gradually approaches zero. However, if the rotor is suspended in a solution of high-molecular-weight DNA, the direction of rotation reverses owing to the relaxation of the previously stretched DNA molecules; ultimately, of course, the rotor stops owing to friction. The exponential decay rate at which the rotor comes to rest can be characterized by a time constant τ_R called a *retardation time* (i.e., the time required to reach $1/e$ the total angular displacement occurring during reverse motion). When τ_R is extrapolated to zero DNA concentration, it is called the *relaxation time*, τ_R°, and is a true molecular parameter that can be related to M. Due to the incompleteness of the theory at present, M for a linear DNA molecule is best determined from the empirical equation

$$M = 1.56 \times 10^8 (\tau_{R,W}^\circ)^{0.60} \tag{9}$$

in which

$$\tau_{R,W}^\circ = \frac{\eta_{H_2O}}{\eta} \tau_R^\circ \tag{10}$$

with η_{H_2O} and η being the viscosity of water and of the solvent, respectively, and all measurements being made at 25°C. For circular DNA molecules, the constant and exponent will differ, but to date they have not been evaluated.

The great usefulness of this method is that, with solutions containing molecules of different sizes, the observed value of M corresponds to that of the *largest* molecules present (unlike $[\eta]$ measurements, which yield an

TABLE 13-2
Size of various DNA molecules determined by viscoelastic measurements.

DNA	Molecular-weight viscoelasticity	Other methods*
E. coli phage T7	$25 \pm 2 \times 10^6$	25×10^6
E. coli phage T2	$109 \pm 6 \times 10^6$	111×10^6
E. coli	2.7×10^9	2.6×10^9
Bacillus subtilis	2.0×10^9	$1\text{–}4 \times 10^9$
Drosophila melanogaster	$41 \pm 3 \times 10^9$	—
Drosophila americana	$79 \pm 10 \times 10^9$	—

SOURCE: Data from L. C. Klotz and B. H. Zimm, *J. Mol. Biol.* 72(1972): 779–800; R. Kavenoff and B. H. Zimm, *Chromosoma* 41(1973):1–27; R. Kavenoff, L. C. Klotz, and B. H. Zimm, *Cold Spring Harbor Symp. Quant. Biol.* 38(1973):1–8.

*A dash indicates that the measurement has not been made by any other method.

average value). Hence, with very fragile molecules, the system is unaffected by the degradation of a fraction of the molecules.

The instrumentation for viscoelasticity measurements is not yet commercially available so that work has been confined to laboratories of Bruno Zimm and his colleagues. To date, the method has been used successfully to evaluate M for the DNAs of large bacteriophages, of several bacteria, and of the chromosomes of several species of the fruit fly *Drosophila* (Table 13-2). This technique should find even greater applications in the future.

Selected References

Eigner, J. 1968. "Molecular Weight and Conformation of DNA," in *Methods in Enzymology,* vol. 12B, edited by L. Grossman and K. Moldave, pp. 386–429. Academic Press.

Uhlenhopp, E. L., and B. H. Zimm. 1973. "Rotating Cylinder Viscometers," in *Methods in Enzymology,* vol. 21, edited by C. H. W. Hirs and S. N. Timasheff, pp. 483–491. Academic Press.

Yang, J. T. 1961. "The Viscosity of Macromolecules in Relation to Molecular Conformation." *Advan. Protein Chem.* 16:323–400. Viscometry as applied to proteins.

Problems

13-1. A DNA has an intrinsic viscosity $[\eta]$ of 50 dl/g. When heated to 95°C in 0.01 M NaCl, $[\eta]$ drops to 20. When heated to 90°C in 0.5 M NaCl, $[\eta]$ drops to 3. Explain the effect of NaCl concentration. What differences might be expected if the heating was performed with 1 M formaldehyde in the NaCl solutions?

13-2. If bacteria are treated with low concentrations of some detergents, the suspension becomes visibly viscous. If the suspension is briefly sedimented, a pellet results that contains all of the DNA and RNA. The supernatant fluid is still viscous. What is the probable explanation for this viscosity?

13-3. Which in the following pairs will produce the greater viscosity when suspended in H_2O: (a) two spheres of identical mass but different radii; (b) a solid sphere and a porous sphere through which solvent can flow; (c) a rigid rod and a flexible rod; (d) a sphere and a sphere with a linear branch (e.g., a lollipop)?

13-4. Which will produce the greater relative viscosity, a molecule in H_2O or in 20% glycerol?

13-5. Which of the following pairs will be more resistant to degradation by hydrodynamic shear: (a) a linear DNA molecule and a circular molecule, both having the same total length; (b) a native linear DNA molecule and a denatured molecule in 1 M NaCl, both having the same total length: (c) denatured linear DNA in 1 M NaCl and 0.01 M NaCl, both having the same molecular weight; (d) a covalent circle and a twisted circle of DNA, both having the same molecular weight?

13-6. What might you conclude about the structure of a protein molecule if $[\eta]$ increases when placed in 6 M guanidinium chloride? If it decreases? What is the principal source of uncertainty in drawing these conclusions?

13-7. In general, the addition of various salts to H_2O increases viscosity, that is, $\eta_r > 1$. However, in a few cases, in dilute solution $\eta_r \leq 1$. Propose a possible explanation.

13-8. Under certain conditions of ionic strength and pH, a particular polynucleotide has a fairly small variation of η_{sp}/c with c, and $[\eta]$ is relatively independent of shear. On changing pH, the concentration and shear dependence, as well as the actual value of $[\eta]$, increase markedly. What is a possible effect of the pH change?

13-9. Normally, native DNA solutions show a relatively small dependence of $[\eta]$ on ionic strength, owing to the rigidity of the molecule. However, a particular DNA sample isolated from bacteria has the property that, at low ionic strength, $[\eta]$ becomes substantially smaller. Furthermore, if the solution is diluted before lowering the ionic strength, η_r is higher than that observed if the ionic strength is decreased before dilution. This is not true of all DNA preparations from these bacteria. What is a possible cause of this effect with the particular sample?

13-10. A sample of supercoiled DNA is subjected to a treatment that introduces single-strand breaks at a constant rate of one every 30 minutes. On the average, it takes ten single-strand breaks before a double-strand break forms. Assume that the measurement can be made sufficiently rapidly that the viscosity does not change significantly during the measurement. Draw a curve showing η_r versus time.

PART V

Spectroscopic Methods

CHAPTER 14

Absorption Spectroscopy

Molecules absorb light. The wavelengths that are absorbed and the efficiency of absorption depend on both the structure and the environment of the molecule, making absorption spectroscopy a useful tool for characterizing both small and large macromolecules.

SIMPLE THEORY OF THE ABSORPTION OF LIGHT BY MOLECULES

Light, in its wave aspect, consists of mutually perpendicular electric and magnetic fields, which oscillate sinusoidally as they are propagated through space (Figure 14-1).

The energy E of the wave is

$$E = \frac{hc}{\lambda} = h\nu \tag{1}$$

in which h is Planck's constant, c is the velocity of light, λ is the wavelength, and ν is the frequency. When such a wave encounters a molecule, it can be either *scattered* (i.e., its direction of propagation changes) or absorbed (i.e., its energy is transferred to the molecule). The relative probability of the occurrence of each process is a property of the particular molecule encountered. If the electromagnetic energy of the light is absorbed, the molecule is said to be *excited* or in an *excited state.* A molecule or part of a molecule that can be excited by absorption is called a *chromophore.*

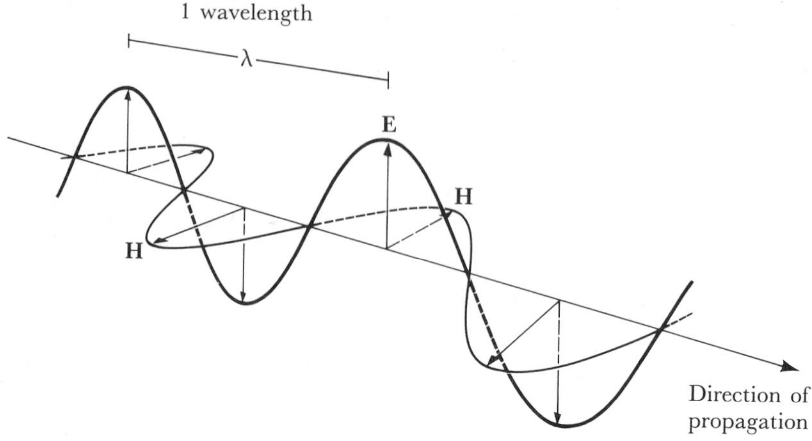

FIGURE 14-1
Propagation of an electromagnetic wave through space. The **E** and **H** vectors are mutually perpendicular at all times.

This excitation energy is usually converted into heat (kinetic energy) by the collision of the excited molecule with another molecule (e.g., a solvent molecule). With some molecules it is reemitted as *fluorescence,* which is discussed in more detail in Chapter 15. In both cases, the intensity of the light transmitted by a collection of chromophores is less than the intensity of the incident light.

An excited molecule can possess any one of a set of discrete amounts (quanta) of energy described by the laws of quantum mechanics. These amounts are called the *energy levels* of the molecule. The major energy levels are determined by the possible spatial distributions of the electrons and are called *electronic energy levels;* on these are superimposed *vibrational levels,* which indicate the various modes of vibration of the molecule (e.g., the stretching and bending of various covalent bonds). (There are even smaller subdivisions called *rotational levels,* but they are of little importance in absorption spectroscopy and will not be discussed.) All these energy levels are usually described by an *energy-level diagram* (Figure 14-2). The lowest electronic level is called the *ground state* and all others are excited states.

The absorption of energy is most probable only if the amount absorbed corresponds to the difference between energy levels. This can be expressed by stating that light of wavelength λ can be absorbed only if

$$\lambda = \frac{hc}{E_2 - E_1} \tag{2}$$

in which E_1 is the energy level of the molecule before absorption and E_2 is an energy level reached by absorption.

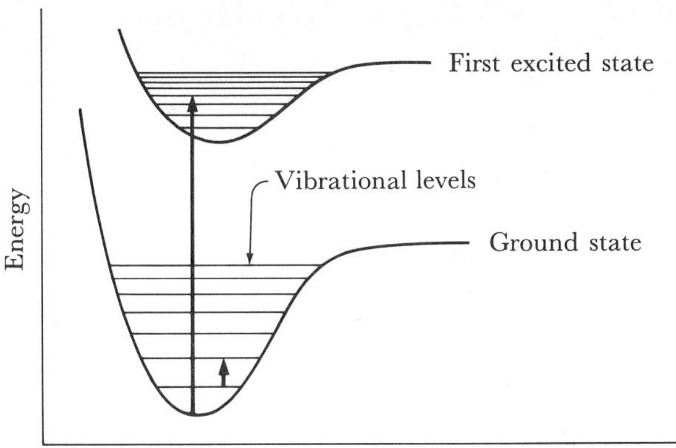

FIGURE 14-2
Typical energy-level diagram showing the ground state and the first excited state. Vibrational levels are shown as thin horizontal lines. A possible electronic transition between the ground state and the fourth vibrational level of the first excited state is indicated by the long arrow. A vibrational transition within the ground state is indicated by the short arrow.

A change between energy levels is called a *transition*. Mechanically, a transition between electronic energy levels represents the energy required to move an electron from one orbit to another. Transitions are represented by vertical arrows in the energy-level diagram.* A plot of the probability of absorption versus wavelength is called an *absorption spectrum* and absorption spectroscopy refers to the gathering and analysis of absorption data. If all transitions were between only the lowest vibrational levels of the ground state and the first excited state, then an absorption spectrum would consist of narrow, discrete lines. However, because transitions are possible from the ground state to any of the vibrational and rotational levels of the first excited state and because the lines have finite width, a spectrum appears to be a relatively smooth curve (see Figure 14-6). For most molecules, the wavelengths corresponding to transitions between the ground state and any vibrational level of the first excited state fall in the range of ultraviolet and visible light. Low-energy transitions are also possible between vibrational levels within a single electronic level. These transitions produce

*All transitions do not occur with high probability; those that do are determined by the so-called *selection rules* of quantum mechanics, which will not be discussed here.

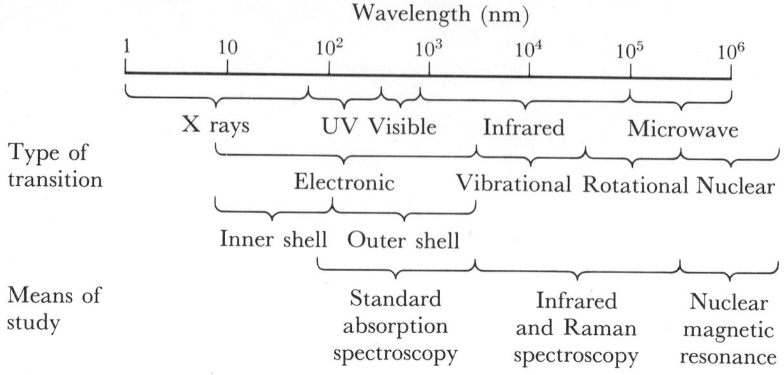

FIGURE 14-3
The part of the electromagnetic spectrum that is relevant to physical biochemistry.

radiation in the *infrared* range. Figure 14-3 shows the part of the electromagnetic spectrum relevant to the work of biochemists and the transitions producing radiation in different frequency ranges.

The probability of absorption at a single wavelength is characterized by the *molar extinction coefficient* at that wavelength. This is most easily defined in terms of how it is measured. If light of intensity I_0 passes through a substance (which may be in solution) of thickness d and molar concentration c, the intensity I of the transmitted light obeys the Beer-Lambert law:

$$I = I_0 10^{-\varepsilon dc}, \quad \text{or} \quad \log_{10}\left(\frac{I}{I_0}\right) = -\varepsilon dc \tag{3}$$

in which ε is the molar extinction coefficient. Absorption data are reported either as *% transmission* ($100 \times I/I_0$) or, more commonly, as the *absorbance,* A, ($\log I/I_0$). When $d = 1$ cm, A is commonly called OD_λ or *optical density,* in which the subscript λ tells the wavelength at which the measurement was made. Optical density is convenient because it equals $\varepsilon \times c$. In some cases, if c is high, ε appears to be a function of c and it can be said that Beer's law* is violated. This can result from scattering or from structural changes (e.g., dimerization, aggregation, or chemical changes) at high concentrations (Figure 14-4).

*The Beer-Lambert law is almost universally called Beer's law. Although this is not strictly correct (there being another law of Beer), this convention will be followed herein.

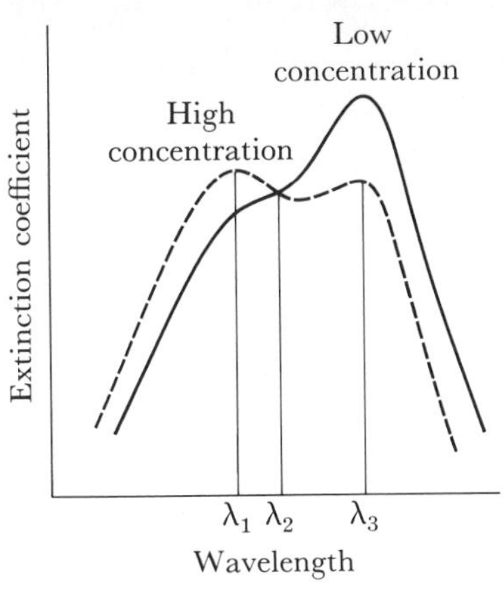

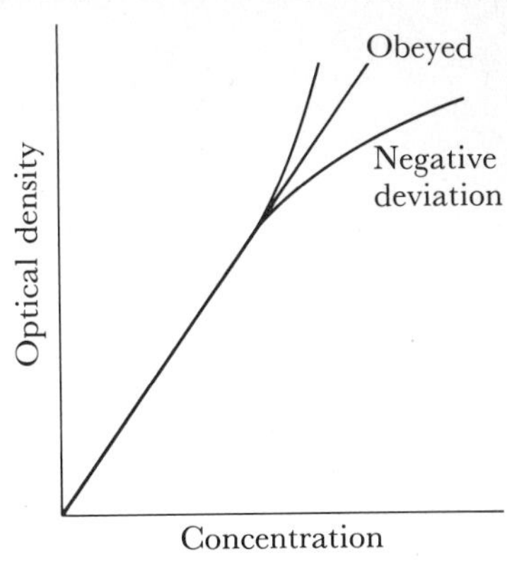

FIGURE 14-4
Positive and negative deviation from Beer's law and the causes. At the left is a spectral shift associated with increasing concentration—often a result of polymerization. Note that at one wavelength, λ_2, there is no change in molar extinction coefficient with change in concentration. This wavelength is called the isosbestic point. At the right is a curve showing deviation from Beer's law. At λ_1, the deviation is positive and, at λ_3, it is negative. At the isosbestic point, λ_2, it is always obeyed.

INSTRUMENTATION FOR MEASURING THE ABSORBANCE OF VISIBLE AND ULTRAVIOLET LIGHT

Absorbance measurements are made by a *spectrophotometer.* (Because in biochemistry almost all studies of molecules are done with molecules in solution, the discussion that follows refers to samples of that type.) Although they vary in design, all spectrophotometers consist of a light source, a monochromator (for wavelength selection), a transparent sample holder called a cuvette, a light detector, and a meter or recorder for measuring the output of the detector (Figure 14-5). In a typical operation at a single wavelength, a measurement is made of the light transmitted by the solvent alone (which may be a buffer or a solution of small molecules), followed by a measurement of that transmitted by the sample when dissolved in the same solvent; the first value is then subtracted from the second to give the absorbance of the solute. In practice, this subtraction is not done arithmetically; rather, the instrument is adjusted to read zero absorbance when the solvent alone is measured (this is called zeroing the instrument). Then, with the instrument so adjusted, the absorbance of the sample is read directly. To obtain a spectrum this operation is repeated at many wavelengths. Certain instruments, called automatic double-beam

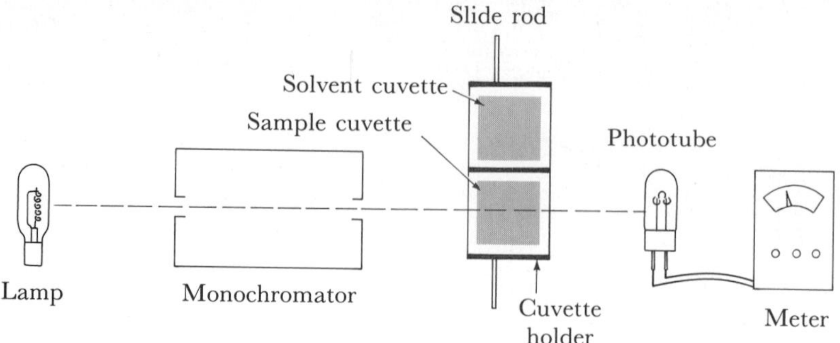

FIGURE 14-5
A spectrophotometer. Light from a lamp passes through a monochromator for wavelength selection. Sample and solvent are contained in two cuvettes in a cuvette holder. Light passes through a cuvette and falls on a phototube whose output is recorded on a meter. The cuvette holder is on a slide so that each cuvette can be separately placed in the beam.

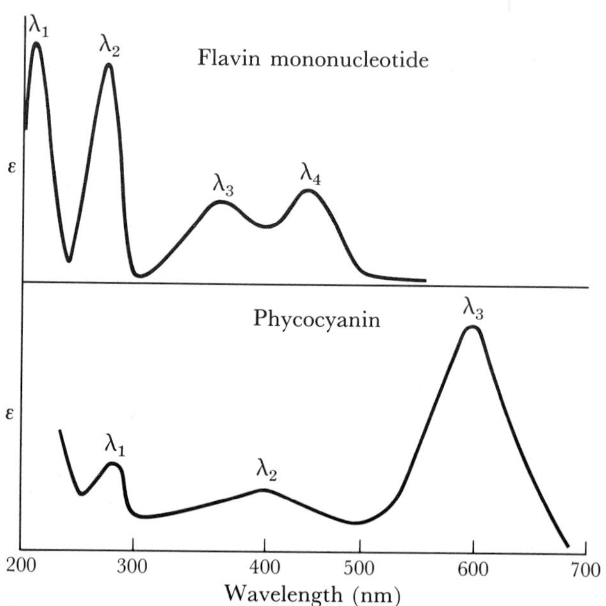

FIGURE 14-6
Spectra of two biological molecules, flavin mononucleotide and phycocyanin, indicating that spectra can be very different, which often allows the identification of compounds from their spectra.

recording spectrophotometers, scan a range of wavelengths and simultaneously measure the absorbance of the sample and solvent (contained in separate cuvettes) and electronically subtract the two values at each wavelength. The spectrum is then plotted on a chart recorder. These instruments are very expensive but essential if a great deal of spectral analysis is to be done.

PARAMETERS MEASURED IN ABSORPTION SPECTROSCOPY

Figure 14-6 shows UV-visible spectra for two biological molecules. The parameters usually measured are OD or ε. The wavelength corresponding to a peak of maximum absorption is called λ_{max}, and it is at this wavelength that ε is usually measured. Some of the absorption bands consist of multiple peaks and the wavelengths corresponding to the peaks having smaller molar extinction coefficients are frequently recorded. These wavelengths are sometimes also called λ_{max}, or it is stated that a substance has absorption maxima at $\lambda_1, \lambda_2 \ldots \lambda_n$.

Sometimes the width of a band is measured, although this is not common.

A useful list of λ_{max} and ε for common biological chromophores is given in Table 14-1.*

FACTORS AFFECTING THE ABSORPTION PROPERTIES OF A CHROMOPHORE

The absorption spectrum of a chromophore is primarily determined by the chemical structure of the molecule. However, a large number of environmental factors produce detectable changes in λ_{max} and ε. Environmental factors consist of pH, the polarity of the solvent or neighboring molecules, and the relative orientation of neighboring chromophores. It is precisely these environmental effects that provide the basis for the use of absorption spectroscopy in characterizing macromolecules.

The general features of these environmental effects are the following:

---

*Note that wavelength is expressed in nanometers (nm). In the older literature and today in certain disciplines, the unit is millimicrons (mμ). Some of the old instruments are labeled in angstrom units (Å). 1 nm = 1 mμ = 10 Å = 10^{-9} meters.

TABLE 14.1
Absorption maxima (λ_{max}) and molar extinction coefficients (ε) for various substances at neutral pH encountered in biological studies.

Molecule	λ_{max}(nm)	ε at λ_{max}($\times 10^{-3}$)
Tryptophan*	280	5.6
	219	47.0
Tyrosine*	274	1.4
	222	8.0
	193	48.0
Phenylalanine*	257	0.2
	206	9.3
	188	60.0
Histidine*	211	5.9
Cysteine*	250	0.3
Adenine	260.5	13.4
Adenosine	259.5	14.9
Guanine	275	8.1
Guanosine	276	9.0
Cytosine	267	6.1
Cytidine	271	9.1
Uracil	259.5	8.2
Uridine	261.1	10.1
Thymine	264.5	7.9
Thymidine	267	9.7
DNA	258	6.6
RNA	258	7.4

*Other amino acids show insignificant absorption.

pH Effects. The pH of the solvent determines the ionization state of ionizable chromophores. An example is shown in Figure 14-7, which illustrates the effect of pH on the tyrosine spectrum.

Polarity Effects. For polar chromophores, it is frequently true (especially if the molecule contains O, N, or S) that λ_{max} occurs at a shorter wavelength in polar hydroxylic solvents (H_2O, alcohols) than in nonpolar solvents. An example can be seen in Figure 14-8.

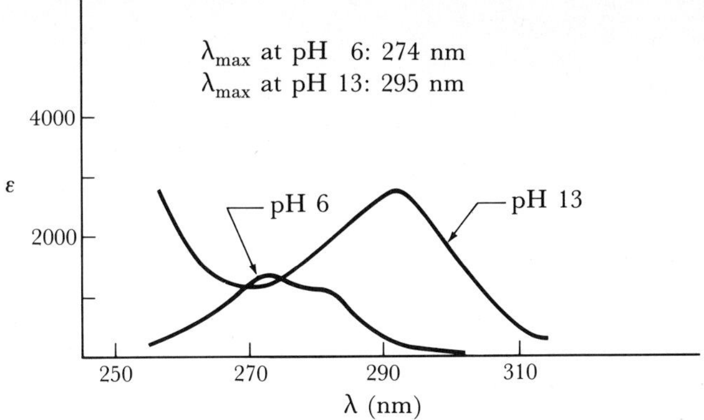

FIGURE 14-7
Absorption spectrum of tyrosine at pH 6 and 13. Note that both λ_{max} and ε are increased when the phenolic OH is dissociated.

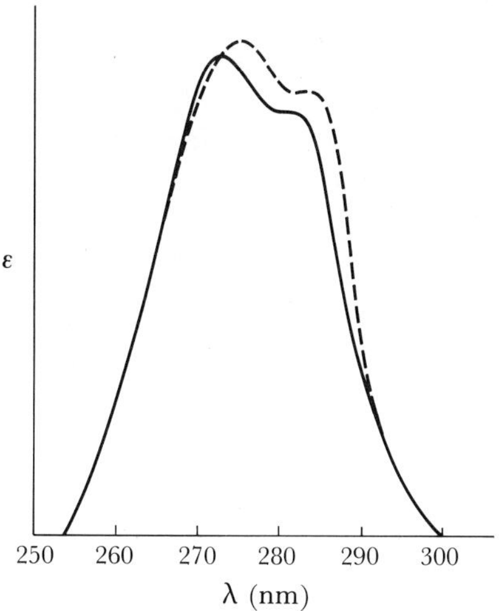

FIGURE 14-8
Effect of solvent polarity on the spectrum of tyrosine. Solvents: H_2O (solid line) and 20% ethylene glycol (dashed line). Notice the increase in λ_{max} in the less polar solvent.

Orientation Effects. Geometric features frequently have strong effects on λ_{max} and ε. The best known is the *hypochromism* of nucleic acids. That is, the extinction coefficient of a nucleotide decreases when the nucleotide is in a single-stranded polynucleotide in which the nucleotide bases are in proximity. There is a further decrease with a double-stranded polynucleotide because the bases are arranged in an even more orderly array. This is shown in Figure 14-9.

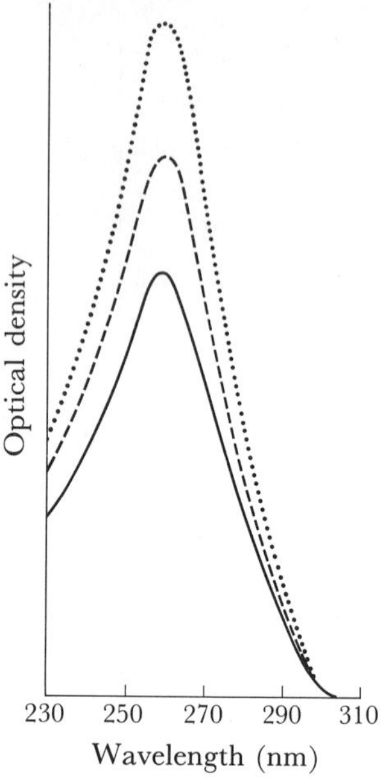

FIGURE 14-9
Spectra of T7 DNA as a double-stranded DNA (solid line), as a single-stranded DNA (dashed line), or after hydrolysis to free nucleotides (dotted line), showing the decrease in optical density (hypochromicity) that accompanies the formation of a more ordered structure. All spectra were obtained at the same concentration.

As a result of a large number of studies with biological compounds and macromolecules whose structures are well known in a variety of conditions, a set of empirical facts has been assembled that may be called the working rules of biochemical absorption spectroscopy. They are given in Table 14-2. Examples of the application of these rules will be given in parts of the following section.

APPLICATIONS OF ABSORPTION SPECTROSCOPY USING VISIBLE AND ULTRAVIOLET LIGHT

Absorbance measurements are made for many purposes: to determine the concentration of a substance, to assay certain chemical reactions, to identify materials, and to determine the structural parameters of macromolecules. Examples of each follow.

Example 14-A. Concentration measurement.
The most common use of absorbance measurements is to determine concentration. This can be done if the extinction coefficient is known and Beer's law is obeyed. For example, for double-stranded DNA, an $OD_{260\,nm} = 1$ corresponds to 50 μg/ml. Because Beer's law is obeyed to at least OD = 2 (which approaches the limit for most spectropho-

TABLE 14-2
Empirical rules for the interpretation of the absorption spectra of biological macromolecules.

1. If the amino acids tryptophan, tyrosine, phenylalanine, and histidine are shifted to a less polar environment, λ_{max} and ε increase. Hence:
   a. If the spectrum of an amino acid in a protein in a polar solvent shows that λ_{max} and ε are higher than they are for the free amino acid in the same solvent, then that amino acid must be in an internal region of the protein ("buried") and surrounded by nonpolar amino acids.
   b. If the spectrum of a protein is sensitive to changes in the polarity of the solvent, the amino acid showing the change in λ_{max} and ε must be on the surface of the protein.
2. For amino acids, λ_{max} and ε always increase if a titratable group (e.g., the OH of tyrosine, imidazole of histidine, and SH of cysteine) is charged. Hence:
   a. If no spectral change is observed for one of these chromophores and if the pH is such that titration of a free amino acid would have occurred, the chromophore must be buried in a nonpolar region of the protein.
   b. If the spectral change as a function of pH indicates that the ionizable group has the same pK as it would if free in solution, then the amino acid is on the surface of the protein.
   c. If the spectral change as a function of pH indicates a very different pK, then the amino acid is likely to be in a strongly polar environment (e.g., a tyrosine surrounded by carboxyl groups).
3. For purines and pyrimidines, ε decreases as their ring systems become parallel and nearer to one another (more stacked). The value of ε decreases in the following series: free base > base in an unstacked single-stranded polynucleotide > base in a stacked single-stranded polynucleotide > base in a double-stranded polynucleotide.

tometers), concentration is easily calculated—that is, OD = 0.5 corresponds to 25 μg/ml, OD = 0.1 to 5 μg/ml, and so forth.

Sometimes the material being measured consists of light-absorbing particles in suspension rather than in solution—for example, bacteriophages whose absorbance is determined almost entirely by their DNA content. Bacteriophages not only absorb but also scatter light (by Rayleigh scattering) and therefore appear to have an artificially high absorbance. However, a scattering correction can be made by measuring the absorbance at a series of wavelengths far from the λ_{max}. Because scattering varies as λ^{-4}, a plot of measured absorbance versus λ^{-4} can be made and the linear part of this curve (where all observed absorption is due to scattering) can be extrapolated to λ_{max} to correct the measured absorbance for the amount due to scattering. An example of such a correction is shown in Figure 14-10. This is a successful method for determining the nucleic acid content of bacteriophages and viruses, using the relation that OD_{260} (corrected for scattering) = 1 means 50 μg DNA/ml.

The concentration of bacteria is also often determined by spectrophotometry, although at the wavelengths used, scattering accounts for all of the apparent absorbance. In this case, instead of correcting for scattering, a calibration curve is constructed by comparing observed optical density with viable cell count. This is a precise way to measure cell concentration or dry mass per unit volume, as shown in Figure 14-11.

Example 14-B. Assay of chemical reactions.
Many chemical reactions can be assayed if one of the reactants changes in absorbance during the course of the reaction.

An example from enzymology that had a tremendous impact on molecular biology in studies of the lactose operon is the measurement of the activity of the enzyme β-galactosidase. This enzyme can cleave *o*-nitrophenylgalactoside (ONPG) to form *o*-nitrobenzene, which can be detected by its absorbance at 420 nm. The OD_{420} is then a measure of the amount of hydrolysis of ONPG. Because, for a certain range at least, the reaction rate is proportional to enzyme concentration, the amount of enzyme can be determined from the slope of a plot of OD versus hydrolysis time (Figure 14-12).

Another example is the enzymatic hydrolysis of polynucleotides. Rule 3 in Table 14-2 states that ε(free base) $>$ ε base in a polynucleotide). Hence, if a polynucleotide is hydrolyzed to mononucleotides, ε_{260} will increase. Therefore, if a nuclease is added to a polynucleotide sample and the OD_{260} is measured as a function of time, the OD_{260} will increase and this increase is an indicator of hydrolysis. Like that of β-galactosidase, the amount of nuclease can be determined from the slope of the curve of OD_{260} versus time.

A third example is the means of determining the dose rate of an x-ray machine. If a solution of $FeSO_4$ in H_2SO_4 is x-irradiated, the Fe^{2+} ion is converted into Fe^{3+}, which can be detected by its absorbance at 305 nm. From OD_{305}, the total dose received can be calculated.

Example 14-C. Identification of substances by spectral measurement.
Most substances have characteristic spectra and can be identified thereby. This can be done either by measuring a complete spectrum or by measuring the ratio of absorbance at different wavelengths. For example, in the early work in determining the base composition of DNA, the DNA was hydrolyzed and the bases were separated by chromatography. The individual bases could be identified by obtaining

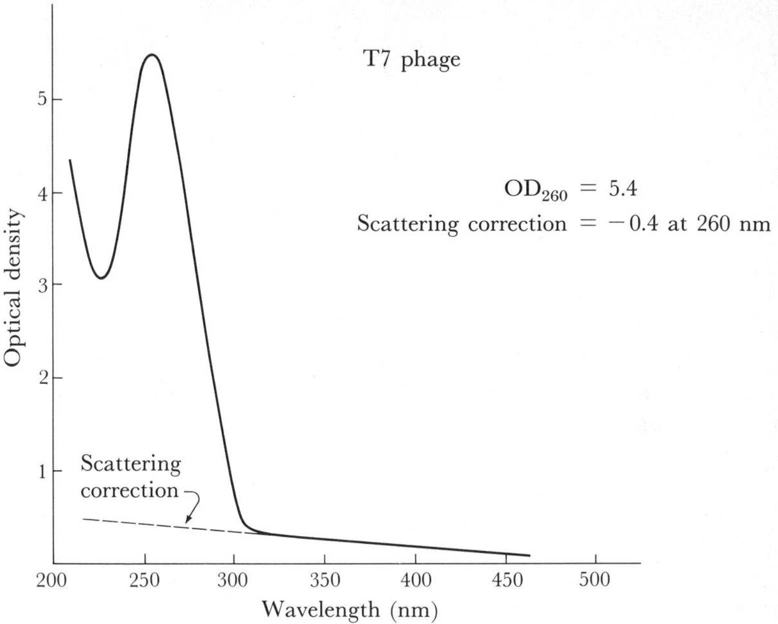

FIGURE 14-10
Spectrum of *E. coli* phage T7 showing the λ^{-4} Rayleigh scattering correction.

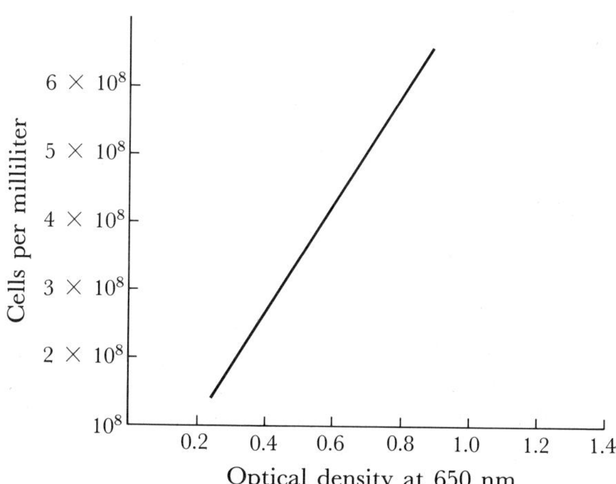

FIGURE 14-11
Concentration of bacteria determined from measurement of optical density.

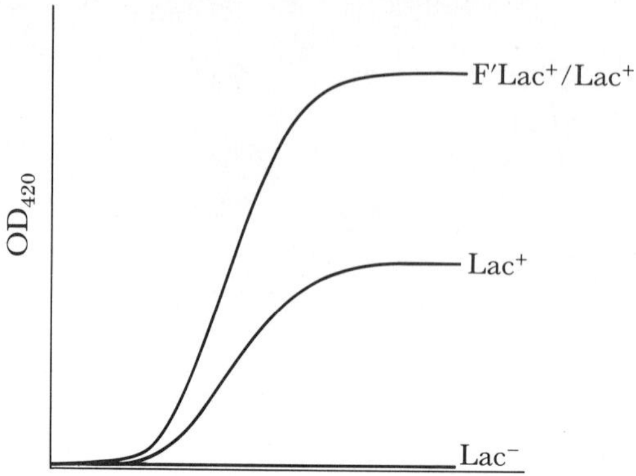

FIGURE 14-12
Synthesis of β-galactosidase assayed by OD_{420} as a measure of hydrolysis of *o*-nitrophenyl galactoside (ONPG). Three cultures, one Lac^-, one Lac^+, and a diploid containing two copies of the *lac* genes were growing in a glycerol-salts medium. At $t=0$, an inducer of β-galactosidase synthesis was added. At various times, samples were removed and treated with toluene to release the enzyme and kill the cells; then ONPG was added. Thirty minutes later, the OD_{420} was measured. The experiment shows that the Lac^- cell makes no enzyme and that the diploid makes roughly twice as much as the haploid.

their spectra and noting their λ_{max}: adenine, 260.5 nm; thymine, 264.5; guanine, 275.0, and cytosine, 267.0. To avoid determining the complete spectrum (which would normally be done with a recording spectrophotometer), the bases could be distinguished by measuring the ratio OD_{250}/OD_{280}. These values are: adenine, 2.00; thymine, 1.26; guanine, 1.63; and cytosine, 0.31. After the bases had been identified, the amount of each was determined as in Example 14-A by measuring optical density because the values of ε at λ_{max} (13.4×10^3, 7.9×10^3, 8.1×10^3, and 6.1×10^3, respectively) are known. (For example, if the OD_{260} of adenine were 0.65, the molar concentration would be $0.65/(13.4 \times 10^3) = 4.9 \times 10^{-5}$ M.)

Example 14-D. The helix-coil transition of double-stranded DNA: denaturation and renaturation.

The OD_{260nm} of DNA increases if the DNA is heated through a particular temperature range (Figure 14-13). This so-called *hyperchro-*

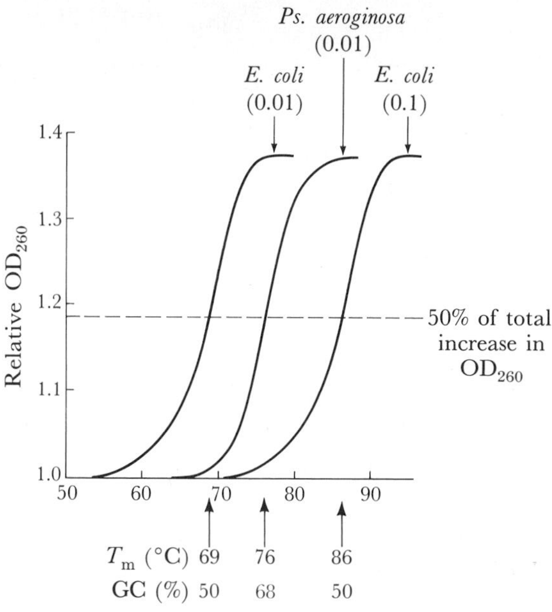

FIGURE 14-13
Optical density of three DNA solutions as a function of temperature: *E. coli* DNA (50% GC) in 0.01 M PO_4, pH 7.8, and in 0.1 M PO_4, pH 7.8; *Pseudomonas aeroginosa* DNA (68% GC) in 0.01 M PO_4, pH 7.8. The temperature at which the absorbance change is 50% complete is T_m, the melting temperature. Note that T_m increases with ionic strength and with GC content.

micity is a measure of denaturation or a helix-coil transition and results from the unstacking of the DNA bases associated with the continuous separation of the two polynucleotide strands (see rule 3 in Table 14-2). This simple optical assay allows the determination of the stability of DNA in relation to temperature, pH, ionic strength, and added small molecules, and in a variety of polar and nonpolar solvents. Some of the most important properties of DNA were elucidated from this powerful assay in the following ways. First, the thermal stability of DNA increases with guanine-cytosine content (Figure 14-14). Hence, a guanine-cytosine base pair is probably hydrogen-bonded more strongly than an adenine-thymine pair. Second, if the temperature is raised (but not to the point of maximum absorbance) and then lowered to a temperature below which no increase in OD is observed, the OD immediately drops to its original value, indicating that, if strand separation is not complete, the native structure is restored (Figure 14-15). A special

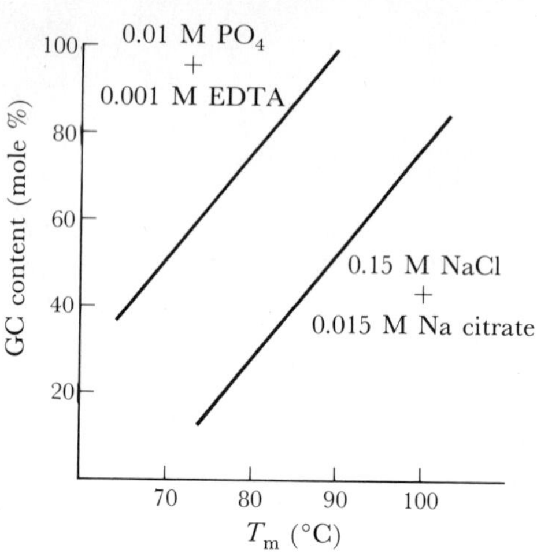

FIGURE 14-14
Plot of T_m versus GC content of various DNA molecules. The values of T_m for two different ionic strengths are shown. Note that the curves have the same slope. [From J. Marmur and P. Doty, *J. Mol. Biol.* 5(1962):109–118.]

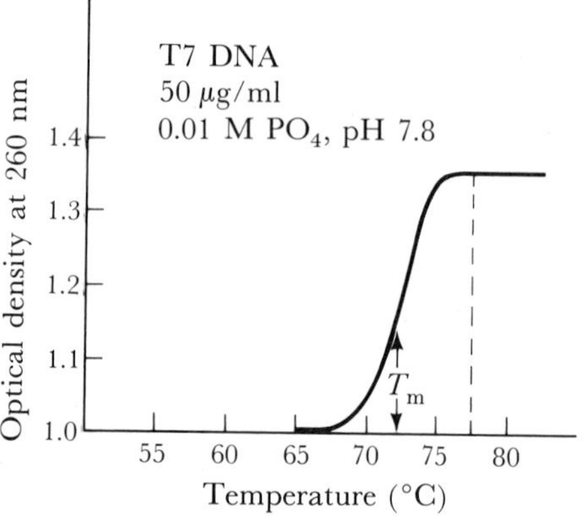

FIGURE 14-15
Difference between melting curves of a DNA solution obtained by measuring OD_{260} at the indicated temperature (solid line) or by heating the DNA to the indicated temperature, cooling to 25°C, and then measuring the OD (dashed line). The temperature at which strand separation occurs is at the intersection of the dashed line and the x-axis.

case is that of an interstrand cross-link in which the absorbance always drops to the normal value. Third, strand separation does not occur until well past the optical-transition region (e.g., 77.5°C, Figure 14-15) because separation of the last few base pairs has only a tiny effect on the total absorbance change. Fourth, if DNA is heated past the point of strand separation and then cooled, the increase in absorbance drops from a value of 37% (i.e., the maximum value) to 12% in high ionic strength (0.1), because hydrogen bonds (intra- and interstrand) reform at random. Hence, in high ionic strength, denatured DNA is aggregated. If the ionic strength is low (0.01), the OD remains at the maximum value because charge repulsion of the phosphate groups keeps the strands separated. Fifth, if the DNA with a 12% increase in OD is put at an ionic strength ranging from 0.5 to 1.0 and at a temperature above the midpoint (T_m) of the transition (Figure 14-16) and the DNA concentration is high, in several hours the OD returns to the original value. This is called *renaturation* and represents reformation of the native double-stranded structure (see Chapters 1 and 18).

Example 14-E. Spectrophotometric pH titration of proteins.

Many studies of protein structure require the determination of pK values for proton dissociation from ionizable amino acid side chains, because these values give an indication of the location of the amino acid in the protein (rule 2 in Table 14-2). This can often be done spectrophotometrically because dissociation often changes the spectrum of one of the chromophores (e.g., tyrosine); see Figure 14-7. Let us consider

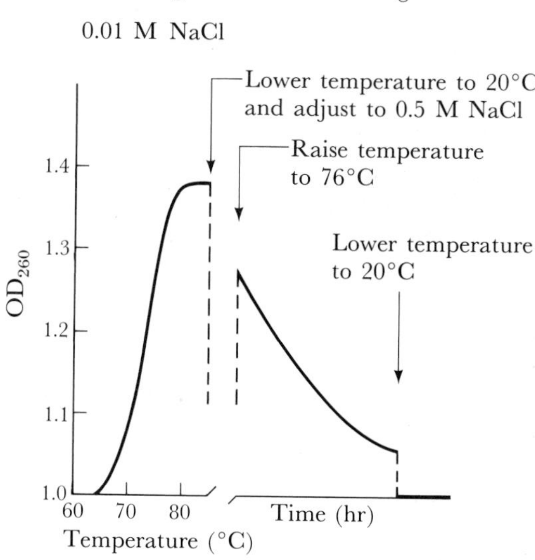

FIGURE 14-16
Detection of the renaturation of T7 DNA by measuring OD_{260}. DNA is first denatured in 0.01 M NaCl. When adjusted to 0.5 M NaCl, random hydrogen bonds form and the OD_{260} drops. When the temperature is raised to 76°C, the random base pairs are disrupted and the OD increases. Correct base pairing begins. Because the temperature is below that for strand separation, the OD decreases. However, DNA is partly denatured at that temperature. When the temperature is lowered to 20°C, all hydrogen bonds reform and renaturation is complete.

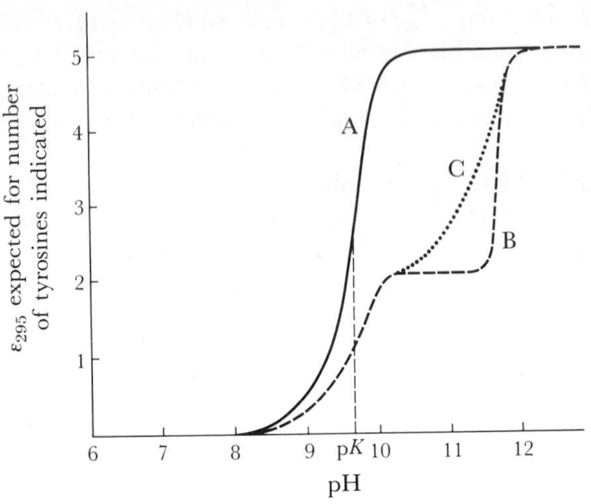

FIGURE 14-17
pH titration curves for tyrosine, using ε_{295} as an assay. The hypothetical protein contains 5 tyrosines. In curve A, all five are on the surface. In curve B, two are on the surface and the remaining three are internal—in a nonpolar environment and therefore not titratable. In curve C, the three internal ones are in a polar environment and accessible to the solvent.

a hypothetical tyrosine-containing protein, and use rule 2 to determine the number of external tyrosines. Suppose that this protein has five tyrosines. If all are on the surface and they are ionized by increasing the pH, the entire tyrosine spectrum will shift to that seen in Figure 14-7 for free tyrosine at high pH (rule 2a). In other words, a plot of OD_{295} (λ_{max} for the ionized form) versus pH would look like curve A in Figure 14-17. If, instead, three tyrosines were internal, and in a nonpolar environment, the curve would be like curve B; the ratio of the first plateau value to the final value would be 2/5 (rule 2b). Note that the curve shows a large rise in OD_{295} at very high pH. This indicates that the internal tyrosines have become exposed to the solvent—that is the protein has unfolded (become denatured).

If the three internal tyrosines were in a polar environment, a curve similar to curve C might be obtained, which indicates that these three have a p*K* value different from that of the exposed groups (rule 2c).

Example 14-F. Determination of some aspects of the conformation of proteins by the solvent-perturbation method and by difference spectroscopy.

Rule 1 in Table 14-2 indicates that the spectra of chromophores depend on the polarity of their environment. This fact can be used in two ways: (1) to determine whether certain chromophores in a protein are internal or external and (2) to determine the polarity of the environment of an internal amino acid.

Point 2 is described first because it is easier to follow. A protein is placed in a polar solvent and the spectrum of an internal chromophore is obtained. This is compared with known spectra of the same chromophore in polar and nonpolar solvents. If the spectrum resembles that of the chromophore in a nonpolar solvent, then the chromophore must be in a nonpolar region in the protein, and vice versa; the extent of the shift in λ_{max} allows an estimate of the degree of polarity.

The determination of whether an amino acid is internal or external by measuring the spectra of a protein in a polar and nonpolar solvent is called the *solvent-perturbation method.* Because one is generally interested in knowing the structure of a protein in aqueous salt solutions (as the protein would be if in a living cell), it is necessary that the nonpolar solvent itself does not introduce conformational changes and this must always be checked by other methods. In fact, proteins are rarely studied in completely nonpolar solvents because most proteins are either insoluble or denatured in these solvents. The usual practice is to use a solvent that is 80% water and 20% a substance of reduced polarity. Some standard mixtures are given in Table 14-3. These solvents of reduced polarity are called *perturbing solvents.*

The most convenient procedure for applying the perturbation method is *difference spectroscopy* because it eliminates the necessity of determining the spectrum twice—that is, in the presence and absence

TABLE 14-3
Solvents commonly used in the solvent-perturbation method.

Liquid additive (20 volumes/100 volumes final solution in H_2O)	
Dimethylsulfoxide	Ethylene glycol
Dioxane	Glycerol
Ethanol	
Solid additive (20 g/100 g final solution in H_2O)	
Arabitol	Polyethylene glycol
Erythritol	Sucrose
Glucose	Urea
Mannitol	

of the perturbant. In standard absorption spectroscopy, a spectrum is determined by measuring the absorbance of a solution and subtracting the absorbance of the solvent (or, as described earlier, by adjusting the spectrophotometer to read zero for the solvent). In difference spectroscopy, the two sample holders contain solutions that are identical except that one contains a perturbant. Instead of measuring the spectra of each, the absorbances at each wavelength are subtracted from one another (of course, it is necessary that both solvents have identical absorbances at all wavelengths used—this is the case for the solvents in Table 14-3).

A nonzero *difference spectrum* will appear only if the spectrum of the sample is affected by the perturbant. The parameters obtained from such spectra are λ_{max} and $\Delta\varepsilon$ at λ_{max}. Figure 14-18 shows the difference spectra for tryptophan and tyrosine (the amino acids most commonly examined in the solvent-perturbation method) in 20% ethylene glycol.

An example of perturbation analysis, using tryptophan as a chromophore, and difference spectroscopy follows. Consider a hypothetical protein with five tryptophans. The difference spectrum between H_2O and 20% ethylene glycol shows a peak at 292 nm, as shown in Figure 14-18. From the OD_{280} and ε_{280} obtained in H_2O, the amount of tryptophan in the sample can be determined; and from $\Delta\varepsilon_{292}$ for free tryptophan in H_2O and 20% ethylene glycol, the expected ΔOD_{292} for the sample—if all tryptophans were perturbed—is known. Let us assume that this value is 1.00. If ΔOD_{292} is measured for this protein, it is found to be 0.60. Because only those on the surface are perturbed, $0.60/1.00 \times 5 = 3$ are on the surface. If the amino acid sequence of the protein is known, the particular tryptophans that are on the surface can be identified by a simple trick. If a protein is treated by various oxidizing agents, the indole group of tryptophan is oxidized and becomes nonabsorbing. Presumably, only those tryptophans on the surface will be

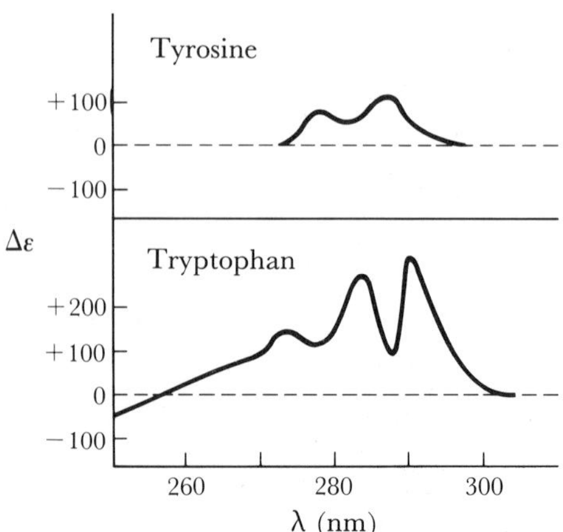

FIGURE 14-18
Difference spectra for tyrosine and tryptophan in 80% H_2O, 20% ethylene glycol. Notice that difference spectra can have negative values of $\Delta\varepsilon$.

oxidizable, but this must be checked. This also can be done by the solvent-perturbation method and difference spectroscopy. If $\Delta OD_{292} = 0$ after oxidation, the oxidized tryptophans must be on the surface. After this has been ascertained, the sequence of amino acids in the oxidized protein can be redetermined and the positions of the oxidized tryptophans noted. Hence, the particular tryptophans that are on the surface can be identified. Such information aids in the elucidation of the three-dimensional structure of the molecule and frequently provides a starting point for x-ray diffraction analysis.

In the preceding example, the ΔOD_{292} expected if all tryptophans are on the surface was calculated from the absorbance and extinction coefficient of the pure protein in the absence of perturbant. This may not be possible if the protein sample contains a contaminating absorbing material. The measurement can, however, be made in another way because if a protein is denatured, all the amino acids are in contact with the solvent. Hence, the maximum ΔOD_{292} can be determined from a difference spectrum of the denatured protein in a perturbing solvent versus the native protein in a polar solvent.

We have so far attempted to distinguish external from internal amino acids. However, this distinction is not always clear because, in some cases, a chromophore is not totally buried but is in a deep crevice so that its spectrum is affected only if the perturbing molecules are below a critical size necessary to enter the crevice. This also provides information about molecular conformation. The substances most commonly used, together with their mean diameters, follow: D_2O (2.2 Å), dimethylsulfoxide (4.0 Å), ethylene glycol (4.3 Å), glycerol (5.2 Å), arabitol (6.4 Å), glucose (7.2 Å), and sucrose (9.4 Å).

Example 14-G. Observation of the helix-coil transition in proteins: denaturation.

Because a buried chromophore becomes exposed to the solvent during denaturation, by monitoring the absorbance of these chromophores, one can observe the helix-coil transition for proteins in the same way that denaturation is studied by examining the hyperchromicity of DNA. For example, if a protein contains tryptophans, some of which are internal, the unfolding as a function of temperature could be detected by measuring $\Delta\varepsilon_{292}$. This could then be used to examine the effects of other agents such as NaCl concentration on the thermal stability. The kind of data obtained for a hypothetical protein is shown in Figure 14-19.

Example 14-H. Detecting the binding of small molecules to proteins. The binding of an enzyme substrate to the active site of an enzyme frequently produces spectral changes in chromophores in or near the active site by affecting the polarity of the region or the accessibility

FIGURE 14-19
Helix-coil transition of a hypothetical protein assayed by perturbation difference spectroscopy at a single wave length using 80% H_2O, 20% ethylene glycol containing two different NaCl concentrations. The reference solution for the difference spectrum is the protein in the ethylene-glycol–NaCl solution at 20°C. Note that this protein is more stable at higher NaCl concentration since a higher temperature is needed for unfolding in 0.1 M NaCl.

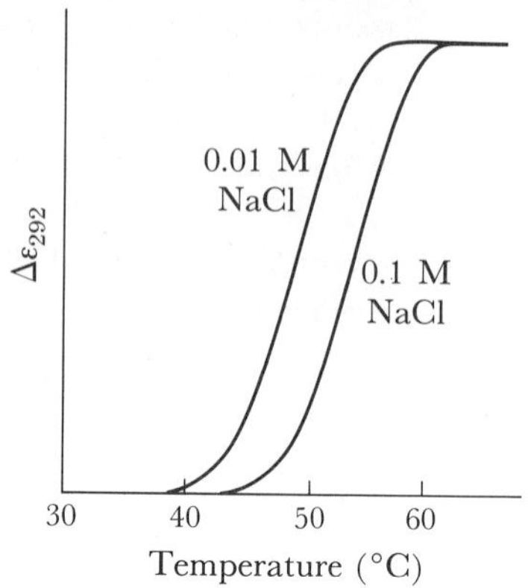

to solvent. By comparing the observed changes with those obtained by solvent perturbation, information about the structure of the active site can be obtained. For example, the addition of various substrates to the enzyme lysozyme produces a shift in λ_{max} for tryptophan to longer wavelength. The magnitude of the change is that expected from the transfer of one tryptophan from a polar to a nonpolar environment. This suggests that a tryptophan is in the binding site. Furthermore, solvent-perturbation studies of lysozyme, such as that in Example 14-E, show that there are four tryptophans on the protein surface (i.e., a certain fraction of the known number of tryptophans are perturbable); if the enzyme-substrate complex is studied, the analysis indicates that only three are on the surface. Hence, one tryptophan is no longer in contact with the solvent when the substrate is added. Again the simple interpretation is that the active site contains tryptophan. This simple optical analysis is important because it is performed in solution and therefore confirms that the structure determined by x-ray diffraction analysis (which is done with dry samples), that is, there is a crevice in the molecule containing tryptophan in which substrate is bound, is probably valid for lysozyme in solution.

In many cases, chromophores in the enzyme, the substrate, or both, change their absorption during complex formation. Hence, a spectral assay (i.e., $\Delta\varepsilon$ as a function of substrate or enzyme concentration) can be used to determine dissociation constants. This is in fact a very useful method applicable to the binding of any material (e.g., small molecules or metal ions) to a protein whenever spectral changes are observed.

Example 14-I. Protein-protein association.

Spectral changes can accompany protein-protein association either because chromophores on the surface become inaccessible to the solvent by being buried in the region in which binding takes place or because a conformational change that buries or exposes a chromophore in another part of the molecule can accompany binding. Hence, as in Example 14-H, the spectral changes can be used to monitor interaction and thereby determine conditions, kinetics, and so forth (Figure 14-20).

This can be studied especially well using difference spectroscopy, in which case the difference spectrum is produced between two solutions that are identical except for concentration.

Example 14-J. Solvent perturbation of nucleic acids.
A change in solvent from H_2O to 50% D_2O causes characteristic spectral changes in mononucleotides but not in base pairs. Thus, spectral changes of a DNA sample in 50% D_2O can be used to determine the fraction of bases that are not base-paired. This is of great value in the study of such substances as transfer RNA (tRNA). In tRNA the sequence does not allow all bases to be engaged in hydrogen bonds; three-dimensional models in which there is partial hydrogen bonding can be constructed (Figure 14-21). Different models require different fractions to be hydrogen bonded so that some of the possible models can be ruled out by spectral data. (See Chapter 17 for a discussion of the role of nuclear magnetic resonance in distinguishing possible tRNA models by measuring the number of hydrogen-bonded base pairs.)

REPORTER GROUPS

A molecule of interest may often have many chromophores but none will be in or near a region that participates in the biological function of the molecule. This situation can sometimes be corrected by the addition of an

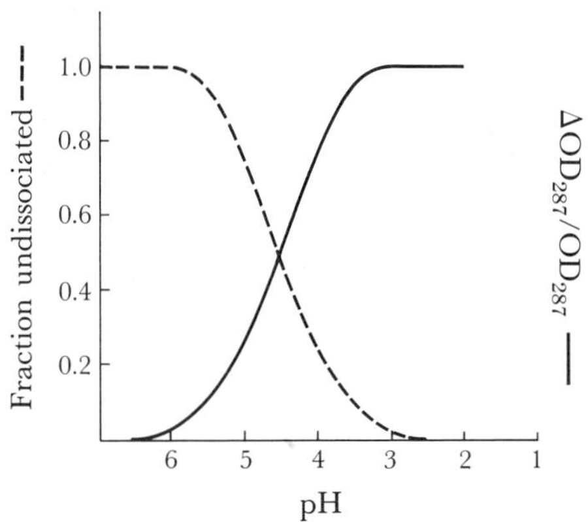

FIGURE 14-20
Detection of the pH-induced dissociation of ovomacroglobulin into its subunits by difference spectroscopy. The OD_{287} at various pH values was measured against OD_{287} at pH 7.

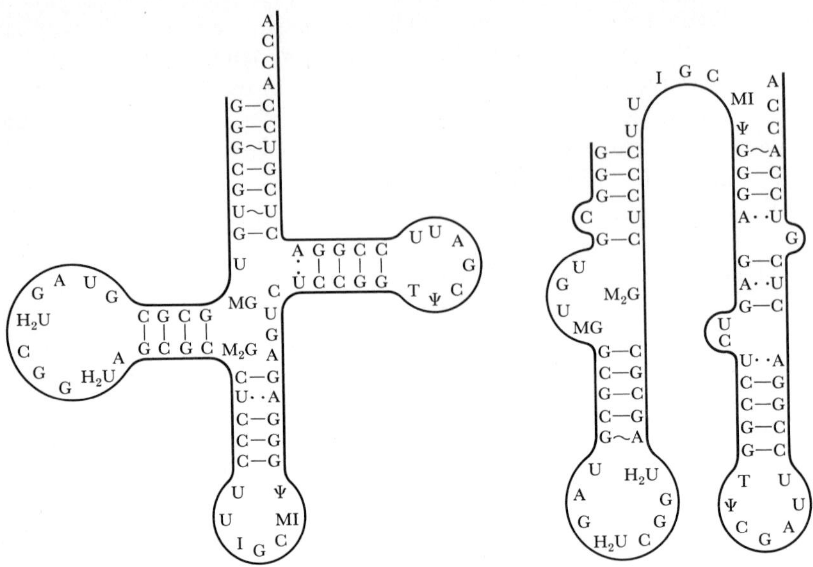

FIGURE 14-21
Two of several possible structures of yeast alanyl tRNA arranged so that a large fraction of the bases participate in hydrogen bonding. The hydrogen bond of a GC pair is indicated by —; that of an AU pair is indicated by · ·; the weaker hydrogen bonds found in other base pairs are indicated by ~. An NMR analysis yielding the number of base pairs of various types indicates that the cloverleaf structure (left) is the correct one. The symbols are: G, guanosine; C, cytidine; A, adenosine; U, uridine; I, inosine; H_2U, dihydrouridine; T, ribosylthymine; MG, methylguanosine; M_2G, dimethylguanosine; Ψ, pseudouridine; MI, methylinosine. [The base sequence is that reported by R. W. Holley, J. Apgar, G. A. Everett, J. T. Madison, M. Marquisee, S. H. Merrill, J. R. Fenswick, and A. Zamer, *Science* 147(1965):1462–1465.]

artificial chromophore in the relevant region. Such a chromophore is called a *reporter group.* For successful use, the reporter group must have a spectrum distinguishable from the remainder of the macromolecule, it must react at a single position, and its insertion must not affect the interaction. Useful reporters are dimethylaminoazobenzene and arsanilic acid. An example of the use of reporter groups follows.

Example 14-K. Conformation of carboxypeptidase A.

The enzyme carboxypeptidase A contains a tyrosine in the active site and a zinc atom at a defined position outside the active site. Arsanilic acid can be coupled to the tyrosine in the active site by diazotization. The absorption spectrum for free arsanilazotyrosine is significantly altered by binding zinc. In the protein in solution, the spectrum is that of the zinc complex. Hence, the protein must be folded so that the zinc site and the active site are close together. This is an especially interesting example because the x-ray diffraction analysis of crystals shows that the zinc is not near the active site. However, the spectrum of the reporter *when in the protein crystal* is also without bound zinc. Hence, the structure of the protein in solution is not the same as that in the crystals used for x-ray analysis.

The similarity of this procedure to the method of extrinsic fluorescence (see Chapter 16) should be noted.

ABSORPTION OF POLARIZED LIGHT

The plane of polarization of an electromagnetic wave is defined as the plane of the **E** vector (Figure 14-1; see Chapter 16 for a more complete description of polarization and the production of plane-polarized light). A typical beam of light is nonpolarized because it can be thought of as a collection of waves whose planes of polarization are randomly oriented. However, plane-polarized light can be obtained either by passing light through various materials or by the reflection from a surface at a critical angle.

All chromophores have at least one preferred axis of absorption of plane-polarized light. For example, the probability of absorption by the symmetric planar ring of hexamethylbenzene

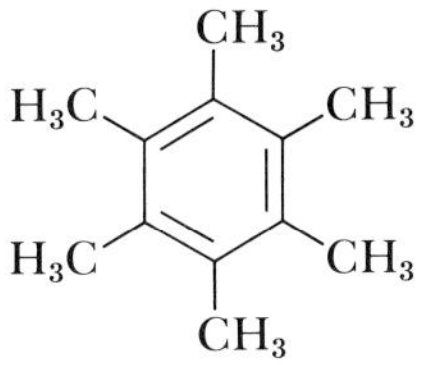

CH_3
H_3C CH_3
H_3C CH_3
CH_3

is maximum if the **E** vector is in the plane of the ring and zero if perpendicular to this plane. If the molecule has axes of symmetry of different lengths, as in the planar molecule naphthalene,

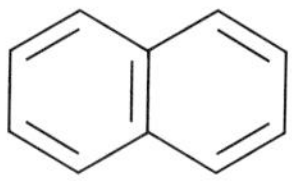

absorption occurs only if the plane of the **E** vector is in the plane of the ring, but the probability of absorption depends on whether the **E** vector is parallel to the long or the short axis of the molecule. This effect is called *dichroism* and requires that ε be defined for particular directions. It is useful in determining the orientation of molecules in biological systems and is described in greater detail in the section concerning polarization microscopy in Chapter 2. In studies of macromolecules, it has been used most frequently in infrared spectroscopy to determine the orientation of particular bonds with respect to molecular axes. This will be described in the next section.

INFRARED SPECTROSCOPY

Transitions between vibrational levels of the ground state of a molecule (Figure 14-2) result from the absorption of light in the infrared (IR) region: from 10^4 to 10^5 Å (Figure 14-3). These vibrational levels and, hence, IR spectra are generated by the characteristic motions (bond stretching, bond bending, and more complex motions) of various functional groups (e.g., methyl, carbonyl, amide, etc.). The value of IR spectral analysis comes from the fact that the modes of vibration of each group are very sensitive to changes in chemical structure, conformation, and environment and, from this point of view, IR is not different from visible and ultraviolet spectroscopy. It is thought of as being different principally because it has a somewhat different technology and because it is used to examine chemical groups not accessible to ultraviolet and visible light absorption spectroscopy.

Technology of Infrared Spectroscopy

Infrared spectrophotometers are in principle no different from ultraviolet and visible light spectrophotometers. The source of radiation is an object heated to 1500 to 1800 K; a monochromator is used for wavelength selection, but a thermocouple (instead of a photocell) is the detector.

The principal complication of IR spectroscopy is that it is usually not possible to use aqueous solutions because of the powerful absorption of infrared by water (because it has a high extinction coefficient and is at a concentration of 55 molar). This problem can be circumvented in part by the use of D_2O or of H_2O-D_2O mixtures. Chloroform is also sometimes used because it dissolves many polar molecules, but chloroform-induced conformational changes present many difficulties. The most common method with macromolecules is to use thin, fairly dry films. They are prepared by dissolving the macromolecule in a volatile solvent, placing the solution on a flat plate, allowing the solvent to evaporate, and lifting the film from the plate. These films can, if desired, be stretched to prepare films in which all molecules are oriented in the same direction. They are useful in studies with polarized light in which the orientation of particular groups with respect to the molecular axis can be determined. Of course, the possibility that the structure of a macromolecule in a dry film is not the same as that in solution must always be considered.

Information in Infrared Spectra

First, it should be noted that infrared spectra are conventionally plotted in terms of wave numbers ($1/\lambda$) or frequency (ν) rather than wavelength. In frequency terms, each band in an IR spectrum can be characterized by the frequency of an absorption maximum (ν_{max}), its band width at half height ($\Delta\nu_{1/2}$), the optical density at ν_{max} (A_{max}), and the band shape.* In oriented films, the dichroic ratio (R), that is, the ratio of the band area if the electric vector is parallel to and perpendicular to the axis, is measured. By extensive studies on many monomers with known structure, the identity of the groups and the type of vibration corresponding to each band in a spectrum is well known. For simple compounds, IR spectra differ from spectra for visible and UV light in that they generally consist of fairly narrow lines (Figure 14-22). However, for macromolecules, each bond type exists in such great numbers and in so many different configurations that each band is shifted—to an extent that depends on where it is in the molecule—so that all bands overlap and therefore the spectrum appears to contain a few relatively broad bands (Figure 14-22).

Applications of Infrared Spectroscopy

Brief descriptions of relatively common applications of IR spectroscopy follow.

Determination of Fractional amounts of α, β, and Random-coil Structures in Proteins from the Intensities of Various Amide Bands. This is possible because the amide I band has $(1/\lambda)_{max} = 1650$, 1685, and 1637 cm^{-1} for the α, β, and random-coil structure, respectively.

Identification of Exchangeable Hydrogen. Many bands change frequencies when deuterium is substituted for hydrogen. In general, the functional group responsible for a given band (i.e., carbonyl, hydroxyl, amino) is known; thus, by observing the bands that have shifted, the groups in

*The wave number is $1/\lambda = \nu/c$, in which c is the velocity of light in vacuum. The units of ν and $1/\lambda$ are sec^{-1} and cm^{-1}, respectively.

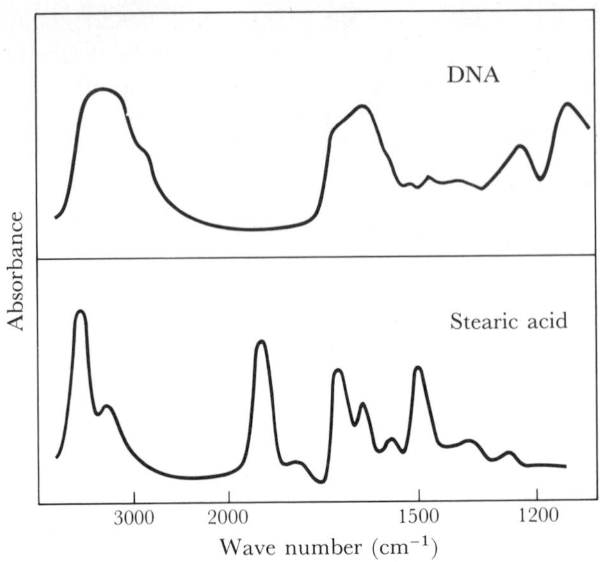

FIGURE 14-22
Infrared spectra for DNA and for stearic acid. Note the broad bands in the DNA spectrum.

which exchange is possible can be identified. Because some functional groups (e.g., hydroxyl) normally exchange rapidly, a delayed shift in ν_{max} indicates a slow exchange; this generally means that the group is buried.

Identification of the Number of Hydrogen Bonds and the Functional Groups Engaged in Hydrogen Bonding and Measurement of Their Breakage During Denaturation. This can be done, for example, by dissolving the macromolecule in D_2O, denaturing the sample, and observing which bands corresponding to deuterated groups appear during denaturation.

Identification of Tautomeric Forms by the Appearance of Unexpected Bands. Suppose that a molecule contains a hydroxyl group, but its IR spectrum indicates that a carbonyl is present. This can be taken as strong evidence for tautomerization. If the molecule is part of a macromolecule, the relative intensities of the hydroxyl and carbonyl bands may change compared with the free molecule—one band might even disappear. This would indicate the chemical structure of the molecule in the polymer. This has been of special importance in understanding nucleotide structure and the effect of various nucleotides on mutation frequency.

Interaction between Small Molecules, Such as Riboflavin and Adenine, and Protein–Ligand Binding. This produces characteristic shifts in ν and intensity, as in ultraviolet absorption spectroscopy.

Determination of the Ratio of AU to GC Pairs in Transfer RNA. The two types of base pairs give bands at different ν. This has been used to distinguish the structures shown in Figure 14-21.

Titration of Protein Carboxyls. Some "buried" carboxyls titrate in a pH region that is far from normal and in a range of titration of other ionizable groups (see Example 14-D for the analogous experiment with UV light). Such titrations can be followed by spectral changes in D_2O.

Determination of the Orientation of Hydrogen bonds in Stretched Films of Proteins and Polypeptides by Measuring the Orientation of the Constituent C═O and NH— Groups, Using Polarized IR. This is possible because the maximum absorption by these groups occurs when the **E** vector is parallel to the group. This can be used to identify an α helix because the theoretical ratio of the absorbance when the **E** vector is parallel to and perpendicular to the axis of a protein (the "dichroic ratio") is 44 at 3300 cm^{-1} (the stretching vibration for NH—). If the measured ratio is near this value, the structure has a high probability of being α-helical. The theoretical value itself will rarely be achieved because the polypeptide chains cannot be perfectly oriented in a film. Also, from the measured dichroic ratios at 3300 cm^{-1} and 1660 cm^{-1} (the stretching vibration for C═O), the maximum angle that these groups could make with the axis can be calculated. This kind of information is frequently of great value in the interpretation of x-ray diffraction patterns.

RAMAN SPECTROSCOPY

The scattering of light normally occurs without a change in frequency (i.e., elastic scattering); however, a small fraction, usually in the infrared range, is scattered inelastically with a frequency shift. This is called Raman scattering. The frequency shift differs from that encountered in fluorescence (Chapter 15) in that excitation to a higher *electronic* state does not occur. The frequency change is caused either by excitation to a higher *vibrational level* or by the addition of the vibrational energy of the molecule to the electromagnetic energy of the light wave.

If, in the scattering process, the light excites the scattering center to a higher vibrational level, energy is lost and the frequency decreases. On the other hand, if the scattering center is at a higher vibrational level (e.g., by previous collision with solvent molecules), it can transfer its vibrational energy to the incident light and thereby increase the frequency. At the temperatures normally used, there are fewer vibrating than nonvibrating molecules, making a decrease in frequency more common. Hence, Raman spectroscopy examines vibrational transitions, as does IR. Because of the very low intensity of the scattered light, it has been a relatively uncommon technique owing to a lack of high intensity light sources. Furthermore, the frequency changes are so small that highly monochromatic light is necessary. Both of these problems have recently been solved by the use of lasers in modern instruments. This is an important step forward because the advantage of Raman over IR spectroscopy is that it is possible to work in H_2O solution. Figure 14-23 shows a typical Raman spectrum.

Raman spectroscopy has not yet had widespread use. A few applications that have been reported follow.

Proof of the Zwitterion Structure of Various Amino Acids. Characteristic spectral changes accompany the ionization of amino and carboxyl groups.

Distinguishing Adenosine Monophosphate, Diphosphate, and Triphosphate and Their Ionized Forms in Solution. This has been useful in studying certain enzyme reactions because the Raman spectrum of a reaction mixture indicates the relative proportions of each molecular species during the reaction or at equilibrium.

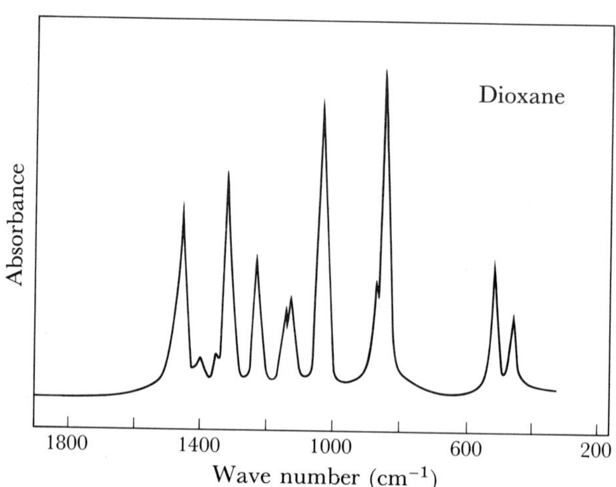

FIGURE 14-23
Raman spectrum of dioxane.

Identification of α-helical, β-sheet, and Random-coil Structures in Poly Amino Acids. As in IR spectroscopy, characteristic amide bands are observed, the intensities of which are proportional to the relative amount of each structure.

Determination of the number of S–S Bonds in Proteins. The S–S and SH bonds each give characteristic ν_{max}.

Determination of the Number of Paired and Unpaired Bases in RNA. Each base is distinguishable by its Raman spectrum so that base composition can be determined by the relative intensity of peaks. There is also a spectral change on deuteration so that sites of slow and rapid hydrogen-deuterium exchange are identifiable. A base showing slow exchange is thought to be in a base pair because hydrogen bonding delays exchange (see Example 14-F).

Selected References

Beaven, G. H., E. A. Johnson, H. A. Willis, and R. G. J. Miller. 1961. *Molecular Spectroscopy.* Heywood.

Bremer, J. M., A. J. Pesce, and R. B. Ashworth. 1974. *Experimental Techniques in Biochemistry,* ch. 7. Prentice-Hall.

Donovan, J. W. 1973. "Ultraviolet Difference Spectroscopy: New Techniques and Applications" and "Spectrophotometric Titration of the Functional Groups of Proteins," in *Methods in Enzymology,* vol. 27, edited by C. H. W. Hirs and S. N. Timasheff, pp. 497–525; 525–548. Academic Press.

Edisbury, J. R. 1965. *Practical Hints of Absorption Spectrometry.* Helger-Watts.

Herskovitz, J. T. 1967. "Difference Spectroscopy," in *Methods in Enzymology,* vol. 11, edited by C. H. W. Hirs, pp. 748–775. Academic Press.

Horton, H. R., and D. E. Koshland. 1967. "Environmentally Sensitive Groups Attached to Proteins," in *Methods in Enzymology,* vol. 11, edited by C. H. W. Hirs, pp. 856–870. Academic Press. A description of the use of reporter groups.

Timasheff, S. N. 1970. "Some Physical Probes of Enzyme Structure in Solution," in *The Enzymes,* vol. 2, edited by P. D. Boyer, pp. 371–443. Academic Press.

Problems

14-1. A solution at a concentration of 32 μg/ml of a substance having a molecular weight of 423 has an absorbance of 0.27 at 540 nm measured in a cuvette with a 1-cm light path. What is the molar extinction coefficient at 540 nm? Assume that Beer's law is obeyed.

14-2. A solution of a molecule (A) has an $OD_{260} = 0.45$ and $OD_{450} = 0.03$. A solution of a second molecule (B) has $OD_{260} = 0.004$ and $OD_{450} = 0.81$. Two milliliters of A are mixed with 1 milliliter of B. The resulting optical densities of the mixture are $OD_{260} = 0.30$ and $OD_{450} = 0.46$. Is there an interaction between A and B? Explain. What assumption is made to justify this conclusion?

14-3. Suppose that you have just prepared two DNA samples, one native and one denatured, and have dialyzed each against 0.01 M NaCl. You then add a very small amount of an enzyme to each and in so doing mix up the samples. How can you determine the identity of each sample by an absorbance measurement? You may assume that a small part of each sample is consumed in the testing and that the enzyme will not interfere with the test. Design a second test that does not require the introduction of an agent or condition that causes denaturation.

14-4. Suppose that you would like to know how much cytochrome *c* is contained in one *E. coli* cell. The molar extinction coefficient at its absorption maximum is known. How would you make this measurement?

14-5. A particular molecule has a molar extinction coefficient of 348 at 482 nm. A solution of this molecule has an $OD_{482} = 1.6$. When diluted 1:1, 1:2, 1:3, 1:4, 1:5, and 1:6, the values of OD_{482} are 1.52, 1.42, 1.05, 0.84, 0.70, and 0.61, respectively. What is the molarity of the original solution?

14-6. The molar extinction coefficients of substance A at 260 and 280 nm are 5248 and 3150, respectively. In isolating A, a reagent B is used whose molar extinction coefficients at 260 and 280 nm are 311 and 350. After isolating A, $OD_{260} = 2.50$ and $OD_{280} = 2.00$. What is the concentration of A?

14-7. The OD_{260} of denatured DNA (measured at a temperature at which the maximum OD_{260} is reached—see Figure 14-13) is invariably 37% higher than that of native DNA (e.g., at 20°C). However, in a solution of 6 M sodium trifluoroacetate, the OD_{260} of a DNA sample is only 16% higher than at 20°C. Propose an explanation for this.

14-8. A protein is placed in a 20% ethylene glycol solution and its difference spectrum determined against the solution without ethylene glycol. At all wavelengths $\Delta\varepsilon = 0$. What can you conclude about the structure of the protein?

A second protein is also studied in both 20% dimethylsulfoxide and 20% sucrose. In sucrose, $\Delta\varepsilon = 0$. In dimethylsulfoxide, there is a characteristic difference spectrum of tryptophan. The observed value of $\Delta\varepsilon$ is doubled if the protein is heated to 60°C in dimethylsulfoxide. In 20% sucrose, after heating to 60°C, the value of $\Delta\varepsilon$ is equivalent to eight tryptophans. What can you say about the protein?

14-9. In performing a pH titration of a protein at 295 nm, it is found that ε_{295} increases sharply at pH 9.6. At pH 11.7, ε_{295} increases again, the latter increase being one-third that of the first rise. If the pH is then reduced to 6 and then gradually increased, the curve for ε_{295} versus pH is nearly identical with that of the protein before titration except that the increase in ε_{295} is only two-thirds that originally observed. The protein is known to have eight tyrosines. What can you say about the protein structure?

14-10. DNA is isolated from an *E. coli* bacterial culture. The mean GC content of *E. coli* is 50%. The DNA is dissolved in a mixture of 0.15 M NaCl and 0.015 M Na citrate. A melting curve is obtained. Instead of a smooth curve, there are two steps having T_m of 80° and 88°C. The first step accounts for 20% of the total increase. What possible explanation can be given for this?

DNA from another bacterium is isolated in the same way but by accident is heated to 75°C for 1 minute. It was later used in a melting experiment. The melting curve was surprising in that, although the major transition was between 82° and 88°C, there was a gradual increase in OD_{260} of 10% in the temperature range of 32° to 55°C. Explain this finding.

CHAPTER 15

Fluorescence Spectroscopy

In the preceding chapter, we saw that certain types of information about both the properties of macromolecules and their interactions with other molecules could be obtained from studies of absorption spectra. The underlying principle is that the absorption spectrum of a chromophore is significantly affected by the physical and chemical environment. In this chapter, we consider a more sensitive spectroscopic probe, fluorescence.

With some molecules, the absorption of a photon is followed by the emission of light of a longer wavelength (i.e., lower energy). This emission is called fluorescence (or phosphorescence, if the emission is long-lived). As is true for absorption spectroscopy, there are many environmental factors that affect the fluorescence spectrum; furthermore, fluorescence efficiency is also environmentally dependent. Because these parameters of fluorescence are more sensitive to the environment than are those of absorbance and because smaller amounts of material are required, fluorescence spectroscopy is frequently of greater value than absorbance measurements (although absorption spectroscopy is simpler to perform). With macromolecules, fluorescence measurements can give information about conformation, binding sites, solvent interactions, degree of flexibility, intermolecular distances, and the rotational diffusion coefficient of macromolecules. Furthermore, with living cells, fluorescence can be used to localize otherwise undetectable substances.

As with other physical methods, the theory of fluorescence is not yet adequate to permit a positive correlation between a fluorescent spectrum and the properties of the immediate environment of the emitter; hence, once again the utility of the procedure is based on establishing empirical principles from studies with model compounds.

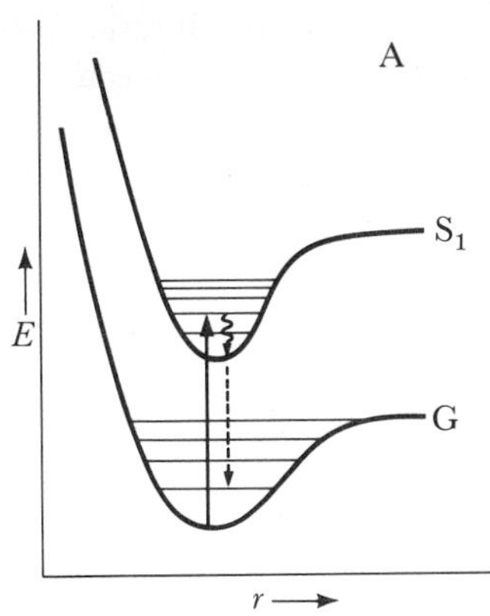

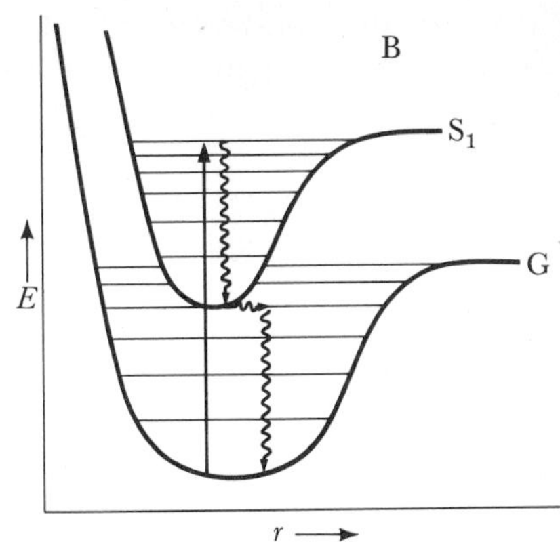

FIGURE 15-1
Energy-level diagram of two chromophores; G and S_1 indicate the ground and first excited states, respectively (heavy lines). The vibrational levels are the thin lines. A. This molecule is capable of fluorescing by the transition (solid arrow) indicated in the diagram. After excitation, there are vibrational losses (wavy arrow) to the lowest level of the excited state and then emission from this state (dashed arrow). B. This molecule fails to fluoresce because the vibrational levels of G are higher than the lowest level of S; hence, there can be a nonradiative transition (horizontal wavy arrows) from S_1 to a vibrational level of G followed by nonradiative losses to the bottom of G (vertical wavy arrow).

SIMPLE THEORY OF FLUORESCENCE

In Chapter 14, it was seen that a molecule can possess only discrete amounts of energy. The potential energy levels of the molecule are described by an energy-level diagram (Figure 15-1). This figure shows two electronic levels, the lower or ground state (G) and one upper or first excited state (S_1) and some of the vibrational levels of each (see Chapter 14 for a discussion of vibrational levels). As explained in Chapter 14, light energy can be absorbed only when the molecule moves from a lower to a higher energy level.* Such transitions are indicated on an energy-level diagram by vertical lines. If the molecule is initially unexcited (i.e., at its lowest energy level or ground state, G) and the absorbed energy is greater than that required to reach the first electronic excited state, S_1, the excess energy can be absorbed as vibrational energy and the molecule will be at

*Not all transitions are possible. Allowable transitions are defined by the selection rules of quantum mechanics.

one of the vibrational levels shown in the figure. This vibrational energy is rapidly dissipated as heat by collision with solvent molecules (if the excited molecule is in solution), and the molecule drops to the lowest vibrational level of S_1. The excited molecule then returns to G either by emitting light (fluorescence) or by a nonradiative transition (described in the discussion of quantum yield). Because energy is lost in dropping to the lowest level of S_1, the emitted light will have less energy (i.e., longer wavelength) than the absorbed light. Therefore, fluorescent light always has a longer wavelength than the exciting light. However, in returning to G, the molecule may arrive at one of the vibrational levels of G instead of the absolute ground state; this vibrational energy will also be dissipated as heat. Hence, if there are many absorbers, the light emitted will have many wavelengths (all of course greater than that of the absorbed light); the probability of dropping from the first excited state to each vibrational level of the ground state determines the shape of the fluorescence spectrum. A typical absorption and fluorescence spectrum is shown in Figure 15-2.

As stated earlier, the excited molecule does not always fluoresce. The probability of fluorescence is described by the *quantum yield, Q*; that is, the ratio of the number of emitted to absorbed photons. (A photon is a unit of light having energy $E = h\nu$, in which h is Planck's constant and ν is the frequency of the light wave.) Several factors determine Q; some of these are properties of the molecule itself (internal factors) and some are environmental. The internal factors derive mostly from the distribution of vibra-

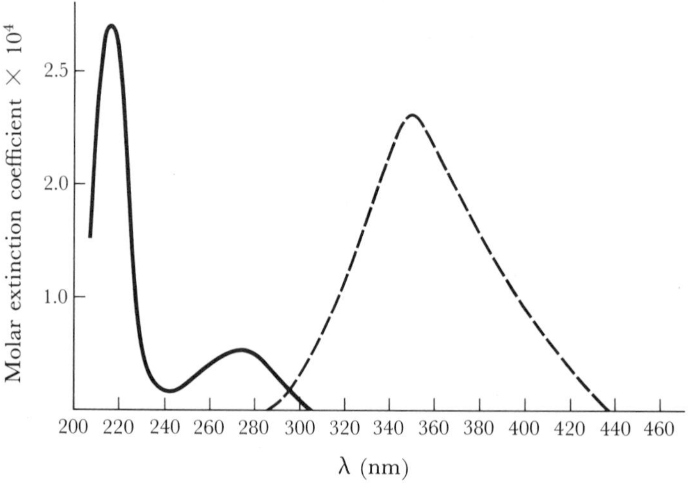

FIGURE 15-2
Absorption and fluorescence spectrum of tryptophan. Solid line: molar extinction coefficient as a function of wavelength. Dashed line: emission spectrum in arbitrary units.

tional levels between G and S. For example, if a vibrational level (V_G) of the ground state has the same energy as a vibrational level of a lower order (V_S) of the first excited state, there can be a nonradiative transition (which will not be explained here) from V_S to V_G (Figure 15-1B), followed by the conversion of the energy of V_G into heat. This is what usually happens with flexible molecules because they have very high vibrational levels of G (see Figure 16-1B). In fact, this is the most common way for excitation energy to be dissipated and accounts for the fact that fluorescent molecules (fluors) are rare and that those which do fluoresce are almost invariably fairly rigid aromatic rings or ring systems.

The internal factors are not generally of interest to biochemists concerned with the properties of macromolecules; environmental factors are more important. The effect of the environment is primarily to provide radiationless processes that compete with fluorescence and thereby reduce Q; this reduction in Q is called *quenching.* In biological systems, quenching is usually a result of either collisional processes (either a chemical reaction or simply *collision* with exchange of energy) or a long-range, radiative process called *resonance energy transfer* (which will be discussed in a later section). These three factors are usually expressed in an experimental situation involving solutions as an effect of the solvent or dissolved compounds (called *quenchers*), temperature, pH, neighboring chemical groups, or the concentration of the fluor. How to make use of these environmental effects in studying macromolecules will be discussed in a later section.

INSTRUMENTATION FOR MEASURING FLUORESCENCE

Figure 15-3 shows a standard arrangement for measuring fluorescence. A high-intensity light beam passes through a monochromator for the selection of an excitation wavelength (i.e., a wavelength efficiently absorbed by the fluor). The exciting light beam then passes through a cell containing the sample. To avoid detecting the incident beam, use is made of the fact that fluorescence is emitted in all directions so that observation of the fluorescence can be made at right angles to the incident beam. The fluorescence then passes through a monochromator for wavelength analysis and finally falls on a photosensitive detector (usually a photomultiplier tube). Many modern instruments have scanning systems and chart recorders that automatically vary the wavelength detected and plot the emitted intensity as a function of the wavelength of the emitted light.

The intensity of light is a measure of the energy E per unit area per unit time. Because the response of the usual light detectors, (i.e., photomultiplier tubes), is wavelength dependent and because $E = hc/\lambda$, the ratio of the outputs (electrical currents) produced by a photomultiplier

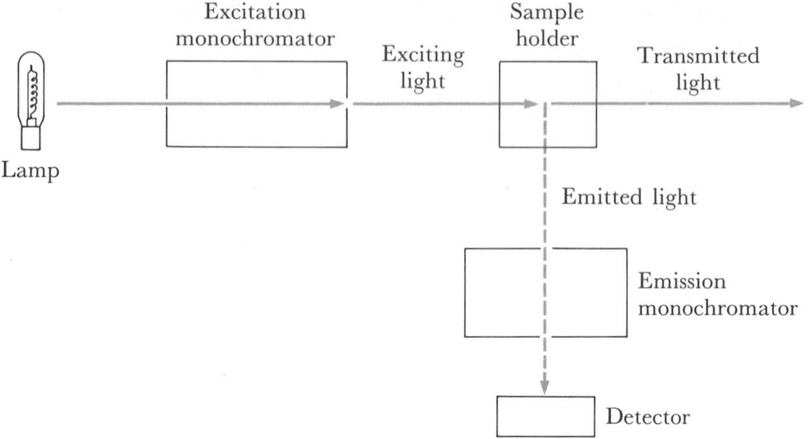

FIGURE 15-3
A spectrofluorometer. Fluorescence is emitted in all directions by the sample, but most systems look at only that emitted at 90° with respect to the exciting light. Note that this differs from an absorption spectrophotometer only in that, with that instrument, the transmitted light would be detected. Modern automatic spectrofluorometers have a chart recorder and plot either detector output versus emitted wavelength or detector output at a single wavelength or a collection of wavelengths versus wavelength of the exciting light.

when two different wavelengths fall on it is not the same as the ratio of the intensities. However, in most experiments, a simple measurement of the photomultiplier output is sufficient because usually only *relative* intensities at each wavelength are being measured—for example, the measurement of fluorescence in the presence and absence of an agent if the agent does not affect the efficiency of absorption of the exciting light. However, in some experiments, it is necessary to measure Q or to determine the absolute energy distribution of a fluorescence spectrum; both of these require measuring absolute intensity. The usual methods for measuring fluorescence intensity require calibrating the photomultiplier system with a *thermopile,* an instrument whose ability to measure the energy of incident light is wavelength independent. This calibration may be supplied by the manufacturer of the spectrofluorometer; if not, another instrument, called a quantum counter can be used.

It is important to know that the distinction between a corrected spectrum and an uncorrected one is not often made in the presentation of fluorescence spectra in journal articles. It is common to plot a spectrum as the photomultiplier output versus wavelength. This is an *uncorrected* spectrum. Plotting fluorescence intensity or quantum yield produces a *corrected* spectrum. Invariably, when photomultiplier output is plotted, it is incorrectly called fluorescence or fluorescence intensity. It is probably safe to assume that a given spectrum is uncorrected unless stated otherwise.

To measure Q requires the counting of photons because

$$Q = \frac{\text{photons emitted}}{\text{photons absorbed}} \tag{1}$$

Note that Q is a dimensionless quantity. Because the energy, E, of one photon is related to the frequency, ν, of the light by the relation $E = h\nu$, a measurement of the number of photons requires measuring the energy of the radiation and correcting for frequency. Measuring the absolute quantum yield is a difficult and tedious process and is rarely done in biochemistry. (Procedures for doing so are given in the Selected References near the end of this chapter.) The usual method for determining Q requires a comparison with a fluor of known Q; two solutions are prepared—one of the sample and one of the standard fluor—and, with the same exciting source, the integrated fluorescence (i.e., the area of the spectrum) of each is measured.

The quantum yield, Q_X, of a sample X is

$$Q_X = \frac{I_X Q_S A_S}{I_S A_X} \tag{2}$$

in which Q_S is the quantum yield of the standard, I_X and I_S are the integrated fluorescence intensities of the sample and the standard, respectively, and A_X and A_S are the percentage of absorption of each solution at the exciting wavelength. Usually, the solutions are adjusted so that $A_X = A_S$.

Once again, it is necessary to understand a convention used in presenting data. Very often published spectra are plotted as Q versus λ, or fluorescence intensities at a particular wavelength are stated as a value of Q. In fact, it is rare that Q has been measured. This inconsistency derives from the fact that, in biochemistry, one is usually measuring the relative intensities in two different situations and, of two spectra (e.g., a quenched and unquenched), the relative values of Q have the same ratio as the fluorescence intensities.

INTRINSIC FLUORESCENCE MEASUREMENTS FOR STUDYING PROTEINS

Two types of fluors are used in fluorescence analysis of macromolecules—intrinsic fluors (contained in the macromolecules themselves) and extrinsic fluors (added to the system, usually binding to one of the components). Intrinsic fluorescence will be discussed first.

For proteins, there are only three intrinsic fluors—tryptophan, tyrosine, and phenylalanine (listed in order of Q, from larger to smaller). The fluorescence of each of them can be distinguished by exciting with and observing at the appropriate wavelength. In practice, tryptophan fluorescence is most commonly studied, because phenylalanine has a very low Q and tyrosine fluorescence is frequently very weak due to quenching. The fluorescence of tyrosine is almost totally quenched if it is ionized, or near an amino group, a carboxyl group, or a tryptophan. In special situations, however, it can be detected by excitation at 280 nm.

The principal reason for studying the intrinsic fluorescence of proteins is to obtain information about conformation. This is possible because the fluorescence of both tryptophan and tyrosine depends significantly on their environment (i.e., solvent, pH, and presence of a quencher, a small molecule, or a neighboring group in the protein).

The use of measurements of intrinsic fluorescence in proteins is based on empirical "rules" obtained from studies of model compounds whose structure and conformation are well known. The rules in common use are presented in Table 15-1. How some of these rules are used in practice is described in the following examples.

TABLE 15-1
Empirical rules for interpreting fluorescent spectra of proteins.

1. All fluorescence of a protein is due to tryptophan, tyrosine, and phenylalanine unless the protein is known to contain another fluorescent component.
2. The λ_{max} of the tryptophan fluorescence spectrum shifts to shorter wavelengths and the intensity of λ_{max} increases as the polarity of the solvent decreases.
   a. If λ_{max} is shifted to shorter wavelengths when the protein is in a polar solvent, the tryptophan must be internal and in a nonpolar environment.*
   b. If λ_{max} is shifted to shorter wavelengths when the protein is in a nonpolar medium, either the tryptophan is on the surface of the protein or the solvent induces a conformational change that brings it to the surface.*
3. If a substance known to be a quencher (i.e., it quenches the fluorescence of the *free* amino acid), such as the iodide, nitrate, or cesium ions, quenches tryptophan or tyrosine fluorescence, the amino acids must be on the surface of the protein. If it fails to do so, there are several reasons:
   a. The amino acid may be internal.
   b. The amino acid may be in a crevice whose dimensions are too small for the quencher to enter it.
   c. The amino acid may be in a highly charged region and the charge might repel the quencher. For example, the iodide ion (a negative quencher) fails to quench tryptophan fluorescence if the tryptophan is in a negative region; the Cs^+ ion is ineffective if the fluor is in a positive region. The neutral quencher, acrylamide, disregards the charge.

TABLE 15-1
(continued)

4. If a substance that does not affect the quantum yield of the free amino acid affects the fluorescence of a protein, it must do so by producing a conformational change in the protein.
5. If tryptophan or tyrosine are in a polar environment, their Q decreases with increasing temperature, T, whereas, in a nonpolar environment, there is little change. Hence, deviation from a monotonic decrease in Q with increasing T indicates that heating is inducing a conformational change because the polarity of the regions to which the tryptophans are being exposed must be changing. An increase with T of the temperature dependence of Q when the protein is in a polar solvent such as water indicates that more tryptophan molecules are being exposed to the solvent—that is, the protein is unfolding.
6. The Q for both tryptophan and tyrosine are decreased if the α-carboxyl group of these amino acids is protonated.
7. Tryptophan fluorescence is quenched by neighboring protonated acidic groups. Hence, if the pK measured by monitoring tryptophan fluorescence is the same as the pK for a known ionizable group (e.g., the imidazole of histidine or the SH bond of cysteine), then that group must be very near a tryptophan. This rule applies only if it can be shown independently that the pH change does not introduce a conformational change.
8. If a substance binds to a protein and tryptophan fluorescence is quenched, either there is a gross conformational change as a result of binding or some tryptophan is in or very near the binding site. Furthermore, because a decrease in the polarity of the solvent causes a shift in λ_{max} to shorter wavelengths, such a shift associated with binding indicates that water is excluded in the complex.
9. If the absorption spectrum of a small molecule overlaps the emission spectrum of tryptophan and the distance is small, there is quenching. Hence, if the binding of such a molecule to a protein quenches tryptophan fluorescence, tryptophan must be in or near the binding site.

NOTE: Two important points must be made about these rules: (1) fluorescence is so sensitive to environmental factors that other interpretations must always be sought for the results described in these rules; (2) if a protein contains several tryptophans, as is usually the case, each may have a different quantum yield. Therefore, the absolute magnitude of changes cannot be used to determine the fraction in a given environment—for example, internal versus external.

*A shift of λ_{max} in a macromolecule is, in standard terminology, in relation to λ_{max} of the free amino acid in water.

INFORMATION ABOUT THE CONFORMATION AND BINDING SITES OF PROTEINS OBTAINED FROM STUDIES OF INTRINSIC FLUORESCENCE

In the following examples, relatively strong conclusions can be made about structural features of proteins, using simple measurements of fluorescence. It should be noted that in many cases measurement of relative fluorescence is sufficient. ("Free amino acid" always refers to an amino acid dissolved in water.)

Example 15-A. Conformational change in a hypothetical enzyme induced by the binding of a cofactor.

The tryptophans of a hypothetical enzyme have a much greater fluorescence intensity and a shorter λ_{max} than does free tryptophan (Figure 15-4). Hence, they must be in a very nonpolar environment (rule 2 in Table 15-1). When the cofactor of the enzyme is added, the fluorescence intensity decreases and λ_{max} becomes longer. Hence, the conformation of the enzyme must have changed (rule 4) either to make the tryptophan available to the polar solvent, water, or to bring it near charged groups. The addition of either the iodide or cesium ion has no effect on the fluorescence spectrum of the enzyme either before or after the cofactor is bound. Hence, the tryptophan has not been brought to the surface (rule 3) and has probably been moved to a more polar, internal region of the protein.

Note that, because of these changes in fluorescence, the value of one of the parameters (e.g., intensity at a particular wavelength, the value of λ_{max}, or Q) could be used as an assay for binding. Hence, a plot of

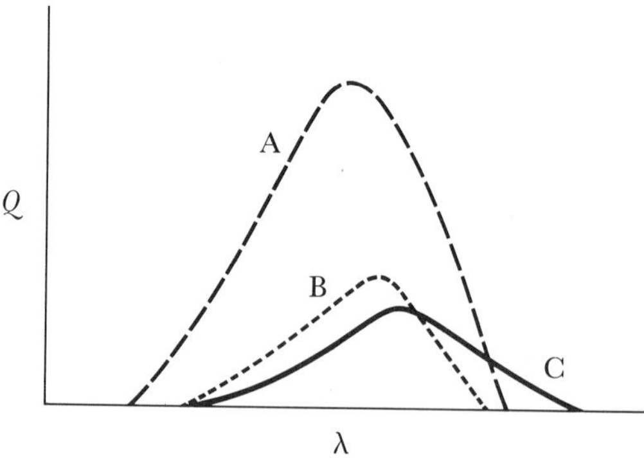

FIGURE 15-4
Fluorescence spectrum of a hypothetical protein in solution: (A) without added cofactor; (B) with added cofactor. Curve C is the spectrum of free tryptophan in water.

fluorescence versus cofactor concentration can be used to determine both the stoichiometry of binding and the dissociation constants by the standard methods of enzyme chemistry.

Example 15-B. Properties of the active site of an enzyme.
A hypothetical enzyme is known to contain a single tryptophan and λ_{max} is nearly the same as that for the tryptophan, suggesting that the tryptophan is in a polar environment (rule 2). The Q is low, suggesting that a quencher might be nearby. Acid titration of the enzyme decreases Q and the pK of the transition is that of a carboxyl. Hence, the low Q is caused by proximity to a carboxyl (rule 7). The addition of the substrate shifts λ_{max} to shorter wavelengths and increases Q. The shift in λ_{max} indicates that the tryptophan is in a less polar environment (rules 2 and 4). This could mean either that there has been a conformational change moving the tryptophan to a nonpolar region or that the tryptophan is in or near the binding site and that the substrate binding excludes water. The increase in Q suggests that the quenching effect of the carboxyl has been eliminated. A likely but not exclusive interpretation is that tryptophan is in the active site and the substrate binds either to the tryptophan or to the carboxyl.

Example 15-C. Studies on the denaturation of a protein.
The Q of tryptophan of a hypothetical protein drops slowly and continually with temperature from 20° to 63°C. Hence, the tryptophan molecules are in a polar environment (rule 6). Between 63° and 65°C, it increases markedly (Figure 15-5) so that, in the range from 63° to

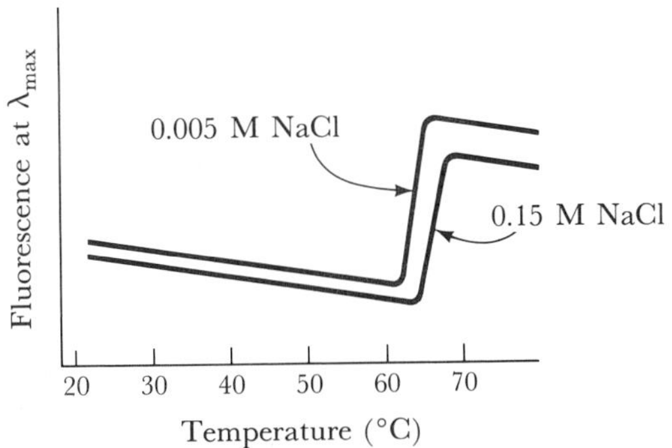

FIGURE 15-5
A helix-coil transition of a protein measured by changes in fluorescence. Note that, in 0.15 M NaCl, the protein is more stable in that a higher temperature is required for the transition.

65°C, there must be a conformational change resulting in a change in the exposure of the tryptophan molecules to the solvent (rule 6a). This change in Q can be used as an assay for the helix-coil transition so that various parameters (e.g., salt concentration, pH, etc.) affecting the process can be studied.

Example 15-D. Location of tryptophans in a hypothetical enzyme.
An enzyme is known to have five tryptophans. The Q is higher than that for free tryptophan, indicating that some are in nonpolar regions (rule 2). If the protein is heated from 20° to 55°C, the decrease in Q is only 35% that found for free tryptophan, suggesting that 0.35×5 or ~2 tryptophans are on the surface (rule 5). If a high concentration of iodide is added, 30% of the fluorescence is quenched, again consistent with two tryptophans (rule 3). In neither case is 40% quenched, because the three internal tryptophans are in a slightly nonpolar environment and therefore contibute *more* than three-fifths of the value of Q. If a substrate of the enzyme is added, there is no change in Q, suggesting either that there is no tryptophan in the binding site or that, if tryptophan is in the binding site, the bound molecules must create a polar environment. Iodide quenches only 15% of the fluorescence when the substrate is bound; hence, $0.15/0.30 = 1/2$ of the tryptophans on the surface are not exposed to the solvent when the substrate is bound. Hence, $1/2 \times 2 = 1$ tryptophan is in the active site.

If the enzyme is titrated, 18% of the fluorescence is quenched. The pK associated with this quenching is that of histidine. Hence, there must be a histidine near one of the tryptophans (rule 7). If the substrate is bound and then the enzyme is titrated, there is no change in Q, indicating that in the presence of substrate, the histidine cannot be titrated. Hence, the binding site contains both a histidine and a tryptophan and the substrate may bind directly to the histidine.

In Example 15-D, the assumption was made that the Q for all tryptophans is nearly the same. This certainly need not be the case (e.g., one or more could be near carboxyls). If this assumption were incorrect, the calculations of the number of each tryptophan would be invalid. However, the value of this naïve analysis is to provide a working hypothesis for the structure of the protein; this hypothesis must then be tested in other ways.

Example 15-E. Assay by iodide quenching of the dependence of conformation of a protein on NaCl concentration.

The conformation of many proteins in solution is affected by the NaCl concentration of the solvent. This can be studied by examining the degree of quenching produced by the iodide ion. For example, consider a protein for which, as the NaCl concentration is decreased to

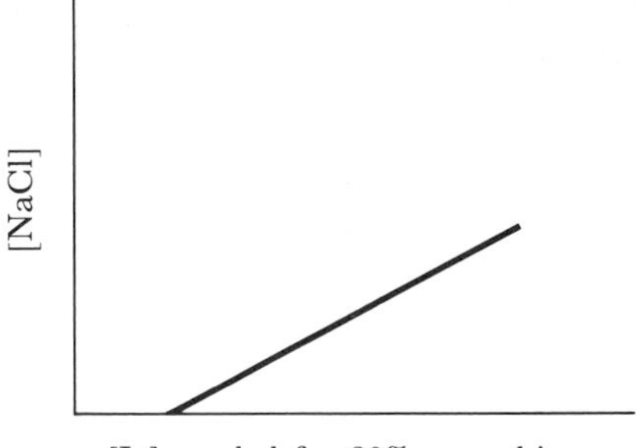

FIGURE 15-6
Plot of the concentration of iodide needed to reduce tryptophan fluorescence of a protein by 30% as a function of NaCl concentration.

zero, the concentration of iodide required to produce a given degree of quenching decreases (Figure 15-6). Hence, at low salt concentration, the structure of the protein becomes more open so that more of the tryptophans become available for collision with iodide. Furthermore, if in the absence of NaCl the maximum quenching is only 50% of the quenching of free tryptophan produced by iodide, then it might be concluded that, if no NaCl is present, the structure cannot be a random coil because 50% of the tryptophans are still not exposed to the solvent. Note that iodide quenching is a general method for studying the helix-coil transition induced by any agent that does not affect the ability of iodide to quench.

EXTRINSIC FLUORESCENCE

Nature does not always supply the investigator with a fluorescent group in the appropriate place in a macromolecule. However, in many cases, a fluor can be introduced into the molecule to be studied either by chemical coupling or by simple binding (as in the use of reporter groups in absorption spectroscopy). The use of such added molecules in fluorescence analysis is called the method of extrinsic fluorescence. Several requirements of the fluor must be met for using extrinsic fluorescence: (1) the fluor must be tightly bound at a unique location; (2) its fluorescence should be sensitive to environmental conditions; and (3) it should not itself affect the features of the macromolecule being investigated. These three criteria must always be verified. For proteins, the most common extrinsic fluors are 1-anilino-8-naphthalene sulfonate (ANS); 1-dimethyl-aminonaphthalene-5-sulfonate (DNS) and its chlorinated derivative, dansyl chloride; 2-*p*-toluidylnaphthalene-6-sulfonate (TNS); rhodamine; fluorescein; and the isothiocyanates of rhodamine and fluorescein. For nucleic acids, various acridines (acridine orange, proflavin, acriflavin) and ethidium bromide are used (Figure 15-7).

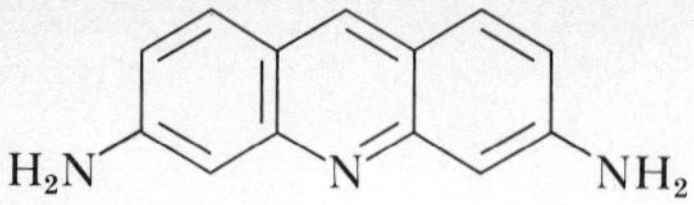

Proflavin

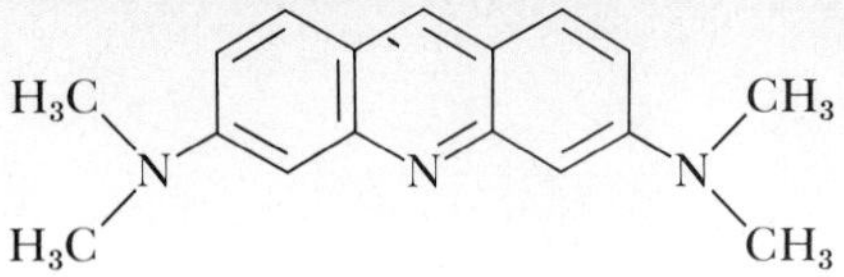

Acridine orange

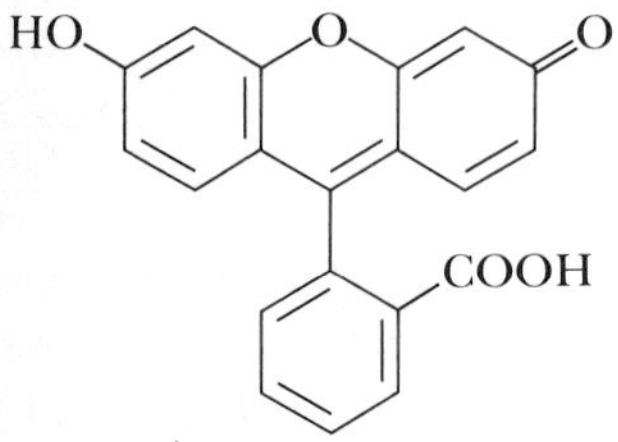

Fluorescein

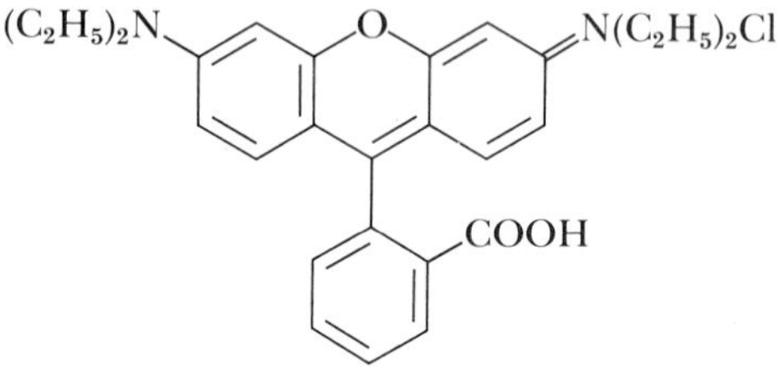

Rhodamine B

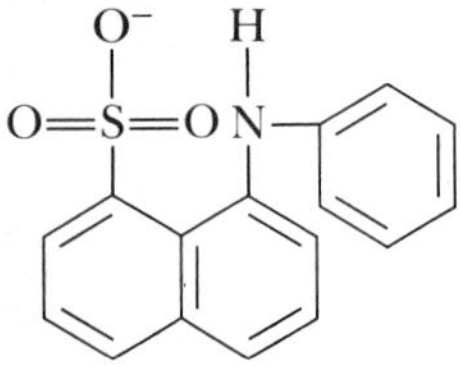

1-Anilino-8-naphthalene sulfonate (ANS)

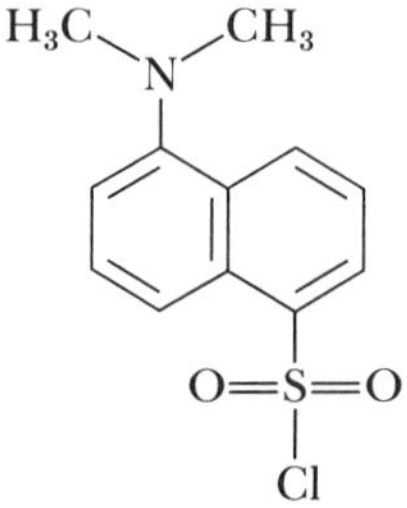

Dansyl chloride

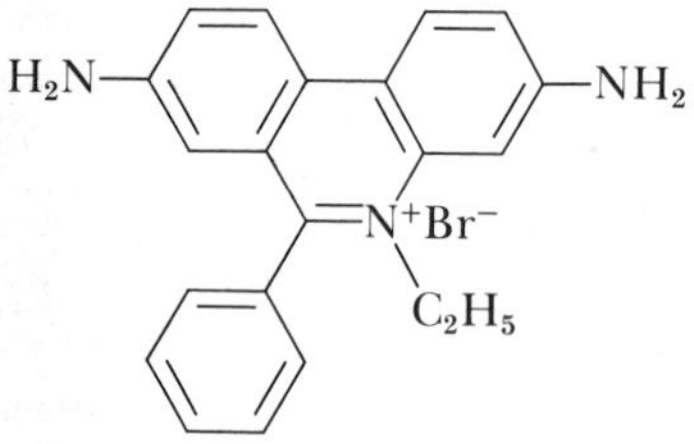

Ethidium bromide

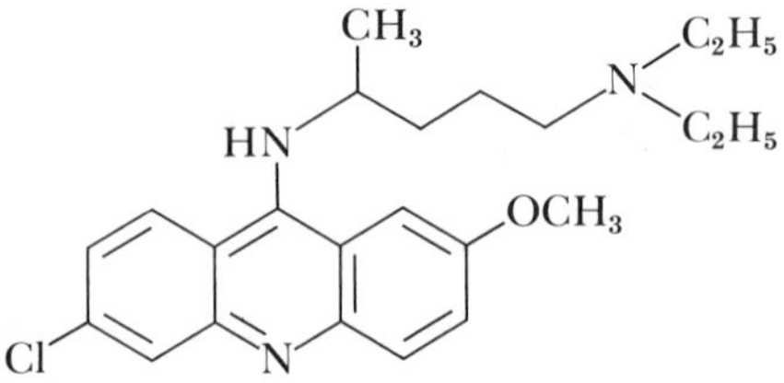

Quinacrine chloride

FIGURE 15-7
Structures of common extrinsic fluors.

ANS, DNS, and TNS have the valuable property that in aqueous solution they fluoresce very weakly. However, in a nonpolar environment, Q increases markedly and the spectrum shifts toward shorter wavelength, both effects increasing as polarity decreases (Figure 15-8). Studies with model compounds have provided a rough relation between Q and the degree of nonpolarity. Dansyl chloride and the isothiocyanates have the useful property of reacting with specific amino acids in proteins. Hence, these substances are widely used to detect nonpolar regions in proteins because, on binding to such a region, their fluorescence increases.

How these extrinsic fluors can be used is shown in the following examples.

Example 15-F. Determination of the properties of the heme binding site in hemoglobin.

Hemoglobin is a complex of a small prosthetic group with the protein apohemoglobin. The extrinsic fluor ANS fluoresces when added to solutions of apohemoglobin but not with hemoglobin. The addition of heme to the apohemoglobin-ANS complex eliminates the fluorescence by displacing the ANS, so that ANS and heme must bind at the same site, which must be nonpolar. From the exact value of Q and the spectral shift, the degree of nonpolarity can be estimated.

A study of this sort can give valuable clues in the interpretation of x-ray diffraction patterns. For example, one would know that certain configurations of amino acids (e.g., a highly polar cluster) could not be the point of interaction with a prosthetic group.

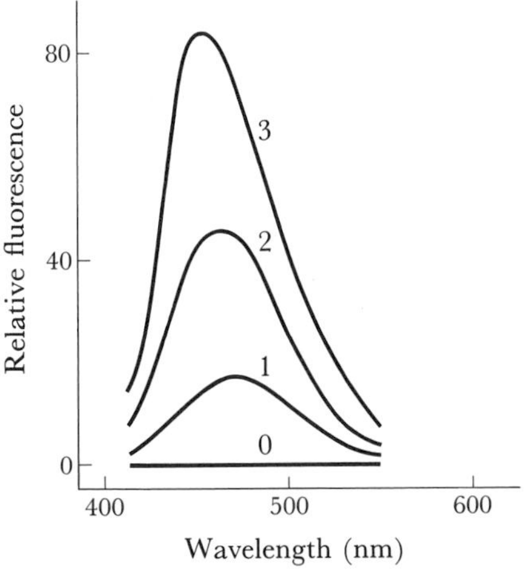

FIGURE 15-8
Effect of the addition of bovine serum albumin (BSA) to ANS: 0 means no BSA; 1, 2, and 3 indicate increasing amounts of BSA.

Example 15-G. Detection of a conformational change in an enzyme when the substrate is bound.

TNS fluoresces when added to α-chymotrypsin and therefore must bind to it. TNS does not competitively inhibit the enzymatic hydrolysis of various substrates so that it cannot bind to the substrate-binding site. However, the addition of a substrate decreases the fluorescence. This decrease could be a result of decreased binding (TNS fluoresces only if bound) or of decreased polarity of the TNS binding site. By using fluorescence as an assay of binding and studying the fluorescence as a function of TNS concentration (in the presence and absence of substrate), it was shown that the binding constants are the same in both conditions. Therefore, there is not less TNS bound and the decrease in fluorescence indicates a decrease in the polarity of the binding site. The change in polarity could result from a polar group moving near the binding site or from the uncovering of a polar group—in either case, a conformational change is associated with substrate binding. Note also that, with this fluorescence assay, the binding parameters of the substrate and the effects of various agents both on binding and the conformational change can be determined.

Example 15-H. Presence of bound fatty acids in bovine serum albumin (BSA).

If different samples of BSA are prepared and ANS is added, ANS fluorescence is observed, but the values of Q and λ_{max} differ from one sample to the other. The variation in Q is too great to be due to variations in the polarity of the binding site and hence is due to the number of binding sites—that is, less ANS is bound in the samples whose fluorescence is weak. If the BSA samples are treated with a lipid solvent, the quantum yields and spectral shifts become the same for all samples. The extraction with the lipid solvent apparently makes some nonpolar sites unavailable to the ANS. However, analysis of the solvent demonstrates that fatty acids are extracted from the samples whose fluorescence is weak. Hence, so-called pure samples of crystalline BSA may contain bound fatty acids. Clearly, ANS fluorescence can be used both to assay the purity of BSA samples and to study the binding of fatty acids to BSA—for example, to determine the binding constants.

Example 15-I. Determining the strandedness of polynucleotides.

The fluor acridine orange shows an increase in Q and a shift in λ_{max} when bound to polynucleotides (Figure 15-9). When saturating amounts of acridine orange are added, the values of λ_{max} are significantly different for double- (green fluorescence) and single- (red fluorescence) stranded polynucleotides. Hence, the structures can be distinguished in

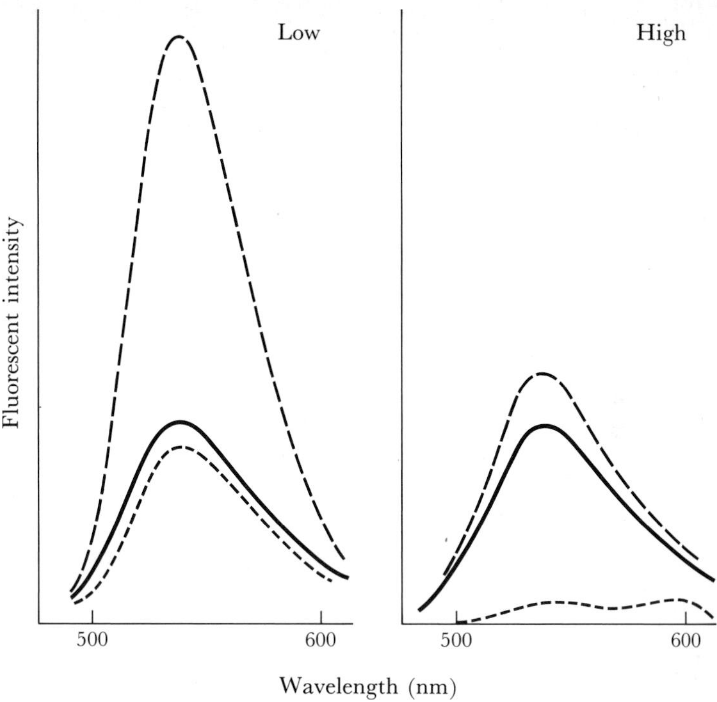

FIGURE 15-9
Effect of a double-stranded (long dashes) polynucleotide (e.g., DNA) and a single-stranded (short dashes) polynucleotide (e.g., denatured DNA or RNA) on the fluorescence of acridine orange. The solid line is the fluorescence spectrum for free acridine orange. Low and high refer to ratios of dye molecules to nucleotides. Note that DNA enhances fluorescence without affecting λ_{max} and that the enhancement decreases as more dye is added. Single-stranded material quenches fluorescence and shifts the spectrum toward the red. This accounts for the fact that, if a high concentration of acridine orange is added to a cell, the nuclear material fluoresces green whereas the RNA-containing cytoplasm appears orange.

this simple way. If a sample contains both double- and single-stranded polynucleotides, the fluorescence spectrum will have two peaks, one for each value of λ_{max}.

More examples of the use of extrinsic fluorescence will be given in the sections on energy transfer and polarization.

EXTRINSIC FLUORESCENCE AND ENERGY TRANSFER

Consider a system containing two fluors (numbered 1 and 2) whose absorption and emission spectra are as shown in Figure 15-10. A convenient wavelength (e.g., λ_3) is selected for detecting the fluorescence of fluor 2 to determine the *excitation* (or *action*) spectrum for producing fluorescence. By excitation spectrum is meant the wavelengths that, when absorbed, will produce the fluorescence. It is clear that, in general, the excitation spectrum will be the same as the absorption spectrum (Figure 15-10C).* However, on occasion the excitation spectrum will be very different from the absorption spectrum (Figure 15-10D) in that fluorescence is produced by absorption of much shorter wavelengths. This is almost invariably due to the fact that under certain circumstances energy absorbed by one molecule (a donor) can be transferred to another fluor (an acceptor) at some distance away. This is called *resonance energy transfer,* a necessary but not sufficient condition of which is that the emission spectrum of the donor overlaps the absorption spectrum of the acceptor (Figure 15-10).†

Note that, if energy transfer occurs, there is a decrease in the amount of the fluorescence of fluor 1 when excitation is by a wavelength in the absorption spectrum of fluor 1. This is clearly a kind of quenching, especially if the acceptor molecule is not a fluor. However, energy transfer can be easily detected if the entire fluorescence spectrum is determined, because it will be found (Figure 15-10E) that the fluorescence spectrum resulting from excitation at λ_1 contains a new band at long wavelengths. In other words, energy transfer is to be suspected whenever the observed emission spectrum cannot be accounted for by the wavelength used to excite a particular fluor.

So far, energy transfer between different fluors has been considered, but energy transfer can also occur between identical fluors because the absorption and emission spectra of a fluor usually overlap. Such transfer is indicated when it is observed that, as increasing amounts of a fluor are added to a sample, the fluorescence intensity decreases without a decrease in the absorbance at the exciting wavelength—that is, Q decreases. (For this to

---

*This statement assumes that the absorption spectrum is due to a single electronic transition. This is not always the case, but the assumption is acceptable for our purpose.

†A simplistic explanation is that the donor emits a photon, which is then absorbed by the acceptor, and the probability of energy transfer is merely the product of the probabilities of emission and of absorption. However, although emission and reabsorption is certainly a possibility, this is not what occurs in resonance energy transfer and the probability of transfer is described by an expression called the *overlap integral.* [This difference is described in some detail in a paper by T. Förster, *Disc. Faraday Soc.* 27(1959):7–17.]

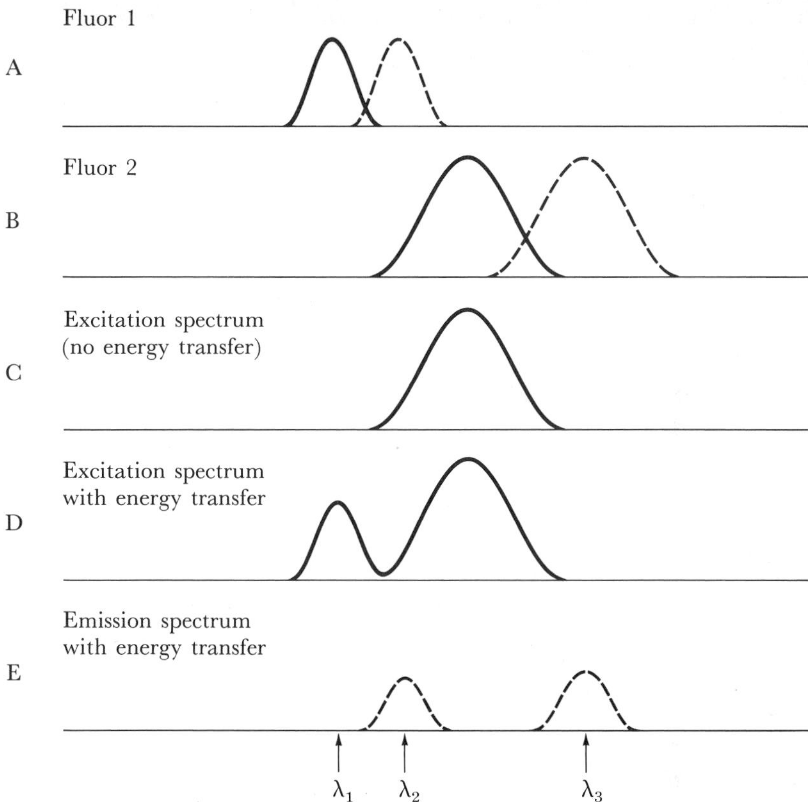

FIGURE 15-10
Spectra for two fluors (the solid curves represent absorption and the dashed lines emission): (A and B) absorption and emission spectra of fluors 1 and 2, respectively; (C) excitation spectrum for emission at λ_3, in which there is no energy transfer between fluors 1 and 2; (D) same as part C, but with energy transfer; (E) emission spectrum with excitation at λ_1 when there is energy transfer between fluors 1 and 2.

occur not only must the usual conditions for energy transfer be satisfied, but also Q must be less than 1.) This kind of energy transfer is clearly a kind of quenching.

The importance of the energy-transfer phenomenon to biochemistry is that the efficiency of transfer is a function of the separation of the fluors and can therefore be used to measure molecular distances. The efficiency of transfer, $\mathscr{E}$, is described by the following equation:

$$\mathscr{E} = \frac{R_0^{\ 6}}{(R_0^{\ 6} + R^6)} \tag{3}$$

in which R is the distance between donor and acceptor, and R_0 is a constant related to each donor-acceptor pair that can be calculated from certain parameters of the absorption and emission spectra of each. The utility of energy transfer becomes apparent if equation (3) is rewritten as

$$R = R_0\left(\frac{1 - \mathscr{E}}{\mathscr{E}}\right)^{1/6} \tag{4}$$

(See Chapter 17 for a discussion of spin-labeling, a method of measuring molecular distance by nuclear magnetic resonance.)

To understand how $\mathscr{E}$ is measured, let us refer to Figure 15-10, which shows the absorption and emission spectra of a donor and an acceptor. We will consider two methods, both of which utilize three wavelengths, λ_1, λ_2, and λ_3, selected as follows: λ_1 is chosen so that absorption by the donor is efficient but inefficient by the acceptor; λ_2 is a wavelength in the emission spectrum of the donor but not in that of the acceptor; and λ_3 is a wavelength emitted only by the acceptor.

In the first method of measuring $\mathscr{E}$, the quenching effect on the donor fluorescence is determined—that is, the quenching of the fluorescence at λ_2 if the donor is excited by λ_1. Let $f_{1,2}$ be the fluorescent intensity at λ_2 if excited at λ_1 and f^{D} and $f^{\mathrm{D,A}}$ represent the fluorescence if only the donor (D) is present or if both donor and acceptor (A) are present, respectively. The fraction of the donors that remains excited is $1 - \mathscr{E}$, or

$$\mathscr{E} = 1 - \frac{f_{1,2}^{\mathrm{D,A}}}{f_{1,2}^{\mathrm{D}}} \tag{5}$$

In the second method, the intensity of emission at λ_3 is measured. In this case, the relevant equation is

$$\mathscr{E} = \left(\frac{\varepsilon_1^{\mathrm{A}} C^{\mathrm{A}}}{\varepsilon_1^{\mathrm{D}} C^{\mathrm{D}}}\right)\left(\frac{f_{1,3}^{\mathrm{D,A}}}{f_{1,3}^{\mathrm{A}}}\right) - 1 \tag{6}$$

in which $\varepsilon_1^{\mathrm{A}}$ and $\varepsilon_1^{\mathrm{D}}$ are the molar extinction coefficients of acceptor and donor, respectively, at λ_1; C^{A} and C^{D} are the concentrations of acceptor and donor, respectively; and $f_{1,3}^{\mathrm{A}}$ and $f_{1,3}^{\mathrm{D,A}}$ are the fluorescence intensities at λ_3 when excited by λ_1, if either the acceptor or the donor-acceptor pair, respectively, are present.

After $\mathscr{E}$ has been measured, R can be evaluated if R_0 is known. (R_0 is usually calculated from various spectral parameters but sometimes is determined by measuring $\mathscr{E}$ in a system in which R is known.)

To use energy-transfer measurements for distance determination, the following conditions must be met: (1) there must be a single donor and acceptor; (2) the value of R_0 for the donor-acceptor pair must be known and it should be near that of the distance being measured; (3) the relative orientation of the donor and acceptor should be known so that it is clear whether a low transfer efficiency is a result of great distance or of unfavorable orientation (see Chapter 14 for a discussion of orientation effects and absorbance); and (4) the addition of the fluors should not alter the structure of the macromolecule being studied.

The validity of the method for determining distance has been demonstrated by studies of a linear polymer in which oligomers of known length of poly-L-proline serve as spacers between a dansyl and an α-naphthyl group (Figure 15-11). These molecules were excited by a wavelength absorbed by the α-naphthyl group and the fluorescence of the dansyl group was measured as a function of the number of proline residues. Efficiency of transfer was calculated as previously described. When $\mathscr{E}$ was plotted as a function of the length of the polyproline chain, it was found to obey the $1/R^6$ law (Figure 15-12).

The kind of information obtained from energy transfer measurements can best be seen in a few examples.

Example 15-J. Proximity of tryptophan to the active site of carbonic anhydrase.

The enzyme carbonic anhydrase contains a zinc atom in the active site. If a donor, *m*-acetylbenzenesulfonamide, is bound to this zinc atom, excitation of this fluor causes tryptophan fluorescence. Hence, a tryptophan must be near the active site.

Example 15-K. Spacing between hydrophobic groups of bovine serum albumin (BSA).

As increasing amounts of ANS are added to BSA, the induced fluorescence of ANS first increases and then decreases. The decrease is caused by energy transfer between nearby ANS molecules. The magnitude of the effect gives a rough idea of the spacing between nonpolar regions.

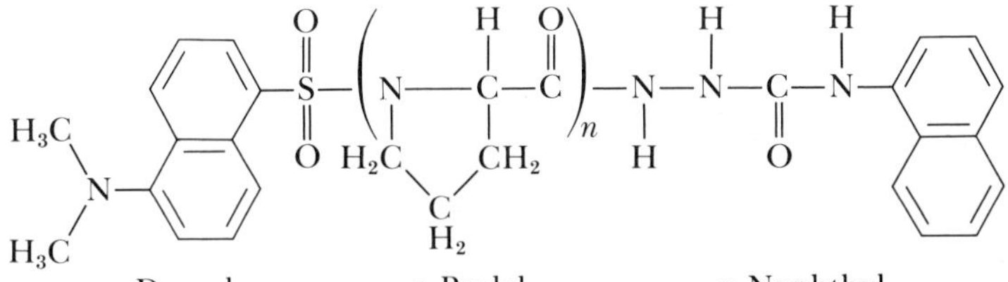

O H O H H
S N C C n N N C N
H_3C O H_2C CH_2 H O
N C
H_3C H_2
Dansyl L-Prolyl α-Naphthyl

FIGURE 15-11
Formula for poly-L-proline separated by a dansyl (acceptor) and a α-naphthyl (donor) group (*n* varied from 1 to 12).

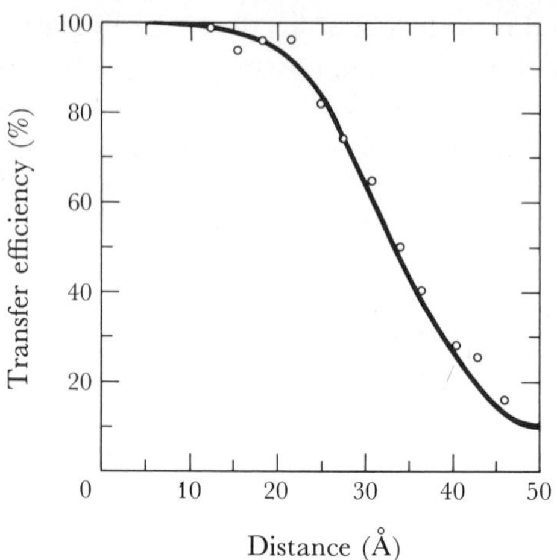

FIGURE 15-12
Efficiency of energy transfer as a function of distance in dansyl-(L-prolyl)$_n$-α-naphthyl, in which n varies from 1 to 12. The distances (in Å) for each value of n were independently determined from other types of measurements. The solid line corresponds to a $1/R^6$ dependence. [From L. Stryer and R. P. Haugland, *Proc. Nat. Acad. Sci.* 58(1957): 719–726.]

EXCIMERS

On occasion, an excited molecule can form a complex with an identical unexcited molecule and the complex will exist until the excited molecule emits. This can be described by the following scheme, in which A is an absorber, A* is an excited absorber, and D is a dimer:

$A + \text{photon} \rightarrow A^*$	(excitation)
$A^* + A \rightarrow D^*$	(formation of excited dimer)
$D^* \rightarrow A + A + \text{photon}$	(fluorescence of dimer)

Such a complex (D*) is called an *excimer* and is recognized by the production of a new fluorescent band at a longer wavelength than the usual emission spectrum. It is distinguishable from resonance energy transfer in that the excitation spectrum is identical with that of the monomer; with energy transfer, the excitation spectrum is composed of the absorption spectra of the donor and the acceptor (Figure 15-10).

Excimer formation is a concentration-dependent phenomenon and can be used to indicate high local concentrations of a fluor. It has had very little use in biochemistry to date but will certainly find applications in the future. For example, the presence of a vesicle or some unit that can concentrate certain substances could be studied by studying the excimers of an extrinsic fluor.

POLARIZATION OF FLUORESCENCE

The intensity of polarized light transmitted by a polarizer depends on the orientation of the polarizer—the transmission is maximum when the plane of polarization is parallel to the axis of the polarizer and zero when it is perpendicular (see Chapter 16, Figure 16-1B). For a light beam that is only partially polarized, a polarization, P, can be defined as

$$P = \frac{I_{\parallel} - I_{\perp}}{I_{\parallel} + I_{\perp}} \tag{7}$$

in which $I_{\parallel}$ and $I_{\perp}$ are the intensities observed parallel and perpendicular to an arbitrary axis. Polarization can vary between -1 and $+1$. It is zero when the light is unpolarized; otherwise it is called *partially polarized.* The polarization of fluorescence can be determined using the standard experimental arrangement for measuring fluorescence (Figure 15-3) with a simple modification (Figure 15-13): a polarizer is placed in the path of the exciting

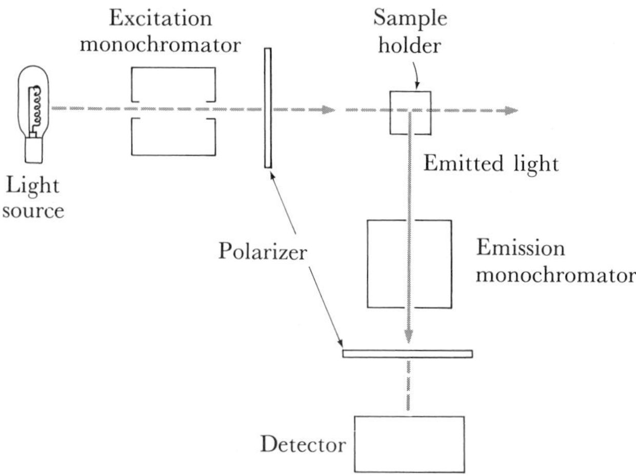

FIGURE 15-13
Experimental arrangement for measuring the polarization of fluorescence.

light in order to excite the sample with polarized light and a second polarizer is placed between the sample and the detector with its axis either parallel or perpendicular to the axis of the excitation polarizer. If the exciting light is polarized, it is usually observed that the fluorescence is either only *partially* polarized or completely unpolarized. The cause of this loss of polarization is explained next, and it will be seen that the magnitude of the change in polarization gives information about the physical state of the emitter.

To begin with, it must be remembered that the absorption of polarized light by a chromophore is maximal when the plane of polarization is parallel to a particular axis of the chromophore, called the electric dipole moment (see Chapter 16). In general, the chromophores will be randomly oriented (this is certainly true in solution); therefore, the probability of absorption of the exciting polarized light is proportional to $\cos^2\theta$, in which θ is the angle between the plane of polarization and the electric dipole moment. Furthermore, the plane of polarization of the emitted light is not determined by the absorption dipole moment but by the transition dipole moment (which is generally not parallel to the absorption dipole moment), and the probability of emission of fluorescence with the plane of polarization at an angle ϕ with respect to the transition dipole moment is proportional to $\sin^2\phi$. The result of these probabilities is that, if the absorbers are randomly arranged (*but stationary*) and the two dipoles are not parallel, the polarization, P, of the fluorescence is $< 1/2$. (In fact, even if the absorbers were perfectly aligned with the plane of polarization of the exciting light, P would be < 1 because of $\sin^2\phi$.) The fact that P is always < 1 is called *fluorescence depolarization.*

Many other factors can increase the extent of depolarization; the two most important from the point of view of biochemistry are (1) the motion of the absorber and (2) energy transfer between like chromophores. If the emitter is rotating very rapidly (i.e., if there is Brownian motion) so that there is a substantial change in orientation during the lifetime of the excited state, the polarization will be further reduced. This is an important phenomenon because the extent of this type of depolarization is affected by temperature, solvent viscosity, and the size and shape of the molecule containing the emitter. The second factor, energy transfer between identical chromophores, results from the fact that, although resonance energy transfer occurs with highest probability between molecules having parallel dipoles, it also occurs when they are nonparallel, with resulting depolarization. Because the efficiency of energy transfer falls off with the sixth power of the distance between the donor and acceptor, this type of depolarization is highly concentration dependent.

The *intrinsic polarization,* P_0, is that which would be observed if the absorber were immobilized and far from all other molecules (i.e., if there were no effect of either molecular motion or energy transfer). In practice, this is the polarization observed if the fluor is in a solvent of very high viscosity and at very low concentration. Note that low concentration can mean either a low concentration of the macromolecule-fluor complex or a large separation between the fluors bound to a single macromolecule. At very high viscosity and high concentration, the effect of energy transfer is primarily detected and, at low viscosity and low concentration, the effect of molecular motion. These effects can be used to obtain information about macromolecules not readily obtained by other techniques.

Applications of Fluorescence Polarization to Proteins and Nucleic Acids

The principal applications in the study of proteins make use of the changes in polarization due to the changes in mobility of either the entire molecule or a part of the molecule.*

To observe depolarization due to motion, it is necessary that the lifetime of the excited state be sufficiently long that reorientation can occur before emission. For both proteins and nucleic acids, the lifetimes of intrinsic fluors are usually too short, although, with small proteins ($M < 20{,}000$), some information can be obtained from studies of the fluorescence depolarization of tryptophan. In general, polarization analysis utilizes extrinsic fluors.

Like other procedures, the method of fluorescence polarization uses a set of rules that govern the imterpretation of data. The basic rule and some of its general consequences are given in Table 15-2.

The utility of the fluorescence polarization method is described in the following examples.

Example 15-L. Measurements of binding using fluorescence polarization as an assay.

If an antigen labeled with a fluor reacts with an antibody, the polarization of the fluorescence increases owing to the decreased mobility of the antigen in the antigen-antibody complex (consequence 3 in Table 15-2). This has been used as an assay for the binding of a variety of fluorescein-labeled antigens—for example, ribonuclease, bovine serum albumin, and the insulin β chain—to their respective antibodies. From such simple titrations, equilibrium constants and stoichiometry have been obtained that have enhanced our understanding of the mechanism of the antigen-antibody reaction.

The binding of a fluorescent substrate by an enzyme enhances polar-

*Note that the kind of *mobility* under discussion is *rotation* because translational motion will not cause depolarization. However, if a molecule is free to move and moves by collision with solvent molecules (i.e., by Brownian motion), it is very unlikely that it could move without rotating also. For this reason, mobility can also be thought of in the general sense as simply the ability to move.

TABLE 15-2
Basic rule used in interpreting fluorescence polarization data and some of the consequences of this rule.

Rule

With increasing mobility, polarization decreases.

Consequences

1. If free in solution, extrinsic fluors show virtually no polarization because they are rapidly and freely rotating. However, if coupled to a low-molecular-weight molecule, their polarization becomes greater owing to somewhat reduced mobility.
2. If an extrinsic fluor is bound to a macromolecule, the polarization can become substantial owing to greatly decreased mobility. The degree of polarization depends not only on the mobility of the entire macromolecule but also on the mobility of the binding site. For example, if a fluor intercalates between the bases of DNA, it is held rigidly in place so that its mobility is that of the DNA molecule; if it is bound to a long amino acid side chain, its polarization will reflect both the mobility of the entire protein and the freedom of movement of the side chain with respect to the polypeptide chain.
3. If a fluor is bound to a macromolecule and then the macromolecule either aggregates (e.g., dimerizes) or binds a large molecule, the polarization increases owing to the reduced mobility of the structure.
4. If a fluor is bound to a macromolecule and the macromolecule undergoes a conformational change (e.g., a helix-coil transition), the polarization decreases as the molecular structure becomes more disordered and increases as it becomes more ordered.

ization by reducing the mobility of the fluor (consequence 2 in Table 15-2). This is a sensitive assay for enzyme-substrate binding and can be used to determine those parameters whose measurement depends on knowing precisely the amount of bound substrate (e.g., binding constants, stoichiometry).

Example 15-M. Detection of the association and dissociation of proteins. The polarization of DNS-derivatives of α-chymotrypsin, chymotrypsinogen, and lactic dehydrogenase increases when the proteins self-associate because the increased size of the macromolecule reduces the mobility (consequence 3 in Table 15-2). Hence, self-association and dissociation can be assayed by measuring changes in polarization. The effects of pH, ionic strength, and composition of the solvent can be examined with this assay because they have little or no effect on the polarization itself (i.e., if the enzymes are so dilute that no association is possible, no changes in polarization are observed).

The hybridization of beef muscle (M) and beef heart (H) lactic acid dehydrogenase (LDH), known tetramers, has been studied elegantly using fluorescence polarization of DNS derivatives of each. Certain conditions (e.g., low pH) lead to a *decrease* in both polarization and enzyme activity. Because decreased polarization indicates increased mobility or decreased molecular size, these conditions presumably cause dissociation of the tetramer to monomers. Having reached their presumably monomeric state, both H-LDH and M-LDH can be treated in such a way that polarization and enzyme activity increase, indicating reassociation. This will also happen if the dissociated H-LDH and M-LDH are mixed before reassociation, indicating that in the mixture one of them does not hinder the other from reassociating. If the reassociated material is then subjected to low pH again and the polarization is measured as a function of time to determine the kinetics of dissociation, it is found that the kinetics of dissociation of the reassociated mixture indicate that more than two components are present. Fractionation of the reassociated mixtures has indicated that hybrid tetramers are present (e.g., a tetramer having 3 H subunits and 1 M subunit) and that each of these has different dissociation kinetics. Note that the value of P is merely used as an assay of the dissociation-reassociation reaction and that this is a general way to detect hybridization.

Example 15-N. Studies of the helix-coil transition in proteins.
For some proteins, there is no simple optical method for observing the helix-coil transition. If an extrinsic fluor (e.g., ANS) can be coupled to a protein without affecting the conformation, the helix-coil transition can be observed by measuring the polarization of the fluorescence because, as the protein unfolds, the polarization will decrease owing to the increasing flexibility of the protein (consequence 2 in Table 15-2).

The polarization method can also be used to detect S–S bonds in a denatured protein. The principle is that *the limiting state of a structureless random coil should have maximum flexibility* and therefore minimum polarization. Hence, a fully denatured molecule containing S–S bonds will not have minimum polarization until the S–S bonds are broken. This means that, if the addition of an agent known to break S–S bonds (e.g., mercaptoethanol) decreases the polarization, S–S bonds must have been present. Polarization is probably the most sensitive way to detect these subtle changes.

Example 15-O. Structure of a fluor-DNA complex: an example of the determination of orientation from fluorescence polarization.

The fluor acridine orange binds tightly to DNA. Because it is an effective mutagen, the structure of the complex is of some interest. If bound to DNA, acridine orange becomes inaccessible to certain chemical treatments (e.g., diazotization), which suggests that it is somehow within the DNA double helix. If bound to DNA, acridine orange fluoresces when it is excited not only by wavelengths in its own absorption band but also by light absorbed only by the DNA bases, thus indicating energy transfer from the bases. The quantum yield of this energy transfer is very high, so that the acridine orange (which is a planar structure) must be very near the bases. The plane of polarization of the fluorescence of acridine orange is known with respect to both the plane of its rings and its long axis so that, if the system were immobilized, the orientation of the acridine orange molecules with respect to the DNA could be determined by measuring the plane of polarization of the fluorescence with respect to the DNA molecules (using light absorbed by acridine orange but not by the bases). The fluorescence of acridine orange is highly polarized, indicating greatly reduced mobility. If it were bound in such a way that it is not free to move with respect to the DNA helix, this low mobility would be expected because DNA is so highly extended that it has a very low diffusion coefficient. The inaccessibility to diazotization, the proximity to the bases, and the apparent rigidity of the binding suggests that a weak binding to an external group such as the phosphates of the polynucleotide chain is not likely. When the complex of acridine orange and DNA is oriented by flow through a capillary tube (see Chapter 13) and the plane of polarization of acridine orange fluorescence is determined, it is found that the plane of the bound acridine orange ring is perpendicular to the helix axis. Because the bases are also perpendicular to the helix axis and the resonance energy transfer observed previously indicates that the acridine orange molecules and the bases are very close together, a reasonable interpretation of the data is that acridine orange is intercalated between the base pairs. This is consistent with the weak binding of acridine orange to single-stranded DNA and with numerous other physical measurements.

Other Applications of Fluorescence Polarization

Polarization of fluorescence can be used to determine the orientation of molecules in *rigid* systems and the hydrodynamic properties of certain solutions of molecules. Some examples of these uses follow.

Example 15-P. Orientation of chlorophyll in chloroplasts by polarization of intrinsic fluorescence.

The polarization of the red fluorescence of chlorophyll in a single chloroplast can be measured with a fluorescence polarization microscope (see Chapter 2). From measurements of artificially immobilized chlorophyll, the angles between the plane of polarization of the fluorescence to the axes of the chlorophyll is known. Hence, from the observed plane of polarization of the fluorescence of the chloroplasts, the orientation of the chlorophyll molecules in the chloroplasts can be determined.

Example 15-Q. Orientation of DNA in chromosomes by polarization of extrinsic fluorescence.

In Example 15-O, it was shown that the fluor acridine orange intercalates between the base pairs of DNA and has its plane perpendicular to the helix axis. This dye also binds intracellularly to eukaryotic chromosomes. The polarization of acridine orange fluorescence with respect to the axis of stretched chromosomes has been measured with a fluorescence polarization microscope and used to determine the orientation of DNA in the chromosomes.*

Example 15-R. Determination of viscosity within living cells.

If a fluorescent molecule is reorienting very rapidly compared with its lifetime, its fluorescence will show little or no polarization (consequence 1 in Table 15-2). However, if it is in a very viscous medium so that its movement is greatly impaired, polarization will be observed. If the lifetime of the fluorescence is roughly known (and indeed lifetimes are measurable), the viscosity can be determined. Note that here we are not concerned with the properties of a macromolecule but merely with determining the viscosity of a liquid that might be inaccessible to direct viscometry. Intracellular viscosity has been measured in both mouse ascites cells and in the bacterium *E. coli,* by allowing them to take up a fluorescent amino acid, aminonaphthylalanine. A small polarization appears; because it is known that this amino acid does not bind to any large molecules in the cells, the polarization can be used to calculate intracellular viscosity. This is of interest because it allows estimation of the rates of diffusion within cells and determination of whether certain diffusion-limited reactions may participate in various regulatory pathways.

Polarization Analysis with Pulsed Excitation

Fluorescence polarization analysis can be used in a quantitative way to determine rotational diffusion coefficients. For this purpose, the standard method of measuring polarization of fluorescence has two disadvantages: (1) it measures an average value, which is undesirable if a molecule has

---

*J. W. MacInnes and R. B. Uretz, *Science* 151(1966):689–691.

two rotational coefficients (e.g., a rodlike molecule can rotate about its long axis or tumble end-over-end); and (2) it requires measurement of the intrinsic polarization, which is accomplished by studying the polarization in solutions of increasing viscosity and extrapolating to infinite viscosity. Such extrapolation often fails because the substances added to increase viscosity sometimes affect conformation. These problems are avoided by the *nanosecond pulse technique,* in which excitation is accomplished with a nanosecond (10^{-9}) pulse of polarized light, and the intensities of the emitted light in directions parallel ($I_{||}$) and perpendicular ($I_{\perp}$) to the direction of excitation are recorded. With a rotating molecule, there is an initial difference in these intensities, which decays in a few nanoseconds as the molecule tumbles owing to Brownian motion. The decay rate is related to the rotational diffusion coefficient.

The data are analyzed in the following way. The emission anisotropy, A, which is defined below, is measured as a function of time;

$$A(t) = \frac{I_{\perp}(t) - I_{||}(t)}{I_{\perp}(t) + 2I_{||}(t)} = A_0\, e^{-3t/\rho} \tag{8}$$

in which A_0 is the anisotropy at the instant of excitation and ρ is the rotational relaxation time. A plot of log A versus t is usually linear and has a slope of $-3/\rho$. If the substance has two rotational relaxation times, this will show up as a log A versus t curve consisting of two connected lines (Figure 15-14). An example of the utility of this method follows.

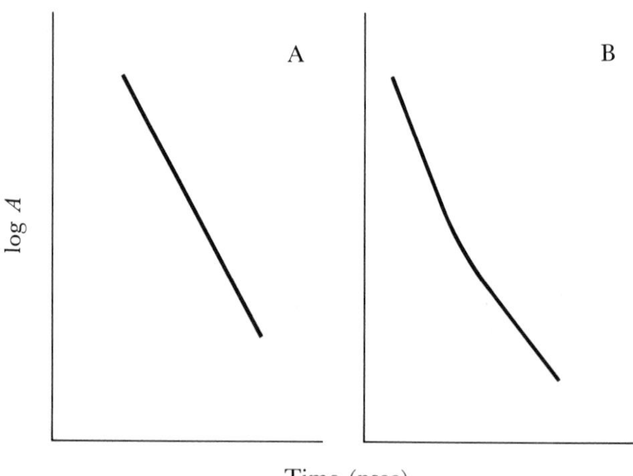

FIGURE 15-14
Plots of logarithm of the emission anisotropy versus time for a molecule having (A) one and (B) two rotational relaxation times.

Example 15-S. Determinants of ρ for chymotrypsin.
An anthraniloyl group can be coupled to the active site of α-chymotrypsin. Using the nanosecond pulse technique, a single value of ρ of 52 nsec is found. This value of ρ could reflect either flexibility of the site of binding of the anthraniloyl group or rotation of the entire protein (consequence 2 in Table 15-2). If it is assumed that the anthraniloyl group is moving virtually independently of the protein, ρ can be calculated (from the rotational relaxation time of anthranilic acid) to be approximately 1 nsec, which is incompatible with the measured value. If it is assumed that the protein is a rotating unhydrated sphere, then $\rho = 3\eta V/kT$, in which η is the solvent viscosity, V is the volume of the sphere, k is the Boltzmann constant, and T is the absolute temperature. Using this equation and determining V from the molecular weight, a value of $\rho = 22$ nsec, which is also too small, is calculated. Hence, chymotrypsin must have a larger V and must be either hydrated or nonspherical. This high value of ρ also shows that the active site must be rigid.

SPECIAL USES OF FLUORESCENCE IN BIOLOGY AND BIOCHEMISTRY

Fluorescence is of great use in obtaining information other than the conformation of macromolecules or the arrangement of molecules within cells (e.g., the localization within cells and tissue or the quantitative assay of minute amounts of material). Examples of some of these applications follow.

Example 15-T. Fluorescence microscopy.
Many fluors are localized intracellularly and can be detected by fluorescence microscopy. A specimen is illuminated with an exciting wavelength and observed through a filter that excludes the exciting light and transmits the fluorescence. Acridine orange binds to nucleic acids and fluoresces green and orange if bound to DNA and RNA, respectively. This has been used with eukaryotes to observe nucleic acids and chromosomes and to detect RNA in the nucleus; it has been used with prokaryotes to localize DNA. See Chapter 2 and Figures 2-19 and 2-21 for details.

Example 15-U. Fluorescent antibody method.
An antibody to a particular substance (e.g., a viral antigen or a cell-wall component) is conjugated to a fluorescent dye. If the antibody is incubated with cells containing the antigen and then washed away,

examination of the cells by fluorescence microscopy will show fluorescence only where the antigen is present. This method is widely used to detect tumor antigens and to identify intracellular viruses. It is described more fully in Chapter 2 and an example is given in Figure 2-20.

Example 15-V. Assay of S–S bonds and SH groups.
In 1 N NaOH, the fluorescence of fluorescein mercuric acetate is quenched by disulfide bonds in protein. This is a relatively accurate means to assay these bonds. A protein with a known number of S–S bonds and SH groups (e.g., ribonuclease) must be used for calibration (Figure 15-15). At neutral pH, the fluorescence is quenched instead by SH groups but not S–S bonds.

Example 15-W. Enzyme assays.
Fluorescence assays in which nonfluorescent substances are converted into fluorescent ones, or vice versa, exist for numerous hydrolytic en-

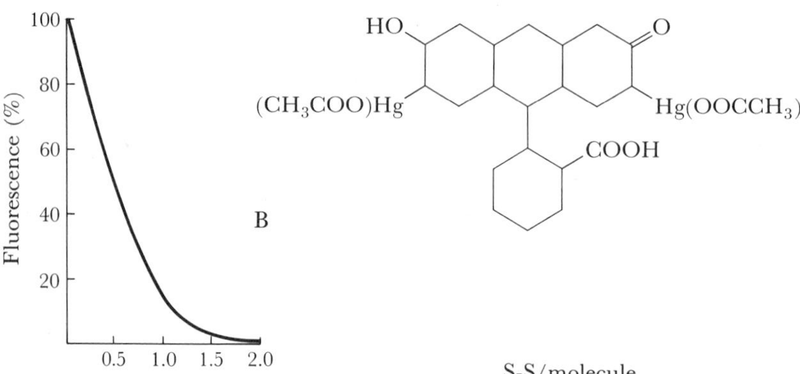

S-S/molecule

Protein	FMA titration	Chemical method
Human serum albumin	18	17
Lysozyme	3.9	4
Trypsin	5.9	6
Insulin	3.0	3
Chymotrypsinogen	5.3	5

C

FIGURE 15-15
A. Measurement of S–S bonds by the quenching of fluorescence of fluorescein mercuric acetate. B. The structure of the fluor. C. A comparison of the results with other methods. [From F. Karush, N. R. Klinman, and R. Marks, *Analyt. Biochem.* 9(1964):100–114.]

zymes (e.g., cholinesterase, lipase, hyaluronidase, and β-galactosidase), oxidative enzymes (e.g., aryl hydroxylases, peroxidases, and oxidases), transaminases, dehydrogenases, isomerases, kinases, and decarboxylases. An example of the extraordinary sensitivity of such assays can be seen in the work of Boris Rotman (*Proc. Nat. Acad. Sci.* 47(1961):1981–1991), who measured the activity of a *single* molecule of β-galactosidase. Purified β-galactosidase was diluted and dispersed into droplets of roughly 10^{-9} ml—each containing a known concentration of the nonfluorescent substance fluorescein di-(β-D-galactopyranoside). Hydrolysis of this substance yields fluorescein. The fluorescence of each droplet was measured with a microscope equipped with a photomultiplier and suitable apertures so that a single drop could be observed. In these droplets it was possible to detect fluorescein at 2×10^{-6} M, or 1.7×10^{6} molecules per droplet. Because one enzyme molecule hydrolyzes roughly 2×10^{5} molecules of substrate per hour, the fluorescein could be detected after approximately 10 hours. This method was then used to determine the number of enzyme molecules in a single *E. coli* bacterium.

Example 15-X. Quantitative measurement of DNA.

The fluor ethidium bromide binds tightly to DNA; in so doing, the quantum yield increases substantially. This increase is linear throughout a wide range and the measurement of fluorescence intensity of an ethidium bromide solution containing small amounts of DNA can be used in a quantitative way to measure DNA concentration.

This enhancement of Q is also being used to detect DNA that has been electrophoresed in polyacrylamide and agarose gels (see Chapter 9, Figure 9-11). If a gel containing DNA is soaked in an ethidium bromide solution, it will take up the fluor. If the gel is then exposed to exciting light, the DNA bands become visible as regions of intense fluorescence.

Selected References

Chen, R. F., H. Edelhoch, and R. F. Steiner. 1973. "Fluorescence of Proteins," in *Physical Principles and Techniques of Protein Chemistry,* part A, edited by S. J. Leach, pp. 171–244. Academic Press.

Kronman, M. J., and F. M. Robbins. 1970. "Buried and Exposed Groups in Proteins," in *Fine Structure of Proteins and Nucleic Acids,* vol. 4, edited by G. D. Fasman and S. N. Timasheff, pp. 271–416. Dekker.

Pesce, A. J., C. G. Rosen, and T. L. Pasby. 1971. *Fluorescence Spectroscopy.* Dekker.

Stryer, L. 1968. "Fluorescence Spectroscopy of Proteins." *Science* 162:526–540. A good review of energy transfer.

Timasheff, S. N. 1970. "Some Physical Probes of Enzyme Structure in Solution," in *The Enzymes,* vol. 2, edited by P. D. Boyer, pp. 371–443. Academic Press.

Udenfriend, S. 1962, 1969. *Fluorescence Assay in Biology and Medicine,* vols. 1 and 2. Academic Press.

Weber, G. 1952. "Polarization of the Fluorescence of Macromolecules: 1. Theory and Experimental Methods; 2. Fluorescent Conjugates of Ovalbumin and Bovine Serum Albumin." *Biochem. J.* 51:145–155; 155–167. The classic papers on fluorescence depolarization.

Weber, G., and F. W. J. Teale. 1966. "The Interaction of Proteins with Radiation," in *The Proteins,* vol. 3, edited by H. Neurath, pp. 445–452. Academic Press.

Problems

15-1. A protein causes ANS to fluoresce. If the protein concentration is increased before adding ANS, the fluorescence decreases. Give two possible explanations for this decrease.

15-2. Iodide quenching of tryptophan fluorescence can be used to determine whether tryptophans are exposed to the solvent.

If a protein is known to contain only one tryptophan and iodide fails to quench, what possible explanations might be given to account for the lack of quenching?

If the protein contains eight tryptophans and iodide quenches 25% of the fluorescence, it is tempting to assume that two tryptophans are accessible to the solvent. State several factors that could make this conclusion invalid.

15-3. When excimers are formed, the excitation spectrum is that of the monomer because the monomer is excited before dimerization. If many dimers are present and the dimer is excited, do you expect the excitation spectrum to match the absorption spectrum of the monomer? Explain.

15-4. Iodide quenching decreases fluorescence intensity. Would you expect there to be a change also in the shape of either the excitation or emission spectrum?

15-5. Give several possible mechanisms for an increase in the quantum yield of a fluor when bound to another molecule. Why might there sometimes be a shift in the excitation and/or emission spectra? Could the shift be to either longer or shorter wavelengths?

15-6. Is the excitation spectrum of a fluor always the same as the absorption spectrum? Explain.

15-7. A protein containing ten tryptophans shows fairly strong fluorescence of tryptophan. A small molecule known to bind tightly to the protein produces virtually no change in the fluorescence, even though it is known that there are two tryptophans in the binding site. Give several possible explanations for this.

15-8. When the fluor acridine orange is bound to DNA and the mixture is irradiated with light absorbed by the acridine orange, irreversible chemical changes (e.g., broken purine rings, single-strand breaks) occur in the DNA. For these changes to occur, molecular O_2 must be present; hence, the process is called photosensitized oxidation, or photo-oxidation. These reactions occur with many acridine derivatives; the efficiency of the reaction is roughly proportional to the quantum yield of fluorescence. Propose a mechanism for photo-oxidation.

15-9. A fluor is covalently coupled to a protein. The polarization of the fluorescence is measured as a function of the ionic strength of the suspending buffer; it is found to decrease markedly as the ionic strength increases. What effect does increasing ionic strength have on the protein?

15-10. A protein has a fluor F covalently attached at a unique site. When tryptophan is excited, tryptophan fluorescence is very weak but the fluor emits strongly. Explain each of the following possible observations. (*Note:* These are *alternative* and mutually exclusive observations.)

Iodide quenches the fluorescence of F but not of tryptophan.

Iodide quenches the fluorescence of F *and* of tryptophan.

Adjustment to pH 9 eliminates the fluorescence of F and enhances that of tryptophan (assume that there is no pH effect on free tryptophan or unbound F).

Adjustment to pH 9 eliminates all tryptophan fluorescence and markedly enhances the fluorescence of F.

CHAPTER 16

Optical Rotatory Dispersion and Circular Dichroism

Another set of techniques for elucidating both the conformation of a macromolecule or macromolecular complex in solution and the interactions between macromolecules is described in this chapter. Although absorption spectroscopy is capable of supplying a great deal of useful information of this sort, by studying the absorption of *polarized* light—that is, the technique of optical rotatory dispersion (ORD) and circular dichroism (CD) spectroscopy, both of which satisfy the criteria of speed and applicability to solutions—even greater information can be obtained (although at the cost of somewhat greater instrumental and theoretical complexity than in absorption spectroscopy). These methods measure the wavelength dependence of the ability of an optically active chromophore to rotate plane-polarized light (ORD) and the differential absorption of right and left circularly polarized light (CD). The physical basis of ORD and CD is the same and, in fact, they are merely two different ways of looking at the same interaction of polarized light with *optically active* molecules. Because a very large fraction of biological molecules contain optically active centers, ORD and CD have great applicability to their study.

In this chapter, it will be seen that, because the ORD and CD spectra of proteins and nucleic acids result primarily from the spatial asymmetry of the constituent amino acids and nucleotides, respectively, in the backbones of the macromolecules, ORD and CD are useful in structural studies with proteins, nucleic acids, and nucleoproteins. However, once again the theory connecting the spectra with molecular structure is not yet fully developed; thus, the working rules used to interpret spectra (like those of absorption spectroscopy) are for the most part empirical.

SIMPLE THEORY OF ORD AND CD

In the preceding chapter, it was explained that light is an electromagnetic wave consisting of an oscillating electric (E) field and a magnetic (H) field, both of which can be represented by mutually perpendicular vectors (Figure 16-1A). The *plane of polarization* is defined as the plane of the **E** vector. Because a light source usually consists of a collection of randomly oriented emitters, the emitted light is a collection of waves with all possible orientations of the **E** vectors. Plane-polarized light is usually obtained by passing light through an object (e.g., a Polaroid screen or a Nicoll prism) that transmits light with only a single plane of polarization (Figure 16-1B).

Suppose that two plane-polarized waves differing in phase by one-quarter wavelength (i.e., when one sine curve crosses the axis of propagation, the other is at a maximum or minimum), whose **E** vectors are perpendicular to one another, are superimposed. As the waves propagate forward, the resultant **E** vector rotates so that its tip follows a helical path (Figure 16-2). This is, of course, also true of the magnetic-field vector. Such light is called *circularly polarized* and is defined as *right* circularly polarized if the **E** vector rotates clockwise to an observer looking at the source.

If a right (R) and a left (L) circularly polarized wave, both of equal amplitude, are superimposed, the result is plane-polarized light, because at any point in space the **E** vector of each will sum as shown in Figure 16-3A. Similarly, plane-polarized light can be decomposed into R and L components. If the amplitudes of the two circularly polarized waves are not the same, the tip of the resultant **E** vector will follow an elliptical path and such light is said to be elliptically polarized (Figure 16-3B). A parameter called the ellipticity, θ, is often used to describe the elliptical polarization. This is the angle whose tangent is the ratio of the minor and major axes of the ellipse shown in Figure 16-3B—that is, $\theta = \tan^{-1} b/a$ [see equation (6)].

When a beam of light (i.e., a propagating electromagnetic field) passes through matter, the electric (**E**) vector of the propagating wave interacts with the electrons of the component atoms. This interaction has the effect of reducing the velocity of propagation (also called *retarding* the light) and in decreasing the amplitude of the **E** vector. Reducing the velocity

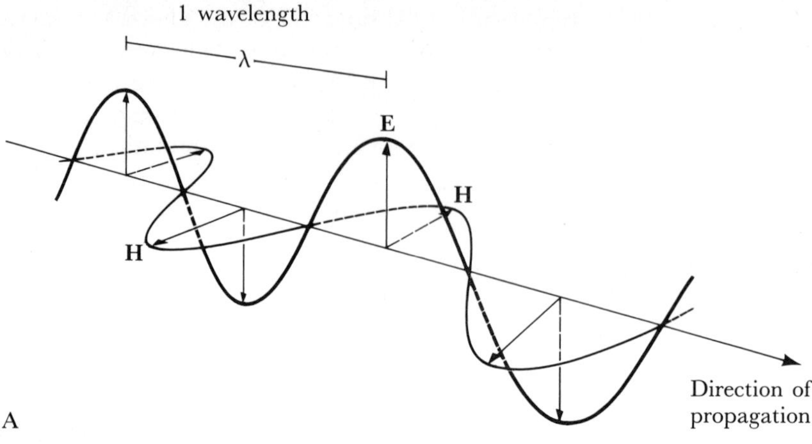

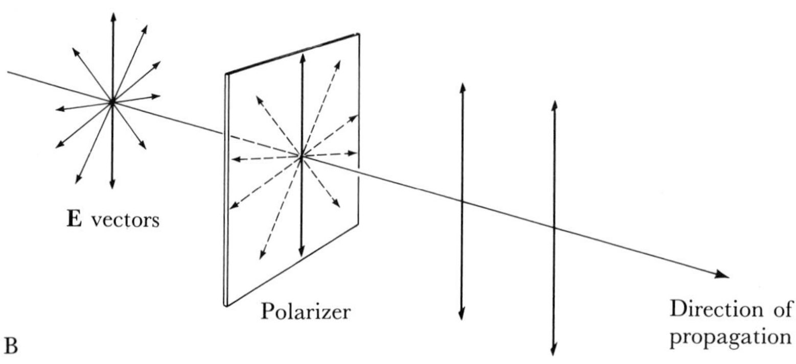

FIGURE 16-1
A. Propagation of an electromagnetic wave. The path of the tip of the **E** vector is indicated by the heavy line; that of the **H** vector by the light line. The **E** and **H** vectors and the direction of propagation are mutually perpendicular. The plane of the **E** vector is called the plane of polarization. B. Production of plane-polarized light. A collection of waves falls on the polarizer, which passes only those components of the **E** vector that are parallel to the axis of the polarizer.

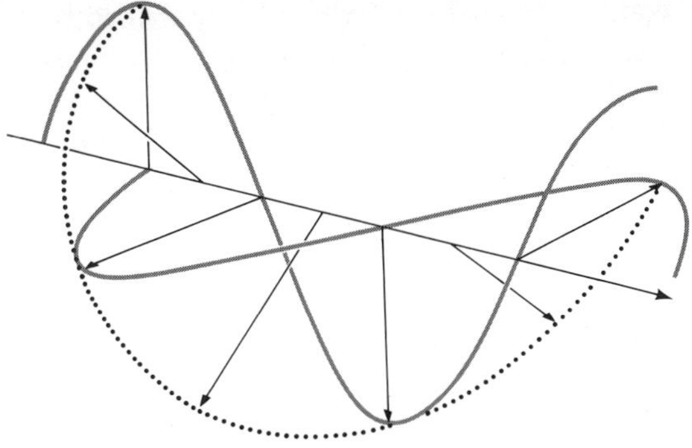

FIGURE 16-2
Generation of circularly polarized light. The **E** vectors of two electromagnetic waves are one-quarter wavelength out of phase and are perpendicular. The vector that is the sum of the **E** vectors of the two components rotates so that its tip follows a helical path (dotted line).

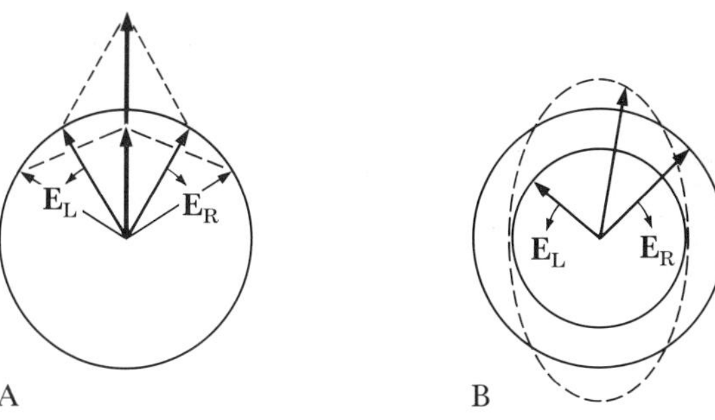

FIGURE 16-3
Diagrams showing how right and left circularly polarized light combine: (A) if the two waves have the same amplitude, the result is plane-polarized light; and (B) if their amplitudes differ, the result is elliptically polarized light—that is, the head of the resultant vector will trace the ellipse shown as a dashed line. The lengths of the major and minor axes of the ellipse are a and b (see page 445).

of propagation is called *refraction* and is described by the *index of refraction, n,* and decreasing the amplitude of the **E** vector is *absorption* and is described by the *molar extinction coefficient,* ε. Both n and ε depend on wavelength in a way that reflects the electronic structure and geometry of the molecules.

For most substances, simple refraction and absorption is the only detectable result of such an interaction even if the light is polarized. However, the behavior of some molecules is sensitive to the plane of polarization of the incident light. Such a molecule or chromophore is called *optically active* and is characterized by having distinct indices of refraction, n_L and n_R, and molar extinction coefficients, ε_L and ε_R, for left and right circularly polarized light, respectively. Optical activity is a characteristic of many organic and almost all biological molecules. The property that determines whether a chromophore is optically active is its *asymmetry.* If a molecule is asymmetric *in the sense that it cannot be superimposed on its mirror image,* it is optically active. Examples of optically active structures are given in Figure 16-4. The physical basis of optical activity is difficult to explain without the formalism of electrodynamics and quantum mechanics and will not be explained here. It can be found in the Selected References near the end of this chapter.

Let us consider the interaction of an optically active substance with polarized light. As indicated earlier, a plane polarized wave can be thought of as a mixture of L and R circularly polarized light. The entire interaction can be thought of in terms of these circularly polarized components and, in fact, it is easier to do so. If a substance retards both L and R equally

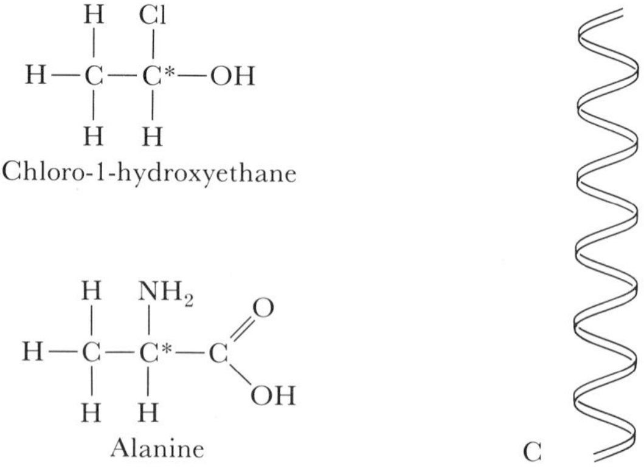

FIGURE 16-4
Examples of several optically active structures. In parts A and B, the asymmetric carbon is identified with an asterisk.

(i.e., if the indices of refraction for L and R circularly polarized light, n_L and n_R, are the same), the L and R waves will recombine on leaving the substance to form plane-polarized light, with the plane of the transmitted beam being the same as that of the incident beam. However, if n_L and n_R are unequal, the transmitted L and R components are each retarded to a different extent so that on leaving the material the phases of the two sine waves differ (Figure 16-5). Henceforth, at any point in space, the **E** vectors of the L and R waves combine to form a beam of plane-polarized light whose angle differs from that of the plane of polarization of the incident wave; hence, the plane of polarization of the resulting wave will be rotated. For any substance that interacts with light in this asymmetric way, the extent of the rotation produced by a sample of a given volume depends on the number of chromophores with which the wave interacts—that is, on the concentration of the molecules multiplied by the path length, d, and on the wavelength, λ, of the light—because n is always a function of λ.

Quantitatively, the observed angle of rotation, α_λ, expressed in degrees, can be described by

$$\alpha_\lambda = \frac{180\, d}{\lambda}(n_L - n_R) \tag{1}$$

For most work, the terms *specific rotation,* $[\alpha]_\lambda$, and *molar rotation,* $[M]_\lambda$, are used. These are defined as

$$[\alpha]_\lambda = \frac{\alpha_\lambda}{dc} \tag{2}$$

in which α_λ is the observed rotation in degrees, d is the light path in decimeters, and c is the concentration in grams per milliliter; and

$$[M]_\lambda = \frac{\alpha_\lambda M}{100\, dc} \tag{3}$$

in which M is the molecular weight in grams per mole. If the substance under study is a polymer, as is often the case in biochemistry, it is more common to use the *mean residual rotation,* $[m]_\lambda$, in which

$$[m]_\lambda = \frac{\alpha_\lambda M_0}{100\, dc} \tag{4}$$

and M_0 is the mean residue molecular weight (i.e., the molecular weight of the polymer divided by the number of monomers).

A curve that shows the wavelength dependence of optical rotation, expressed either as α, $[\alpha]$, $[M]$, or $[m]$, is called an optical rotatory dispersion spectrum.

So far, only the *retardation* of R and L waves has been considered, but it is also of interest to know what happens to the *intensity* of each as these waves pass through matter. If the substance is optically *inactive,* the absorption of each is equal. If, on the other hand, the material is optically active, then in the range of wavelengths in which absorption occurs (the absorption band) there will be, for each wavelength, differential absorption of the L and R circularly polarized light. This difference is usually expressed in terms of the extinction coefficients for L and R light, ε_L and ε_R; that is,

$$\varepsilon_L - \varepsilon_R = \Delta\varepsilon \tag{5}$$

in which $\Delta\varepsilon$ is called the *circular dichroism,* or CD. It is positive if $\varepsilon_L - \varepsilon_R > 0$ and negative if $\varepsilon_L - \varepsilon_R < 0$. An important point is that, if a given optically active molecule has positive CD, then its mirror image will have a negative CD of precisely the same magnitude. Because differential absorption of the L and R waves means that the amplitudes of the transmitted waves will differ, the result, as shown in Figure 16-3B, is elliptically polarized light. Experimentally, it is usual to measure $\Delta\varepsilon$, but for historical reasons the ellipticity, θ, is plotted; θ is related to $\Delta\varepsilon$ by the equation

$$\theta = 3300\ \Delta\varepsilon \tag{6}$$

A curve showing the dependence of θ on wavelength is called a *CD curve* or *CD spectrum.*

FIGURE 16-5
Rotation of the plane of polarization: (A) both right and left circularly polarized light are retarded equally so that the resultant **E** vector remains in the same plane; (B) the left circularly polarized light is retarded more than the right so that the resultant **E** vector changes orientation (because the amount of retardation is proportional to the distance the light travels, the **E** vector will rotate clockwise with increasing optical path length); (C) physical significance of the retardation of a light wave by passing through media of different indices of refraction and traveling equal distances *ad.* As a consequence of the lower velocity of light in refractive media and the invariant frequency of light, more cycles of vibration are squeezed in a given path length in the refractive media than in air ($n=1.00$). Note that all three beams are in phase up to position *b.* In this example, if $n=2.00$, the beam emerges at *c* in phase with the unretarded beam. If $n=1.83$, it is exactly out of phase. [From E. Slayter, *Optical Methods in Biology*, Wiley-Interscience, 1970.]

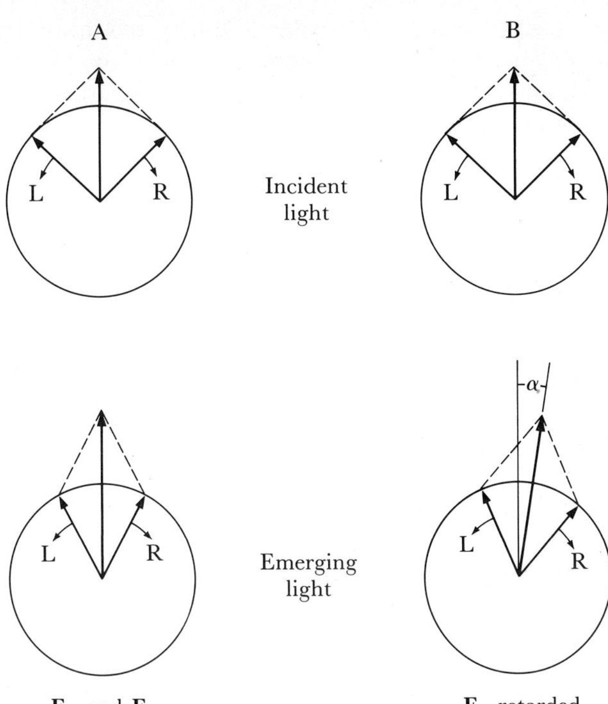
A
B
L
R
Incident light
L
R
L
R
Emerging light
α
L
R
E_L and E_R retarded equally
E_L retarded more than E_R

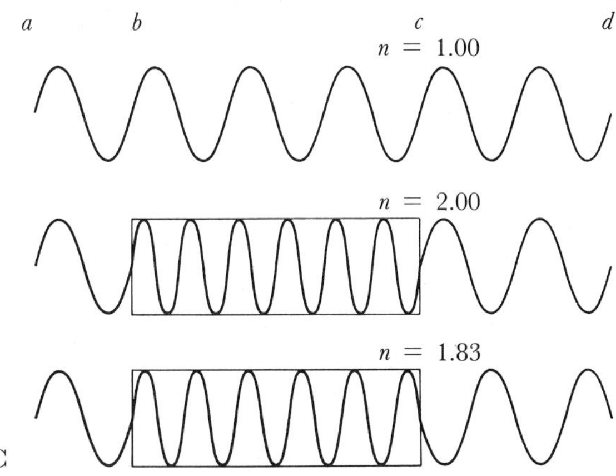
a
b
c
d
n = 1.00
n = 2.00
n = 1.83
C

For most work, the terms molar ellipticity and mean residue ellipticity are used. Unfortunately, in contrast with molar rotation and mean residual rotation in optical rotatory dispersion [equations (3) and (4)], the same notation is used for both of these terms:

$$[\theta]_\lambda = \frac{M\theta_\lambda}{10\ dc} \tag{7}$$

in which θ_λ is the observed ellipticity in degrees, M is the molecular weight *or* mean residue molecular weight, d is the path length in centimeters, and c is the concentration in grams per milliliter.

It is useful to see the relation between an ordinary dispersion curve (i.e., the index of refraction, n, versus wavelength, λ), the ORD curve (which is, in fact, $\Delta n = n_L - n_R$ versus λ), a standard absorption curve (ε versus λ), and a CD curve ($\Delta\varepsilon = \varepsilon_L - \varepsilon_R$ versus λ). The types of curves obtained by examining a single absorption band of an optically active chromophore are shown in Figure 16-6. The solid curve of Figure 16-6A shows the spectrum for the absorption of nonpolarized light; for example, for a single electronic transition from the ground state to the first excited state. The wavelength corresponding to maximum absorption, λ_0, is often called the absorption peak or λ_{max}. This absorption band can also be characterized by a value, Δ, or the half-width at $1/e$ times the maximum height. The solid line of Figure 16-6B shows the index of refraction, n, for nonpolarized light as a function of wavelength. The principal features of this curve are that (1) $n = 1$ at $\lambda = \lambda_0$; (2) the curve to the left of λ_0 can be superimposed on that to the right by means of a 180° rotation; (3) the wavelengths corresponding to maximum and minimum n are approximately $\lambda_0 + 0.9\Delta$ and $\lambda_0 - 0.9\Delta$, respectively; and (4) at wavelengths outside of the absorption band, n approaches 1 asymptotically. If the chromophore is optically active, the curves for CD and ORD, respectively, are essentially the same except that the y-axes are $\Delta\varepsilon = \varepsilon_L - \varepsilon_R$ and $\Delta n = n_L - n_R$. The ORD curves are usually presented with the y-axis being some measure of rotation (i.e., $[\alpha]$ or $[M]$), but it must be remembered that the rotation is a result of the different indices of refraction of L and R circularly polarized light.

An ORD curve of the type shown in Figure 16-6B is called a *Cotton effect.* Such curves can be negative or positive, as shown in Figure 16-6D. (If an optically active molecule has a positive Cotton effect, its mirror image will have a negative Cotton effect of precisely the same shape.) Each Cotton-effect curve consists of two extremes called a *peak* and a *trough;* the magnitude of Δn at the wavelength corresponding to the maximum of a peak is always the same as that for a trough, except with the opposite sign. Two other parameters used to describe a Cotton effect are the amplitude and breadth, as shown in Figure 16-6D. The parts of the Cotton effect that tail

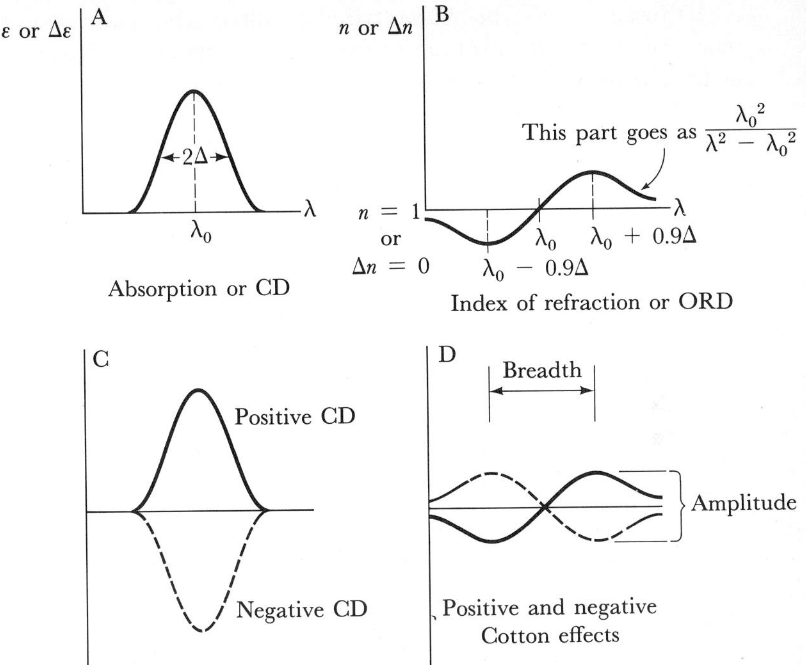

FIGURE 16-6
Relation between (A) an absorption spectrum (ε versus λ) and circular dichroism (Δε versus λ) and between (B) a dispersion curve (*n* versus λ) and optical rotatory dispersion (Δ*n* versus λ) for an optically active substance (note that the symbol Δ is used to denote both the y-axis and the band width—a common notation, unfortunately; (C) the CD curve drawn as a solid line is called positive CD, and the dotted line is negative CD; (D) the ORD curve drawn as a solid line is called a positive Cotton effect, and the dotted line is a negative Cotton effect. If the substance is not optically active, the *n*-versus-λ and ε-versus-λ curves are of the same type, but Δε and Δ*n* equal zero at all λ.

outward from the absorption band are called *plain curves.* These curves are only rarely studied now but have the occasionally useful property that they extend very far from the absorption band and can therefore be used to determine approximately some of the parameters of a molecule in a solvent that absorbs strongly in the region near λ_0. For proteins and polypeptides, the principal Cotton effect occurs in the wavelength range near 200 nm and is caused by the absorption of the peptide bond. For nucleic acids, it occurs in the 250-to-275-nm range and is caused by electronic transitions of the nucleotide bases.

CD bands can also be positive or negative, as shown in Figure 16-6C, and a positive CD band always corresponds to a positive Cotton effect.

A CD band can also be characterized by the height—that is, the magnitude of θ at λ_0—which is to some extent a measure of the degree of asymmetry. However, with CD it is more common to refer to the *rotational strength, R,* of a band—the area under the $\Delta\varepsilon$-versus-λ curve. Rotational strength is difficult to evaluate precisely and is normally calculated approximately for a Gaussian band as

$$R \sim 1.23 \times 10^{-42}[\theta_0]\frac{\Delta}{\lambda_0} \tag{8}$$

in which λ_0 is the wavelength of the peak of the CD curve, $[\theta_0]$ is $[\theta]$ at λ_0, and Δ is the half-width of the band at 1/e the height. It can also be measured with a curve analyzer.*

The rotational strength describes the *intensity* of a CD band and physically tells something about the motion of the electrons when the absorbing center is raised from the ground state to an excited state by the absorption of light; it is not necessary to understand in detail the factors that determine the magnitude of R in order to interpret CD data. The main rule is the R is not zero in an optically active substance and that it generally increases with increasing asymmetry.†

RELATIVE VALUES OF ORD AND CD MEASUREMENTS

It should be apparent by now that ORD and CD are both manifestations of the same underlying phenomenon. In fact, as might be expected, ORD and CD spectra can be generated from one another. This is done by means of a mathematical conversion called the general Kronig-Kramers transformation, which can be found in the specialized texts referred to near the end of the chapter.

The degree to which ORD and CD have been used to obtain information about macromolecules has been determined primarily by the availability of the necessary instrumentation. At first, only ORD instruments were available. Furthermore, before about 1960, these instruments were operable only in a wavelength region well above the absorption bands for proteins and nucleic acids. The result was that only the plain curve was

---

*A curve analyzer is an electronic instrument that can (1) measure the area of a curve and (2) decompose a curve into a collection of Gaussian curves whose sum is the original curve.

†A corollary derived from this statement is that sometimes a substance either without optical activity or with a low R can achieve very high R by interacting with a polar asymmetric molecule.

available for study. These ORD curves were usually analyzed by a relation called the Drude equation:

$$[\alpha]'_\lambda = \frac{A}{\lambda^2 - \lambda_0^2} \tag{9}$$

in which $[\alpha]'_\lambda = \alpha_\lambda/dc$ [see equation (1)], and λ is the wavelength at which $[\alpha]'_\lambda$ is measured; λ_0 was called the *dispersion constant* (often written λ_c) and later proved to be the wavelength of the center of the absorption band; A was called the *rotatory constant* and was later shown to equal $2\ R\lambda_0^2/(0.696 \times 10^{-42}\ \pi)$, in which R is the rotational strength. It was clear that this equation could not be correct because it predicted infinite rotation at $\lambda = \lambda_0$; nonetheless, it was all that was available at the time and some information was actually obtainable from the determination of λ_0 and A. Various empirical studies showed that λ_0 and A could be related in certain ways to conformation of macromolecules. During this period of rather inadequate instrumentation, other equations were used to analyze the data. The most notable of these was the Moffitt equation, a multiparameter equation that attempted to relate the shape and magnitude of the plain curve to the fraction of amino acid residues in a protein that were in the α-helical and random-coil configurations. In later years, the instrumentation of ORD was improved so that the range of measurements extended to the far ultraviolet and allowed studies in regions of absorption—that is, observation of the Cotton effect. As will be seen in later parts of this chapter, analyses of Cotton effects have provided a large amount of information.

However, in the late 1960s, instruments for measuring CD became available and the simplicity of the CD curve compared with the Cotton-effect curve (see Figure 16-6) makes CD analysis vastly superior to ORD analysis; at the present time, ORD is rarely used.

The principal advantage of CD analysis is its greater ability to resolve bands due to different optically active transitions. This is best seen by examining curves consisting of more than one band. First, referring to Figure 16-6, which shows the ORD and CD for a particular optically active substance, it should be noted that at no point is a flat baseline reached in an ORD curve; the curves are asymptotic. The CD curve, however, has a defined zero baseline outside of the absorption band. Furthermore, CD bands are narrow, allowing fairly good resolution of nearby bands.

Figure 16-7A shows the ORD and CD spectra that would result if the molecule contained two optically active centers giving rise to two Cotton effects very near one another. Note that the CD spectrum clearly shows two elements, whereas the ORD curve is somewhat complex. Even more striking is Figure 16-7B, which shows the results for a molecule with a strong Cotton effect at one wavelength and a very weak one at a more

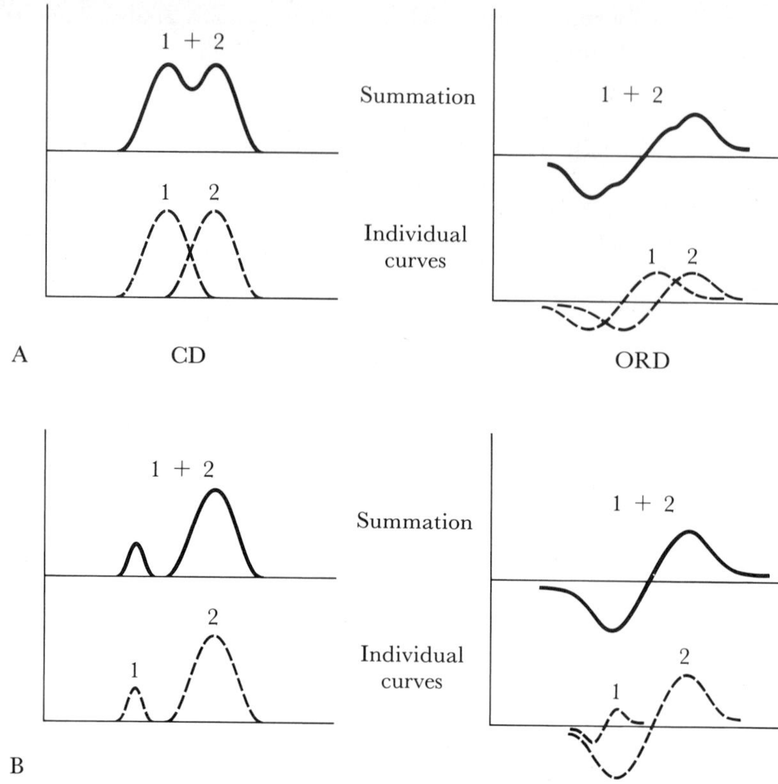

FIGURE 16-7
Comparison of the ability of ORD and CD to indicate that two optically active bands exist. In each case, the solid upper curve is the sum of the two dashed lower curves. In part A, the two curves have the same sign and amplitude but different λ_{max}. In part B, the amplitudes differ. In part C, the signs differ but the amplitudes are nearly the same. Part D is a comparison of absorbance (upper solid line), CD (dashed line), and ORD (lower solid line) of a system containing four bands. Note that the CD curve tells which bands are optically active and distinguishes the sign.

distant wavelength. Because ORD spectra totally overlap, the weak band is almost undetectable, whereas the CD spectrum with its narrow bands shows both bands quite clearly. In Figure 16-7C, the effect of the juxtaposition of positive and negative curves can be seen. These examples should serve to convince the reader that the resolution of CD spectra is superior to ORD analysis.

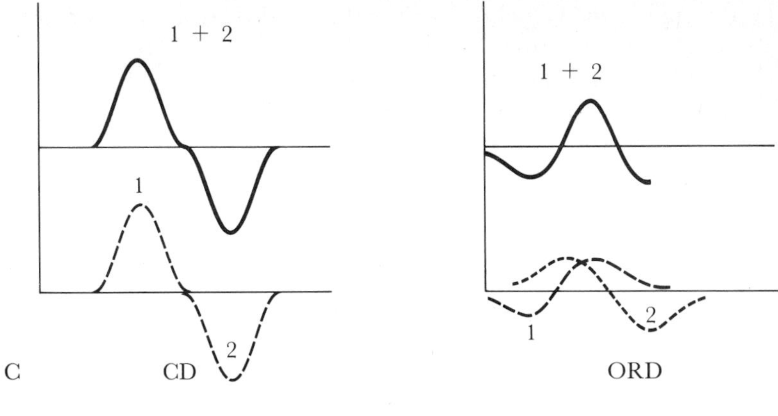

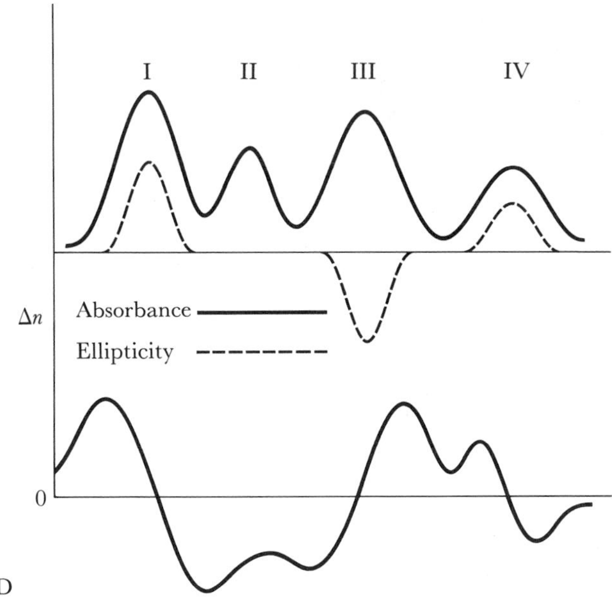

A second diagram (Figure 16-7D) serves to reinforce this conclusion in addition to indicating that CD analysis also provides greater information than simple absorption spectroscopy with unpolarized light. Note that the absorption spectrum merely indicates that the system consists of four chromophores (i.e., there are four bands), whereas the CD spectrum shows that three of the four (I, III, and IV are due to optically active centers.

The sign of the CD also describes the handedness of the chromophore. It can also be seen that the ORD fails to give clear evidence for whether band II is optically active because of the overlapping of the plain curves of the Cotton effects for bands I and III.

ORD measurements are presented in some of the examples of this chapter because many important results have been obtained in this way, despite the complexity of the spectra. However, in reading the recent scientific literature, the student will find that CD is now the method more frequently used.

TECHNIQUES FOR MEASURING ORD AND CD

Although films and solids are occasionally used, solutions are generally used for ORD and CD measurements. The solution is contained in a vessel called a cell. The basic instrumentation requires a light source whose wavelength can be varied, a system for polarizing the light, a system for measuring the polarization after the light has passed through the cell, and a detector by which the amount of light can be measured.

Figure 16-8 shows a simple model for an ORD instrument. A light

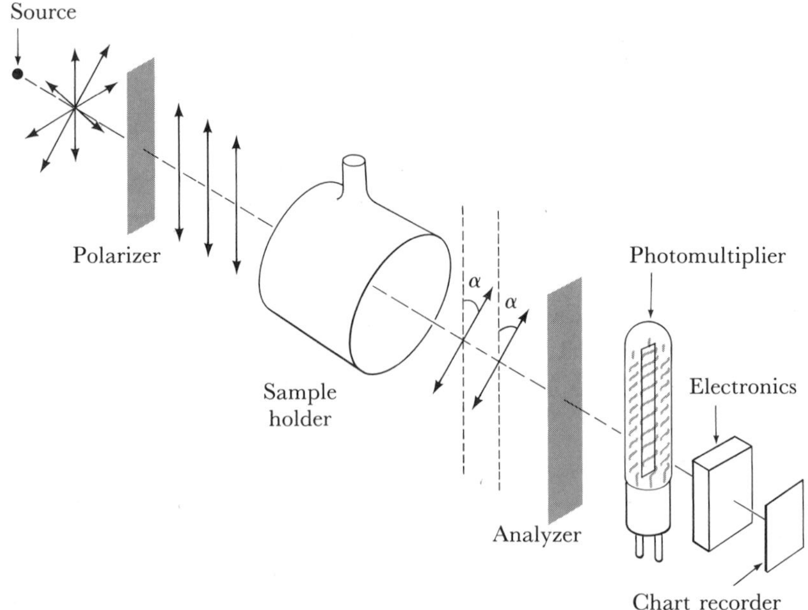

FIGURE 16-8
An ORD instrument.

source plus a monochromator is used to select the illuminating wavelength. This light passes through a polarizer, which produces plane-polarized light. An analyzer (which is, in fact, just another polarizer) is used to determine the angle of polarization. When the plane of the analyzer is parallel to that of the polarizer, the maximum intensity of light passes through. The transmitted light then falls on a photomultiplier tube, which converts the intensity into an electric current. If the cell is filled with a sample that rotates light, the analyzer must be rotated to allow the maximum passage of light. The angle at which the transmitted intensity is maximal defines the observed rotation in degrees, α_λ [see equation (1)]. In practice, modern automatic instruments simultaneously vary the wavelength, determine the rotation, and make a plot of α_λ versus λ. Different commercial instruments vary slightly in the way this is done, but the principle is as described.

For CD measurement (Figure 16-9), in principle two light sources are needed, one for L and the other for R circularly polarized light, each provided with a monochromator for wavelength selection. However, commercial instruments utilize a simple trick for generating L and R light from a single source. Plane-polarized light passes through a crystal that is subjected to an alternating electric field. This crystal (called an electrooptic modulator) has the remarkable property that the polarity of the field determines whether the L or R component of the light is transmitted. Because the field is alternating current, the beam continually modulates from the production of L to the production of R light. This beam then passes through the sample cell and falls on a photomultiplier. The output of the photomultiplier is then processed electronically in a fairly complex way to provide a voltage that is proportional to the ellipticity. This is automatically plotted as a function of wavelength to give the CD spectrum.

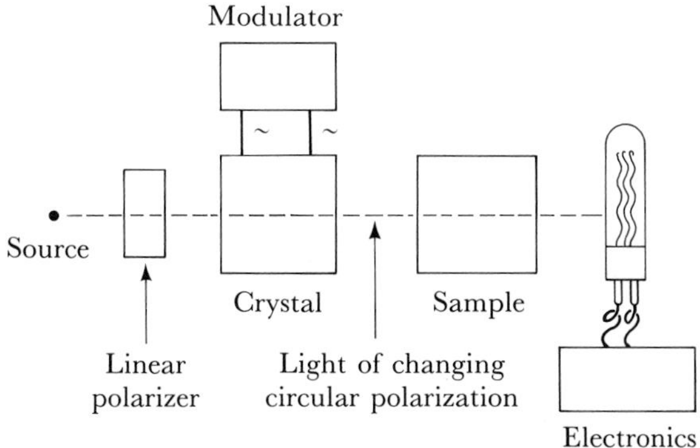

FIGURE 16-9
A CD instrument.

INTERPRETATION OF ORD AND CD CURVES

The theory of optical activity is not yet capable of yielding the precise structure of a protein from its CD spectrum, although, as will be seen later, it is somewhat better for nucleic acids. The complications are that very frequently it is not that the chromophore is asymmetric but that the chromophore is asymmetrically perturbed by neighboring groups. Furthermore, with proteins, there is the additional complication that, because the peptide bond (which is the principal element whose spectrum is detected by CD) exists in many conformations depending on its precise location in the protein, the spectrum is a result of an average of the various conformation parameters. Hence, in practice, an empirical approach of obtaining an ORD or CD spectrum for molecules whose structure is accurately known from x-ray diffraction is used, and the spectrum is related to the structural features of the molecule. This spectrum is then compared with the spectrum of a protein of unknown structure. The principal problem of this approach is that the (rarely proven) assumption must be made that the structure of a macromolecule in solution (remember that ORD and CD are determined in solution) is nearly the same as that of a fiber, crystal, or dry powder (as is used in x-ray analysis) prepared from the same solvent.

As a result of these problems, a set of working "rules" that are used to analyze spectra has been developed. Some of these rules are presented in Table 16-1. Rule 1 is primarily used to make an estimation of the basic conformation of a macromolecule (e.g., the fraction of a protein that is helical) or it may be used to confirm that a structure determined by x-ray

TABLE 16-1
Empirical rules for interpreting ORD and CD spectra.

1. An ORD or a CD spectrum is additive—that is, it is the simple sum of the spectra of its components (as in Figure 16-7). This is not always strictly true but is a good approximation.
2. The amplitude of an ORD curve or the rotational strength of a CD curve is a measure of the degree of asymmetry. An agent that increases or decreases these parameters usually does so by increasing or decreasing asymmetry (although other spectral features usually accompany the change in asymmetry).
3. A chromophore that is symmetric can become optically active when it is in an asymmetric environment (e.g., a helix). This may or may not be accompanied by a change in λ_0.
4. The value of λ_0 and the magnitude and sign of $\Delta\varepsilon$ at λ_0 allows the chromophore to be identified because it is always very near the value of λ_0 obtained from simple absorption spectroscopy.

analysis is valid in solution. Rules 2 and 3 are primarily used to assay interactions between macromolecules and small molecules. These applications will be seen in many of the examples later in this chapter.

APPLICATION OF ORD AND CD ANALYSIS TO PROTEIN AND POLYPEPTIDE STRUCTURE

Because of the lack of adequate theory discussed earlier, the approach to the elucidation of the secondary structure of a protein has been to determine empirically ORD or CD curves for model polypeptides. (A model polypeptide has only a single conformation and its structure is known from x-ray scattering.) Then an attempt is made to construct from these "standard" curves a weighted sum that is the same as the observed curve of the sample.

For proteins, the principal standards are three forms of poly-L-lysine: α helix, β form, and the random coil, whose ORD and CD spectra are shown in Figure 16-10A. If it is assumed that no other conformations exist and that the amino acid side chains have no effect on the spectra, then it is possible to calculate the expected curve for a protein containing a mixture of the three conformations by simple graphic addition. An example is shown in Figure 16-10B, which shows the observed and calculated CD curve for myoglobin (the structure of which is accurately known from x-ray studies). As can be seen, the analysis is fairly good, although the fit is not exact. Such graphic addition for calculating a curve for a protein *with known structure* is simple. However, the more interesting problem of determining the fractional contribution of the three forms in a protein of unknown structure is a substantial task that must be done by curve fitting, using electronic computers. This is done by continually adjusting the fraction of each in the summation until the observed curve is obtained.

The problem with such an analysis is that, as noted for myoglobin, the calculated curve is rarely a perfect fit. As might be expected, this is due to violations of the assumptions stated in the preceding paragraph. The fact is that the spectra are somewhat affected by the presence of aromatic side chains, disulfide bridges, and prosthetic groups, and by chain length, and the peptide bond may exist in conformations other than α, β, or random coil* (e.g., the various kinds of β turns described in Chapter 1). The observations leading to this realization are:

---

*It is also probably the case that no part of a native protein is ever in a true random-coil configuration because of a variety of subtle interactions between the side chains and because short regions of a protein chain bounded by structured regions can never have complete flexibility because the ends of the short regions are not completely free to move.

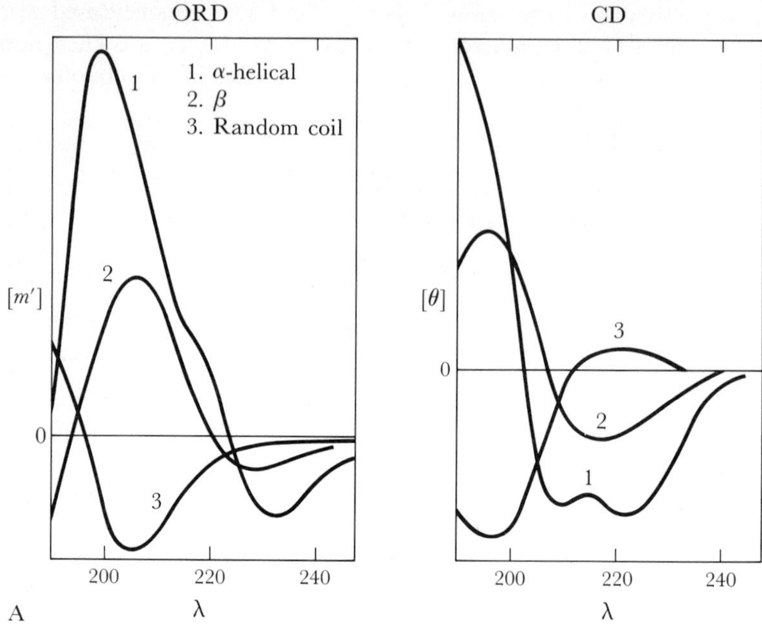

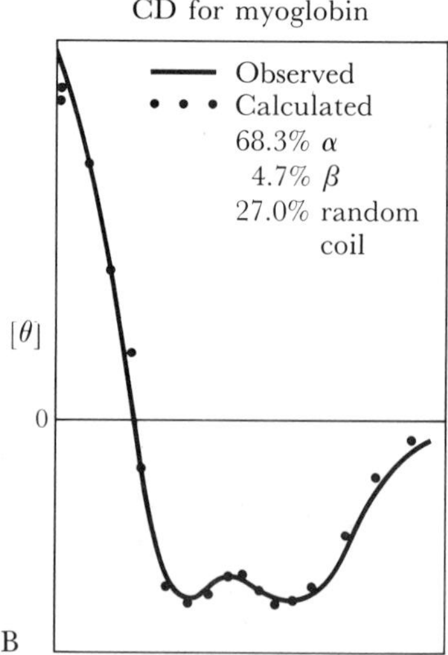

FIGURE 16-10
A. ORD and CD spectra for poly-L-lysine in the α-helical, β, and random-coil conformations. B. Observed CD curve for myoglobin and a curve calculated from the poly-L-lysine data of part A. [From N. Greenfield and G. Fasman, *Biochemistry* 8(1969):4108–4115.]

1. The spectra for proteins claimed to be "true" α helices are not all the same; this is caused by a small effect of nonaromatic side chains on the rotatory strength of the peptide bond and by an occasional distortion of the α helix due to hydrogen bonding between two amino acids separated by *one* amino acid rather than three (the α-helical situation).
2. Long-chain homopolypeptides do not have the same rotational strength per residue as short ones. This is a problem because α-helical regions of proteins are usually from three to twenty amino acids in length; yet all model compounds that have α helicity are very long.
3. The random-coil "standards" are probably not totally in the random-coil configuration. This is also the case for random-coil segments of proteins, which probably contain β turns.
4. By x-ray diffraction, structures have been found in proteins that are unknown in the synthetic polypeptides used as model compounds.
5. Phenylalanine, tyrosine, histidine, and tryptophan side chains can contribute to CD spectra if in certain configurations. Polymers of these amino acids give spectra that are different from the spectra of polypeptides of nonaromatic amino acids.
6. Proline and glycine can form left-handed helical structures, as in collagen.
7. Cystine disulfide bridges give CD bands, like those in insulin and ribonuclease, which cannot be due to peptide absorption.
8. Nonprotein prosthetic groups (e.g., heme in hemoglobin) affect spectra.

By this time, the reader will have certainly decided that ORD and CD analyses are best dispensed with insofar as determining protein structure is concerned. However, as Figure 16-10 illustrates for myoglobin, these analyses are useful in some cases. In fact, for a general description of the helical content of proteins the ORD-CD analysis is adequate, as shown in Table 16-2. Furthermore, recent studies using CD analysis have begun to clarify some of the problems and to yield information that will make spectral features more easily interpreted. The important point to realize is that an experimenter cannot always defer experimentation until precise x-ray analysis is available, and the information gained from CD studies, if coupled with other methods such as fluorescence, nuclear magnetic resonance, and sedimentation, can often lead to a reasonable description of the conformation of a macromolecule.

The next important use of ORD and CD in the study of proteins is to assay conformational *changes* for which it is exquisitely sensitive, as will be seen in the following section.

TABLE 16-2
Helical content of proteins determined by CD compared with x-ray analysis.

Protein	α helix (%) by CD	α helix (%) by x-ray analysis
Myoglobin	77	77
Lysozyme	29	29
Ribonuclease	18	19
Papain	21	21
Lactic dehydrogenase	31	29
α-Chymotrypsin	8	9
Chymotrypsinogen	9	6

SOURCE: Data from Y. H. Chen, J. T. Yang, and H. M. Martinez, *Biochemistry* 11(1972):4120–4131.

NOTE: Similar data for β structure show poorer agreement because the CD analysis does not always adequately detect the various kinds of β turns.

ASSAY OF CHANGES IN CONFORMATION BY CD ANALYSIS

Although CD analysis rarely gives absolute information about structure and it is always necessary to make comparisons with standards whose structures are known from x-ray diffraction, CD is extraordinarily sensitive to *changes* in conformation—if a CD spectrum changes in any way, there must be a conformational change. Hence, even if the structure is not known at all and the CD curve is virtually uninterpretable, CD analysis can still be a sensitive assay for any interaction or agent that causes a conformational change. Furthermore, this can be made semiquantitative by looking at changes of particular features of a spectrum. Hence, experiments can be designed with a CD change as the variable. How this is done is best seen by example.

Example 16-A. Changes in enzyme structure caused by substances that bind to the enzyme.

The CD spectra of many enzymes change when the enzyme interacts with its substrate, a coenzyme, or an inhibitor. This can be used as an assay of binding. For example, binding constants can be determined by measuring rotational strength as a function of concentration of the bound molecules. Because each bound molecule will change R by a definite amount, the number of bound molecules can be determined from the magnitude of the total change. In cases in which the changes are small, they can be amplified by introducing at or near the active site an optically active chromophore or one that becomes optically active when coupled to the protein and studying the changes in CD of that chromophore. Active sites can also be identified in this way by introducing a chromophore at various sites and determining at which site the CD is affected by substrate binding. This is much like the reporter-group method of absorption spectroscopy (Chapter 14). Suppose that it is known that there is a histidine in the active site but that there are four histidines in the protein. A chromophore can be introduced (often with difficulty) next to one of the histidines and the CD of the chromophore can be examined before and after substrate binding. If there is no effect on the CD of the chromophore when adjacent to histidines 1, 2, and 4, but there is for 3, this can be taken as evidence that 3 is in the active site.

Example 16-B. Protein denaturation.

The denaturation of proteins is always accompanied by CD changes, indicating the loss of α and β structure and the enhancement of the random-coil spectral components. Hence, denaturation can be followed by plotting ellipticity at a particular wavelength as a function of the denaturing conditions. Figure 16-11 shows an example for the thermal denaturation of myoglobin.

Example 16-C. Studies of the folding of proteins.

The forces responsible for determining the three-dimensional structure of a protein are of great interest. This has frequently been studied by

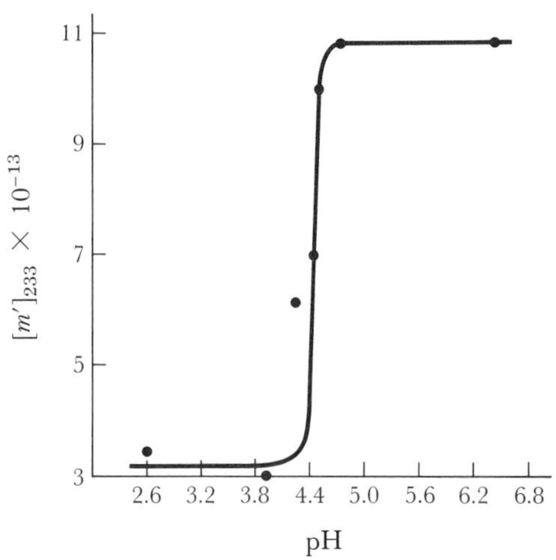

FIGURE 16-11
Detection of the denaturation of myoglobin by acid titration, using specific rotation at 233 nm as an assay. [From P. Appel and W. D. Brown, *Biopolymers* 10(1971):2309–2322.]

examining helix-coil transitions. CD adds another dimension to this analysis if the amino acid sequence is known, because it is possible to look directly at the environment of aromatic amino acids and of disulfide bridges because they have CD bands at characteristic wavelengths. Hence, it is possible to detect changes in the neighborhood of particular amino acids caused by the destruction of hydrophobic forces, or hydrogen bonds, the titration of particular amino acids, and so forth. An interesting example is that of a tyrosine in pancreatic ribonuclease far removed from the active site of the enzyme—its CD is changed when the inhibitor 3′-cytidylic acid, which binds to the active site, is added. This shows that binding to one part of a protein can cause a conformational change in a distant region.

ORD AND CD MEASUREMENTS FOR POLYNUCLEOTIDES, NUCLEIC ACIDS, AND NUCLEOPROTEINS

For all polymers containing nucleotides, the optically active groups that are observable are the purines and the pyrimidines because the bonds in the sugars and the phosphoester linkages do not absorb in the wavelength range that is usually studied. Purines and pyrimidines themselves are examples of symmetric chromophores that become somewhat optically active when attached to a sugar by means of an *N*-glycosidic bond; furthermore, the optical activity increases substantially when they become part of a helical structure.

ORD and CD are extraordinarily sensitive probes of conformational changes in polynucleotides. For example, for many years, the most commonly used assay of conformational change in polynucleotides has been what is known as hypochromicity—the decrease in optical density that accompanies the formation of an ordered structure (see Chapter 14). The optical density at 258 nm of polyadenylic acid (poly A) is 26% smaller than that of the monomer adenylic acid. However, the CD for the same substances changes, as shown in Figure 16-12. Note the reversal of the sign and the increase in the value of θ at 260 nm. Clearly, these are huge changes compared with the change in absorbance. Changes such as these can be used to study (1) the loss of helicity of single-stranded polymers by various agents such as high temperature or extremes of pH (note in Figure 16-12 the large change in the CD spectrum of poly A when denatured); (2) the transition from single- to double-stranded polynucleotides and vice versa; (3) structural changes introduced by the binding of cations, peptides, proteins, and so forth; and (4) the effect of the charging of tRNA with an amino acid.

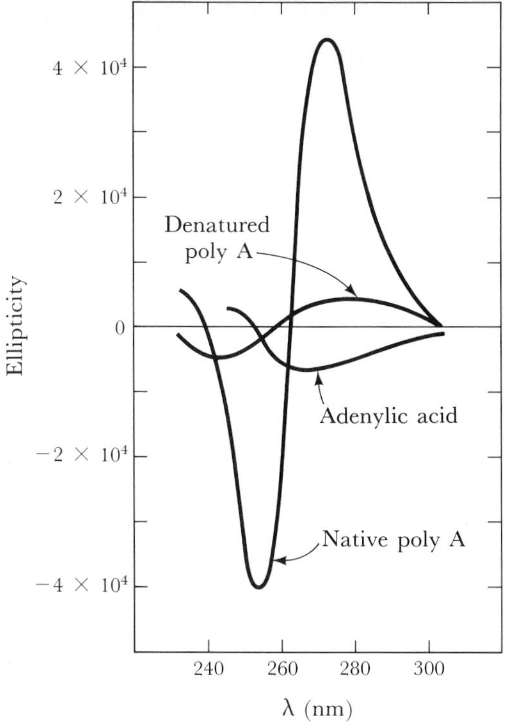

FIGURE 16-12
Circular dichroism of adenylic acid, denatured polyadenylic acid (poly A), and native poly A. Note the change in sign when adenylic acid is polymerized and the great increase in ellipticity when in the native structure.

For single-stranded polynucleotides, a reasonable theory (that of Ignacio Tinoco) exists relating ORD and CD spectra to structure. This theory has been nearly confirmed in all of its details and actually allows the calculation of a CD spectrum from molecular structure with some accuracy. The most important element of the theory is that the optical activity of oligonucleotides is a result of an interaction between two *adjacent bases—the nearest neighbors*—that are *stacked* one above the other (Figure 16-13). Second neighbors (i.e., two bases separated by the nearest neighbor of each) do not make a significant contribution. This has the result that the CD spectra of all nearest-neighbor pairs are different so that it is possible to calculate by direct summation the CD spectrum of a single-stranded polynucleotide if the nearest-neighbor frequencies are known. Furthermore, with computer

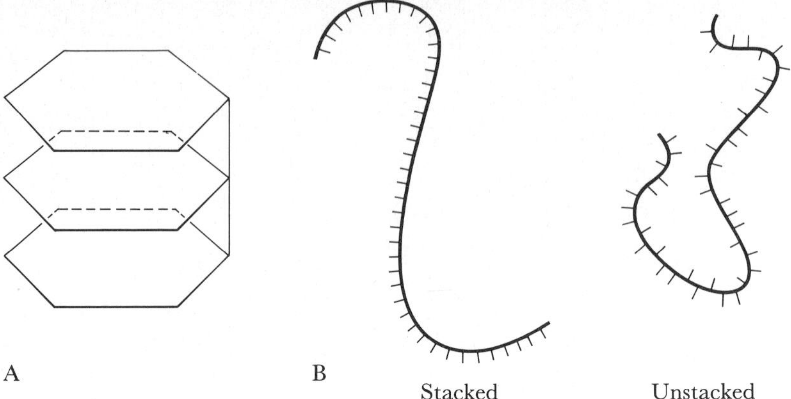

FIGURE 16-13
A. Three stacked bases. B. Structures of stacked and unstacked polynucleotides. The stacked polynucleotide is more extended because the stacking tends to decrease the flexibility of the molecule.

techniques, it is possible to determine the nearest-neighbor frequencies from a CD spectrum, which is of tremendous practical value in determining base sequences. The precision of CD analysis is no greater than that of the standard enzymatic methods, but to obtain a CD spectrum requires a few hours, whereas the enzymatic analysis would take about a week provided that pure enzymes were available.

The theory for double-stranded polynucleotides is not yet complete, although progress is rapidly being made. For double-stranded DNA, the CD spectra seems to be independent of base composition in the wavelength range usually studied, although there is very recent evidence for such effects in the far (vacuum) ultraviolet.

EXAMPLES OF ORD AND CD ANALYSIS IN STUDIES OF POLYNUCLEOTIDE STRUCTURE

Several examples of the use of ORD and CD in determining the structural features of polynucleotides follow.

Example 16-D. Structure of polycytidylic acid (poly C).
The ORD spectrum of poly C shows $[m']_{292} = 35{,}160$, which by rules 2 and 3 (Table 16-1) means that poly C is very asymmetric—for polynucleotides, this implies helicity. The forces responsible for the helicity are easily identified by spectral analysis. For example, the addition of

formaldehyde—which titrates amino groups and thereby eliminates hydrogen bonding—leaves $[m']_{292}$ virtually unchanged. Hence, the helicity is not stabilized by hydrogen bonds. However, in 90% ethylene glycol, a reagent known to interfere with interactions between nonpolar (hydrophobic) groups, there is $[m']_{292} = 7{,}223$, a fivefold decrease; this means that the asymmetry (i.e., the helicity) is sharply reduced. The $[m']_{292}$ of the monomer, cytidylic acid, is approximately 8,000 in both H_2O and 90% ethylene glycol. Hence, the structure is stabilized by the so-called hydrophobic forces. Stabilization by a hydrophobic force means that the interacting molecules tend to become sufficiently near that water is excluded; hence, this result suggests that the bases are arranged one above the other and that they are relatively near—that is, the stacked-base conformation (Figure 16-13B).

Suppose that the effect of these reagents was examined instead by simply determining absorbance changes. The OD_{260} increases by about 10% when formaldehyde is added, but this could be owing to the loss of helicity or to a change in the absorbance of the cytosine itself (because formaldehyde titration of free bases causes an increase in absorbance). Hence, no statement about the role of hydrogen bonding could be made. The OD_{260} increase when ethylene glycol (a weakly polar solvent) is added is 30%. This would suggest hydrophobic effects but, without the ORD results, that this is an effect on helicity could not be proved.

Example 16-E. Strong evidence for base stacking.

Di- and trinucleoside phosphates show *strong* ORD and CD bands compared with the monomers, indicating that even short oligomers have a very asymmetric and probably a helical conformation. The rotational strength is vastly reduced by organic solvents so that in aqueous solution there must be helical arrangement stabilized by hydrophobic forces, as in Example 16-A.

Tinoco's theory provides a basis for understanding these data because CD spectra of dinucleotides agree almost precisely with those calculated on the basis of the stacked arrangement. From the magnitude of $[M]$ and $[\theta]$, the fractions of the bases that are stacked under various conditions can also be determined.

Uracil oligonucleotides show little or no optical activity; hence they are rarely in an asymmetric configuration so that uracil does not stack.

The studies on stacking confirmed and actually proved an early hypothesis that a hydrophobic force is the principal force in stabilizing the double-stranded DNA structure. Many studies involving hyperchromicity and viscosity measurements in solvents of different polarity had shown that the rigid character of double-stranded DNA was lost as polarity decreased. The ORD and CD studies proved that the absorbance changes were due to changes in helicity.

Example 16-F. Structure of various forms of polyadenylic acid (poly A). Single-stranded poly A at neutral pH has very strong CD bands and therefore possesses a highly ordered structure (Figure 16-14). A possible structure is a double-stranded helix held together by hydrogen bonds between the 6-amino group and the OH group of the ribose on the opposite strand. However, poly-N^6-hydroxyethyladenine, which cannot form hydrogen bonds (Figure 16-15), has a similar CD spectrum; hence the hydrogen bond is not likely to be a part of the structure and the structure is probably not a double helix. Because the CD of the dinucleotide is that predicted for stacking and because, with increasing numbers (as many as 7 or 8) of A in oligo A, the rotational strength of the CD band increases, poly A at neutral pH is probably an extended semirigid helical molecule whose structure is stabilized by base stacking.

At acid pH, the rotational strength (area) of the CD band is one and one-half times as great as that at neutral pH (Figure 16-14), indicating greater asymmetry at acid pH. Furthermore, λ_0 for the CD band shifts toward shorter wavelength compared with that at neutral pH and approaches λ_{max} of the simple absorption spectrum of a double-

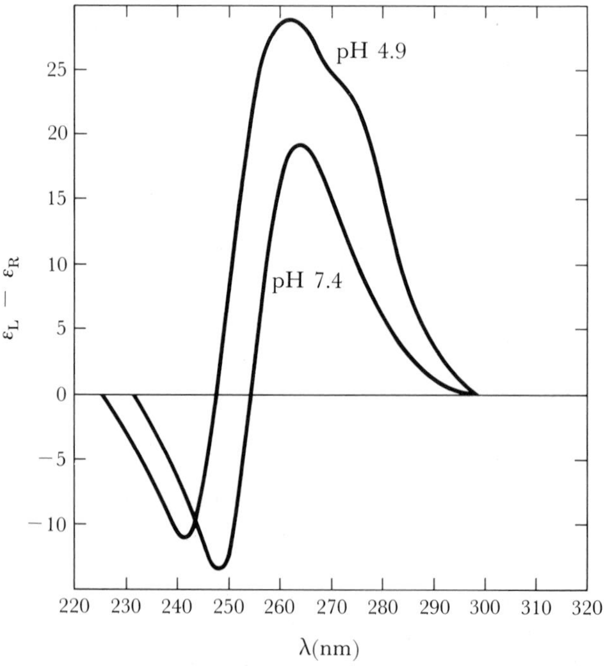

FIGURE 16-14
CD spectra of poly A at acid and neutral pH. [From J. Brahms, *Nature* 202(1964):797–798.]

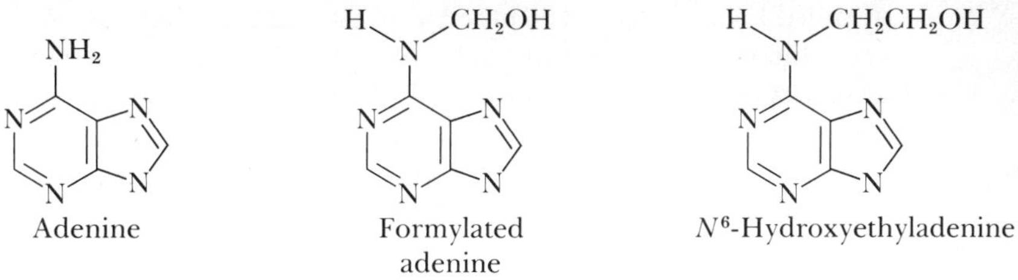

Adenine

Formylated adenine

N^6-Hydroxyethyladenine

FIGURE 16-15
Chemical structures of adenine, adenine reacted with formaldehyde (formylated adenine), and N^6-hydroxyethyladenine. Formylation blocks the NH group of adenine that participates in hydrogen bonding. This group is already blocked in N^6-hydroxyethyladenine.

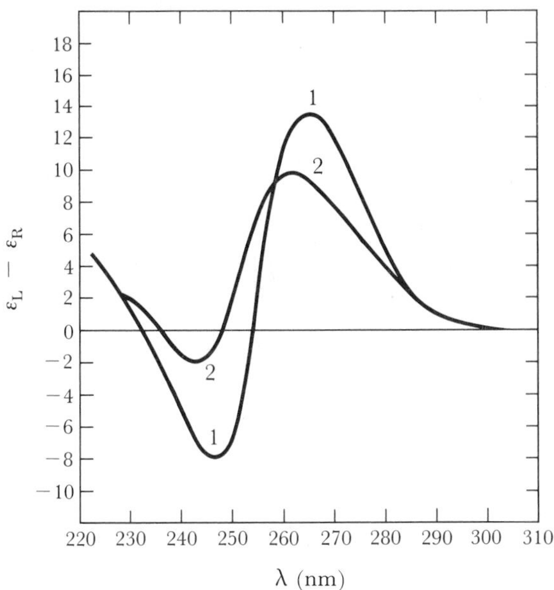

FIGURE 16-16
Circular dichroism of the poly A : poly U complex. Curve 1 is the sum of the spectra of poly A and poly U measured separately. Curve 2 is for the complex. Note that curve 2 is shifted to shorter wavelengths and that there is less negative CD. This is a characteristic of double-stranded polymers. [From J. Brahms, *J. Mol. Biol.* 11(1965):785–801.]

stranded helix. By rule 4 (Table 16-1), this suggests that poly A is also a double-stranded helix at acidic pH. Poly-N^6-hydroxyethyladenylic acid does not show either of these CD changes when shifted to acid pH. Because this polynucleotide cannot form hydrogen bonds, it is likely that the poly A structure at acid pH requires hydrogen bonding. This evidence suggests that acidic poly A is a hydrogen-bonded, double-stranded helix. This agrees with the results of x-ray analysis of poly A fibers prepared from acidic solution (i.e., that it is a double-stranded helix) and confirms that the structure determined by x-ray analysis also exists in solution.

Example 16-G. Structure of the poly A:poly U complex.
If poly A is mixed with equimolar amounts of poly U, a polymer is formed whose rotational strength is less than that obtained by summation of the CD spectra of poly A and poly U. This lack of simple additivity shows that the spectrum of at least one of the components has changed (rule 1, Table 16-1). Furthermore, λ_0 is shifted to shorter wavelenths, as it is for acidic poly A. Hence, again by rule 4, a reasonable conclusion is that poly A:poly U is a double-stranded helix. Note the simplicity of this observation (Figure 16-16).

Example 16-H. Structure of single-stranded polydeoxynucleotides.
The polydeoxynucleotides generally show the same spectral features as the ribose forms except that the rotational strength is much lower. Hence, these structures are less helical and probably less stacked. Furthermore, the OH group present in ribose but not in deoxyribose must somehow be a part of the helical structure.

Example 16-I. Structure of RNA in ribosomes.
The ORD spectrum of ribosomes can be easily resolved into two components—that of the RNA and that of the protein—because the two spectra are in different wavelength ranges. The calculated ORD of the RNA is essentially the same as that of ribosomal RNA free in solution so that it may be concluded that there is probably no conformational change in the RNA resulting from assembly into ribosomes. This suggests that perhaps ribosomes are assembled by proteins building onto the RNA.

Selected References

Adler, A. J., and G. D. Fasman. 1967. "Optical Rotatory Dispersion as a Means of Determining Nucleic Acid Conformation," in *Methods in Enzymology,* vol. 12B, edited by L. Grossman and K. Moldave, pp. 268–302. Academic Press. Dated but excellent.

Adler, A. J., N. J. Greenfield, and G. D. Fasman. 1973. "Circular Dichroism and Optical Rotatory Dispersion of Proteins and Polypeptides," in *Methods in Enzymology,* vol. 27D, edited by C. H. W. Hirs and S. N. Timasheff, pp. 675–735. Academic Press. An excellent, up-to-date review.

Brahms, J., and S. Brahms. 1970. "Circular Dichroism of Nucleic Acids," in *Fine Structure of Proteins and Nucleic Acids,* vol. 4, G. D. Fasman and S. N. Timasheff, pp. 191–270. Dekker. An excellent, up-to-date review.

Jirgensons, B. 1969. *Optical Rotatory Dispersion of Proteins and Other Macromolecules.* Springer. Dated but good.

Kauzman, W. 1957. *Quantum Chemistry.* Academic Press. This contains both the classical and the quantum mechanical theory of optical activity.

Mommaerts, W. F. H. M. 1967. "Ultraviolet Circular Dichroism in Nucleic Acid Structural Analysis," in *Methods in Enzymology,* vol. 12B, edited by L. Grossman and K. Moldave, pp. 302–329. Academic Press. A good description of both experiment and theory.

Problems

16-1. Figure 16-2 shows how right circularly polarized light is generated. How should the figure be altered to show the production of left circularly polarized light? Of elliptically polarized light?

16-2. If two substances are found whose absorption spectra are identical and whose CD curves are identical except that one curve is positive and the other negative, what can probably be said about the structural relation between the two substances?

16-3. The structure of a protein has been determined by the x-ray diffraction of a protein crystal. It is found to contain 31% α helix, 58% β sheet, and 11% random coil. From CD analysis, the values are 60% α helix, 35% β sheet, and 5% random coil. What can you conclude about the structure?

16-4. Draw the expected and observed CD curves for the protein of Problem 16-3.

16-5. Why are amino acid side chain effects easier to detect by CD than by ORD?

16-6. If a rotation of $\pi/12$ radians is observed for a certain substance, how can you be sure that it is not really $4\frac{1}{12}\pi$ radians?

16-7. In what wavelength range would you expect to find the CD spectrum for tyrosine in a protein?

16-8. Ethidium bromide is known to intercalate between the base pairs of DNA. This binding reduces the rotational strength of the CD curve for the DNA and substantially increases it for the ethidium bromide. Why?

16-9. The CD spectra of native DNA in 0.01 M NaCl and in 0.5 M NaCl are only slightly different. After boiling the DNA (to denature it), the rotational strength decreases substantially. In which solvent would the decrease be greater? Explain.

16-10. The CD spectra of a variety of phages differ markedly from that of free DNA. Furthermore, the spectra differ from one phage to the next. What can you conclude about the structure of DNA in phages?

16-11. The rotational strength of oligonucleotides increases with the number of nucleotides up to about 10; after that the changes are insignificant, if present at all. Explain.

16-12. The CD spectra of polyamino acids are not all the same—both rotational strength and λ_{max} differ. What are some of the factors that account for this?

CHAPTER 17

Nuclear Magnetic Resonance

Nuclear magnetic resonance (NMR) is another spectroscopic method capable of yielding information about the structure of biological polymers, about interactions between molecules, and about molecular motion. Its special advantages are that (1) the theory is sufficiently good that, in principle, the detailed arrangement of individual atoms, could be calculated from NMR spectra; (2) hydrogen atoms (which are beyond the resolution of x-ray diffraction analysis) can be located; and (3) different atoms (e.g., H, N, C, and P) can be examined separately. For small molecules (e.g., molecular weight <500), NMR has been highly successful in determining structure. However, for macromolecules, the potential has not yet been realized because of the huge number of spectral lines, which are often poorly resolved and for which identification of the atom producing the line is frequently very difficult, and because of the large number of possible interactions for each atom.

How to approach the problem of structure determination is an important part of this chapter. Unfortunately, the discussion will be limited because of the great complexity of the subject. The intent is that the reader will be able to understand some of the biochemical literature employing NMR and will in the future know when to turn to NMR for aid in the solution of biochemical problems.

It will also become apparent that another difference between NMR and other methods for studying macromolecules is that the limitation of NMR is not primarily the complexity of the theory but often in instrumentation.

This point is made because it seems that the great advances in the utility of the method have come principally from improvements in instrumentation.

BASIC THEORY OF NMR

For the present purpose, a nucleus consisting of a single proton will be considered (although most of the concepts apply to more complex nuclei). In addition to charge and mass, a proton possesses angular momentum, or *spin.* A spinning charge generates a magnetic field and can be thought of as a tiny bar magnet oriented along the spin axis. The strength of the magnetic field is expressed as a magnetic moment, μ. Like a bar magnet, which has a north and south pole, μ has a direction. In an external magnetic field, a macroscopic bar magnet will become oriented with its magnetic moment along the lines of magnetic force of this field. However, in dealing with atomic and nuclear particles, the rules of quantum mechanics come into play, and it is found that, if a proton is in a magnetic field of strength H_0, the proton magnetic moment assumes one of two angles with respect to the direction of H_0—that is, the angles θ and $180 - \theta$, in which θ is in degrees (Figure 17-1). These two orientations of the magnetic mo-

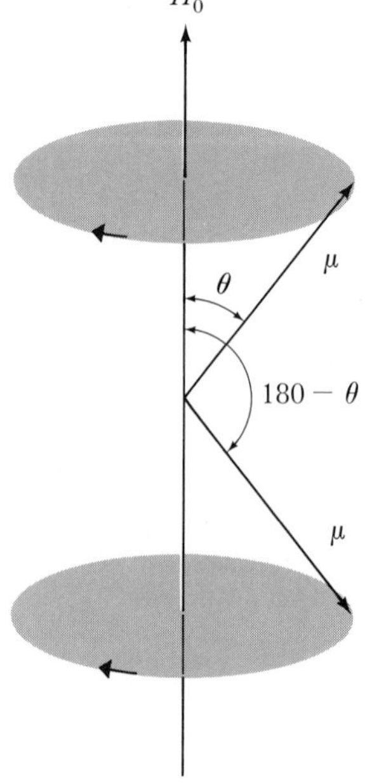

FIGURE 17-1
Precession of the magnetic moment in the two spin states ($I = \pm 1/2$) induced by the magnetic field H_0.

ment are referred to as being *aligned with and against* H_0, respectively. The potential energy of each of the orientations is $-\mu H_0 \sin\theta$ and $+\mu H_0 \sin\theta$, respectively, so that the difference in the energy levels is $2\mu H_0 \sin\theta$. As discussed in Chapter 14, transitions between electronic and vibrational levels can occur by the absorption of electromagnetic radiation (in the ultraviolet to infrared range) having energy equal to the difference (ΔE) between the two energy states. This is also true of nuclear energy levels except that the radiation is in the radiofrequency range. Because $\Delta E = h\nu$, in which ν is the frequency of the absorbed radiation and h is Planck's constant,

$$\nu = \frac{2\mu H_0 \sin\theta}{h} \tag{1}$$

In the jargon of NMR workers, absorption of radiation of this frequency causes the proton to *flip* from one orientation to the other.

To generalize to nuclei other than the single proton requires making use of the quantum mechanical spin number, I, which is a characteristic of each nucleus. (The factors determining the values of I can be found in the Selected References near the end of the chapter.) The number of different orientations and therefore energy levels of a given nucleus in a magnetic field is $2I + 1$, in which I is always an integral multiple of 1/2. Hence, for protons, which have two levels, $I = 1/2$. (In this chapter, only nuclei with $I = 1/2$ will be considered.) The magnitude of μ also varies with the nucleus. It is difficult to predict the values of I and μ for a given nucleus so that in all cases they have been determined by actual measurement.

The parameters for several nuclei of interest to biologists are shown in Table 17-1. Nuclear energy levels differ from electronic and vibrational

TABLE 17-1
NMR data for nuclei of importance in biological systems.

Isotope	Natural abundance (%)	Spin (I)	Sensitivity*	Relative sensitivity†	Approximate range of chemical shifts (ppm)
^{1}H	99.98	1/2	1.000	1.00	12
^{2}H	0.0016	1	0.0096	1.5×10^{-6}	12
^{13}C	1.1	1/2	0.016	0.0002	350
^{15}N	0.37	1/2	0.001	0.00037	1,000
^{19}F	100	1/2	0.83	0.83	500
^{31}P	100	1/2	0.066	0.066	700

*Relative to ^{1}H, in terms of equal number of nuclei.
†Product of (natural abundance) × (sensitivity). For ^{2}H, ^{13}C, and ^{15}N, in which enrichment is possible (to >99% for ^{2}H and ^{15}N), the sensitivity is the more meaningful number.

levels in that the energy levels are equally spaced and transitions can occur only between adjacent levels. This means that, *for a given nucleus, there is only a single transition frequency for each value of H.*

An understanding of the resonance phenomenon can also be gained by considering the theory from the viewpoint of classical electrodynamics. If a spinning nucleus is placed in a magnetic field, H_0, the basic equations of electromagnetic theory show that the magnetic moment of the nucleus is subjected to a torque. Because the nucleus is spinning and therefore possesses angular momentum, the law of conservation of angular momentum requires that the magnetic nucleus precess about the field direction in the same way that a gyroscope responds to a gravitational field (Figure 17-1). Note that θ is not changed. The angular frequency of precession, ω_0 (called the Larmor frequency), is

$$\omega_0 = \frac{2\mu}{h} H_0 \sin\theta \tag{2}$$

To flip the nucleus from one orientation to another, θ must be changed to $180 - \theta$. Increasing H_0 cannot do this because this increase affects only ω_0 and increases the rate of precession. However, if a small magnetic field, H_1, is added perpendicular to H_0 (Figure 17-2), the magnetic dipole will then be subjected to a second torque, tending to induce precession about this field direction. If $H_1 \ll H_0$, this torque is much smaller than that due to H_0 and, because the magnetic moment is precessing and therefore continually changing direction with respect to H_1, the torque due to H_1 will have little effect on the orientation of μ. To be effective, H_1 would have to act continuously and therefore must also rotate in the plane indicated

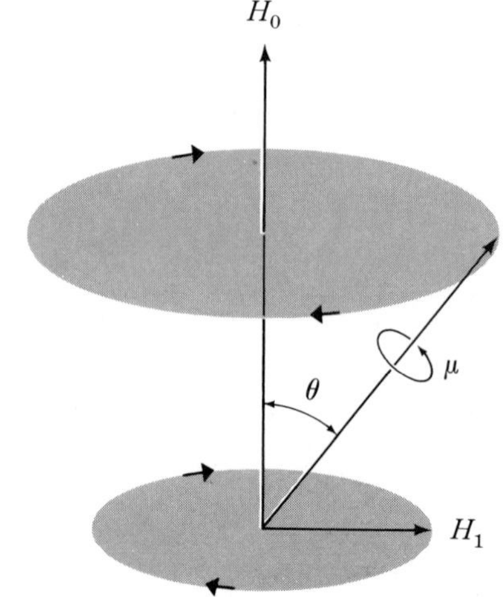

FIGURE 17-2
The magnetic moment μ rotates about H_0. At the Larmor frequency, the perpendicular field H_1 induces a torque on μ; if H_1 rotates at the Larmor frequency it can exert the torque continuously.

in Figure 17-2 and *at the same frequency as the rotating dipole*—that is, ω_0. If H_1 rotates at a different frequency, the force acting on the dipole will be continually changing in magnitude and direction and the precession due to H_0 will experience only a periodic wobbling (called nutation, with a gyroscope). Because there is only a finite number of orientations of μ with respect to H_0 (i.e., because the system is quantum mechanical), only if the frequency of oscillation of H_1 is ω_0 (the so-called resonance condition) can energy be transferred to reorient the magnetic moment. Note that this is exactly what is expressed by equation (1) if the oscillating magnetic field, H_1, is simply the magnetic field of the electromagnetic radiation used to raise the nucleus from one energy level to the next.

THE CHEMICAL SHIFT

In the preceding discussion, it has been stated that only one transition can occur for a given nucleus in a particular magnetic field so that the spectrum should consist of only one resonance line. However, this is not the case because the magnetic field seen by the nucleus is not simply the applied field, H_0. For example, all nuclei are surrounded (shielded) by electrons and these electrons are also induced to circulate by H_0. Because moving electrons themselves generate a magnetic field, the nucleus sees an effective field, $H_{\text{eff}} = (1 - \alpha)H_0$, in which α is called the screening constant. Other factors (such as the magnetic field due to nearby nuclei) also affect the value of H_{eff}. What this means is that *the observed resonance frequency depends on the environment.* It is this environmental effect that makes NMR spectroscopy valuable (exactly as in other forms of spectroscopy) because, in a molecule, each nucleus of the same type will have a resonance frequency that depends on the chemical group in which the nucleus resides. For example, the hydrogen protons in a methyl group will have a different resonance frequency from that of an amino hydrogen proton, and, furthermore, the protons of toluene will differ from those of acetic acid. This shift in resonance frequency due to chemical environment is called the *chemical shift.*

In optical spectroscopy, every transition has a defined frequency. However, because nuclear resonance frequency is dependent on H_0, there is no natural basic scale unit. Furthermore, H_0 is difficult to measure accurately. Hence, the following approach has been adopted for reporting the resonance condition or the chemical shift. A reference material is added to the sample and its resonance line is assigned an arbitrary value of H or ν_0 (whichever is varied in the particular procedure used to obtain a spectrum), and chemical shifts are expressed as displacements from this value. An even better system is to use a dimensionless displacement from the reference standard—called parts per million (ppm). If H is varied and ν is constant,

$$\text{ppm} = \frac{H_s - H_{ref}}{H_{ref}} \times 10^6 \tag{3}$$

in which H_s and H_{ref} are the magnetic fields producing resonance for the sample and the reference material, respectively. If ν is varied,

$$\text{ppm} = \frac{\nu_s - \nu_{ref}}{\nu_{ref}} \times 10^6$$

This dimensionless scale has the advantage that chemical shifts are independent of the actual value of H_0 or the frequency of the radiofrequency signal, and spectra obtained with different NMR spectrometers are comparable. Nonetheless, it is still fairly common to see data reported as a frequency or field shift at fixed field or frequency, respectively.

Figure 17-3 shows an NMR spectrum for protons, clearly indicating the large number of resonances for a single nuclear type and showing the effect of the chemical group in which the proton resides.

Many factors affect the chemical shift, of which the following are important in biochemistry.

Intramolecular Shielding

This effect, which produces shifts on the order of 10 to 20 ppm, is the type briefly mentioned earlier—that is, nearby electrons are induced to move by the external magnetic field; this generates a field at the nucleus opposite in direction to that of the applied field so that $H_{eff} < H_0$. Somewhat larger effects are found if the nucleus is in an organic ring or a conjugated ring system. In rings, the π electrons are "delocalized" and free to move in a circular path in the plane of the ring. If a magnetic field is applied that is perpendicular to the plane of the ring, these electrons circulate in such a way that a magnetic field that is antiparallel to the applied field is induced in the center of the ring and a parallel field is induced external to the ring (Figure 17-4). This is analogous to what takes place when current flows through a wire. The direction of the resulting chemical shift therefore depends on where the nucleus is located with respect to the carbon atoms of the ring. The shift is in the opposite direction if the applied field is in the plane of the ring so that the chemical shift depends also on orientation. This type of chemical shift is called the *ring-current shift* and occurs in aromatic amino acids, purines, pyrimidines, porphyrins, and flavins.

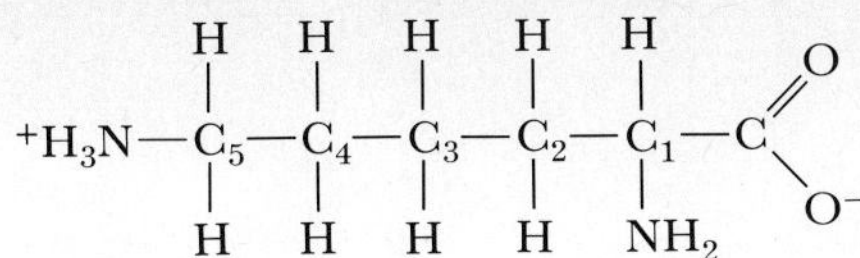

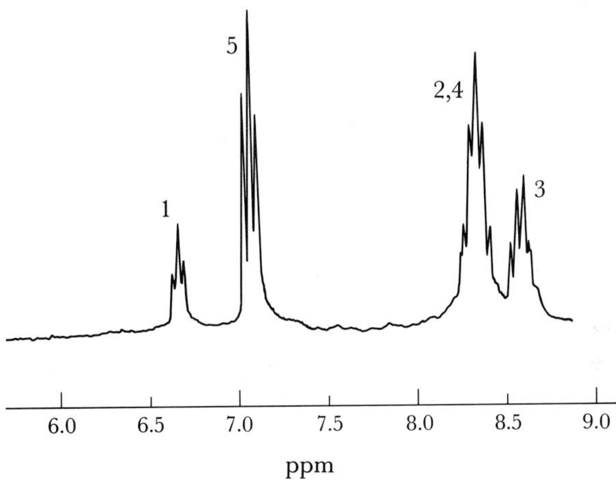

FIGURE 17-3
NMR spectrum of lysine in D_2O in zwitterion form. The structure of lysine is shown with its carbon atoms numbered. The numbers next to the peak indicate the carbon atom whose protons produce the cluster of lines [From F. A. Bovey, *High Resolution NMR of Macromolecules,* Academic Press, 1972.]

Paramagnetic Effects

An unpaired electron (e.g., in free radicals or in certain metal ions) produces a large chemical shift because the electron itself has a significant magnetic moment. The shift is from about 20 to 30 ppm and can be detected when the unpaired electron is as much as 10 Å from the nucleus producing the resonance; furthermore, because the magnitude is inversely related to the cube of the distance, these effects can be used as a molecular yardstick.

Intermolecular Effects

When two molecules interact, the electron distribution of one molecule can affect the chemical shifts of nuclei in the other. If the nucleus is in a polymer, such effects can be between different regions or residues that are

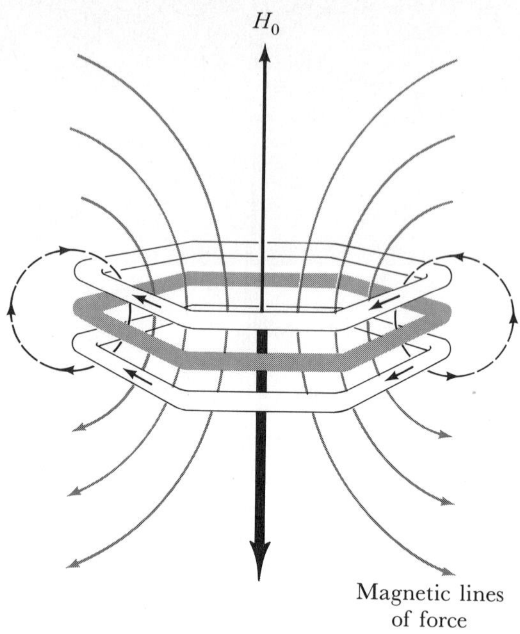

FIGURE 17-4
Electron density, ring currents, and magnetic lines of force about a benzene ring. The solid thin arrow indicates the applied field H_0. The solid thick arrow is the magnetic field induced by the circulation of electrons. [From R. A. Dwek, *Nuclear Magnetic Resonance in Biochemistry,* Clarendon Press, 1973.]

distant within the linear sequence of molecules but nearby in three-dimensional space. That is, the folding and tertiary structure of a polymer can bring together distant chemical groups, thereby creating new environments so that new chemical shifts occur; these are generally smaller than the intramolecular effects described in the preceding paragraph. These effects are of great importance in biochemistry because they can be used to monitor the binding of a ligand to a macromolecule (e.g., interactions between an enzyme, a cofactor, and a substrate) and to describe the three-dimensional structure of a region of a polymer.

Chemical Exchange

If a nucleus is in a molecule that is rapidly undergoing reversible chemical or physical changes, the environment is continually changing. Hence, the effect on chemical shift depends on the time scale of the exchange. Let us consider a ligand that can be either bound to another molecule or free in solution—clearly, two distinct environments. If the ligand exchanges very slowly between the bound and unbound states, the greater fraction of a population of ligands will be either bound or unbound rather than changing from one state to the other. Therefore, there will be two chemical shifts for a nucleus in the ligand—that of the bound and that of the free environment (Figure 17-5A). If exchange is more rapid, there will be more nuclei entering and leaving each environment and these will have intermediate chemical shifts. This will have the effect of broadening the resonance lines (Figure 17-5B). As exchange becomes more rapid, there will be fewer bound or free and more in the intermediate state. Hence, the lines corresponding to the two states will not be present, and a single intermediate and broad band will appear (Figure 17-5C). If the exchange is very rapid, all nuclei will be in the intermediate state, and a narrow resonance line will result whose chemical shift is the average of that of the bound or free ligand (Figure 17-5D). This is a useful phenomenon for studying exchange reactions and the mechanism of enzyme reactions. For example, the number of nuclei in each state can be determined because the area of a resonance line is proportional to the number of nuclei producing the signal (see page 488).

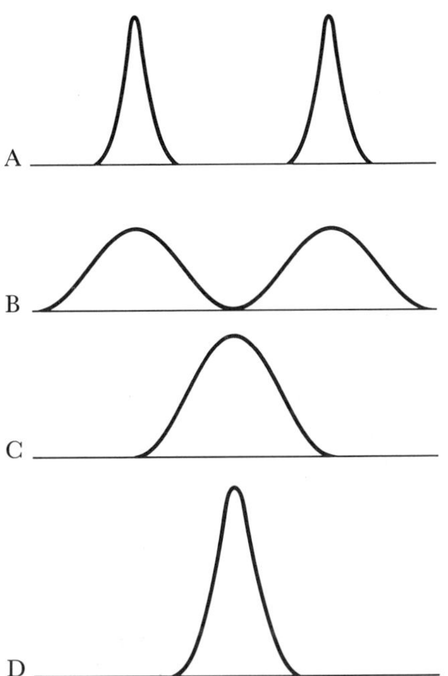

FIGURE 17-5
Shapes of lines as a function of exchange rate.

SPIN-SPIN INTERACTIONS

In all of the preceding discussion, a single proton in a particular environment was assumed to produce a single resonance line. However, a peak will usually appear to consist of a cluster of lines (see Figure 17-6). This is called *line splitting* and results from *spin-spin coupling.* A simplified explanation of this follows. In any atom, the magnetic field of its nucleus will tend to orient its valence electron (which itself has a magnetic moment) so that the electron spin is antiparallel to that of the nucleus. If the atom is in a diamagnetic compound so that its valence electron is paired with an electron of a second nucleus, the valence electron of the second nucleus will be antiparallel to that of the first and therefore parallel to the spin of the first nucleus. The orientation of the second electron affects the orientation of the second nucleus, so that the first nucleus creates a magnetic field at the second. Similarly, the second nucleus affects the first. This mutual interaction is called *spin-spin coupling* and, according to the laws of quantum mechanics, increases the number of resonances. In general, if a nucleus is coupled to n identical nuclei of spin I, its absorption line will be split into $2nI + 1$ components. It is important to realize that spin-spin coupling comes about only if nuclei are coupled by covalent bonds (i.e., by means of paired electrons); it is not a result of simple proximity as in binding, adsorption, or protein folding.

Spin-spin coupling can best be understood by a simple example—that of the spectrum of the protons of acetaldehyde:

$$CH_3{-}C{\overset{\displaystyle O}{\underset{\displaystyle H}{\lessgtr}}}$$

This molecule has two classes of protons—the three methyl protons and one OH proton. The molecule is small enough for the two sets of protons to be close enough together that they can mutually affect one another and cause splitting. Because each proton has roughly equal probability of having spin $+1/2$ and $-1/2$ (i.e., of tending to be oriented parallel (($\uparrow$) or antiparallel ($\downarrow$) to the applied field), the spin arrangements of the protons are like those shown in Figure 17-7. Note that, in the methyl group in which there are three identical protons, there are four distinct situations—all parallel; two parallel and one antiparallel; one parallel and two antiparallel; and all antiparallel. Note that each of the second and third configurations can be produced by three different arrangements of the three spins. Because there are two possible spin arrangements of the OH proton, the OH proton can alter the field in the region of the methyl protons in two ways. Hence, the OH proton results in the methyl proton resonance being split into two peaks. Because the parallel and antiparallel

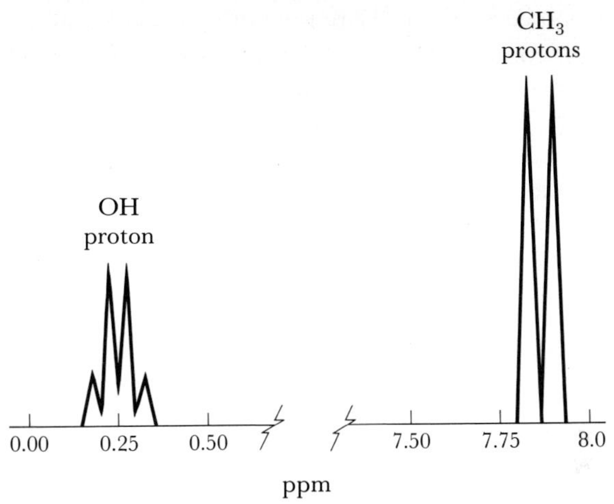

FIGURE 17-6
NMR spectrum for acetaldehyde, showing spin-spin splitting of the OH and methyl protons.

arrangements are equally probable, and because the area of a peak is proportional to the number of nuclei producing the signal (see page 488), the areas of the two peaks of the methyl doublet are the same. Conversely, because there are four possible arrangements of the methyl protons, the OH proton line is split into four components (called a quartet). However, two of the methyl configurations are three times as probable as the others (because they are a result of three possible arrangements); hence, the ratio of the areas of the four components is 1:3:3:1. The spectrum of acetalde-

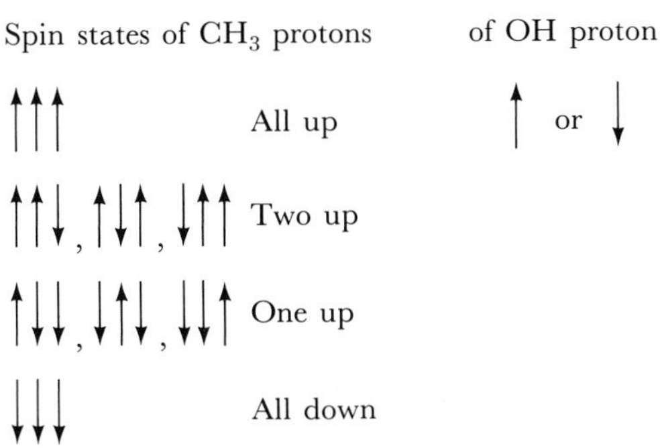

FIGURE 17-7
Possible spin arrangements in acetaldehyde for the CH_3 protons and the OH proton.

hyde is shown in Figure 17-6; the multiplets are easily seen.* The presence of multiplets is sometimes a help in identifying the nuclei producing particular lines. On the other hand, they may hinder analysis by decreasing resolution. This could be a consequence either of unresolved components having the appearance of a broad line or of components of one cluster of lines overlapping a second cluster, thus making it difficult to distinquish peaks. This is especially true with macromolecules.

RELAXATION PROCESSES

At the resonance frequency, nuclear magnetic moments are reoriented. Because the transition from the upper to the lower energy level occurs with the same probability as that from the lower to the upper, equal populations of the nuclei in the two energy levels would result in no net absorption of energy. However, because the difference in energy levels is very small (approximately 10^{-2} calories), the distribution of nuclei in lower and upper levels is strongly dependent on temperature. At absolute zero, all nuclei are in the lower state. In the range usually used for measurement (0°–25°C), thermal excitation has raised many to the upper state; however, there is still a small excess of nuclei in the lower state (1 in 10^5). When radiofrequency radiation of the resonance frequency is applied, energy is absorbed and the populations of nuclei in the upper and lower states are equalized. Once equal, no further absorption can be detected. For absorption to be continuous (as indeed it is), there must be a mechanism to restore the initially unequal distribution. The general term for any process that restores a system to its initial state is *relaxation.* In optical spectroscopy (involving electronic and vibrational levels), the status quo is restored either by degradation of the absorbed energy to heat (by collisional processes) or by fluorescence. There are basically two kinds of relaxation processes in NMR: *spin-lattice,* or *longitudinal,* relaxation and *spin-spin,* or *transverse,* relaxation. These complex processes will not be discussed in detail; in rather oversimplified terms, they result from the fact that the *motion* of molecules or the relative motion of parts of molecules produces fluctuating local magnetic fields, causing changes in resonance conditions and the dissipation of energy as heat owing to the induced motion of electrons.

*The statement that a cluster of lines is a methyl quartet often appears in the literature about NMR. It is important to realize that the four lines are not methyl resonances but the splitting of a resonance of another proton because of its being coupled with a methyl group.

These processes are characterized by relaxation times T_1 and T_2. For the depth of analysis in this chapter, there are only two important results concerning relaxation: (1) $\Delta\nu_{1/2} = 1/\pi T_2$, in which $\Delta\nu_{1/2}$ is the resonance line width at half-height, and (2) T_2 depends on molecular motion. The second point is that rapid motion decreases the ability to transfer energy by spin-spin relaxation (because the nuclei are rarely at the correct relative orientation); hence, with rapid motion, T_2 is long and line width is narrow.

INSTRUMENTATION OF NMR

An NMR spectrometer needs both a fixed and a rotating radiofrequency magnetic field, a sample holder, and a detector of some sort. The basic instrument is shown in Figure 17-8. The sample is placed in a tube (A) that is situated between the poles (B) of a powerful electromagnet (usually 10^4–10^5 gauss). A coil of wire (C) in a plane perpendicular to the field H_0 of the electromagnet surrounds the sample. A radiofrequency transmitter generates a fixed, high frequency (approximately 10^8 Hertz or cycles per second) oscillating magnetic field in the coil of wire. The frequency in the coil is selected to be near the Larmor frequency corresponding to H_0. To

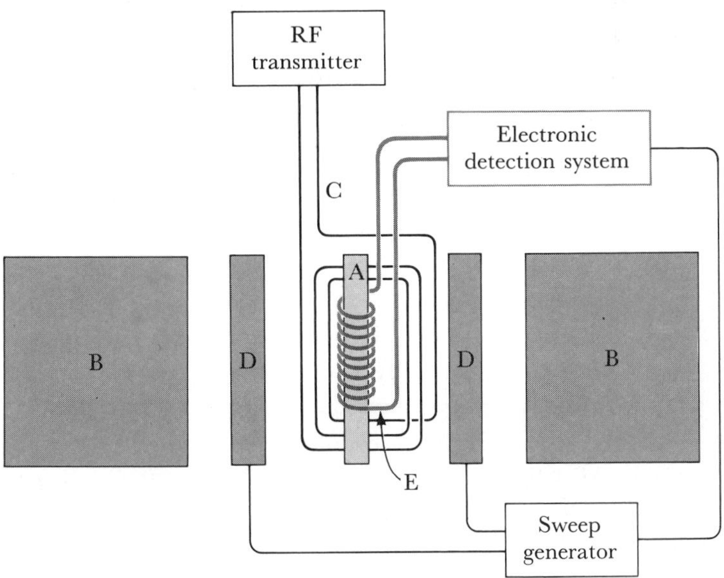

FIGURE 17-8
A sweep-type NMR spectrometer. See text for details.

achieve resonance, the magnetic field of a small accessory magnet (D) is increased slowly. (In some instruments, there is no accessory magnet and H_1 is varied by changing the frequency.) At resonance, the nuclear magnetic moments flip, and this sudden change in the magnetic field induces a current in a small coil (E) that is at right angles to both H_0 and H_1. Variations of this design exist, the most important being the Fourier transform NMR spectrometer, which will be discussed later.

DATA OBTAINED FROM AN NMR SPECTRUM

Four parameters of NMR spectra are used to obtain information about the molecules under study: the line position (i.e., its chemical shift), the area of the line, the band width, and the splitting. It is worth reviewing the information supplied by each parameter.

1. The *position* of the line or its chemical shift is determined by local magnetic fields (from other nuclei or unpaired electrons) and induced magnetic fields (produced by the surrounding electrons), resulting in the magnetic field in the region of a nucleus being different from the applied field. In other words, in a particular molecule, all nuclei of the same type (i.e., chemically identical) need not be in the same environment and hence can have different chemical shifts.
2. The area or intensity of a line is proportional to the *number* of nuclei in a given chemical environment. This means that a solution of molecules having a concentration that is twice that of another solution will have a peak with twice the area and that, if a molecule contains nuclei in two functional groups (e.g., methyl) in identical environments, or magnetically equivalent, the resonance will also have double area. This property differs from all other types of spectroscopy—that is, the area of an NMR line is *independent* of the electronic environment of the nucleus; there is no such thing as absorbance probability or quenching.
3. The *band width* (i.e., width at half-height) in spectra of macromolecules is primarily determined by molecular motion.
4. The *splitting* or clustering of lines is caused by the interactions of one nucleus that is covalently coupled with another. It is, however, often confused with a collection of nearby lines corresponding to different nuclei. The criterion for a multiplet is the relative areas of the components—for example, 1:1 for a doublet, 1:3:1 for a triplet, 1:3:3:1 for a quartet, 1:4:6:4:1, for a quintet, and so forth.*

---

*The relative intensities are proportional to the coefficients of the binomial expansion.

RULES FOR INTERPRETING NMR SPECTRA

Like those of other spectroscopic methods, the theory of NMR is not yet adequate for calculating the molecular parameters of complex molecules. However, if peaks are identified (how this is done is discussed in a later section), certain properties of the molecule can be deduced from measurements of the four parameters given in the preceding section. In practice, a collection of "rules" is used, some empirical and some theoretical. Those that are applicable to biological molecules are given in Table 17-2.

TABLE 17-2
Rules used in the interpretation of NMR spectra.

1. The magnitude of the chemical shift for protons in a particular chemical group (e.g., methyl, ethyl, hydroxyl) varies with the molecule of which it is a part (e.g., a particular amino acid or nucleotide).
2. If a compound that contains a proton having a particular chemical shift is in a polymer, the chemical shift generally changes owing to the proximity of other molecules or groups.
   a. The largest shifts (20–30 ppm) are caused by the presence of unpaired electrons (i.e., paramagnetic centers).
   b. Shifts of approximately 2 ppm are usually the result of ring-current fields. The magnitude of the ring-current shift is roughly in the following order: flavins and porphyrins > tryptophan and nucleotides > histidine > tyrosine and phenylalanine.
   c. Electric fields from many charged groups (such as in aspartate, glutamate, lysine, arginine, and histidine) cause shifts. The shift is small if the group is on the exterior of a folded macromolecule and hence interacting primarily with the solvent but is larger if the group is internal.
3. A change in the chemical shift of a particular proton following treatment of the macromolecule by a physical or a chemical agent means a change in the structure of the macromolecule in the region surrounding the particular proton. If the treatment includes a change in pH, the shift may result from a change in the ionization state rather than a three-dimensional change.
4. Splitting implies that two sets of nuclei are covalently coupled to one another. The number of lines of the multiplet gives a rough indication of what group it might be. For example, a quartet often means that a methyl group is nearby. The separation of the components of a multiplet is related to the angle between the groups containing the nuclei. From a comparison with compounds whose structure is known from x-ray analysis, the bond angles for simple, small molecules can sometimes be determined. In a macromolecule, any agent that alters the separation of the components of a multiplet must alter the bond angle; this indicates what may be a subtle conformational change.
5. In a multicomponent, rapidly self-associating (e.g., dimerizing) system, the magnitude of the chemical shift, the width of the line, and the number of lines (Figure 17-5) is a result of the relative amount of time spent in each chemical

TABLE 17-2
(continued)

or physical state. In such a system, a change in the equilibrium or the rate can change the width of a particular line or produce a splitting like that shown in Figure 17-5. Such changes can be used to study equilibria.

6. The width of a line is a measure of the relative mobility of a nucleus. If the nucleus is moving very rapidly, as it might in a small molecule, the peaks are very narrow. If the nucleus moves more slowly, as in a macromolecule diffusing or in a rigid part of a macromolecule, the lines are broader. Often band width can be used to estimate the mobility of a functional group or residue in a polymer, if the change in band width is not due to the cause given in rule 5 or to a change in position of a very close, unresolved and unrelated line.
7. The binding of a ligand to a site in a macromolecule usually affects the spectra of *both* the ligand and the macromolecule. Although such factors as the redistribution of a local charge, changes in orientation at a distance from the binding site, and so forth, may cause spectral changes, in many cases, the changes are effected by nuclei in the binding site. The usual effect is a shift in line position accompanied by a broadening of the line because there is likely to be decreased freedom of motion in the binding region.

TECHNICAL PROBLEMS OF NMR AND THEIR SOLUTIONS

The first problem of NMR spectroscopy is obviously to obtain a spectrum. For proton resonance spectroscopy (still the most frequently done) of macromolecules in which aqueous solvents are required, it is necessary to prepare the sample in D_2O rather than in H_2O to eliminate a powerful H_2O-proton resonance that obscures most of the resonances of the protons of interest (because H_2O is 55 molar). The necessity of using D_2O, unfortunately, eliminates the possibility of observing protons bonded to nitrogen or oxygen because these exchange rapidly with the solvent deuterons and are usually not observable (because of both the low sensitivity of the deuteron resonance and the fact that the lines are in distant parts of the spectrum). Hence, carbon-bound hydrogens are normally observed. To express all data as parts per million, a reference standard is frequently added. A common one is tetramethylsilane. Often, small amounts of H_2O are added as a standard, or the DHO, which contaminates all D_2O samples, can be used.

With special techniques using Fourier transform NMR, it is possible to suppress the H_2O resonance and work totally in H_2O solution. In this case, D_2O is used as a standard. This method has not yet had widespread use but will probably ultimately supplant most if not all of the work in D_2O.

For other nuclei (e.g., ^{13}C, discussed later), H_2O can be used as a solvent because the frequency range used is far from the H_2O resonance.

A second problem in NMR spectroscopy is the low sensitivity of detection. This difficulty results from the weakness of an NMR signal compared with the background noise of the instruments used. The method to reduce this problem is to sweep through the spectrum many times in succession, sum the results, and average the spectra (each sweep is usually called a *scan*). Because the noise is random, it will tend to cancel out, whereas the signal from the resonance will be enhanced. In practice, this is done with a small computer called a multichannel pulse-height analyzer. The success of this procedure is shown in Figure 17-9. After one scan, the signal is not discernible but, after several hundred, the resonances become clear. By this procedure, a particular proton can be detected at a concentration of 10^{-3} M. Other nuclei have lower sensitivity (see Table 17-1) and require higher concentration. Fourier transform NMR, which increases the signal-to-noise ratio tenfold, allowing detection of protons at 10^{-4} M, is described in a later section.

Resolution is the third major problem because of the huge number of peaks (100–1,000) in the spectra of many macromolecules. However, because the chemical shift is a linear function of the size of H_0, the resolution problem has been attacked by constructing stronger magnets. The largest fields (~ 50 kgauss) are currently generated by superconducting magnets. At the same time, of course, it is necessary to increase the fre-

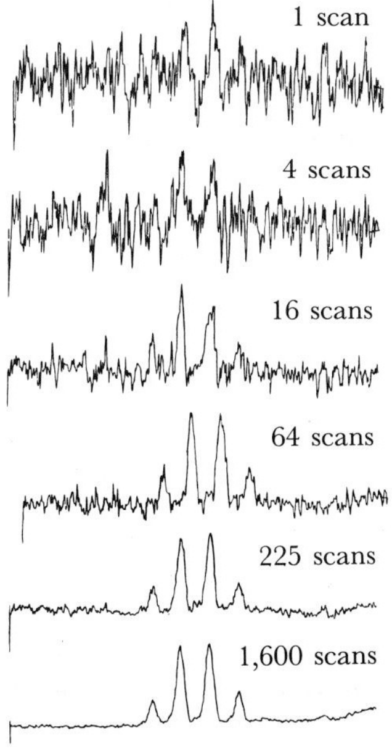

FIGURE 17-9
The methylene quartet of ethylbenzene as a function of the number of scans. [Courtesy of Varian Associates.]

quency of the electromagnetic radiation; radiofrequency generators of 3×10^8 Hertz are now available but not yet in common use. With nuclei other than protons (e.g., ^{13}C), the resolution problem is less severe because chemical shifts are much larger.

After a high-resolution spectrum has been obtained, the final problem is to identify the peaks—that is, to identify the nucleus producing a particular peak. For small molecules, this is not difficult; for large molecules, it is rarely simple, although the methods described in the following section are easy to understand.

ASSIGNMENT OF LINES FOR PROTEINS AND POLYNUCLEOTIDES

The identification of the peaks in polymers is usually done by the following methods:

1. The spectra of amino acids, nucleotides, and certain functional groups have been tabulated. How these spectra vary with pH and certain solvents is also fairly well known. Many of these substances have sufficiently characteristic sets of lines at particular positions that they can be recognized in complex spectra by simple inspection. Examples of three such spectra are shown in Figure 17-10. This is probably one of the most common methods for peak identification.
2. If a protein has a known sequence, it can be enzymatically fragmented and the spectra of some of the purified fragments can be determined separately. In this way, peaks can sometimes be shown to come from particular regions of the molecule. The limitation of this method is that the chemical shifts are not always the same in the fragments because the interaction with other parts of the molecule may not be

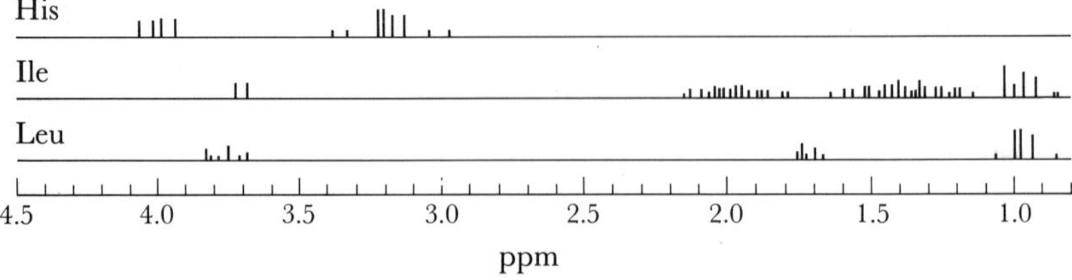

FIGURE 17-10
Spectra of histidine, isoleucine, and leucine in the zwitterion form. The relative heights of lines correspond to areas in true spectra.

the same. This method is usually not of great value for very large structures but is applicable to oligopeptides.

3. Chemical reactions with a particular residue or the binding of a ligand to the residue may result in the loss or displacement of particular lines. This is usually accompanied by significant conformational changes so that the effect on the resonance may be indirect. Occasionally, however, the situation is more favorable and peaks can be identified in this way.
4. Some polymers may be selectively deuterated by the growth of an organism in a medium containing a particular deuterated residue. Proton resonances from the deuterated residue will vanish because of the weakness of the deuteron signal. Inversely, an organism can be grown in a completely deuterated medium but with a single protonated residue; then, only peaks corresponding to the protonated species will be seen. This method is reliable but is very expensive owing to the high cost of deuterated compounds.
5. Spectra have been obtained for proteins and polynucleotides whose three-dimensional structure is accurately known from x-ray diffraction analysis. The spectra of samples can be compared with them. This is a highly useful and fairly common method except for the reservation that the structure derived from x-ray analysis is that of a crystal, powder, or otherwise dry material, whereas NMR spectra are obtained in solution. Also, there are not very many macromolecules whose structures are accurately known.

In summary, no single method is ever used to identify a peak; usually several must be used—along with a fair amount of trickery or reasoning.

DETERMINATION OF THE CHEMICAL STRUCTURE OF SMALL MOLECULES

The most straightforward application of NMR is the determination of the structure of simple organic molecules. This is a common technique of organic chemistry but is rarely used in biochemistry. It is based on the fact that, for simple molecules, the chemical shifts for functional groups such as methyl, amino, and so forth, are fairly well defined. Therefore, if a chemical formula is known, the NMR spectrum can give some indication of the functional groups. The multiplets of spin-spin coupling also show the proximity of functional groups (as in Figure 17-6). This method will not be discussed here because it can be found in many standard texts on organic chemistry.

USE OF NMR TO STUDY PROTEIN STRUCTURE

NMR has the potential for supplying a great deal of information about protein structure because the parameters of the spectra are sensitive to changes in both sequence and conformation. For example, in the absence of any interaction, the spectrum of a polypeptide or protein would be the sum of the spectra of the constituent amino acids, all of which are well known. However, this is not the case, as seen in Figure 17-11, which shows the spectrum of the enzyme lysozyme in its native and denatured (i.e., random coil) forms and that calculated from the amino acid composition. It is clear that there are many differences between the observed and calculated spectra; this is the usual situation. Although each protein has its own differences, there are a few general features of the spectra that seem to be characteristic of most proteins. For example, it is generally observed that large chemical shifts occur if a proton is in an amino acid that is either amino-terminal, carboxyl-terminal, or next to the nitrogen atom of a peptide bond, and small shifts are produced if the proton is next to either the carbon atom of a peptide bond or a carbon or nitrogen atom of the nearest neighbor amino acid; small shifts are also produced by titration of a carbon or nitrogen atom in the nearest-neighbor amino acid. Furthermore, if a protein is in a native configuration, there are characteristic chemical shifts of certain amino acid protons produced by being in an α helix. Unfortunately, the magnitude of the changes in chemical shift, band width, and intensity due to tertiary structure and the proximity to distant amino acids (e.g., because of ring-current shifts or hydrogen bonds) vary for each protein, and there are no truly general principles that might be stated as empirical rules.

That the NMR spectra of amino acids are affected by the secondary and tertiary structures of proteins theoretically allows the determination of three-dimensional structure; however, this has not yet been done both because of the great complexity of the analysis and because of the following problems.

First, a typical protein has about 850 lines, many of which overlap. Because line width is affected by the mobility of the molecule (Table 17-2, rule 6), roughly increasing with decreasing diffusion coefficient and therefore increasing molecular weight, useful spectra (in which lines can be resolved) are obtained at present only for molecules whose molecular weight is less than 25,000. Second, the *unambiguous* identification of lines requires that the complete amino acid sequence be known. This is not always available and is not even sufficient owing to the interactions resulting from the tertiary structure. Observing changes in line position during the titration of particular amino acids or the effects of selective deuteration is an aid in identification but does not always work. The net result is that in a complex protein a significant number of lines remain unidentified. Nonetheless, a great deal of information can be obtained.

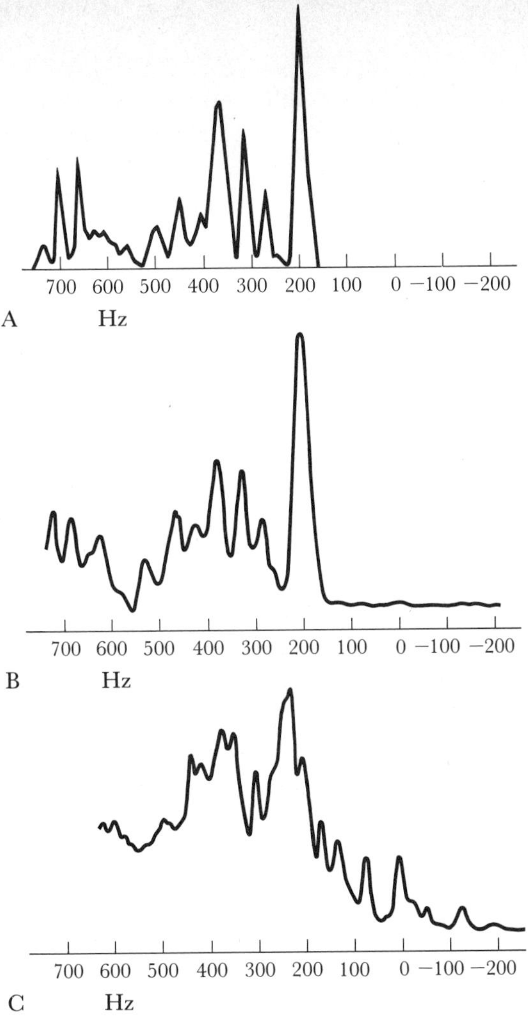

FIGURE 17-11
Part of the proton spectra of lysozyme: (A) calculated; (B) measured spectrum of denatured lysozyme; and (C) measured spectrum of native lysozyme. [Redrawn with permission from C. C. McDonald and W. C. Phillips, *J. Amer. Chem. Soc.* 91(1969):1513–1521. Copyright by the American Chemical Society.]

Currently, NMR is used successfully in five ways to study proteins: (1) to determine the fraction of amino acids in the α-helical configuration in order to confirm that the structure derived from x-ray diffraction is valid in solution; (2) to monitor helix-coil transitions; (3) to determine the conformation of selected regions of the protein (e.g., around a particular amino acid); (4) to observe the binding of small molecules and metal ions to selected regions of the protein (by observing either the ligand or the protein spectra); and (5) to study paramagnetic active sites in electron-transfer proteins.

The types of observations made and the method of analysis of the data can best be seen in a few examples.

Example 17-A. Fraction of residues of poly-γ-benzyl-L-glutamate in the α-helical configuration.

The proton of the α-carbon of some amino acids has two different chemical shifts, depending on whether the carbon atom is in an α helix or a random coil (Figure 17-12). Because the area of a line is proportional to the number of protons giving that resonance and because there is one α-carbon proton per unit of the α helix, the relative areas of the two lines can be used to calculate the fraction of amino acids in α-helical configuration. The fraction is simply

$$\frac{A_H}{A_H + A_R}$$

in which A_H and A_R are the areas of the helix and the random-coil lines, respectively.

Example 17-B. Unfolding of proteins.

The denaturation of proteins produces substantial changes in NMR

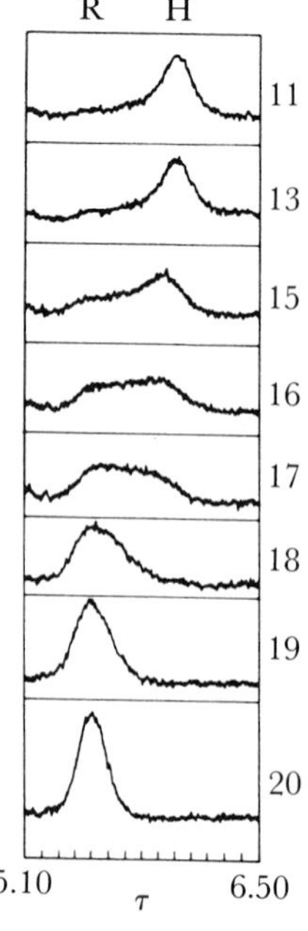

FIGURE 17-12
Selected region of the spectrum of poly-γ-benzyl-L-glutamate in $CDCl_3$ with added trifluoracetic acid (TFA): the peak represents the α-carbon proton only; H and R refer to the line positions when in the α-helical and random-coil configurations, respectively; the numbers indicate the percentage of TFA added. [Redrawn with permission from J. A. Ferretti, B. W. Ninham, and V. A. Parsegian, *Macromolecules* 3(1970):30–42. Copyright by the American Chemical Society.]

spectra. Lines become narrower because freedom of motion is increased and the chemical shifts of identical amino acids approach one another. By following line width and position as a function of pH, temperature, the concentration of various denaturing agents, and so forth, it is possible to follow the unfolding of various parts of the protein independently.

For example, let us consider the thermal denaturation of a hypothetical protein containing several histidine (His) lines and one tryptophan (Trp) line (Figure 17-13). The subscripts N and R denote line positions for native and random-coil configuration, respectively. At 25 °C, the protein is in the native configuration, and various histidines have different chemical shifts owing to their various environments. The tryptophan line must represent many residues in the same environment because its area is greater than any of the histidine lines (remember that area is proportional to the number of protons). At 37 °C, the histidine lines have shifted to the position they have in a random coil. Hence, the region of the protein containing the histidines must have relatively low thermal stability compared with that containing the tryptophan. At 45 °C, a line appears at Trp_R. Because the Trp_N line remains, the tryptophans must be in at least two different parts of the protein; that is, one that is disrupted at 45 °C and one that is not. From the relative

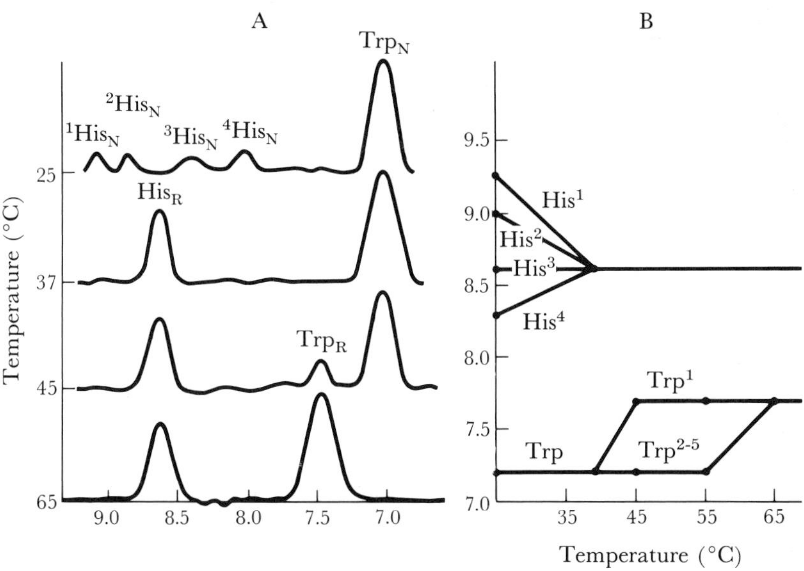

FIGURE 17-13
A. Spectra for a hypothetical protein as a function of temperature. Only the relevant lines have been drawn. In a real protein, as many as 50 additional lines might be in this region. B. Plot of line position as a function of temperature for each line of part A.

areas of Trp_R and Trp_N (1:5), one-sixth of the tryptophans are in a region that is a random coil at 45°C. Note that the Trp_R line is as wide as the Trp_N line, suggesting that the Trp must have been in a highly flexible region of the molecule. By 65°C, the Trp_N line is gone so that all regions containing tryptophan are denatured.

Note that this NMR study provides more information than would a hydrodynamic analysis of denaturation. For example, if this protein had been studied by viscometry (Chapter 13), one would have simply observed a continuous change in viscosity over the same temperature range and this would be a reflection of the *average* state of the molecule. However, the NMR analysis gives information about the state of *particular* regions of the molecule.

Example 17-C. Properties of the active site of ribonuclease A.
The structure of the active site of pancreatic ribonuclease A has been elucidated by examining the spectrum of the C-2 histidine protons when the inhibitor 3′-cytidine monophosphate (3′-CMP), which binds to the active site, is added at various concentrations and over a range of pH values. This enzyme has several histidine residues but only three (12, 48, and 119)* show significant changes in chemical shift when 3′-CMP is added (Figure 17-14A and B), and histidine-119 shows by far the greatest. Clearly, each histidine is in a new environment when 3′-CMP is added. The problem is to determine which histidines participate in direct bonding. A study of the chemical shift of the C-2 proton peak of imidazole (the aromatic part of histidine) as a function of phosphate concentration (i.e., using imidazole instead of histidine and phosphate instead of 3′-CMP) shows that the curve relating chemical shift and concentration matches that for histidine-119. Hence, the chemical shift for histidine-119 *may* be a result of binding to the phosphate group of 3′-CMP (Table 17-2, rule 7). This can be verified by further studies with the inhibitors 2′-CMP and 5′-CMP, whose binding constants to ribonuclease differ. The chemical shift of histidine-12, as well as the pH dependence of the shift, is the same for all three inhibitors even though their binding differs. In other words, histidine-12 is not responsive to factors that affect the binding of the inhibitors, suggesting that histidine-12 is not the site of binding of the phosphate. On the other hand, both the chemical shifts and the pK of binding for histidine-119 differ for the three inhibitors—that is, histidine-119 is sensitive to the position of the phosphate and therefore probably forms a direct bond with it. However, the changes in histidine-12 are sufficiently great that it is probably in

*These numbers and their alternate designations, His-12, His-48, and so forth, refer to the position of the amino acid in the complete amino acid sequence.

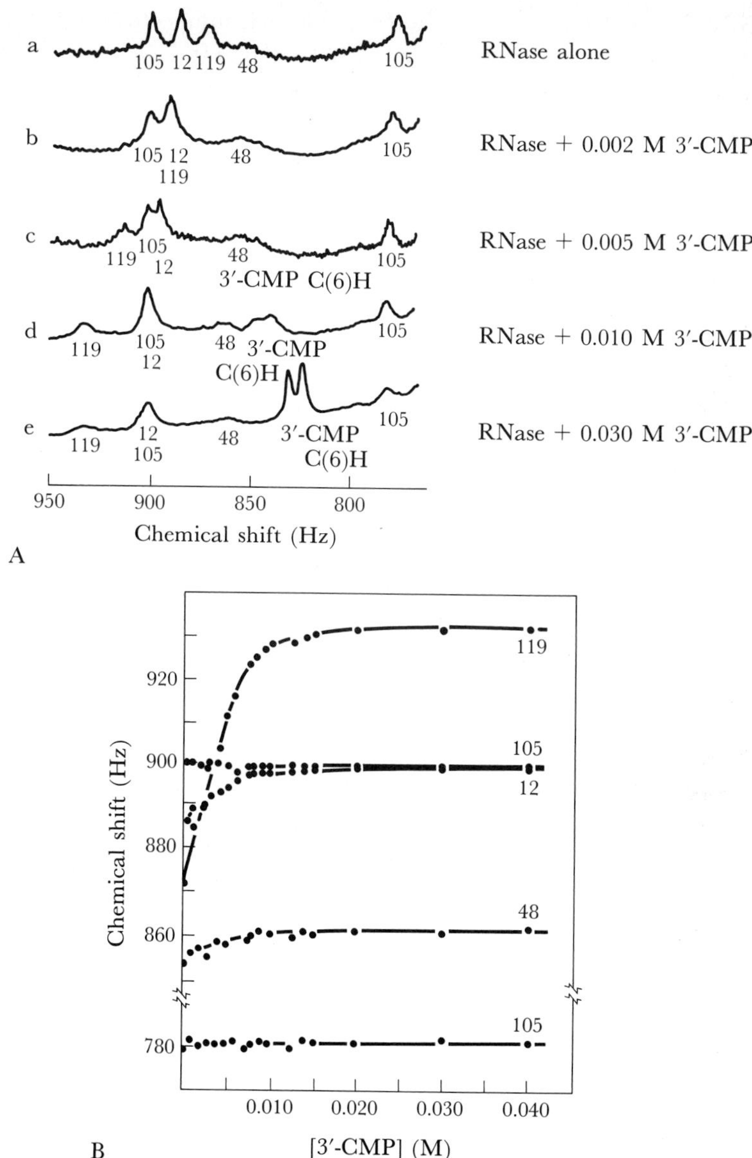

FIGURE 17-14
A. A part of the proton spectrum showing peaks corresponding to various histidines in RNase as a function of the amount of added 3′-CMP. Numbers refer to the particular histidine. B. Plot of the data in part A. [Data from D. H. Meadows and O. Jardetzky, *Proc. Nat. Acad. Sci.* 61(1968):406–413.]

or near the active site. The small shift in histidine-48 is probably a result of a conformational change in the protein induced by the binding.

Adjacent to histidine-119 is phenylalanine-120. The lines produced by this residue have not been unambiguously identified; however, because there are five protons in the ring and because there is a peak in the spectrum in the range of ppm corresponding to aromatic amino acids, the peak can be presumed to correspond to a phenylalanine. This peak shows a large change in position when 3′CMP is added, but none when phosphate alone is added. This suggests that binding also includes an interaction between the pyrimidine ring of 3′-CMP and phenylalanine. Note that this is just suggestive because it is not clear that the line is actually due to phenylalanine-120.

Further evidence for the participation of the pyrimidine part can be derived from an examination of the spectrum of 3′-CMP (by rule 7, Table 17-2, these regions of the ligand that take part in binding should also show spectral changes). Indeed, it is found that the ring-proton lines are affected by binding and the shifts are the same for 2′-, 3′-, and 5′-CMP. This agrees with the idea that the pyrimidine ring participates in binding and, because its binding is insensitive to the position of the phosphate, it does not bind to histidine-119.

Figure 17-15 shows a postulated structure for the 3′-CMP complex. Although the details might not be correct, it is shown to give an idea of the sophisticated conclusions that may be drawn from NMR data.

Example 17-D. Mode of binding of sulfacetamide to the protein bovine serum albumin (BSA), assayed from the spectrum of the ligand.

The spectrum of sulfacetamide consists of a single large peak due to the methyl group and a cluster of four peaks due to the aromatic ring. When BSA is added, the line widths of the methyl and the aromatic protons increase substantially (Figure 17-16), which indicates a decreased mobility of all of these protons (Table 17-2, rule 6). A trivial explanation of the change in line width is that the increased viscosity of the solution by the protein decreases the proton mobility. However, this explanation cannot be correct because, as shown in the figure, the line width of the aromatic protons increases more rapidly than that of the methyl protons. If the broadening were due solely to the increase in solution viscosity caused by the addition of the BSA, all lines would broaden by the same factor; indeed, this is the case if γ-globulin (which does not bind sulfacetamide) is added. Hence, the change in line width probably results from binding; because the change is greater for the aromatic protons, the aromatic ring is probably the principal site of interaction. Note also that the change in line width can serve as an assay for binding; indeed, further studies of the effects of concentration of both BSA and sulfacetamide, of pH, and of ionic strength on line

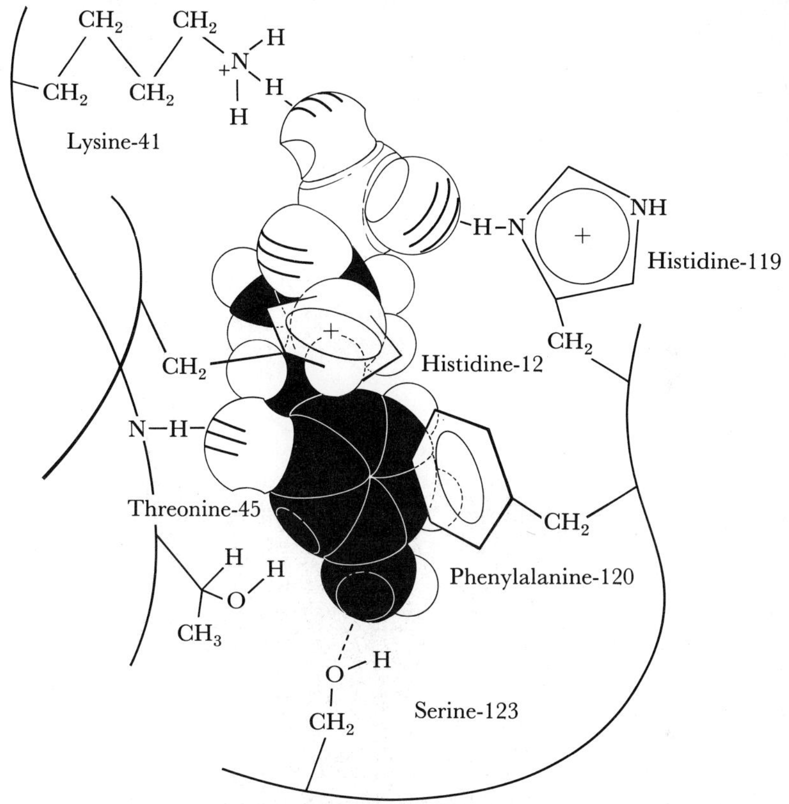

FIGURE 17-15
Postulated structure of the 3′-CMP ribonuclease complex. [From D. H. Meadows, G. C. K. Roberts, and O. Jardetzky, *J. Mol. Biol.* 45(1969):491–511.]

width allows the calculation of the dissociation constant, and thus the determination of the role of ionic forces in binding.

Example 17-E. Effect of a cofactor on enzyme-substrate binding. Yeast alcohol dehydrogenase converts ethanol into acetaldehyde but only if the enzyme cofactor nicotine adenine dinucleotide (NAD) is present. A hint about the role of NAD in this reaction can come from NMR studies. The nicotinamide proton lines are broadened when the enzyme and the NAD are mixed. This broadening is specific to the enzyme (i.e., other proteins do not do this) and is therefore due to enzyme-cofactor binding and not to the viscosity effect mentioned in Example 17-D. If ethanol (the substrate) or acetaldehyde (the product)

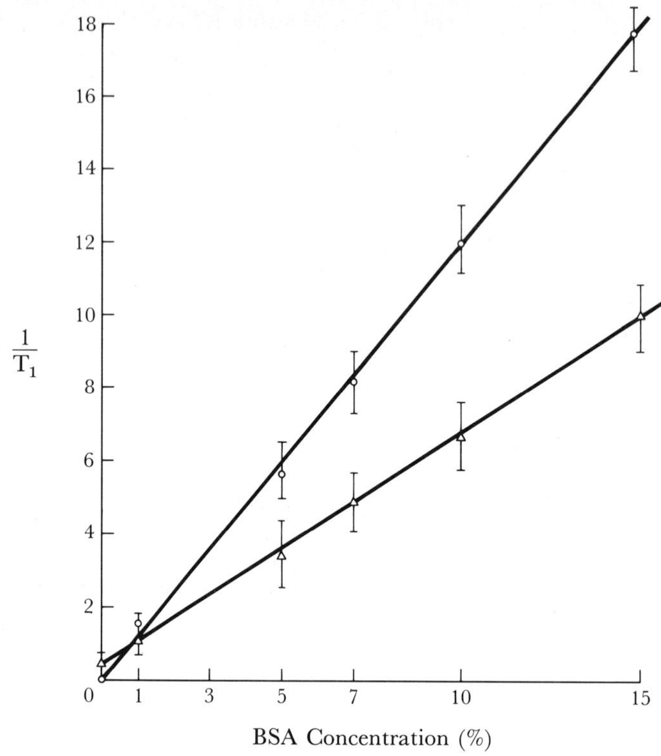

FIGURE 17-16
Relaxation rate ($1/T_1$) of 0.1 M sulfacetamide as a function of bovine serum albumin (BSA) concentration in D_2O: circles represent *p*-aminobenzene sulfonamide protons; triangles represent methyl protons. [From O. Jardetzky and M. G. Wade-Jardetzky, *Mol. Pharmacol.* 1(1965):214–230.]

is added to the enzyme-NAD mixture, the methyl proton lines of both substrate and product are broadened. No broadening of any lines occurs if the enzyme is absent, so that both substrate and product must bind to the enzyme-cofactor complex. If enzyme, substrate, and product are mixed *in the absence of NAD,* there is no broadening of the methyl lines of either ethanol or acetaldehyde. Hence, the NAD must be bound to the enzyme in order for the substrate to bind.

Example 17-F. Binding of a polymer to a protein—the DNA-histone complex.

DNA binds tightly to histone to form nucleohistone; this is of great interest because of its role in chromosome structure.

From studies of amino acid sequences, it is known that most histones have an asymmetric distribution of amino acids (i.e., the amino acids are arranged in three clusters—a basic region, an acidic region, and a nonpolar region). The amino acid protons of these three regions have distinguishable resonance lines. If histones and DNA are mixed, the lines representing the basic region are broadened. This cannot be due to the effect of DNA viscosity because, if so, all lines would be broadened equally. This suggests that the DNA is bound to the basic region of the histones.

Studies of the effect of salt concentration on histones in the absence of DNA show that the lines of the nonpolar amino acids are broadened with increasing salt concentration. This is taken to mean that the histones self-associate by means of the nonpolar regions to generate large structures (Table 17-2, rules 5 and 6). From this, one may reasonably propose that the histone-protein gel (called chromatin) consists of histones self-associated by means of the nonpolar groups with DNA bound to the basic groups.

USE OF NMR TO STUDY POLYNUCLEOTIDE STRUCTURE

To date, NMR has had little use in the study of polynucleotides, partly because (1) DNA is so rigid and has such a low diffusion coefficient that the lines are extemely broad and in fact barely observed (its molecular weight is far above the 25,000 limit for proteins); and (2) most of the structural features of polyribonucleotides were fairly well worked out by the time NMR workers turned their attention to nucleic acids. An exception is in determining the structure of the smaller polynucleotide transfer RNA (tRNA), an example of which follows.

Example 17-G. Base-pairing structure of transfer RNA.
For studying the base pairing of tRNA, it is possible to use an H_2O solution because there are two clusters of lines far from the H_2O lines that give a great deal of information. One cluster consists of guanine N-1 protons of the guanine-cytosine (GC) pairs and the other, uracil N-3 protons of the adenine-uracil (AU) pairs. Because the intensity of a line is proportional to the number of protons present, the areas of the lines in the two clusters are indicative of the number of hydrogen bonds in the molecule. Furthermore, the ratio of the areas of the two clusters indicates the ratio of GC to AU pairs. To obtain information about the distribution of GC and AU pairs in the molecule, one can make use of the fact that tRNA can be enzymatically cleaved at particular sites to generate a discrete set of fragments, each of which apparently maintains the original hydrogen bonds. By examining the spectra of fragments, it is possible to determine for each fragment the number of GC and AU base pairs. Therefore, whenever the base sequence of a particular tRNA

is known, it is possible to construct a three-dimensional model of the tRNA, at least insofar as hydrogen bonding is concerned. One of the most important conclusions drawn from these studies is that the so-called cloverleaf model (Figure 17-17) is the correct description of tRNA in solution.

SPIN LABELING

An immediate aim of NMR spectroscopy is to determine distances—either between residues of a macromolecule or between a ligand and a binding site. The technique of spin labeling allows fairly precise measurements to be made. The principle underlying the method is that a paramagnetic

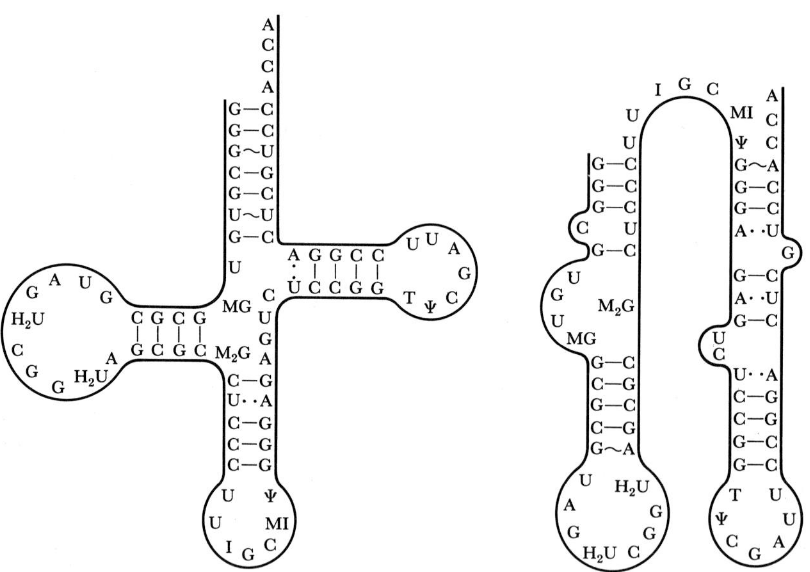

FIGURE 17-17
Two of several possible structures of yeast alanyl tRNA arranged so that a large fraction of the bases participate in hydrogen bonding. The hydrogen bond of a GC pair is indicated by —; that of an AU pair is indicated by · ·; the weaker hydrogen bonds found in other base pairs are indicated by ~. An NMR analysis yielding the number of base pairs of various types indicates that the cloverleaf structure (left) is the correct one. The symbols are: G, guanosine; C, cytidine; A, adenosine; U, uridine; I, inosine; H_2U, dihydrouridine; T, ribosylthymine; MG, methylguanosine; M_2G, dimethylguanosine; Ψ, pseudouridine; MI, methylinosine. [The base sequence is that reported by R. W. Holley, J. Apgar, G. A. Everett, J. T. Madison, M. Marquisee, S. H. Merrill, J. R. Fenswick, and A. Zamer, *Science* 147(1965):1462–1465.]

center (i.e., an unpaired electron), with which there is associated a very large, fluctuating magnetic field generated by the unpaired electron spin, produces a substantial broadening of the lines of any nucleus that is exposed to that fluctuating field. The amount of broadening decreases as the distance between the unpaired electrons and the nucleus of interest decreases or as the time of exposure of the nucleus to the paramagnetic center is decreased. With protons, an effect is seen with distances of as much as 40 Å for ^{13}C nuclei (which have not yet been studied appreciably by the spin-labeling technique), the distance is much greater. In practice, a substance called a spin label (i.e., one that contains a free radical) is introduced into the system by reacting it either with a particular residue of a protein or with a ligand that is bound by the protein. Common substances are bromoacetamide, *N*-(D,L-2,2,5,5,-tetramethylpyrrolidinyl)-maleimide, and various compounds containing a nitroxide group. To obtain meaningful information from the spectra of a spin-labeled system, it is necessary to assume that the introduction of the spin label itself does not affect the conformation of the macromolecule. Usually, this is verified either by demonstrating identical enzymatic activity under a variety of conditions or by the identity of physical properties. These are not powerful criteria and constitute the only weakness of the method. The use of spin labels is best seen by an example.

Example 17-H. Distance between binding site and an amino acid residue of a protein.

If histidine-15 of the enzyme lysozyme is covalently spin-labeled with nitroxyl nitrogen and the enzyme substrate *N*-acetylglucosamine (NAG) is added, the resonance of the acetamido methyl proton of NAG is broader than if the spin label were absent (Figure 17-18). By means of a simple equation, the distance between the unpaired electron in the nitroxyl group and the proton responsible for the resonance can be calculated from the line width. After the length of the spin-labeled compound itself has been corrected for, the true distance between the proton and histidine-15 can be calculated. Distances determined in this way agree with values determined by x-ray diffraction to about 85% to 95%.

FOURIER TRANSFORM NMR

As explained earlier, traditional NMR instruments generate spectra either by choosing a fixed frequency of the electromagnetic field or by fixing the field and varying the frequency. To obtain a good signal-to-noise ratio, the scan must be repeated many times and the data averaged. This is a slow and inefficient method because only a narrow region of the spectrum

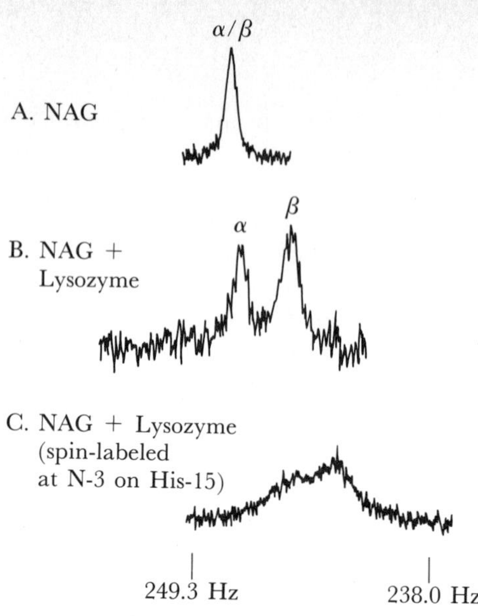

FIGURE 17-18
Methyl lines of *N*-acetylglucosamine: (A) alone; (B) plus lysozyme; and (C) plus spin-labeled lysozyme. [Redrawn with permission from Wien, Morrisett, and McConnell, *Biochemistry* 11(1972):3707–3716. Copyright by the American Chemical Society.]

is being recorded at any one time. Consider the effect of using a fixed field and exciting with a short but very intense radiofrequency *pulse.* Because a pulse in fact consists of a *superposition of a spectrum of frequencies* (the mathematics behind this statement—i.e., the concept of the Fourier transform—can be found in almost any text on wave theory), a pulse can simultaneously excite the entire range of resonance frequencies. This is the principle of Fourier transform NMR spectroscopy and constitutes a great technical advance. This manner of excitation requires substantial changes in instrumentation because, to obtain a spectrum, the signal received (Figure 17-19) must be unraveled. This is done by an on-line digital computer. The details of the receiving and processing system are too complex to be described here but are handled by commercially available instrumentation. Again, repeat scans are made, and the data are averaged to improve the signal-to-noise ratio, as in conventional NMR work; however, the advantage of the Fourier transform method is that it takes roughly one-tenth the time to obtain a given signal-to-noise ratio, thus providing a tenfold increase in sensitivity. What this means is that (1), at concentrations used in conventional NMR, lines are detected that are normally obscured by noise and resolution is improved and (2), to obtain standard sensitivity, one-tenth the concentration is required. The ability to use lower concentrations has two important consequences: (1) other nuclei such as ^{13}C, for which the natural abundance and sensitivity is low (Table 17-1), can be studied; and (2) systems that are highly concentration dependent, such as enzyme reactions or associating polymers, can be studied, using a range of concentrations. In the future, it is likely that all NMR work will use Fourier transform instruments.

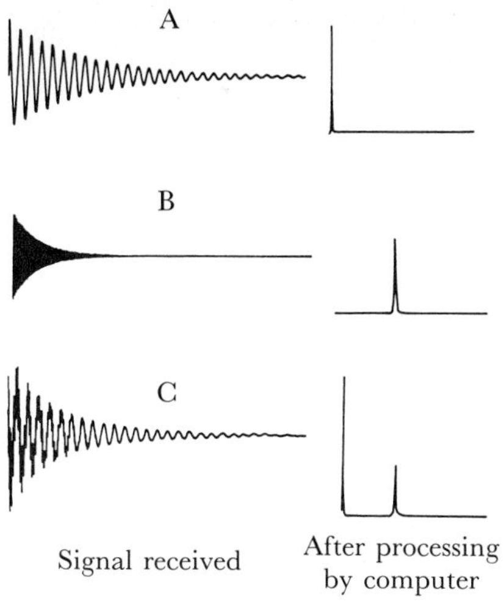

FIGURE 17-19
Signal received in Fourier transform NMR: the left-hand side shows the signals received by the computer; the right-hand side shows the frequency spectrum after computer processing. The parameters to notice are the frequency and the decay time. The frequency determines the position on the frequency scale at the right, and the decay time determines T_2, which is inversely proportional to band width: (A) low frequency, slow decay; (B) high frequency, rapid decay; and (C) a mixture of A and B. [From R. Dwek, *Nuclear Magnetic Resonance in Biochemistry,* Clarendon Press, 1973.]

USE OF NUCLEI OTHER THAN THE PROTON

To produce a resonance line, a nucleus must have nonzero spin. This, unfortunately, eliminates the common isotopes ^{12}C and ^{16}O (an empirical rule is that, if both the atomic number and mass number are even, $I = 0$). A second requirement is that the natural abundance of the isotope and the magnetic moment be high enough that a signal can be detected (the relative sensitivities of nuclei are roughly proportional to the cube of the ratio of the magnetic moments). It is also advantageous that the spin is not $>1/2$ because, for such nuclei, the lines are very broad owing to a phenomenon called quadrupole relaxation (which will not be discussed).

Three nuclei that satisfy these criteria are ^{13}C, ^{19}F, and ^{31}P. They are of special interest because the large polarizability of their electron clouds results in chemical shifts of approximately 400, 400, and 600 ppm, respectively. Because these shifts are about fifty times as great as those for the proton, resolution is much greater and it is possible to look at more subtle environmental differences.

The isotope ^{13}C has tremendous potential because of its wide range of chemical shifts, its greater sensitivity to structural changes, and its simple spectra (i.e., because of its low abundance, it is rarely adjacent to another ^{13}C so that the splitting of lines is unusual). Furthermore, with ^{13}C, information is obtained about the environment of carbon atoms in addition to that of protons. Furthermore, D_2O, which may sometimes have an effect on structure, is not needed. However, because of its low natural abundance (1%), it has been very difficult to detect by conventional means. This problem has to some extent been alleviated by Fourier transform NMR, although the sensitivity is still lower than with protons. The ultimate solution to the sensitivity problem will be the development of techniques for enriching with ^{13}C, although along with the obvious advantage of enrichment will come the disadvantage of splitting. However, even with the current problem of low sensitivity, ^{13}C NMR has already made important contributions to biochemistry.

The isotope ^{19}F has the disadvantage that it is not normally present in biological materials. It has the advantage that it can be introduced at defined sites in a molecule and therefore serves as an external probe. If the number of sites is small, the spectra consist of a small number of lines. In proteins, ^{19}F is used in two ways: (1) by introducing a ^{19}F at a site on the protein and observing the ^{19}F resonances as a function of various agents—pH, temperature, ligands, and so forth; (2) by using a fluorinated ligand and observing the signal from both a bound and an unbound ligand. This can be used to study chemical exchange, determine various parameters of binding, and learn something about the structure of the binding site. The isotope ^{31}P has so far had only limited use with nucleotides, membranes, and phospholipids but should be of greater importance in the future.

Selected References

Becker, E. D. 1969. *High Resolution Nuclear Magnetic Resonance.* Academic Press.

Bovey, F. A. 1969. *Nuclear Magnetic Resonance Spectroscopy.* Academic Press.

Bovey, F. A. 1972. *High Resolution Nuclear Magnetic Resonance of Macromolecules.* Academic Press.

Dwek, R. A. 1973. *Nuclear Magnetic Resonance in Biochemistry.* Clarendon Press.

McDonald, C. C., and W. D. Phillips. 1970. "Proton Magnetic Resonance Spectroscopy," in *Fine Structure of Proteins and Nucleic Acids,* vol. 4, edited by G. D. Fasman and S. N. Timasheff, pp. 1–48. Dekker.

Roberts, G. C. K., and O. Jardetzky. 1970. "Nuclear Magnetic Resonance Spectroscopy of Amino Acids, Peptides, and Proteins." *Advan. Protein Chem.* 24:448–545.

Sheard, B., and E. M. Bradbury. 1970. "Nuclear Magnetic Resonance in the Study of Biopolymers and Their Interaction with Ions and Small Molecules." *Progr. Biophys. Mol. Biol.* 20:187–246.

Problems

17-1. How many proton resonances would you expect to see in (a) liquified methane, (b) liquified ethane, (c) chloroform, (d) formaldehyde, (e) acetone, (f) methanol, (g) ethanol in CCl_4?

17-2. Would you expect the proton spectrum of a mixture of phenylalanine and glycine to be the sum of the spectra of each? Would the total concentration of each be important?

17-3. Explain why ^{13}C spectra contain fewer lines than proton spectra and why the number of lines would increase as the abundance of ^{13}C is increased.

17-4. Suppose that you are trying to determine the mechanism of action of a certain enzyme. You hypothesize that the substrate passes through two intermediates in being converted into the product. The two intermediates are known compounds but extremely difficult to identify in the reaction mixture by standard chemical and physical tests. Explain how NMR might help.

17-5. Suppose that an enzyme has a single histidine. On the addition of the substrate, each histidine line moves from 2 to 4 ppm. Can you unambiguously state that histidine is in the binding site?

17-6. Proton spectra of proteins invariably show pronounced changes on denaturation. The usual changes are line positions and widths. Would you expect the appearance or disappearance of splitting to be one of the changes?

17-7. A particular enzyme reaction involves an enzyme, a cofactor, and a substrate molecule. Design an experimental protocol using NMR to distinguish the following possibilities: (a) the enzyme binds the cofactor before the substrate; (b) the substrate binds the cofactor and the complex binds to the enzyme; (c) the enzyme binds the substrate but there is no reaction until the cofactor is added.

17-8. What differences might be expected between the spectra of water and ice? Explain.

17-9. What kinds of spectral changes might accompany dimerization of a protein?

17-10. A proton spectrum is obtained for an enzyme. About fifty lines show chemical shifts about twice those normally found for proteins. Furthermore, the lines are broader than those normally encountered. What might be the cause, and how could you test your hypothesis?

17-11. Would you expect there to be observable differences in the proton spectrum of a linear and a circular DNA?

PART VI

Miscellaneous Methods

CHAPTER 18

Miscellaneous Methods

This chapter comprises brief descriptions of miscellaneous, but useful, techniques that do not belong with the topics treated in the other chapters.

HYDRODYNAMIC SHEAR AND SONIC DEGRADATION

Both the viscosity and the sedimentation coefficient of a DNA solution decrease when the solution flows rapidly through a narrow orifice (e.g., a hypodermic needle, a fine-tipped pipette, or a narrow bore viscometer) and when it is stirred or shaken too vigorously. These effects are due to breakage of the molecule by hydrodynamic shear forces.

As explained in Chapter 13, if a liquid containing a rodlike molecule is flowing through a tube, the shear force will tend both to orient the rod and to stretch it. If the molecule is stretched to its limit, it will break, somewhere near its middle. Similar shear forces can be generated by the violence of turbulent agitation or violent stirring. By controlling the flow rate of solutions through narrow tubes or the speed of rotation of the blades of a stirrer, it is possible to break a molecule successively into halves, quarters, eighths, and so forth. As might be expected, the critical shear force (i.e., the smallest shear stress that will break the molecule) decreases with increasing molecular weight.

In general, shearing is avoided in the laboratory in order to isolate DNA of maximum molecular weight and to maintain the size of isolated DNA. Table 18-1 gives the effects of various laboratory practices on DNA of various molecular weights. Note that the higher the molecular weight, the more care is required.

TABLE 18-1
Extent of breakage of double- and single-stranded DNA of various molecular weights by common laboratory operations.

Operation	Double-stranded DNA $M = 25 \times 10^6$	Double-stranded DNA $M = 106 \times 10^6$	Single-stranded DNA $M = 12.5 \times 10^6$	Single-stranded DNA $M = 53.0 \times 10^6$
Pipetting				
with 1-ml serological pipettes	Safe	Safe	Safe	Safe
with 0.1-ml pipettes	Safe	Some breakage	Safe	Breaks
Filling centrifuge cells				
with number 22 needles	Safe	Safe	Safe	Breaks
with number 24 needles	Safe	Some breakage	Some breakage	Breaks
Pouring	Safe	Safe	Safe	Breakage with splashing
Mixing with another solution	Safe	Safe	Safe	Breakage if densities differ
Shaking	Safe	Breakage with turbulence	Breakage with turbulence	Breaks
Bubbling air	Safe	Safe	Safe if slow	Breaks

General trend

1. Sensitivity to shear increases with molecular weight.
2. A single-stranded molecule is more sensitive than a double-stranded molecule having the same length.

Determination of critical shear values can also be used as a semiquantitative analytical tool. For example, if a molecule is known to be linear, it is possible to estimate its molecular weight from the shear stress required to halve the molecule. It is important to understand, though, that the actual shear stress is never measured. What is done is to take a series of DNA molecules whose molecular weights are known, stir solutions of them at various stirring speeds, and plot the lowest speed necessary to produce halving versus molecular weight. This provides a standard calibration that can be used to estimate M for a sample by measuring the critical stirring speed for the sample. Figure 18-1 gives an example of such an experiment. A calibration curve could also be prepared by flowing the sample through

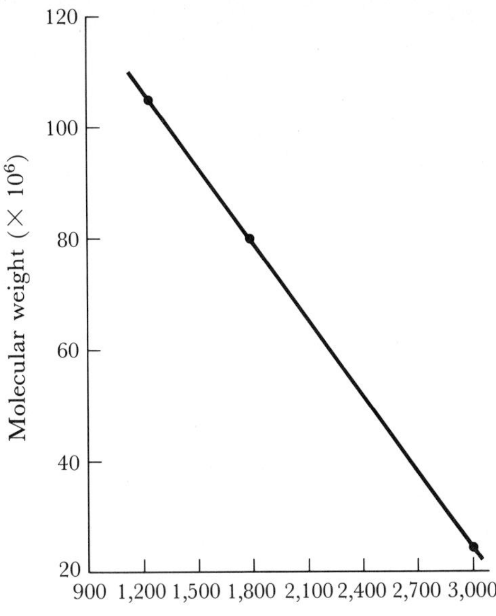

FIGURE 18-1
Typical calibration curve, relating stirring speed that produces 50% breakage for DNA having various molecular weights. DNA solutions are 20 μg/ml in 1 M NaCl. These curves are strongly concentration dependent, the critical speed decreasing markedly with lower DNA concentration. To determine molecular weight, it is necessary to use the same DNA concentration and the same stirring system.

narrow tubes and plotting critical flow rate versus molecular weight; however, the stirring procedure is easier.

If the molecular weight is known, one can tell if the molecule is nonlinear because an anomalous shear sensitivity results. For example, if a circular molecule of molecular weight M were stretched, it would form a loop having length equal to that of a molecule of $M/2$ and would also be twice as thick. Hence, its shear sensitivity would be one-quarter that expected from its molecular weight (Figure 18-2).

To make an estimate of molecular weight from shearing studies, it is necessary to maintain a constant DNA concentration because the resistance to shear increases substantially with concentration ("self protection").

Preparation of half-molecules, quarter-molecules, and smaller fractions is also of great value in isolating specific regions of a DNA molecule

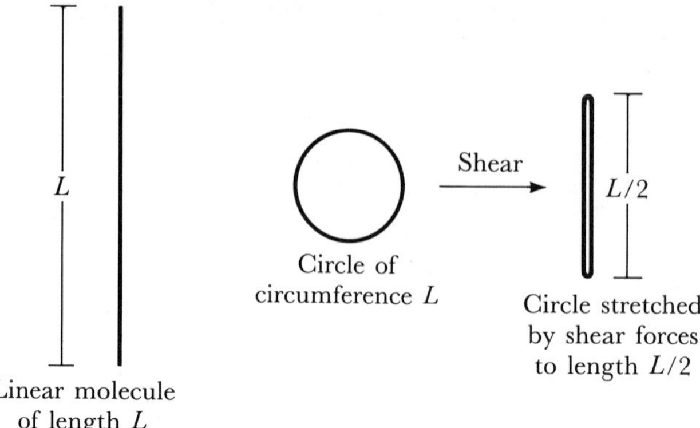

FIGURE 18-2
Shearing of a circle. A circle of circumference L is stretched to a nearly linear structure of length $L/2$, having twice the number of strands.

because frequently the fragments are separable by their buoyant density in solutions of cesium salts (see Chapter 11).

Hydrodynamic shear can also be used to disrupt large structures, even whole cells. For example, one of the most common methods of breaking cells is to expose them to intense sonic fields, although it is not commonly understood that sonication causes breakage by hydrodynamic shear. When a liquid is placed in a high-intensity sonic field (either by putting it in a resonating cavity or, more commonly, by inserting a vibrating probe into the liquid), the liquid *cavitates*—that is, microbubbles rapidly form and collapse. Cavitation is caused by the dissolved gases coming out of solution as bubbles, and the disruption of cells, aggregates, or macromolecules is caused by the violent collapse of these bubbles. (In extremely intense sonic fields, some chemicals decompose to produce free radicals, which can cause other types of molecular damage.) It is not clear whether the collapse of the bubble produces turbulence or local laminar flow, but the mechanism of breakage is certainly by means of hydrodynamic shear; in fact, by controlling the sonic intensity, DNA molecules have been successively broken into halves, quarters, and so forth. Controlled sonication has rarely been used as an analytical tool; its principal use is to rupture cells in preparation for the isolation of internal components. In general, the technique is simple and the only precaution necessary is to avoid heating. Because sonication pours a considerable amount of vibrational energy into water, the temperature rises rapidly. The usual protocol is to chill samples to ice temperature, sonicate for 30 seconds, recool, and repeat. In this way, a low temperature can be maintained.

HYDROGEN EXCHANGE

Many substances dissolved in D_2O or tritiated water (^3H_2O) will exchange some of their hydrogen atoms with D or ^3H. Amino acids, nucleosides, short-chain polypeptides, proteins in the random-coil configuration, and single-stranded nucleic acids rapidly exchange hydrogen atoms bound to nitrogen, oxygen, and sulfur atoms; carbon-bound hydrogen exchanges much more slowly. In proteins, for chemical reasons, exchangeable protons on some amino acid side chains (e.g., serine OH and the NH_2 of glutamine and asparagine) exchange much more rapidly than those in the peptide bond and in the amides of glutamine and asparagine. These two classes are distinguished by their pH dependence—the first class having a minimum at pH 7, the second at pH 3. Each class may be subdivided according to whether the protons participate in the formation of hydrogen bonds. Because the chemical rate of exchange is usually much slower than the making and breaking of hydrogen bonds (which controls the access of the solvent to the groups that are hydrogen-bonded), the rate of exchange of any group is the product of the chemical exchange rate and the fraction of time exposed to the solvent. Thus, the primary effect of a structure such as a hydrogen bond is to reduce the fractional exposure of the group to the solvent because the group is exposed only when local opening takes place. Hence, by measuring the chemical exchange rate for free groups and the rate when in a macromolecule, the fraction open at any given time can be determined.

In double-stranded nucleic acids, the exchange of a hydrogen-bonded proton is also greatly facilitated by an agent that breaks the hydrogen bonds. In summary, a potentially rapidly exchangeable proton exchanges poorly if it is hydrogen-bonded and this can be used to determine the number of hydrogen bonds, the effectiveness of various agents as denaturants, or the dynamic state of a macromolecule. When the technique was first employed, only deuterium exchange was followed because ^3H_2O was not available. Detection of the deuterium was fairly difficult and was done either by density determination in a Linderstrøm-Lang density-gradient column (see Chapter 12), or by infrared spectrophotometry (Chapter 14). However, since ^3H_2O has become available, tritium exchange has totally replaced deuterium exchange because ^3H is easily detected by its radioactivity.

The tritium-exchange technique, which employs gel chromatography with various molecular sieves (Chapter 8, Figure 8-21), is performed by means of the following procedure. Tritiated water is added to a protein or nucleic acid in H_2O solution. At various times, samples are removed and applied to the column. The ^3H_2O is strongly retarded by the column, whereas proteins and nucleic acids pass through rapidly. Using a column whose length is such that the protein or nucleic acid emerges after 10 seconds, the concentration of unbound ^3H is reduced approximately 10^8-fold. Fractions are taken from the column and the radioactivity in each fraction is determined by scintillation counting; the concentration of protein or nucleic acid is determined by spectrophotometry (Chapter 14), or, in some cases, by color tests for proteins. Hence, the number of exchanged hydrogens per unit weight or per molecule (if the molecular weight is known) can be measured as a function of time.

Several important results have been obtained from the ^3H- exchange method. For example, the protons in the hydrogen bonds of double-stranded DNA exchange. From the pH effect, kinetics, and equilibria, it has been shown that the hydrogen bonds in double-stranded DNA are continually breaking and reforming; this process is called *breathing* and provides a potential mechanism for such processes as the initiation of DNA synthesis and DNA-DNA pairing in genetic recombination. A second example is the determination of the number of hydrogen bonds in transfer RNA from a measurement of the fraction of exchangeable hydrogens that exchange rapidly; these data have been used to substantiate one of the proposed structures of tRNA. A third example is a major effort that is now in progress to study the kinetics of protein folding by determining the number of hydrogen bonds as a function of time following the transfer of a denatured protein to conditions resulting in reformation of the native structure.

EQUILIBRIUM DIALYSIS

Equilibrium dialysis is a convenient way to measure the binding of a small molecule to a macromolecule, as long as the small molecule is dialyzable and an assay for the small molecule exists. The principle of equilibrium dialysis follows. A solution of macromolecules is placed inside a dialysis bag (Chapter 7), as shown in Figure 18-3. The bag is suspended in a medium containing a particular concentration of a small molecule that binds to the macromolecule. The small molecules then diffuse into the bag. If no macromolecules were present, the concentration inside and outside the bag would be the same. However, because the macromolecules are present, the concentration inside the bag is greater by virtue of the number of bound molecules, because the concentration of unbound molecules within the bag will always equal the concentration outside the bag. Measurement of the internal and external concentrations as a function of total concentrations of both the small molecule and the macromolecule yields the binding constant, k, as follows:

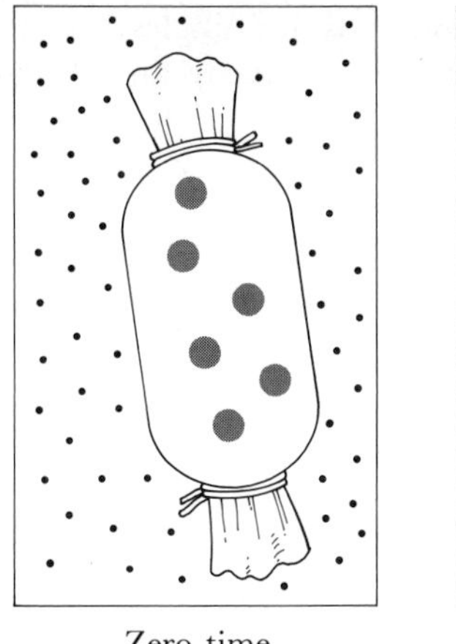

Zero time

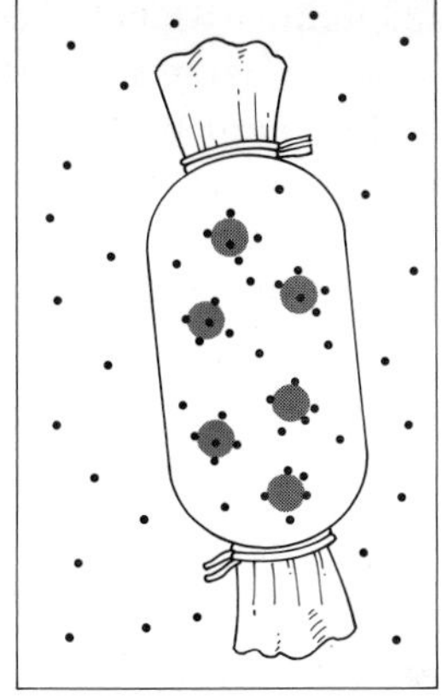

At equilibrium

FIGURE 18-3
Equilibrium dialysis. A dialysis bag filled with macromolecules (shaded circles) is placed in a solution containing dialyzable small molecules (solid circles) that can bind to the macromolecules. At equilibrium, the concentration of *free* small molecules is the same inside and outside the bag. Because the macromolecules bind some of the small molecules, the *total* concentration of small molecules is greater inside the bag than outside.

$$k = \frac{[\text{unbound}][\text{macromolecule}]}{[\text{complex}]}$$

in which the brackets signify concentration and "unbound" means the concentration of the small molecules within the bag. Because [unbound] = [outside of bag] and [complex] = [inside bag] − [unbound] = [inside bag − outside bag], the equation can be written in terms of measurable quantities:

$$k = \frac{[\text{outside}][\text{macromolecule}]}{[\text{inside} - \text{outside}]}$$

To obtain reliable values of k, certain conditions must be met: (1) the nonspecific binding of ligand and macromolecule to the dialysis bag itself must be small and preferably measurable; (2) binding must be strong because of the following effect, which tends to indicate an artifactual negative binding (which, of course, is meaningless). That is, if the macromolecule is very large and does not bind the small molecule at all, its great size takes up so much volume that it excludes the small molecule from the solution within the dialysis bag. Hence, binding must always be great enough that this negative effect is negligible.

Although equilibrium dialysis is a valuable method for determining binding parameters, it can also be used to *detect* binding proteins. This technique was used in the classical experiments by Walter Gilbert on the isolation of the *E. coli* Lac repressor. Because the Lac repressor binds the inducer isopropylthiomethyl galactoside (ITMG), cell extracts were fractionated and different fractions were assayed by equilibrium dialysis for the ability to bind ITMG. By using separation procedures that continually maximized the extent of binding, the Lac repressor was ultimately purified.

HYBRIDIZATION BETWEEN SINGLE-STRANDED POLYNUCLEOTIDES

If double-stranded DNA is denatured, the individual polynucleotide strands become physically separated (Chapter 1). If the ionic strength is low (e.g., 0.01 M), the negative charge of the phosphate groups of the sugar-phosphate chain repel one another. This has two effects: (1) two different strands cannot come into contact because of mutual electrostatic repulsion and (2) an individual strand tends to be highly extended and relatively rigid, because all parts of the backbone repel one another. Hence, at low ionic strength, no two nucleotide bases ever come near enough to form hydrogen bonds. If the ionic strength were suddenly increased (e.g., to 0.5 M), the charge on the phosphates would be neutralized and the strands would rapidly approach the random-coil configuration. However, in the absence of charge repulsion, hydrogen bonds would reform, that is, between guanine and cytosine and between adenine and thymine (for DNA) or adenine and uracil (for RNA). Because of the flexibility of these long polynucleotide strands, the probability of forming intrastrand (same strand) hydrogen bonds is greater than that of forming interstrand (different strand) hydrogen bonds, unless the concentration is very high. The intrastrand bonds will tend to be between very short, complementary tracts of bases (Figure 18-4), because long, complementary sequences will not frequently be found in a single strand. For steric reasons, not all bases will form hydrogen bonds, and this is reflected by the fact that the OD_{260} of the DNA is not restored to the original value for native DNA (Figure 18-5A). Because the hydrogen-bonded regions are very short, they have low thermal stability and can be disrupted at temperatures well below that needed to denature native DNA (Figure 18-5B). Therefore, the intrastrand

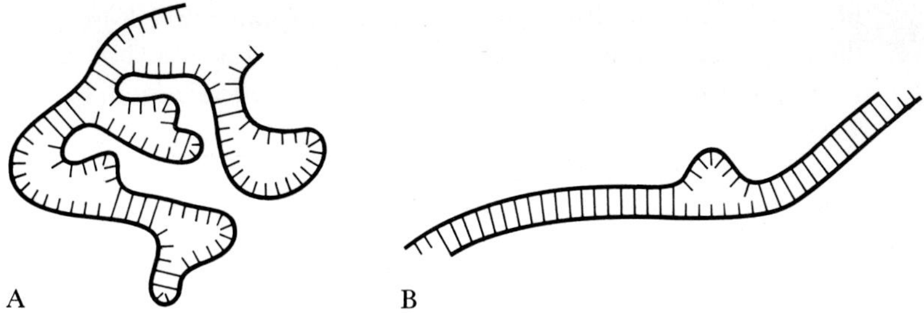

FIGURE 18-4
A. Internally aggregated single-stranded DNA. Base pairs are formed between different parts of the strand, thus making a very compact structure.
B. Renatured DNA, showing one short region of noncomplementarity.

hydrogen-bonded DNA can be converted to single-stranded random coils at temperatures and ionic conditions in which native DNA would be stable, and it is expected that, given sufficient time for the complementary single strands to find one another, native DNA will reform. This is indeed the case, and this process is called *renaturation* or *annealing* (Figure 18-5C). The two-strand molecule need not consist of two identically complementary strands because one strand could be DNA and the other RNA and some regions (ranging from a single base pair to an extended tract) might be noncomplementary (Figure 18-4B). The process by which a two-strand

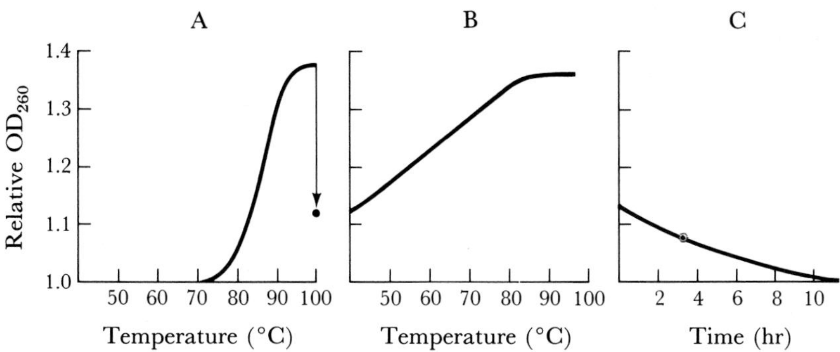

FIGURE 18-5
A. The OD_{260} of DNA in high ionic strength as a function of temperature. After reaching 100°C, the temperature has been lowered to 20°C. The OD_{260} drops, as indicated, to a value of 1.12. B. The DNA that has been heated to 100°C and lowered to 20°C is reheated to give a broad curve of OD_{260} versus temperature. C. The DNA used in part B is heated to 88°C instead, for the times indicated. The OD_{260} drops, indicating reformation of hydrogen bonds (renaturation).

molecule is formed from unlike strands is usually called *hybridization,* although the terms renaturation and hybridization are used interchangeably. A description of the techniques for detecting hybridization, together with several examples of its use, follows.

Detection by Equilibrium Centrifugation in Cesium Chloride Solution

This technique is rarely used now but was so common at one time that it deserves brief mention. For the preparation of DNA-RNA hybrids, use was made of the fact that, for polynucleotides of roughly 50% GC content, the densities in CsCl of DNA and RNA differ by approximately 0.1 g/cc (a very large difference). Hence, a denatured and a renatured mixture could be centrifuged to equilibrium in a CsCl solution and hybridization was detectable by the appearance of a band at the appropriate density. For DNA-DNA hybrids of DNA from the same species of organism, usually one DNA was density labeled with ^{15}N, ^{2}H, or ^{13}C and the other with ^{14}N, ^{1}H, and ^{12}C. If the DNAs were from different organisms having different GC content, the intrinsic densities of the DNAs would not be the same and the hybrid would be detected by the appearance of a band indicating DNA of intermediate density (Figure 18-6). Note that in this technique hybridization is performed between DNAs in solution.

Hybridization on Nitrocellulose Filters

If single-stranded DNA in solutions of high ionic strength is passed through nitrocellulose filters, the DNA is bound by the filters (Chapter 7). If the filters are then dried in a vacuum, the DNA is bound so tightly that it cannot be washed off even at high temperature and low ionic strength. Complementary single-stranded DNA can then be hybridized to these bound DNA molecules. An immediate problem arises in that added single strands will, of course, bind to the filters whether or not DNA is already present on the filters. To avoid this problem, after the DNA is bound to the filters, the filters can be treated in various ways so that no further single-stranded polynucleotide can be bound. One way to do this is to wash the filter with a solution of bovine serum albumin, which seems to saturate the DNA-binding sites. After the filters have been prepared and treated, a small volume of a solution of radioactive single-stranded DNA or RNA is added and the mixture is incubated under conditions appropriate for hybridization—that is, high ionic strength and a temperature between 65° and 70°C. Complementary radioactive DNA or RNA binds to the DNA on the filter (see Figure 7-4 on page 151). The filter is then washed to remove unhybridized material; after it has dried, the radioactivity is counted.

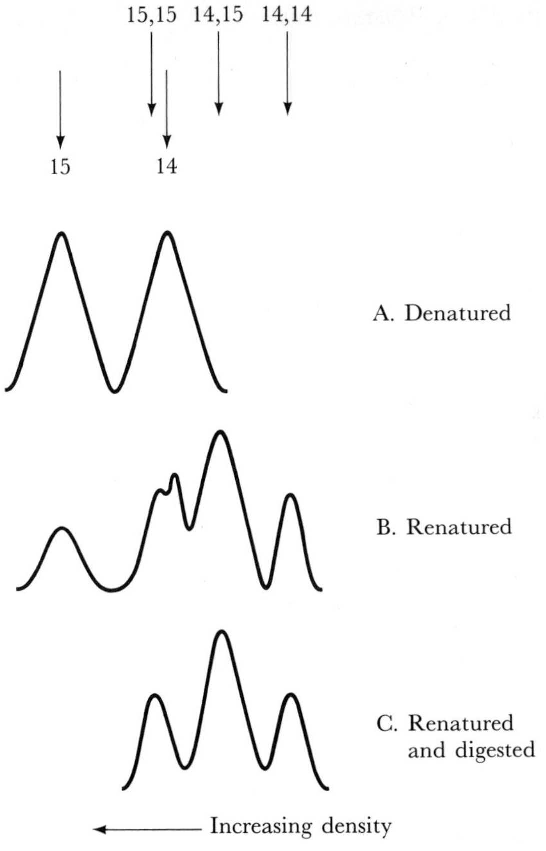

FIGURE 18-6
Detection of hybridization by equilibrium centrifugation in CsCl (Chapter 11). The curves are photometric traces of UV-absorption photographs, showing the DNA concentration distribution in the centrifuge cell: (A) mixture of denatured ^{14}N and ^{15}N phage T7 DNA; (B) after renaturation of the sample in part A (note that there are five peaks—three of renatured DNA and two of unrenatured DNA); (C) result of treatment of the sample in part B with an exonuclease specific for single-stranded DNA—only the renatured DNA remains. The positions of ^{14}N and ^{15}N single-stranded DNA are indicated by the arrows labeled 14 and 15, respectively; the positions of the ^{14}N, hybrid, and ^{15}N double-stranded DNA are indicated by the arrows labeled 1414, 1415, and 1515, respectively.

Examples of the Use of Hybridization

The following examples illustrate the use of hybridization both as an analytic tool and as a preparative one.

Example 18-A. Measurement of genetic relatedness between organisms. Genetic relatedness between two different organisms implies that there should be regions of the DNA that show partial or complete homology—that is, some extended base sequences should be identical or nearly so. To test this, nitrocellulose filters containing single-stranded DNA of organism A are prepared. Then radioactive, single-stranded DNA of very low molecular weight (M is usually reduced by sonication) from organism B is added, and hybridization is allowed to occur. The molecular weight of the B DNA is so low that a large piece of DNA will not be bound by virtue of a small region having homology. The fraction of the added B DNA that binds to the filter is a measure of the fraction of the sequences that are common to the two organisms. The observed fraction must be corrected for the efficiency of hybridization. This is usually done by adding single-stranded A DNA (radioactively labeled with another isotope) and determining the fraction bound. Hence, the fraction of B DNA that is homologous to A DNA is (fraction B bound)/(fraction A bound). The completeness of homology can be estimated from the thermal stability of the hybridized DNA, using certain empirical relations.

This method can also be used to measure the fraction of a DNA sample that is of a particular type. For example, suppose that a bacteriophage infects a host bacterium in a radioactive growth medium such that both newly synthesized phage DNA and bacterial DNA are labeled. The fraction of the label that is phage DNA can be determined by hybridization of the mixture to filters containing phage DNA. As in Example 18-A, a correction must be made for hybridization efficiency.

Example 18-B. Purification of messenger RNA. If a bacteriophage infects a bacterium in growth medium containing ^{3}H-uridine, radioactive phage and bacterial mRNA are synthesized. If the RNA is isolated and hybridized to filters containing single-stranded phage DNA, only phage mRNA is bound to the filters. Washing the filters will free them of all bacterial mRNA, because it is unbound. By placing the washed filter in a buffer and heating to the appropriate temperature for denaturing an RNA-DNA hybrid, phage mRNA will be dissociated from the filter and will be in the buffer. The filter can be removed and the phage mRNA is thereby purified.

Example 18-C. Identification of the DNA template for mRNA.
If mRNA, purified as in Example 18-B, is hybridized with DNA that is genetically deleted for certain sequences and then treated with RNase to digest unbound RNA, the amount of bound RNA will decrease if the mRNA is made partly from the deleted region. Thus, DNA sequences corresponding to an mRNA can be identified.

Example 18-D. Physical mapping of genetic deletions in DNA—an electron microscopic analysis.

If two single-stranded DNA molecules, one of which is deleted for certain sequences, are hybridized, hydrogen-bond formation will be complete except that the sequences in the normal DNA corresponding to the deletion will have nothing with which to form hydrogen bonds. Therefore, a double strand will result of the same length as that of the deleted DNA but with a single-strand loop at the site of the deletion whose length corresponds to the length of the deleted sequence. Such molecules, which are prepared by hybridization in liquid, can be examined with the electron microscope. The position and length of the loop allows the deletion to be mapped physically with respect to the length and ends of the normal DNA (Figure 18-7). This is described in greater detail in Chapter 3 (page 70, Figure 3-16).

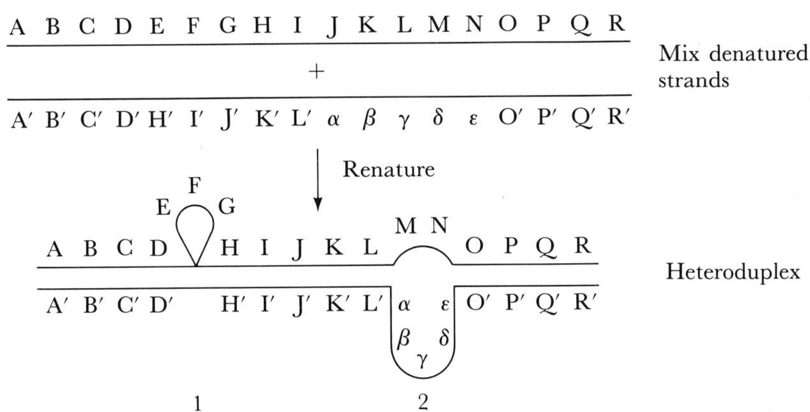

FIGURE 18-7
Diagram of heteroduplex analysis. Two different denatured DNA strands are renatured. Complementary bases (e.g., AA′) reform hydrogen bonds. Regions of DNA for which no complementary regions exist (e.g., EFG, MN, and $\alpha\beta\gamma\delta\varepsilon$) form loops. Note that there are two different types of loops. Type 1 contains one piece of single-stranded DNA and occurs with a true deletion. Type 2 contains two pieces of single-stranded DNA and occurs when each strand contains a different sequence in the same region.

CONCENTRATION OF MACROMOLECULES

In the course of isolating and purifying macromolecules, solutions often become very dilute and must be concentrated. A straightforward procedure is to concentrate the molecules by centrifugation, but often the sedimentation coefficient (Chapter 11) is too small for this to be effective. There are many other successful procedures of which six are described here.

Salting Out with Ammonium Sulfate

Charged solute macromolecules (e.g., macromolecules in polar solvents) are solvated and are thereby rendered soluble. If high concentrations of electrolytes are added, the solvent molecules are bound so tightly by the ions that they are unable to solvate the solute molecules. Hence, the solute molecules come out of solution. This is called salting out and is a useful technique for precipitating macromolecules such as proteins.

In particular, for most, if not all, proteins there is a concentration of ammonium sulfate (a highly soluble, highly pure, inexpensive reagent) above which the protein precipitates.* Hence, solid $(NH_4)_2SO_4$ can be added to a protein solution until precipitation occurs. The precipitate is collected by centrifugation and then redissolved in a smaller volume of whatever buffer is required. This rather general method sometimes fails if the total protein concentration of the starting solution is too low.

Both salting out and centrifugation are based on the removal of the sample from the solution. The following methods are based on the removal of water.

Flash Evaporation

In this technique, the solution is placed in a rapidly spinning flask, which is then evacuated (Figure 18-8). Because of the spinning, the liquid forms a relatively thin film, which significantly increases the surface-to-volume ratio. The vacuum serves to reduce the boiling point of water to room temperature or below. In this way, water can be removed without subjecting the molecules to temperatures that might cause denaturation or dissociation.

---

*The precise amount differs for each protein so that, by varying the amount, a protein mixture can be fractionated.

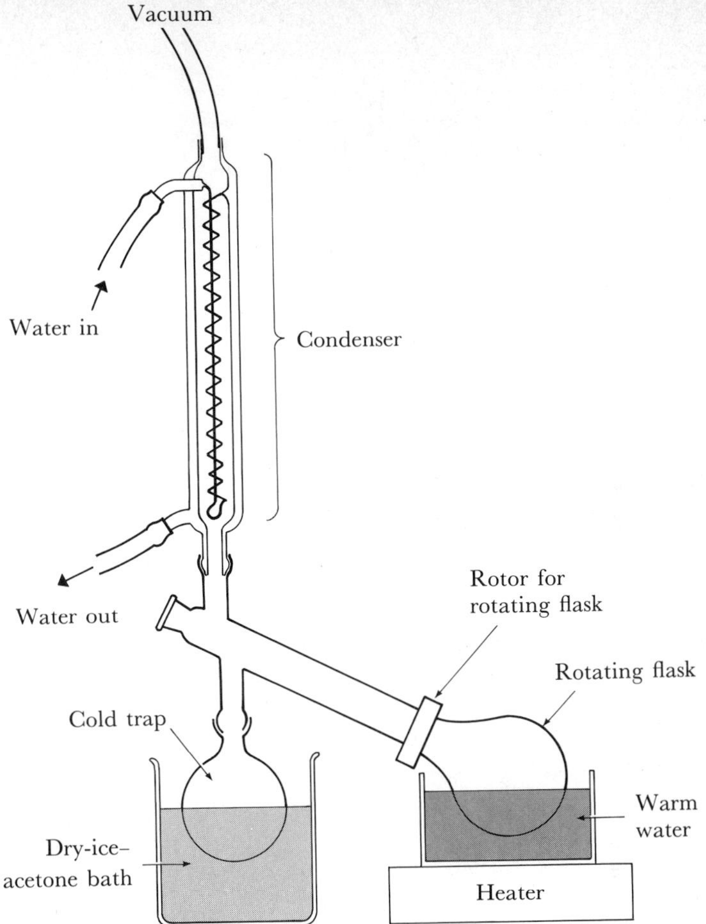

FIGURE 18-8
A flash evaporator. The sample is contained in the rotating flask and is heated by the water bath. The entire system is in vacuum so that the boiling point is depressed. Vapor is reliquified in the condenser and flows into the cold trap where it freezes. A special baffle (not shown) prevents the condensed liquid from flowing back to the rotating flask.

Lyophilization

This technique is based on the fact that, at sufficiently low pressure, ice sublimes. Hence, the sample is frozen and subjected to high vacuum. Note that both lyophilization and flash evaporation can be used to concentrate small molecules as long as they are nonvolatile.

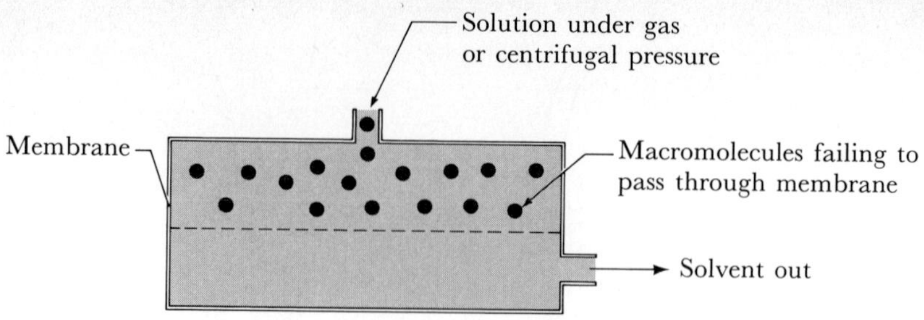

FIGURE 18-9
Concentration of a solution of macromolecules, using a Diaflo or Pellicon membrane. Under pressure, solvent and small solute molecules pass through the membrane. Macromolecules do not and are thereby concentrated.

Pressure Dialysis

Semipermeable membranes (e.g., Diaflo and Pellicon) that pass small molecules but not macromolecules were described in Chapter 7. If a solution is placed in a chamber, one wall of which is such a membrane, the rate of passage of the small molecules (including water) through the membrane increases with increasing pressure. Hence, when pressure is applied to the chamber, water is forced through and the macromolecules are concentrated (Figure 18-9).

Reverse Dialysis

If a solution is placed in a dialysis bag and the bag is allowed to remain in air, water will evaporate from the bag, thus concentrating the solution. If the bag is surrounded by a dry, highly soluble polymer that cannot pass through the membrane, water will leave the bag to dissolve the dry polymer. This tends to be faster than air-drying. The material most commonly used is polyethylene glycol (PEG).

A variation of this procedure is the Minicon multipurpose concentrator made by Amicon Corporation. The sample is placed in a chamber, which is schematically shown in Figure 18-10. A permeable membrane is backed by an absorbent pad. Water and salts pass through the membrane into the pad until the sample volume is such that it is in contact only with an impermeable seal. This is usually set for twentyfold concentration.

Hollow Fiber Membranes

The rate of liquid removal by pressure dialysis (see above) increases with the surface-to-volume ratio. If the solution is contained in hollow

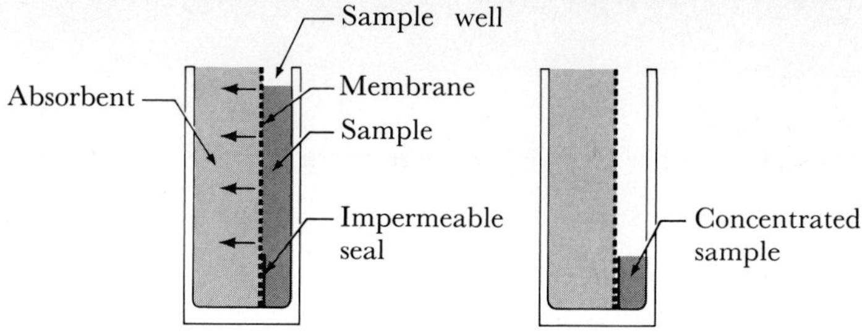

FIGURE 18-10
A Minicon concentrator. The sample is placed in the right-hand compartment. Liquid passes through the membrane and is absorbed by the material in the left-hand compartment. When the level of the liquid reaches the plastic barrier, concentration is complete.

porous fibers (see Chapter 7, Figure 7-10), the surface-to-volume ratio becomes huge. These devices (commercially available as Bio-Fiber, made by Bio-Rad Laboratories) consist of bundles of semipermeable, hollow fibers attached to filling and emptying ports. Pressure is established across the fiber wall and solvent passes through the fiber wall, leaving behind a more concentrated solution.

SEPARATION BY PARTITIONING BETWEEN DEXTRAN AND POLYETHYLENE GLYCOL SOLUTIONS

Concentrated aqueous solutions of the polysaccharides dextran and polyethylene glycol are immiscible. Many biological polymers, cellular components, and even cells show markedly different solubility in these two solutions and will therefore separate by partitioning. The standard procedure is to add the sample to either the dextran or the polyethylene glycol solution, add the other solution, shake for complete mixing, and then allow the two phases to separate. Table 18-2 lists several of the numerous materials that have been separated in this way.

FLOW BIREFRINGENCE

As explained in Chapter 13, elongated macromolecules can be oriented by flow under the influence of a shear gradient; the greater the axial ratio (ratio of length to width), the more easily the molecules are oriented. As explained in Chapter 2, oriented molecules exhibit birefringence—that is, they are capable of polarizing light. When birefringence is produced by a shear gradient, it is usually called flow birefringence. The amount of

TABLE 18-2
Examples of separation and concentration by partititioning.

Common solvent systems	Factors affecting separation
Dextran and polyethylene glycol	Ionic strength and pH
Dextran and hydroxypropyl dextran	Ionic strength
Dextran and methyl cellulose	pH
Dextran sulfate and polyethylene glycol	Particular ions
Dextran sulfate and methyl cellulose	Temperature

Substances separated

Native and denatured DNA
Covalently closed circular DNA and open circles or linear DNA
Native and denatured proteins
Various proteins (useful in enzyme purification)
Protein and nucleic acid
DNA and RNA
Various polynucleotides
Virus and virus-antibody complex
Various viruses and phages
Various microorganisms
Male and female *E. coli*
Different species of *Chlorella*
Different poliovirus strains
Intact and broken chloroplasts
Erythrocytes from various species of animals
Lymphocytes, leucocytes, and platelets
Spores and vegetative cells
Insulin-secreting granules and acid phosphatase particles from pancreatic β cells

SOURCE: P. A. Albertsson, *Partition of Cell Particles and Macromolecules,* Wiley-Interscience, 1971, and P. A. Albertsson, *Advan. Protein Chem.* 24(1970):309–341.

birefringence as a function of the shear gradient can be measured and this yields the rotational diffusional coefficient, a parameter from which the axial ratio can be calculated.

Consider a Couette viscometer (Chapter 13) with a stationary inner cylinder and a rotating outer cylinder. Between these cylinders is an annulus, which is observed by looking down the axis of the instrument. If there is a polarizer between the light source and the instrument and the emerging light passes through a second polarizer (the analyzer) whose axis is perpendicular to the first, no light reaches the observer. If the annulus contains a solution of elongate molecules and the outer cylinder is rotated, these molecules are oriented by the shear gradient and they resolve the light transmitted by the polarizer, so that some light is passed by the analyzer. However, because the system has cylindrical symmetry, there are two regions in which the average orientation is parallel to the direction of the polarizer (see Figure 18-11). At these points (which are opposite one another), there is no resolution of the polarized light and no light is transmitted by the analyzer. Similarly, there are two regions in which the molecules are oriented in the direction of the analyzer and again no light is transmitted. Hence, there are four regions that appear dark against the light background, as shown in Figure 18-11. This is called the *cross* and the angle χ is the *extinction angle.* It can be shown that

$$\chi = 45^\circ - \frac{15G}{\pi\Theta} + \cdots \tag{1}$$

in which G is the shear gradient and Θ is the rotational diffusion coefficient. Hence, by plotting χ versus G, a straight line with slope $-15/(\pi\Theta)$ is obtained. Equation (1) is valid only for low G and a small axial ratio.

This method has been used primarily to determine axial ratio and, in some cases, to monitor helix-coil transitions. In the latter case, when the molecule becomes disordered, it can be less easily oriented and the birefringence decreases.

The flow birefringence apparatus can also be used to measure intrinsic birefringence or dichroism (Chapter 14) of molecules that can be oriented almost totally at low shear forces (e.g., DNA). Most molecules cannot be completely aligned because the required shear gradient would be so great that turbulence would set in and disrupt the orientation. However, this is not the case for very elongated molecules.

Two simple examples serve to indicate how orientation within a molecule can be measured.

Example 18-E. Orientation of nucleotide pairs in DNA.

The bases of DNA are dichroic—that is, they have a preferred direction of absorption of UV light with respect to their axes (Chapter 14).

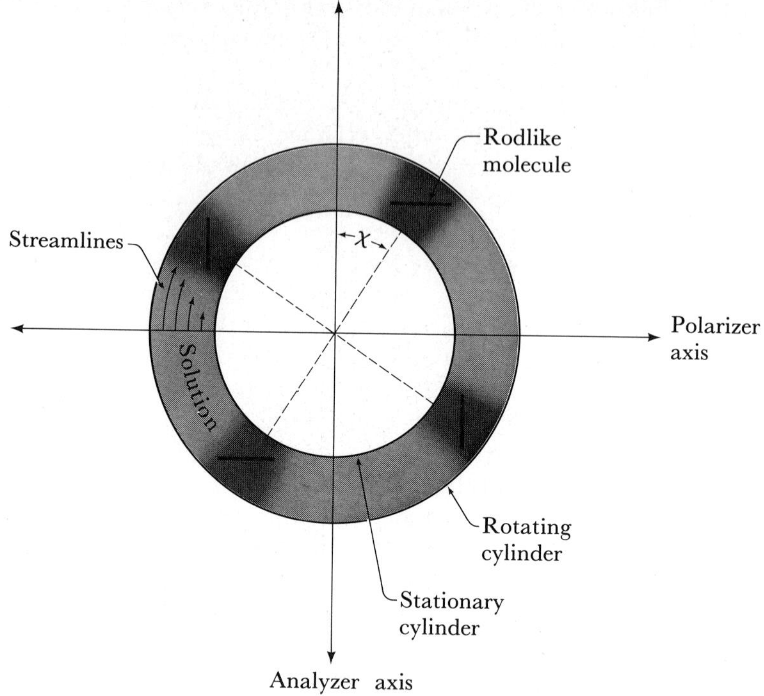

FIGURE 18-11
View down the axis of a flow birefringence instrument. The solution is contained between the inner and outer cylinders. When the outer cylinder rotates, the molecular rods are oriented by the streamlines. At two points, they are parallel to the analyzer; at two other points, they are parallel to the polarizer. No light is transmitted at these four points; the pattern of four dark regions is known as the *cross.* The angle χ formed by the axes of the cross is called the extinction angle.

Hence, if DNA is totally oriented in a flow birefringence instrument and illuminated with polarized UV light, the oriented molecules absorb only if they are situated at certain angles with respect to the plane of polarization. Hence, the orientation of the bases with respect to the DNA axis can be determined.

Example 18-F. Orientation of proflavin molecules bound to DNA. Proflavin also has a preferred axis for the absorption of blue light. By choosing a wavelength absorbed by proflavin but not by DNA, the orientation of the bound proflavin can be determined in the same way as in example 18-E. Such a measurement has shown that the proflavin molecules are parallel to the base pairs.

Selected References

Albertsson, P. A. 1971. *Partition of Cell Particles and Macromolecules.* Wiley-Interscience. The best book on the subject.

Cerf, R., and H. A. Scheraga. 1952. "Flow Birefringence in Solutions of Macromolecules." *Chem. Rev.* 51:185–261. A good review of flow birefringence.

Englander, S. W. 1967. "Measurement of Nucleic Acid Hydrogen Exchange," in *Methods in Enzymology,* vol. 12B, edited by L. Grossman and K. Moldave, pp. 379–386. Academic Press.

Printz, M., and P. H. von Hippel. 1965. "Hydrogen Exchange Studies of DNA Structure." *Proc. Nat. Acad. Sci.* 53:363–370. Experiments on "breathing" of DNA.

Problems

18-1. A substance partitions so that 98% appears in 20% dextran and 2% in 20% polyethylene glycol. If solid polyethylene glycol were added to a solution of the substance to make a final concentration of 30%, what would probably happen to the substance?

18-2. Which of the following would show flow birefringence: a sphere, a prolate ellipsoid, a rigid rod, a flexible rod, a chain of beads, a circular DNA molecule?

18-3. Why do you think the sample vessel in a flash evaporator is made to rotate?

18-4. Actively growing bacilli are rod-shaped, typically with a ratio of length to width of about 4. A stationary culture consists of bacteria with a length-to-width ratio of about 2. If you were going to prepare a cell-free extract by sonication, would you choose a growing or stationary culture? Explain.

18-5. What factors should be considered in selecting a method for concentrating?

18-6. Suppose that, in doing a flow birefringence measurement, you are measuring the position of the cross at various speeds. In the course of increasing the speed, the cross is suddenly observed to fade and move to a larger angle. What has probably happened?

Glossary

absorbance A measure of the decrease in light transmitted by a sample. Defined as $-\log_e(I/I_0)$ in which I and I_0 are the transmitted intensity and the incident intensity, respectively.

active site The region of a protein in which interaction with another molecule takes place.

adenine A base found in DNA and RNA. Abbreviated A for the free base, rA for adenosine, and dA for deoxyadenosine.

alcohol dehydrogenase An enzyme that catalyzes the following oxidation-reduction reaction:

$$\text{acetaldehyde} + \text{NADH} + \text{H}^+ \rightleftarrows \text{ethanol} + \text{NAD}^+$$

α-carbon of an amino acid The carbon atom to which the amino group is attached.

α-phosphate of a nucleotide The phosphate attached to the sugar.

amino acid sequence The linear order of the amino acids in a polypeptide.

anisotropic Having a physical property that is in some way dependent on direction.

antibody A protein synthesized by an animal in response to a foreign substance.

antigen A substance capable of eliciting antibody formation.

antigenic determinant The particular site or chemical group on an antigen against which antibody is directed.

apoprotein A protein requiring a prosthetic group for activity.

AT pair An adenine residue and a thymine residue joined by hydrogen bonds.

bacteriophage A virus that multiplies only in bacteria.

blue shift of a spectrum Any shift to a shorter wavelength. This term is a misnomer in that a shift to shorter wavelengths of a wavelength maximum in the ultraviolet range is called a blue shift even though the shift is actually away from the blue.

boundary In physical biochemistry, this usually means a region in which the composition of a solution changes.

5-bromouracil A base that can substitute for thymine in DNA; in so doing, the density of the DNA increases. Abbreviated BU for the free base and BUDR for the deoxyriboside.

catenane A structure consisting of at least two circles linked as in a chain.

cell, sample A container for a sample.

chemical shift In nuclear magnetic resonance, the change in resonance conditions of a nucleus caused by a chemical or a physical interaction with other nuclei or with electrons.

chlorophyll The principal photoreceptor in plants.

chloroplast A cell organelle that contains chlorophyll.

codon A sequence of three adjacent nucleotides coding for an amino acid.

cofactor A small molecule essential for the activity of an enzyme; usually distinguished from a prosthetic group by being loosely bound.

collagen A major body protein; found in tendon, bone, and skin.

concanavalin A A plant protein of a class called lectins that binds to α-mannosyl groups on the surface of mammalian cells and causes agglutination.

concatemer A linear polymer consisting of a repeating unit of a polymer; usually refers to DNA molecules of greater than unit length.

concentration gradient A system in which the concentration of a substance changes with time or location.

convection The movement of a liquid caused by local changes in density.

copolymer A polymer containing more than one type of monomer.

curve fitting A process by which many known curves are graphically summed in the right proportions to make an observed curve.

cytosine A base found in DNA and RNA. Abbreviated C for the free base, rC for cytidine, and dC for deoxycytidine.

dalton A unit of molecular weight, the mass of a hydrogen atom. In some chemistry texts, this is called an avogram.

dansyl chloride A chemical that reacts with the amino groups of proteins to form a highly fluorescent compound.

density Mass per unit volume.

density gradient A change in density with position.

density gradient centrifugation The centrifugation of a solution that is layered on or contained in a density gradient.

dichroic Having the absorbance depend on the relative orientation of the plane of polarization of incident light and a molecular axis.

difference spectroscopy A technique whereby a spectrum is obtained by subtracting one spectrum from another.

DNA polymerase I An *E. coli* enzyme capable of copying a single strand of DNA or of filling in gaps; it is not the principal enzyme responsible for synthesizing DNA.

DNA polymerase III An *E. coli* enzyme responsible for the duplication of DNA.

DNA-RNA hybrid A double helix consisting of one strand of DNA hydrogen bonded to an RNA strand by complementary base pairing.

double-sector cell A centrifuge cell containing two sample compartments.

DPN Diphosphopyridine nucleotide; also called nicotinamide adenine dinucleotide (NAD).

DPNH The reduced form of DPN; also called NADH.

elution The process of removing material from a chromatographic system.

endonuclease An enzyme that makes internal phosphodiester cuts in the sugar-phosphate chain of a polynucleotide.

exonuclease An enzyme that cuts off nucleotides one by one from the end of a polynucleotide strand by breaking phosphodiester bonds.

ferritin An iron-containing protein used as a marker in electron microscopy.

γ-phosphate of a nucleotide triphosphate The phosphate most distant from the sugar.

GC pair A guanine residue and a cytosine residue joined by hydrogen bonds.

glycoprotein A protein to which a carbohydrate is covalently linked.

guanine A base found in DNA and RNA. Abbreviated G for the free base, rG for guanosine, and dG for deoxyguanosine.

heme The prosthetic group of hemoglobin.

histone A type of protein rich in basic amino acids found in the chromosomes of all eukaryotic cells except sperm.

hydrolysis The breaking of a molecule into two or more parts by the additon of water.

hydrophilic Having a tendency to bind water molecules.

hydrophobic Having a tendency to repel polar molecules such as water but to bind to nonpolar molecules.

inducer A substance that binds to a repressor and reduces or eliminates its activity.

in vitro Used in reference to an experiment done in a cell-free system.

in vivo Used in reference to an experiment done in living cells.

ionic bond A bond formed by the attraction between a positively charged group and a negatively charged one.

ionic strength A measure of the total amount of free charges per unit volume; for example, a divalent ion contributes twice as much to the ionic strength of a solution as a monovalent ion.

isotropic Having a physical property independent of direction.

ligand A small molecule that binds to a larger molecule.

lipoprotein A protein to which a lipid is covalently attached.

lysis The bursting of a cell by the destruction of its cell membrane or cell wall.

lysozyme An enzyme that lyses certain bacteria by cleaving a polysaccharide component of their cell walls.

melting Denaturation by heat.

melting temperature The temperature at which a melting transition is half complete.

nearest neighbor An adjacent monomer in a polymer.

ninhydrin A substance,

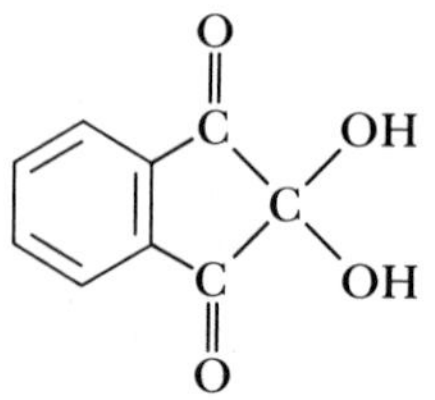

that reacts with amino acids to produce strongly colored compounds that can be easily detected.

nuclease An enzyme that hydrolyses nucleic acids by breaking phosphodiester bonds.

oligonucleotide A small number of nucleotides joined by phosphodiester bonds.

oligopeptide A small number of amino acids joined by peptide bonds.

operator A sequence of bases in DNA to which a repressor binds.

optically active Having the ability to rotate the plane of polarization of light and to absorb differentially right and left circularly polarized light.

peak The maximum of a curve. The position of the peak usually refers to the x-coordinate. The size of the peak is the y-value.

polymerase An enzyme that joins nucleotides in phosphodiester linkages.

polynucleotide kinase An enzyme that transfers the γ-phosphoryl of adenosine triphosphate to a 5′-OH group of a polynucleotide.

promoter A region of a DNA molecule at which RNA polymerase binds and initiates transcription.

prosthetic group A tightly bound, specific, nonpolypeptide unit required for the biological activity of a protein.

red shift of a spectrum Any shift to longer wavelengths. This term is a misnomer in that a shift to longer wavelengths of a wavelength maximum in the infrared range is called a red shift even though the shift is actually away from the red.

relaxation The return of a system, which was initially at equilibrium but which has been disturbed, to the equilibrium state.

replication fork The Y-shaped region of a chromosome or DNA molecule that is a growing point for DNA replication.

repressor A protein product of a regulatory gene capable of binding to an operator and preventing the initiation of transcription.

resolving power The ability to separate two components.

ribonuclease An enzyme that degrades RNA. Often abbreviated to RNase.

ribosome A complex nucleoprotein on whose surface protein synthesis occurs.

RNA polymerase An enzyme that polymerizes nucleoside triphosphates to RNA by copying the base sequence of a DNA strand.

SDS Sodium dodecyl sulfate, a detergent that disrupts most protein-protein and protein-lipid interactions.

single-sector cell A centrifuge cell containing a single sample compartment.

substrate The substance acted on by an enzyme.

Svedberg A unit of sedimentation.

swelling The process by which a dry gel particle imbibes liquid.

thymine A base found in DNA. Abbreviated T for the free base, rT for ribosylthymine, and dT for thymidine.

tRNA The RNA species that recognizes both a codon and an amino acid and carries the amino acid to the growing terminus of a protein. Formerly called sRNA.

uracil A base found only in RNA. Abbreviated U for the free base, rU for uridine, and dU for deoxyuridine.

zwitterion A molecule that can be either positively or negatively charged, depending on the pH.

Answers to Problems

CHAPTER 1

1-1. Because of the alternating sequence of A and T, any region of the chain can form hydrogen bonds with any other region. Hence, the structures could range from a hairpin

```
A T A T A T A
T A T A T A T
```

to various complex branched structures; for example,

```
      A T                           A T
      T A                           T A
A T A   T A T A T             A T A   T A T A T
              A T A                 T         A T A
            T                 T A     A T
        T A                   A T     T A
        A T                           A T
        T A                           T A
```

1-2. Fifteen.

1-3. Linear and circular dimers, trimers, and so forth, because at high concentration the probability of two different molecules interacting with one another approaches that of the two ends finding one another.

1-4. Polyvaline, because the charged amino group of lysine causes mutual repulsion of the monomers. Polylysine shows the greater effect of pH on shape because, in a certain pH range, the net charge varies.

1-5. The DNA would denature because the bases would interact with the solvent rather than with themselves.

1-6. There are six possible linear types:

However, if the subunits were either isotopic or symmetric, structure 1 would be identical with structure 3 and structure 4 would be identical with structure 6; then there would be four types.

There are two possible triangular types:

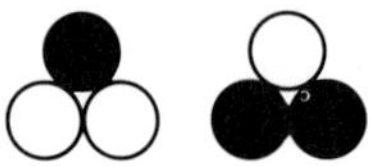

1-7. Eleven.

1-8. There are several possibilities. Each subunit might have a positively charged and a negatively charged region. These regions could interact by a simple charge interaction. If they have hydrophobic regions on the surface, they might interact so that water can be excluded.

1-9. In the first case,

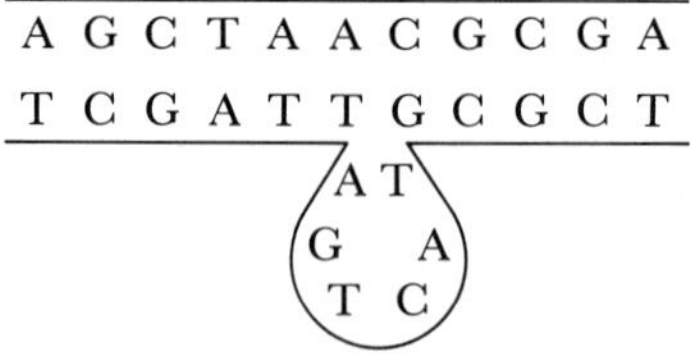

In the second case,

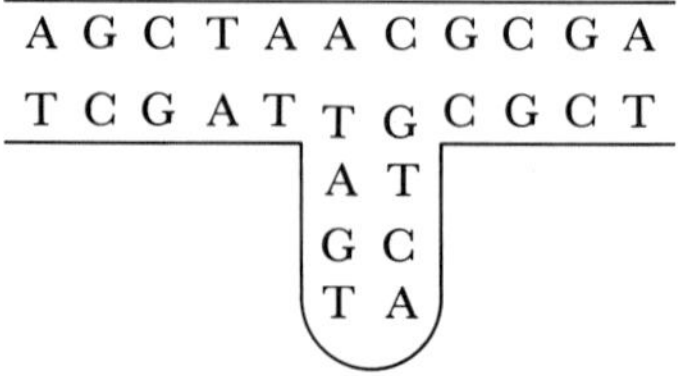

1-10. 3/8 hybrid, 9/16 heavy, and 1/16 light.

1-11. There are only four (↑) peptide bonds because the C—N bond (↓) to the left or proline is not a peptide bond.

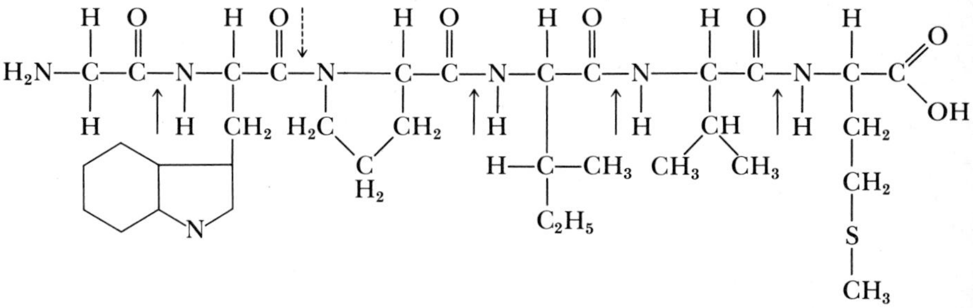

```
      H  O       H  O ¦        H  O       H  O        H  O       H
      |  ||      |  || ↓       |  ||      |  ||       |  ||      |    //O
H2N—C—C—N—C—C—N———C—C—N—C—C—N———C—C—N—C—C
      |    ↑ |   |      |      |   ↑ |    |    ↑ |    |    ↑ |   |    \
      H      H  CH2  H2C      CH2    H    |      H   CH      H  CH2   OH
                         \    /           |          / \         |
                          C          H—C—CH3   CH3  CH3        CH2
                          H2              |                      |
              N                          C2H5                    S
                                                                 |
                                                                CH3
```

CHAPTER 2

2-1. If the virus takes up acridine orange, it can be visualized as a point of light against a black background. The fact that the size of the point is below the limit of resolution is irrelevant because resolution refers to the ability to separate nearby objects. Hence, the virus should not aggregate or the observed count will be too low. A DNA virus is easier to see than an RNA virus because the fluorescence of acridine orange is greater if bound to DNA than if bound to RNA.

2-2. If viewed from the side, they will appear the same.

2-3. Contrast is produced by shifts in phase, which are converted into intensity differences. If the object has regions that differ in absorbance, the contrast effects caused by phase shifts may be either weakened, cancelled, or enhanced by intensity differences.

2-4. Either the polarizer or the analyzer is removed and the sample is rotated. A dichroic object will vary in brightness as it is rotated.

2-5. If a blue filter is placed anywhere in the light path, contrast should be enhanced because red or pink light is usually absorbed by a blue filter.

2-6. It probably consists of either partially oriented fibers or parallel fibers, each of which is internally collapsed.

2-7. Fluorescence microscopy because contrast increases with increasing intensity.

2-8. First, the sample could dry out; second, the sample could curl; and, third, if the sample were in liquid, the liquid surface could act as a lens.

2-9. The cell wall would not be visible. By phase microscopy, the cell itself would be visible if the index of refraction of the cell contents differed from that of the suspending medium. If the cell wall were very thin, its absence might be unnoticed because of the phase halo.

2-10. The first observation shows that the UV light has not broken the chromosome. The second shows that material has been either added or lost; loss is more probable. The third observation confirms the second. The fourth and fifth show that DNA has been lost. Therefore, the UV irradiation has resulted in the depletion of DNA. Other components must be present in the chromosome that are capable of maintaining chromosome structure even when the DNA chain is broken.

2-11. The object either must consist of linear units emanating radially from a single point or a small region or it must consist of concentric rings, each consisting of parallel fibers. If the analyzer were removed, two arms of the cross would disappear.

CHAPTER 3

3-1. Those with dark centers are empty particles.

3-2. 750 × (820/1250) = 492 Å. The viruses with short shadows have probably collapsed.

3-3. If single- and double-stranded DNAs are denoted by thin and heavy lines, respectively, the structure will be:

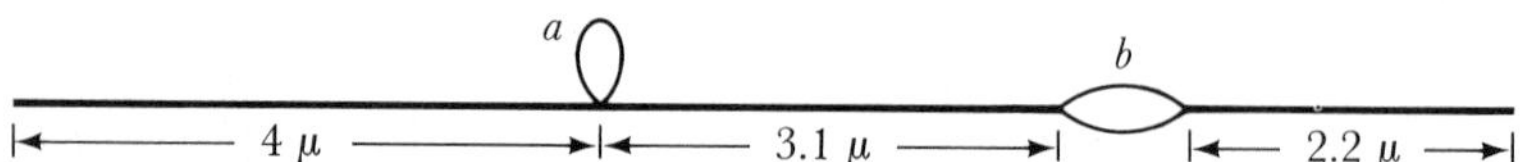

Loop *a* will have a length of 0.5 μ; each strand of loop *b* will be 0.7 μ long.

3-4. The molecule is circular, and one part of it consists of a double-stranded region equal in length to 1% of the total length flanked by two single-stranded regions of equal length that is also approximately 1% of the total. The remainder of the molecule is double-stranded. The short double-stranded region is the terminally redundant region. The total length of the molecule is 99% of the original length.

3-6. No, because the metal film would be less transparent to electrons than the sample, which would therefore be invisible.

3-7. The molecules will overlap so that it will be difficult to follow a single molecule from end to end.

3-8. There are approximately (100/400) × 22 = 5.5 viruses per 100-Å layer. The cell contains roughly 500 such layers, assuming that its shape has not been altered by the processes required for preparing thin sections. Hence, there are approximately 2,750 viruses per cell. The most important assumption in this calculation is that the viruses are uniformly distributed throughout the cell (such an assumption would rarely be satisfied). Another assumption, made for arithmetical simplicity, is that all layers have the same diameter; although this is not possible in a sphere, it introduces a smaller error than the assumption about homogeneity.

3-9. If the slices are perpendicular to the long axis of the cylinder, circles will be observed. If they are parallel to the long axis, stacked discs will be seen. The width of the disc will vary according to the placement of the cut and will be greatest if the cut is on the long axis. If the cut is at an angle, the cylinder will appear short, but discs will still be seen.

CHAPTER 5

5-1. One millimole contains 6×10^{20} methyl groups. If all were labeled with one tritium atom, half or 3×10^{20} would decay in twelve years. Hence, the activity is 2.5×10^{19} decays per year or 8×10^{11} decays per second, or 22 curies. Hence, $6/22 = 27\%$ of the thymidine molecules are radioactive.

A molecule of DNA having a molecular weight of 25×10^6 contains 75,700 bases or 18,940 thymidines of which $0.27 \times 18{,}940 = 5{,}100$ are radioactive.

If a sample registers 1,000 cpm on a counter with 52% efficiency, it contains an amount of radioactivity corresponding to $1{,}000/0.52 = 1{,}923$ cpm $= 8.7 \times 10^{-10}$ curies. This would be contained in $(1/6) \times 8.7 \times 10^{-10} = 1.45 \times 10^{-10}$ millimoles of thymidine $= 1.45 \times 10^{-13}$ moles. If 50% of the base pairs in a molecule of DNA are AT, for every mole of thymidine nucleotide, there are three moles of other nucleotides. Hence, there are $4 \times 1.45 \times 10^{-13} = 5.8 \times 10^{-13}$ moles of nucleotides. The average weight of one mole of nucleotide is 330 grams. Therefore the weight of DNA corresponding to 1,000 cpm is $330 \times 5.8 \times 10^{-13} = 1.9 \times 10^{-10}$ grams.

5-2. ^3H and ^{32}P would be the pair of choice because the levels of a scintillation counter can be adjusted so that only a small fraction of the ^{32}P activity is in the ^3H channel.

5-3. ^3H-thymidine would be the better choice because it is incorporated only in DNA, whereas ^{32}P is also incorporated in RNA, phosphoprotein, and phospholipid.

5-4. There is no ^3H activity in the ^{14}C channel and 20% of the ^{14}C activity is in the ^3H channel. Hence, the 1,620 cpm in channel B represents ^{14}C only. However, $0.2 \times 1{,}620 = 324$ cpm of ^{14}C is in channel A. Therefore, the total ^{14}C activity is $1{,}620 + 324 = 1{,}944$ and the ^3H activity is $1{,}450 - 324 = 1{,}126$. The $^3H/^{14}C$ ratio is $1{,}126/1{,}944 = 0.58$.

5-5. Correcting for counting efficiencies, the ^3H activity is 4,504 dpm and the ^{14}C activity is 2,371 dpm. The $^3H/^{14}C$ ratio is $4{,}504/2{,}371 = 1.90$. The molar ratio of thymidine to uridine is then $1.90/(6/0.5) = 0.16$.

The number of curies of thymidine is $4{,}504/(2.2 \times 10^{12}) = 2.1 \times 10^{-9}$, which is equivalent to $(1/6) \times 2.1 \times 10^{-9} = 3.5 \times 10^{-13}$ moles. In a DNA that is 43%

GC, 0.285 of the nucleotides are thymidine. Therefore, there are $(1/0.285) \times 3.5 \times 10^{-13}$ moles, or 4×10^{-10} grams, of DNA.

The number of curies of uridine is $2{,}371/(2.2 \times 10^{12}) = 1.08 \times 10^{-9}$ or 2.16×10^{-12} moles. The number of moles of RNA nucleotides is $(1/0.28) \times 2.16 \times 10^{-12} = 7.7 \times 10^{-12}$ or 2.54×10^{-9} grams of RNA.

5-6. The DNA with a molecular weight of 20×10^6 contains 6×10^4 atoms; therefore the ratio $^{32}P/^{31}P$ is $1/(6 \times 10^4)$ and the concentration of ^{32}P in the medium must be $10^{-3}/(6 \times 10^4) = 1.6 \times 10^{-8}$ molar. This is 1.6×10^{-11} moles per milliliter.

5-7. A 1% probable error means roughly that N^{-2}, in which N is the total number of counts, should be 0.01. Hence, one must count 10^4 counts and the sample should be counted for at least $(1/752) \times 10^4 = 13.3$ minutes.

5-8. The count rate should be unaffected by volume as long as the two photomultipliers observe the entire sample. The background due to environmental radioactivity will decrease as the volume decreases.

5-9. The channel ratio (B/A) for each quenched standard must be calculated and plotted against the efficiency of detection—for example, (A + B)/1,000. This graph has the following equation: efficiency of detection = channel ratio. Because the observed channel ratio for the sample is 1,211/1,822 = 0.66, the actual number of cpm is (1,211 + 1,822)/0.66 = 4,595.

5-10. The 0.1-ml sample contains $0.04 \times 25 = 1$ μg of leucine. This has 1,251/0.70 = 1,787 cpm = 8.1×10^{-4} μcuries. The specific activity is 8.1×10^{-4} μC/μg. In the original culture, the concentration of radioactivity was $(0.1 \times 50 \times 25)/10 = 12.5$ μC/ml and the concentration of added leucine was $(0.1 \times 25)/10 = 0.25$ μg/ml. Because the specific activity in the protein is much lower than the specific activity of the added ^{14}C-leucine, there must be ^{12}C-leucine in the medium at a concentration of 1.54×10^4 μg/ml. This is so high that the increase in this concentration by the added leucine is negligible. Therefore, to increase the count rate from 1,251 to 5,000 cpm, a fourfold increase, simply requires the addition of four times as much ^{14}C-leucine, or 0.4 ml.

5-11. The ratio of count rate to background is 3.7 for the Geiger counter and 1.9 for the scintillation counter. Because the count rate is expected to be small, the reliability of the measurement will be better if the background is the smaller fraction of the total. The Geiger counter is preferable. Of course, the sample must be counted for a longer period of time than with the scintillation counter to achieve the same percentage of error.

CHAPTER 6

6-1. Grow phages in medium containing ^3H-thymidine. X-irradiate with various doses (including zero). Let the phages adsorb to the bacteria and then inject DNA. Remove the phages from the bacteria by violent agitation. Separate the bacteria, which now contain injected DNA, from the phages. Place the bacteria on a glass slide, dry, and coat with either stripping film or a thin layer made by the dipping method. After exposure and development, count the number (n) of grains over the cells and the number (N) of cells with grains. The value of n obtained for zero dose describes the number of grains corresponding to a whole DNA molecule. If an x-irradiated phage injects all of its DNA, n and N will be independent of dose. If only a fragment is injected, n will decrease, but N will remain nearly the same.

6-2. With 94% efficiency, only $0.94 \times 42 = 39$ rays will be detectable. Because of logarithmic decay, the equation describing the time t to get N rays from $N_0 = 39$ detectable ^{32}P atoms is:

$$39 - N = 39\, e^{-0.693t/\tau_{1/2}} = 39\, e^{-0.693t/14.2} = 39\, e^{-0.049t}$$

Then, to get $N = 12$, $t = 7.5$ days.

6-3. If a ^{3}H-labeled compound is used and an autoradiogram is prepared using either stripping film or the dipping method, the grain density will be maximal near the observed cell boundary but spread across the entire cell area because the cell wall is both above and below the cell when observed. If the compound is contained only in the cytoplasm, the grain density will be either uniform across the cell area or maximal in regions where the cell is thick. If an autoradiogram is prepared of a thin section of the cell, all grains will be peripheral if the compound is localized in the cell wall.

6-4. Cells are grown to stationary phase (i.e., no more cell division) for long periods of time in ^{3}H-thymidine. The cells are spread on an agar surface so that individual cells are well separated. The cells are then allowed to grow in the absence of ^{3}H-thymidine to form a microcolony consisting of several hundred cells and this is overlaid with stripping film. Because of the semiconservative nature of DNA replication, if there is a single chromosome, two cells in the microcolony will be labeled. If the cell contains four chromosomes, eight cells in the microcolony will be labeled.

6-5. The short range of ^{3}H would reduce efficiency so that ^{14}C is preferable. The range of ^{14}C is not so great that resolution would be decreased because the track length of ^{14}C is very short compared with the size of spots on chromatograms.

6-6. A ^{3}H-labeled amino acid should be used, together with either stripping film or dipping. Cells are grown first at 37°C in nonradioactive medium. Labeled amino acid is added, and the culture is divided into two parts, one at 37°C and one at 42°C. After some time has elapsed, the cells are autoradiographed. If the rate at 42°C reflects the fact that 10% of the cells incorporate normally and 90% fail to incorporate, then grains will be found above only 10% of the cells. If the rate per cell is reduced tenfold, all cells will be labeled, but the average number of grains per cell grown at 42°C will be 10% that of a cell grown at 37°C. To count grains, different exposure times are necessary so that the number of grains is small enough that individual grains can be unambiguously counted.

6-7. If cells are prelabeled by growth in ^{3}H-thymidine and then autoradiographed, grains will be formed over all cells. If the small cells contain no DNA, no grains will be found over these cells. A reduced number of grains indicates a lower amount of DNA.

RNA can be studied by growth in a medium containing ^{3}H-uridine and a large amount of unlabeled thymidine. Uridine is a precursor of both RNA and DNA. The thymidine prevents the appearance of ^{3}H in DNA by competing with the thymidine synthesized by means of the uridine-deoxyuridylate-thymidylate pathway.

6-8. This cannot be done by growth in ^{3}H-thymidine because too many grains will result from the huge amount of chromosomal DNA in which the nucleoli are immersed.

6-9. Male and female cells are usually not distinguishable by light microscopy. However, if one type is grown in medium containing ^{3}H-thymidine and the other in nonradioactive thymidine, the two types can be distinguished by autoradiography. Hence, if ^{3}H males are mixed with ^{1}H females, pairs will be observed consisting of one labeled and one unlabeled cell. If one culture of ^{3}H males is mixed with another culture of unlabeled *males* and if there is homosexual pairing, pairs will be found consisting of one labeled and one unlabeled cell.

CHAPTER 7

7-1. Nitrocellulose is soluble in acetone. Hence, fiberglass is better.

7-2. Filtration is preferable if the particle sediments very slowly and if the pellet is hard to resuspend. It is a poor method if the amount of precipitate is so great that the filter would clog or if the precipitate binds to the filter material.

7-3. Study the retention as a function of pore size. If adsorption is occurring, retention is usually independent of pore size.

7-4. At high flow rate, the linear molecules are oriented and pass through the pores end on, whereas circular molecules cannot become thin enough and are retained. At low flow rates, both are retained.

7-5. Sterilize filter, filter holder, and collection vessel.

7-6. The sample is collected on a fiberglass filter and incubated in alkali. The filter is removed and the liquid made acidic. The resulting precipitate is collected on a separate filter. The background should be $0.0001 \times 0.01\% = 0.000001\%$. Nitrocellulose filters are alkali soluble and therefore not usable.

7-7. Note that between 10^7 and 2×10^7 cells remain on the filter. Therefore, if 10^7, 10^6, or 10^5 cells were filtered, none could be removed.

7-8. Use radioactive protein or nucleic acid. Dialyze, empty, and dry tubing. Then count tubing to see if radioactivity is bound to the membrane.

CHAPTER 8

8-1. Gel chromatography.

8-2. By gel chromatography.

8-3. Chromatography on hydroxyapatite.

8-4. The linear molecule would elute first because the circular one occupies a smaller spherical domain and would more easily penetrate the agarose pores. Denatured ribosomal RNA would elute first for the same reason.

8-5. At high concentration, the protein probably aggregated. This larger structure could not penetrate the pores and would pass through with the void volume.

Increasing the diameter is like increasing the number of columns and should be sufficient.

8-6. Two-dimensional chromatography using isopropanol-HCl and acetic acid.

8-7. Because only 5% is lost and enzyme X is 30% of the total, it could not be the case that X was not eluted. Therefore, the activity of X has been lost. There are many possible explanations of which two are the dissociation of subunits and denaturation caused by binding to the charged DEAE-cellulose.

8-8. In the single-strand region because single-stranded DNA binds more tightly than double-stranded DNA. If the phosphate concentration is sufficient to dissociate the double-stranded material from the hydroxyapatite, the molecule will remain bound by the single-stranded piece.

8-9. In the ion-exchange procedure, the enzyme and the Mg^{2+} ions are probably eluted at different ionic strengths. With the gel, they probably separate because of their different sizes. In both procedures, the enzyme will be found in a solution lacking Mg^{2+} and this may cause denaturation. This can be avoided in gel chromatography if the gel has been equilibrated with Mg^{2+} beforehand and if the eluting buffer contains Mg^{2+}.

8-10. Morphine could be coupled to a matrix and affinity chromatography could be performed.

8-11. Gel chromatography at pH 8.5 should be used. Each fraction should be readjusted to pH 5.5 before assaying the protein.

CHAPTER 9

9-1. At high ionic strength, a solution would be too conductive, the current would be too high, and the solution would overheat.

9-2. No.

9-3. pH usually has the greatest effect because the total charge and the sign of the charge can be varied. Reducing ionic strength increases the availability of charged groups to the solvent, although it can have the reverse effect by changing protein conformation. Temperature also can affect conformation but is not always a useful parameter because the alterations are frequently reversible.

9-4. No, because two proteins could easily have the same mobility—for example, if their charge-to-mass ratio were very nearly the same.

9-5. The protein appears to consist of two kinds of subunits, one having twice the molecular weight of the other. If the SDS-treated protein were fractionated by gel chromatography, the relative values of the molecular weights could be ascertained from the relation between elution volumes and molecular weight. After the removal of the SDS from the fractionated subunits, they could be chromatographed one by one with the untreated protein to determine the relative molecular weight of each subunit compared with the intact protein; in this way, it could be known whether the protein consisted of one subunit of each type, two of each type, and so forth.

9-6. The relative distances migrated are the ratio log 26,000/log 1,800 = 1.36. The relative areas reflect the total mass ratio, or $(192 \times 26{,}000)/(64 \times 1{,}800) = 43.3$.

9-7. It would be most useful to vary the pore size of the gel (by varying the concentration or the degree of cross-linking), because in a particular range it might be expected that two proteins having the same mobility but different molecular weights will separate by the molecular-sieve effect. To aid in ascertaining homogeneity, the protein could be analyzed by gel chromatography because two proteins having different molecular weights can have the same molecular volume. Homogeneity can also be checked by sedimentation (see Chapter 11).

CHAPTER 10

10-1. Because the OD increases with dilution, the initial concentration must be between 0.00 and 0.05 μg. Hence, the concentration is approximately 0.03 μg.

10-2. The position of the band is determined both by the diffusion coefficient of each reactant and by the relative concentration of each at equivalence. Because the equivalence point for precipitin formation is not necessarily the same for two proteins having the same diffusion coefficient, two bands will result. One band would appear only if the two proteins had precipitin curves with the same equivalence point. Similarly, two proteins with different diffusion coefficients might not result if the equivalence points were different but such that a balance was achieved.

10-3. The reaction would probably be very different because the three-dimensional structure of the antigenic site of the native protein would probably be greatly changed on denaturation.

10-4. If the amino acid change is in a region of the protein that is not the antigenic site and if it does not affect the structure of this site by virtue of an overall change in the structure of the protein, there will be no effect on antigenicity. If it does these things, antigenicity can be either lost if the changes are great or altered if they are small.

10-5. Use the radioimmunoassay. Couple the drug to bovine serum albumin and immunize a rabbit to prepare antidrug antibody. Prepare the drug in radioactive form. Obtain a standard curve by adding various amounts of the non-radioactive drug to a constant amount of a mixture of the radioactive drug and antidrug. Then add various aliquots of blood to the mixture of the radioactive drug and the antibody and determine the amount of drug from the standard curve.

CHAPTER 11

11-1. The flexible rod would have the higher s because its frictional coefficient would be lower. The hollow and the solid spheres would have the same s.

11-2. If, at time t, the molecule with $s = s_1$ is at a distance r from the meniscus (which is at a distance r_0 from the center of rotation) and that with $s = s_2$ is at a distance R from r_0, the following equations may be written:

$$\ln r - \ln r_0 = s_1 t$$

and

$$\ln R - \ln r_0 = s_2 t$$

If the separation $d = R - r$, these equations can be rearranged to yield:

$$d = (e^{s_2 t} - e^{s_1 t})r_0$$

11-3.

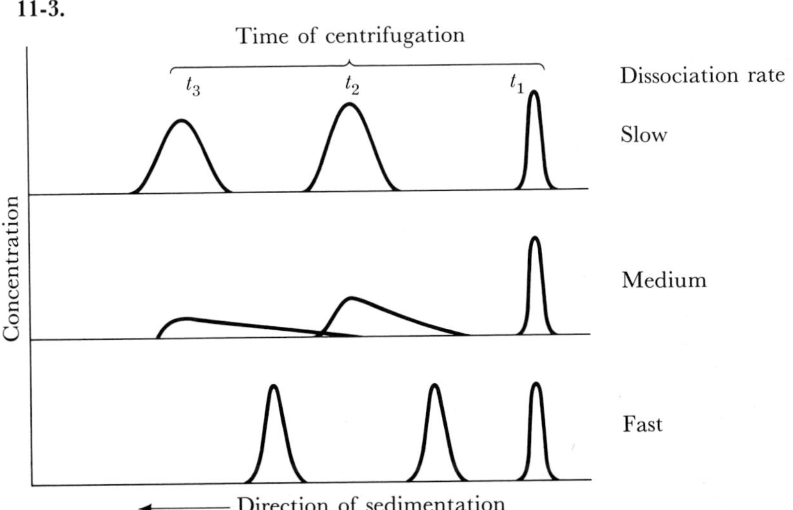

11-4. If the DNA is denatured in formaldehyde and then sedimented in a solvent of high ionic strength, its s will increase as shown in Figure 11-17 but will not decrease if there are interstrand cross-links.

11-5. If the boundary of a homogeneous substance becomes narrower, this usually indicates that the diffusion coefficient has decreased. A decrease in the diffusion coefficient usually means that the molecule has expanded. Such expansion also causes a decrease in s. Hence, in the first case, the reagent has disrupted internal bonds (e.g., disulfide bonds) so that the molecule is less compact. If the boundary broadens, the diffusion coefficient has increased and the molecule becomes more compact. This increases s. Hence, a decrease in s with a broader boundary is not due simply to a shape change. The molecular weight has probably decreased and the reagent probably dissociates subunits.

11-6. No. The bands would be too wide.

11-7.

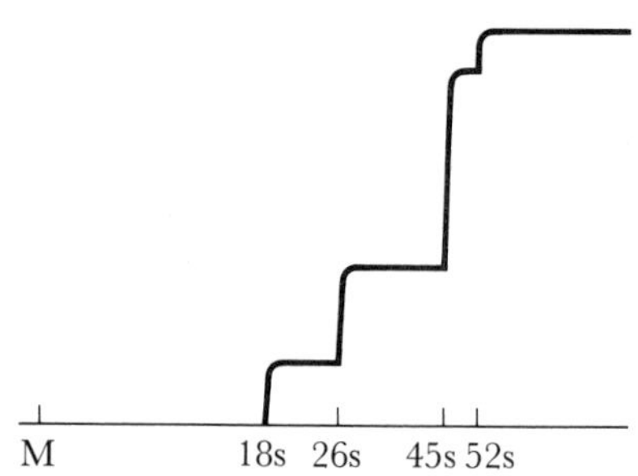

11-8. The number of subunits is three or a multiple of three. For each group of three, there are two subunits having a molecular weight of 5,000 and one with a molecular weight of 15,000.

11-9. The density difference between ^{14}N- and ^{15}N-DNA $= 0.015$ g/ml. Therefore, the density gradient is 0.015/1.32 g/ml/mm. Because *E. coli* is 50% GC, the difference in density of each from that of *E. coli* DNA is $0.098 \times 0.20 = 0.0196$ g/ml. This corresponds to a distance of $(0.0196/0.015) \times 1.32 = 1.72$ mm from *E. coli* DNA.

11-10. Two-thirds of the molecules in the DNA sample are covalent circles; one-third are open circles or linear molecules. If there were an average of one break per molecule, $1/e = 0.37$ of the covalent circles would not receive a break. Therefore, $0.37 \times (2/3) = 0.246$ of the molecules would remain covalent circles and 0.754 would not. The ratio of the area of the denser band to that of the lighter band would be $0.246/0.754 = 0.326$.

11-11. As the number of bound ethidium bromide molecules increases, the supercoiled DNA unwinds until it is no longer supercoiled and then winds up again in the opposite direction. In the form that is not supercoiled, the molecule is least compact and therefore has the minimum value of s.

11-12. The lipoprotein must have a lower density than 1 M NaCl and then sediments in a direction opposite to the centrifugal force.

11-13. In part B, the reagent has precipitated the DNA so that it has formed a pellet before reaching speed. In part C, the reagent has hydrolyzed the DNA to either mononucleotides or small oligonucleotides, neither of which sediments at the centrifugal forces normally encountered.

11-14. At 25 C/mmol, almost every thymidine contains one tritium atom. This increases the molecular weight by two atomic mass units. For a molecule of DNA for which 50% of the base pairs are AT, this would be two atomic mass units per mass of two base-paired nucleotides having a molecular weight of approximately 1,320. Therefore, if the density were originally 1.698, it would increase to 1.698 $\times$ (1,322/1,320) = 1.701, an increase of 0.003 g/ml.

^{14}C has a half-life of roughly 5,000 years or 400 times that of ^{3}H. Therefore, one ^{14}C per thymidine would have a specific activity of 25/400 = 0.62 C/mmol. Therefore, at 1 mC/mmol, there would be 1/0.62 = 1.6 ^{14}C/thymidine. Compared with ^{12}C, this would increase the mass by 3.2 atomic mass units. By a calculation similar to that given for ^{3}H, the density of a 50% AT DNA would be 1.702 g/ml.

For ^{32}P, a similar calculation yields one ^{32}P per eight nucleotides, which would increase the density by 0.0006 g/ml.

11-15. There could be a shape change, sufficient binding of the iodide ion to affect the molecular weight, or a dimerization accompanied by a shape change. A change in molecular weight could be detected by sedimentation equilibrium.

11-16. DNA can be detected by absorption optics. The extinction coefficient of DNA at the absorption maximum (i.e., 260 nm) is much higher than that of protein at 280 nm. For proteins with strongly absorbing groups (e.g., hemoglobin or cytochrome *c*), this is not necessarily the case.

11-17. It is more likely that the material that had pelleted on the tube bottom did not resuspend. The *s* obtained in the second centrifugation is too high because the concentration is lower than that initially placed in the cell.

11-18. The combination of acridine orange and visible light alters the DNA in such a way that a bond in the sugar-phosphate backbone is cleaved at alkaline pH.

11-19. Sediment for different times and show that the distance moved is proportional to the time of sedimentation. A density gradient that is isokinetic for DNA would not be isokinetic for protein because DNA and protein have different densities.

11-20. No, because there are no free ends to experience viscous drag.

11-21. The $s_{20,w}$ of linear DNA should decrease because the molecule increases in length and becomes more rigid. This effect is greater than that produced by the greater mass of the complex.

11-22. For native protein, the effect, if detectable, would be small because the compactness of a protein is usually not an effect of the presence of disulfide bonds.

For denatured protein, it would be detectable because the disulfide bonds prevent the molecule from becoming a random coil.

11-23. Mg^{2+} must be bound to the DNA in CsCl and Mg-DNA is less dense than Cs-DNA.

CHAPTER 12

12-1. The effect would be quite small for proteins because the cause of the effect for amino acids is that Li^+ and K^+ can bind to the free carboxyl groups and alter the density of the molecules. Proteins have few free carboxyl groups.

12-2. Yes, because of the relative mass increases due to the binding of Na^+ and Mg^{2+} to the phosphates.

12-3. The pycnometer holds 9.9249 g of H_2O and therefore has a volume of 9.9249/0.9982 = 9.9428 cc. The weight of the solution in the pycnometer is 11.3251 g and the density is 11.3251/9.9428 = 1.1390 g/cc. The solution consisting of 3.5921 g of solute and 9.9413 g of H_2O has a total weight of 13.5334 g and a volume of 13.5334/1.1390 = 11.8818 cc. The volume of water used to prepare the solution is 9.9413/0.9982 = 9.9592 cc. Hence, if the increase in mass is 3.5921 g, the change in volume is 11.8818 − 9.9592 = 1.9226. Therefore, $\bar{v}$ = 1.9226/3.5921 = 0.535.

12-4. Curvature usually means that the sample is not homogeneous but contains many components. However, thermal convection can cause curvature. If the curve has two components, the sample contains two types of molecules.

12-5.
$$M = \frac{s^{\circ}_{20,\mathrm{w}} RT}{D^{\circ}\,(1 - \bar{v}\rho)}$$
$$= \frac{14.2 \times 10^{-13}}{5.82 \times 10^{-6}} \cdot \frac{(8.3100 \times 10^{7})293}{1 - 0.74(0.9982)}$$
$$= 2.275 \times 10^{5}$$

12-6. Lower speed is preferable because this would allow a shorter time for diffusional spreading to occur. A mixture of two components could be measured as long as the two boundaries are totally resolved.

12-7. Decrease due to increased friction. The flexible rod has greater D because it encounters less friction. D increases as density decreases because denser molecules are more easily moved by collision with solvent molecules.

CHAPTER 13

13-1. At the higher NaCl concentration, intrastrand hydrogen bonds form and the single-stranded DNA is more compact; hence $[\eta]$ is lower. In formaldehyde, the amino groups of the bases are titrated and intrastrand hydrogen bonds cannot form. There would be no effect in 0.01 M NaCl because there are no hydrogen bonds in the absence of formaldehyde. In 1 M NaCl, the single-stranded

DNA would be less compact if formaldehyde were present so that $[\eta]$ would be greater than 3. However, $[\eta]$ would be less than 30 because, in 1 M NaCl, the negatively charged phosphates would be neutralized and a random coil could be assumed.

13-2. The bacteria have become permeable to proteins and the intracellular proteins have come out of the cells.

13-3. (a) the sphere with the larger radius; (b) the solid sphere; (c) the rigid rod; (d) the lollipop.

13-4. The molecule in H_2O because glycerol has a greater viscosity than H_2O.

13-5. (a) the circle because in a shear field it will be deformed to a structure of half length and double thickness; (b) the denatured DNA because it will be very compact owing to intramolecular hydrogen bonds; (c) the DNA in 1 M NaCl because it will be more compact; (d) they will be nearly the same, but the twisted circle will be slightly more resistant because it will be shorter.

13-6. If $[\eta]$ increases, it is probably compact. If $[\eta]$ decreases, it is probably an extended fiber. With a decrease of $[\eta]$ uncertainty arises because the protein might consist of subunits that dissociate in 6 M guanidine chloride.

13-7. The ions might disrupt the quasi-crystalline water lattice (i.e., decrease the interaction between water molecules).

13-8. The polynucleotide has become more extended. If the ionic strength is low, the pH change might break intrastrand base pairing with the result that charge repulsion between the phosphates would produce an extended single-stranded molecule. At high ionic strength, a likely possibility is that the polynucleotide would undergo a transition from being single-stranded to being double-stranded.

13-9. The DNA sample probably contains basic proteins. At low ionic strength, the proteins bind to the DNA and cause intermolecular aggregation, which decreases viscosity. If the DNA is diluted before decreasing the ionic strength, the concentration of DNA and protein can be sufficiently low that aggregation does not occur.

13-10.

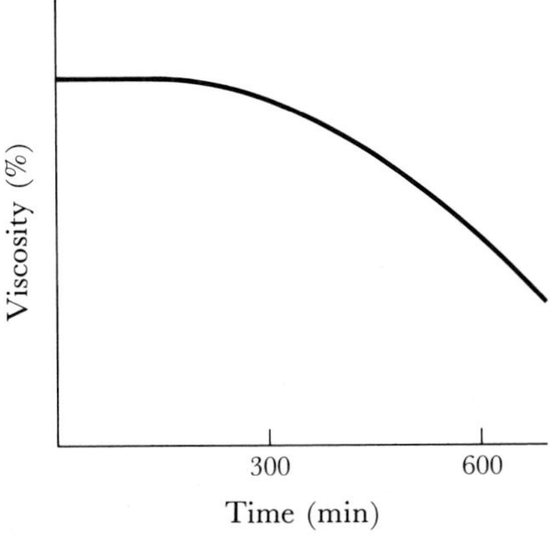

CHAPTER 14

14-1. 32 μg/ml = 0.032 g/1 = 0.032/423 = 0.000076 molar. Therefore, $\varepsilon = 0.27/0.000076 = 3552$.

14-2. By mutual dilution of both A and B, the optical densities of the mixture should be $OD_{260} = 0.301$ and $OD_{450} = 0.460$. There appears to be an interaction between A and B. It has been assumed that Beer's law is obeyed during the dilution.

14-3. Native and denatured DNA can be distinguished by the hyperchromicity at 260 nm resulting from boiling or from the addition of alkali. An alternative is to measure the spectrum because the absorption maxima of native and denatured DNA differ.

14-4. Measure the optical density at the absorption maximum of cytochrome *c*. However, the bacteria scatter strongly so that it is necessary to determine the spectrum at long wavelengths in order that a scattering correction can be carried out.

14-5. Beer's law is not obeyed until the solution has been diluted 1:2. The 1:4 dilution is definitely in the linear range, and its molarity is 0.84/348 = 0.0024. The molarity of the original solution is fivefold greater or 0.12 molar.

14-6. The ratio OD_{260}/OD_{280} of the isolated substance A is less than that of pure A so that some B must be present. If for every part of A there are x parts of B, the following equation may be written:

$$\frac{5248 + 311x}{3150 + 350x} = \frac{2.50}{2.00}$$

Therefore, $x = 10.5$. Then the fraction of the observed OD_{260} due to A is $5248/[5248 + (10.5)311] = 0.616$. Hence, the OD_{260} due to A only is $0.616 \times 2.50 = 1.54$, and the concentration of A is $1.54/5248 = 0.00019$ molar.

14-7. The DNA is partly denatured at 20°C.

14-8. All of the absorbing groups must be internal or at least in crevices inaccessible to ethylene glycol.

The protein has eight tryptophans. None are on the surface. Four are probably in crevices narrower than 9 Å and larger than 4 Å because they are accessible to dimethylsulfoxide but not to sucrose. Four are either internal or in very narrow crevices. Alternately, they may all be internal if dimethylsulfoxide causes partial unfolding.

14-9. Six tyrosines are on the surface. Complete unfolding occurs at pH 11.7. When the protein is restored to pH 6, it refolds but in such a way that only four tyrosines are on the surface.

14-10. There are two classes of DNA: 20% of the DNA has a GC content less than 50%, and 80% has a GC content greater than 50%. This frequently occurs if the bacterium contains an accessory DNA molecule called a plasmid. Alternately, the culture from which the DNA is isolated might be contaminated with another bacterium.

Ten percent of the DNA was denatured by the heating at 75°C.

CHAPTER 15

15-1. The protein dimerizes at high concentration and either the region of contact is the region binding ANS or dimerization causes a conformational change, which increases the polarity of the surface.

15-2. The tryptophan could be internal, in a crevice too narrow for iodide to enter, or adjacent to negatively charged amino acids.

All tryptophans may not have the same quantum yield, and some of the surface tryptophans may be in negatively charged regions.

15-3. No, because the absorption spectrum of a dimer usually differs from that of a monomer.

15-4. No.

15-5. The complex may have different energy levels or the probability of quenching may be reduced. Spectral shifts in either direction are possible because of the different energy levels.

15-6. No, because some of the transitions may not lead to fluorescence.

15-7. The tryptophans may already be quenched. The ligand may remove a quenching factor in the protein yet introduce another. Binding of the ligand may produce a conformational change in the entire molecule so that quenching of the fluorescence of those in the binding site is counteracted by enhancement of the fluorescence of tryptophans that are elsewhere in the molecule and were partially quenched before ligand binding.

15-8. By an energy transfer process, O_2 could be excited. Alternately, the DNA bases could be excited and more susceptible to oxidation by unexcited O_2. The acridine orange and the O_2 could form a complex that, if excited, would be highly reactive.

15-9. As ionic strength increases, the protein unfolds so that the region containing the fluor is more flexible. Alternately, the protein might become very compact or very symmetric, thus allowing more rapid rotation.

15-10. For the first observation, the fluor is sufficiently near the tryptophan that energy transfer occurs. The fluor is on the surface. Therefore, the tryptophan is on the surface. The tryptophan is probably not in a negatively charged region and is protected from quenching by the fluor.

The explanation for the second observation is the same as that for the first except that the fluor does not protect the tryptophan from collisional quenching.

For the third observation, there is a conformational change at pH 9 that moves F away from the tryptophan.

For the fourth observation, the tryptophan and the fluor are very near one another at pH 9.

CHAPTER 16

16-1. The wave that lags by 90° should be advanced by 90°. For elliptically polarized light, the maximum length of the **E** vectors should be different.

16-2. They are probably mirror images.

16-3. The conformation in solution probably differs from that in the crystal. Because the change is very great, there is probably a region that is either extremely hydrophilic or extremely hydrophobic.

16-4.

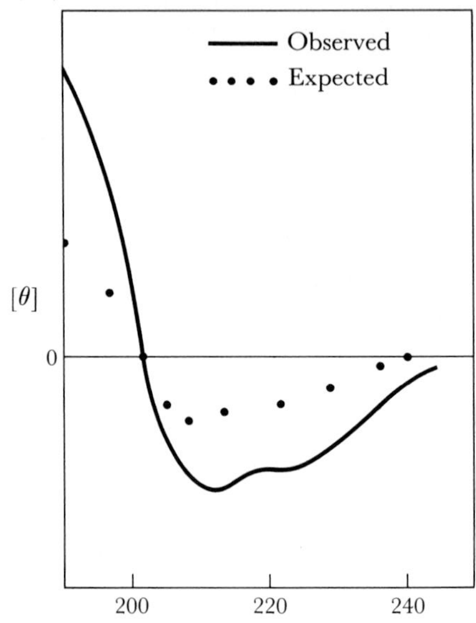

16-5. They can be seen against a baseline that is nearly zero.

16-6. By dilution.

16-7. In the range of the absorption peak.

16-8. The DNA is unwound and the ethidium bromide assumes the helical arrangement of the DNA.

16-9. The decrease would be greater in 0.5 M NaCl because, in 0.01 M NaCl, the molecule would be extended and the bases would all be stacked.

16-10. DNA is tightly folded in the phage head. The manner of folding differs from one phage to the next.

16-11. The fraction of unstacked bases (i.e., those at the ends) becomes small.

16-12. Some form α helices; others do not. In some, there are no side chain interactions; in others, there are and these can be attractive or repulsive, depending on the degree of polarity. Polyproline does not have peptide bonds.

CHAPTER 17

17-1. (a) 1; (b) 1; (c) 1; (d) 1; (e) 1; (f) 2, one for the hydroxyl and one for the methyl hydrogens; (g) 3, one for the hydroxyl, one a quartet for the CH_2 group, and a triplet for the methyl group.

17-2. At all concentrations, the spectra of phenylalanine and glycine would be additive because it is unlikely that the concentrations could ever be high enough for the phenylalanine to induce a ring-current shift in the glycine. However, with phenylalanine and tyrosine, there would be a change at high concentration because phenylalanine and tyrosine have a tendency to stack. Hence, there would be mutual ring-current effects.

17-3. Due to the low abundance of ^{13}C, it would be very rare that two ^{13}C nuclei would be adjacent. Therefore, splitting would not occur. With enrichment, the probability of adjacent nuclei would increase and this advantage would be lost.

17-4. Lines corresponding to the intermediates should appear while the reaction is in progress. The area of the lines with respect to the lines of the initial reactants indicates the relative concentrations of the intermediates and the reactants as a function of time.

17-5. No, because binding of the substrate might introduce a conformational change in the protein that could cause the shift. The shift in line position would indicate that further investigation of the histidine is worthwhile, especially because the shift is fairly large. If there were no shift, it would be very likely that the histidine is not in the binding site.

17-6. No, because splitting requires that the interacting nuclei are covalently coupled and denaturation does not involve changes in covalent bonds.

17-7. First, mix enzyme and cofactor in the absence of substrate and look for displacement of peaks. Second, mix substrate and cofactor and look for displacement. Then add enzyme and look for further displacement or shifts in enzyme peaks. Third, mix enzyme and substrate. If no displacement, add cofactor and look for displacement.

17-8. The line width would differ because the protons in ice are less mobile than in water. Changes in line position could also result if freezing imposed a well-defined orientation between H_2O molecules.

17-9. Lines derived from protons in the region of mutual binding would shift. Most of the line widths would increase owing to the reduced mobility of the dimer compared with that of the monomer.

17-10. It is likely that the protein contains at least one paramagnetic center—probably a transitional metal ion. The addition of a chelating agent should reduce the size of the chemical shift and cause narrowing. This test might not work if the metal is covalently bound. Chemical tests for the presence of the metal would be called for.

17-11. Not for open circles. If the circle were supercoiled, there might be differences because, with a high degree of supercoiling, many of the base pairs would be disrupted, leaving single-stranded regions in the DNA.

CHAPTER 18

18-1. It would probably precipitate.

18-2. All but the sphere, except that, for the circular DNA, high shear stress would be necessary.

18-3. Because a film of liquid is carried upward from the main body of solution, the surface-to-volume ratio is increased, thus increasing the rate of evaporation.

18-4. A growing culture, because the longer cells would be more susceptible to breakage by shear forces.

18-5. Thermal stability, sensitivity to low or high ionic strength, and the materials to which the substance of interest can irreversibly adsorb.

18-6. The molecules have probably been broken in half by the shear stress.

Index